OXFORD READINGS IN FEMINISM

FEMINISM AND THE BODY

OXFORD READINGS IN FEMINISM

Feminism and the Body

Edited by

Londa Schiebinger

OXFORD
UNIVERSITY PRESS

OXFORD
UNIVERSITY PRESS

Great Clarendon Street, Oxford OX2 6DP

Oxford University Press is a department of the University of Oxford.
It furthers the University's objective of excellence in research, scholarship,
and education by publishing worldwide in

Oxford New York

Athens Auckland Bangkok Bogotá Buenos Aires Calcutta
Cape Town Chennai Dar es Salaam Delhi Florence Hong Kong Istanbul
Karachi Kuala Lumpur Madrid Melbourne Mexico City Mumbai
Nairobi Paris São Paulo Singapore Taipei Tokyo Toronto Warsaw

with associated companies in Berlin Ibadan

Oxford is a trade mark of Oxford University Press
in the UK and in certain other countries

Published in the United States
by Oxford University Press Inc., New York

British Library Cataloguing in Publication Data

Data available

British Library Cataloging in Publication Data

(Data applied for)

ISBN 0–19–873191–4

1 3 5 7 9 10 8 6 4 2

Typeset in Minion
by RefineCatch Limited, Bungay, Suffolk
Printed in Great Britain by
Biddles Ltd., Guildford, Surrey

Contents

Part IV. Masculinities

Part V. Restrained Bodies

Notes on Contributors

C. FRED BLAKE is Professor of Anthropology at the University of Hawaii. His research interests focus on how modalities of cultural meaning, such as memory, ideology, and aesthetic compositions, are produced, embodied, and understood in discursive practices that mediate various historical formations. His work mostly deals with China and its diasporic communities.

ANDREW BELSEY is Lecturer in Philosophy at Cardiff University. In addition to recent publications on ethical issues in journalism, he is author of many papers on philosophy, literature, and the history of ideas. With Catherine Belsey he has also written 'Christina Rossetti: Sister to the Brotherhood', *Textual Practice*, 2 (1988).

CATHERINE BELSEY chairs the Centre for Critical and Cultural Theory at Cardiff University. Her books include *Desire: Love Stories in Western Culture* and *Shakespeare and the Loss of Eden*.

JANET BROWNE is Reader in History of Biology at the Wellcome Institute for the History of Medicine, London. She is author of *Charles Darwin: Voyaging; The Secular Ark: Studies in the History of Biogeography;* and editor of the British journal, *History of Science.*

BARBARA BUSH is Principal Lecturer in History at Staffordshire University. She is author of *Slave Women in Caribbean Society, 1650–1938* and *Imperialism, Race and Resistance: Africa and Britain, 1918–1945.*

ALICE DOMURAT DREGER, Assistant Professor of Science and Technology Studies in the Lyman Briggs School of Michigan State University, focuses her research on the biomedical treatment of people born with unusual anatomies. She is author of *Hermaphrodites and the Medical Invention of Sex* and editor of *Intersexuality in the Age of Ethics.* Her essays on the human origins and impacts of science and medicine appear occasionally in the 'Science Times' section of the *New York Times.*

ANNE FAUSTO-STERLING is Professor of Medical Science at Brown University. She is author of *Myths of Gender: Biological Theories about Women and Men* and *Sexing the Body: Gender Politics and the Construction of Human Sexuality.*

SABINE GIESKE is Assistant Professor at the Institut für Europäische Ethnologie und Kulturforschung in Marburg (Germany). Her prize-winning dissertation was entitled 'Johann Gerhard Trimpe (1827–1894), Neubauer und

Weltbürger: Zum gesellschaftlichen und kulturellen Umbruch auf dem Land'. She is editor of *Lippenstift: Ein kulturhistorischer Streifzug über den Mund* and *Jenseits vom Durchschnitt: Vom Kleinsein und Grosssein.*

SANDER L. GILMAN is Henry R. Luce Distinguished Service Professor of the Liberal Arts in Human Biology and Chair of the Department of Germanic Studies at the University of Chicago. He is the author or editor of over fifty books, most recently *Love + Marriage = Death, and Other Essays on Representing Difference* and *Creating Beauty to Cure the Soul: Race and Psychology in the Shaping of Aesthetic Surgery.*

NILÜFER GÖLE teaches in the Department of Sociology at the Bogazici University in Istanbul, Turkey, and is the author of *The Forbidden Modern: Civilization and Veiling.*

LYNN HUNT is Eugen Weber Professor of Modern European History at the University of California, Los Angeles. She is author of *The Family Romance of the French Revolution* and *Telling the Truth about History* (with Joyce Appleby and Margaret Jacob), and editor of *Histories: French Constructions of the Past* (with Jacques Revel). She is now working on the origins of human rights.

MARY C. KENNEDY is Curator of Collections at the Schingoethe Center for Native American Cultures at Aurora University. In addition to gender in prehistory, her interests include the archaeological uses of radiocarbon dating and Native American material culture, past and present.

THOMAS W. LAQUEUR is Professor of History at the University of California, Berkeley. In 1981 he spent a year in medical school on an ACLS training grant. He is author of *Making Sex: Body and Gender from the Greeks to Freud* and editor (with Catherine Gallagher) of *The Making of the Modern Body: Sexuality and Society in the Nineteenth Century.* He continues to work on the history of sexuality and is also engaged in studies on the history of death, of memory, and of human rights.

NELLY OUDSHOORN is Professor of Philosophy of Science and Technology at the University of Twente and Lecturer in the Department of Science and Technology Dynamics at the University of Amsterdam. Her research interests include the social and material shaping of gender, bodies, and technologies. She is the author of *Beyond the Natural Body: An Archeology of Sex Hormones.*

LONDA SCHIEBINGER is Professor of History of Science at Pennsylvania State University. She is author of *The Mind Has No Sex? Women in the Origins of Modern Science*, the prize-winning *Nature's Body: Gender in the Making of Modern Science*, and *Has Feminism Changed Science?* She was awarded a Humboldt Forschungspreis to pursue her current research on gender in the European voyages of scientific discovery.

Alan C. Swedlund is Professor of Anthropology at the University of Massachusetts, Amherst. His research interests are in historical epidemiology of the United States and the history of physical anthropology. Recent publications have focused on infant mortality in late nineteenth- and early twentieth-century America, and the discourses of health reformers, eugenicists, and statisticians.

Jacqueline Urla is Associate Professor of Anthropology and Director of Modern European Studies at the University of Massachusetts, Amherst. She is co-editor (with Jennifer Terry) of *Deviant Bodies* and is completing an ethnography on language revival and cultural politics in the Basque Country of Spain.

Marina Warner is a historian, critic, and novelist. She has published widely on myths and fairy tales, including *Alone of All Her Sex: The Myth and Cult of the Virgin Mary*, *From the Beast to the Blonde*, and, most recently, *No Go the Bogeyman: Scaring, Lulling and Making Mock*. In 2000, she will be Visiting Professor at Stanford University.

Patty Jo Watson, Edward Mallinckrodt Professor of Anthropology at Washington University in St Louis, earned her MA and Ph.D. degrees in Anthropology at the University of Chicago. She has carried out archaeological fieldwork in Iraq, Iran, and Turkey; Arizona, New Mexico, Kentucky, and Tennessee. Her research foci include agricultural origins in Western Asia and Eastern North America, as well as ethnoarchaeology and archaeological theory and method.

Introduction

Londa Schiebinger

In the 1970s, feminists reinserted the body into history, bringing to light issues that had previously been considered too vulgar, trivial, or risqué to merit serious scholarly attention. Nineteenth-century gentlemen scholars—finely turned out in sombre suits and vests complete with pocket watches—thought it improper to have women students, let alone discuss topics so messy and seemingly exasperating as female menstruation or the cultural history of the breast. The mind/ body dualism that long underpinned Western culture made males the guardians of culture and the things of the mind, while it associated females with the frailties and contingencies of the mortal body. Females, subject to unruly humours, unpredictable hormones, and other forces, have been identified so closely with nature that nature itself is often called 'Mother Nature'.

Body studies are larger than feminism; critics such as Norbert Elias and Michel Foucault have enlarged our understanding of how the body is cultural and political as well as biological. Feminism was crucial for opening universities to women in the nineteenth and twentieth centuries but it also opened up university disciplines to new fields of study, including what we now call 'body studies'.[1] Because females have historically been identified with the body, most of the classic papers reprinted here focus on females and female bodies.

A central principle of feminist theory has been to recognize that gender differences are not fixed in the character of the species but arise from specific histories and from specific divisions of labour and power between the sexes. By the same token, it is important to look at the specific history forming feminist body studies and accounting for its theoretical structure. Feminist scholars have sought to unveil the mythos and power wrapped up in the notion that science is value-neutral.[2] American academic feminism came of age in the midst of a virulent sociobiology that, like its nineteenth-century precursors, pro-

1

claimed that 'biology is destiny'. Sociobiologists such as Harvard's E. O. Wilson taught that social order followed natural order: 'men forage for game or its symbolic equivalent in the form of barter and money', while women seek the male with the best genes, and then subsequently bear and raise his young.[3] Sociobiology and, more recently, parts of evolutionary psychology have attempted to rebuff the women's movement: if fundamentals, such as the sexual division of labour, are hard-wired into the species, efforts to counteract them must be foolhardy.[4]

Feminists—since the seventeenth century—have opposed the argument from 'nature' with an argument from 'nurture': in François Poullain de la Barre's resounding formula of 1673, 'the mind has no sex'.[5] In efforts to check the increasingly popular biological determinism, feminists in the 1960s introduced the term *gender*, distinguishing culturally specific forms of masculinity and femininity from biological 'sex', construed as anatomy, physiology, and chromosomes.[6] The primary force of body history ever since has been to show that universal, transhistorical masculine and feminine bodies do not exist apart from culture.

The crucial distinction between nature and nurture was further refined to appreciate two points. First, too strict a demarcation can obscure how 'nurture' (culture) can form 'nature' (the body). Peeling gender (culturally specific forms of behaviours and aspirations) off the body can create the impression that sex and sexual characteristics are natural objects, existing apart from culture and discoverable through science. Philosopher Susan Bordo and others have explored how culture can have a 'direct grip' on the body—etching ideals of thinness into human flesh, disciplining sexual practices, and sculpting bodies themselves.[7] We know, for example, that the relatively greater height of men compared to women—a supposed sexual characteristic—has closed over the past century, as women and girls quit eating what was left over after the men and boys had had their fill. In this case, as in the case of intersexuality discussed below, cultural ideals and practices directly form the body (nature) itself.

Second, driving a strict distinction between nature and nurture, feminists sought to break the stranglehold of arguments from nature—the conservative argument that by nature women are incapable of great things—and in the process a certain constructivism was let loose that tended to dissolve all bodily differences into political and cultural artefacts. Recent developments in medicine, however, have shown the advantage of taking nature seriously. The revolution in women's health research in the United States, epitomized in the 1990 founding of the NIH Office of Research on Women's Health and the 1991 Women's Health Initiative, rested on sharply distinguishing anatomical and physiological differences in the bodies of men and women. Current biomedical research has found previously unsuspected differences between men's and women's bodies: they often do not respond in the same way to particular drugs, their symptoms are not always the same (as, for example, in heart attacks), and they do not suffer the same degree from the same illnesses. In this sense, the feminist revolution in women's health care has led to a new appreciation of sexual differences.[8]

In this volume, I highlight the 'difference dilemma' in feminist studies—the irony that by calling attention to sexual differences we may reinforce them, but that by ignoring differences we may leave invisible power hierarchies in place.[9] I also highlight other strands of scholarship that focus not on bodies themselves but on how it is that sexed and gendered bodies function in modern biopolitics. A hallmark of early feminist theory was the notion that 'the personal is political'.[10] This, in concert with Foucault's work, was key in devising new understandings of politics and power. Several chapters in this volume treat realignments in biopolitics in the shift from the absolutist, corporate states of Europe to enlightened, democratic societies. In the old regime, the state and one's place in it was thought to be sanctioned by divine order: a king was born a king, a peasant was born a peasant. Sexual distinctions were by and large subordinate to birthright. The new 'enlightened' states, by contrast, attempted to build just societies on the laws of nature. The natural destiny of the individual—his or her intrinsic merit and ability—was to establish his or her 'natural' place in society. As a result, the supposed and real characters of bodies took on a new importance, reinforcing divides in political power and social well-being along the fault lines of race and sex.

This volume brings together a variety of interdisciplinary and, to the extent possible, cross-cultural essays on the body. I have chosen each article to illustrate a distinctive approach to body studies as well as to introduce a particular topic; in this sense the reader should consider each chapter as a microcosm opening onto a larger field of scholarship. Even so, many themes and subthemes worthy of attention could not be included because of space restrictions. The history of sexuality, for example, can hardly be separated from body history, but it has received but passing attention here (Robert Nye's new collection *Sexuality* provides a welcome supplement).[11] Another area not treated is women's health and medical care, especially in relation to childbirth, menstruation, menopause, and reproductive technologies, where so much innovative feminist work has been done; this topic is massive and worthy of a separate volume.[12] Much could also be said about the body in relation to religion—the body of the Virgin Mary, anorectic saints, Christ's body as a metaphor for the church, or the church as a lactating mother—which does not appear here.[13] This volume, however, brings together several 'classic' essays in feminist body studies in an effort to highlight dominant themes in current scholarship and to offer new perspectives for future research.

SCIENTIFIC (MIS)REPRESENTATIONS

Early feminist scholarship focused on scientific misreadings of female bodies, exposing the privileged first-born twins of modern science: the myth of the natural body and the myth of value-neutral knowledge. This literature reveals how science, like other human endeavours, is guided by social priorities and cultural ideals, and documents how scientific definitions of woman's 'nature' came to justify her exclusion from the public realms of politics and the learned professions.

We begin our exploration of scientific representation in the eighteenth century with my article, 'Skeletons in the Closet', that analyses the first portrayals of distinctively female skeletons in European anatomy. Although drawn from nature with painstaking exactitude, great debates erupted over the distinctive features of the female skeleton as she emerged on the eighteenth-century stage. Political circumstances drew immediate attention to depictions of the skull as a measure of intelligence and the pelvis as a measure of womanliness. The woman's supposedly narrower cranium seemed to explain her lesser achievement in science and culture, and justified denying her the rights of citizenship in the state, while her wide and capacious pelvis indicated that nature had intended her to be a loving mother confined to the private sphere of hearth and home.[14]

The timing of European anatomists' search for sexual differences and efforts to define the nature of women, I argue, arose both from changes internal to science and also from the loudly discussed 'woman question' about women's proper place in the newly envisioned democratic societies. In the eighteenth century, there was still great optimism that social issues—the intellectual, physical and moral character of women, for example—could be resolved by science. Perhaps the knife of the anatomist could find and define sexual difference once and for all. Perhaps sexual differences—even in the mind—could be weighed and measured.

Thomas Laqueur's classic essay, 'Amor Veneris', zeros in on male and female genitalia. The rise of feminist body studies in the 1970s took place within a vigorous women's health movement that put specula in women's hands and made it respectable to probe below the waist, an area heretofore shrouded in laced undergarments and embarrassed whispers.[15] In the same way that feminist historians brought to light forgotten and 'insignificant' women—seamstresses, prostitutes, and the like—body scholars spotlighted heretofore unspeakable body parts and functions, including breasts, the vagina, menstruation, and menopause. Laqueur's delightful essay discusses the historical discovery of the clitoris in 1559 (by Renaldus Columbus, the 'Christopher Columbus' of this new-found bit of female anatomy).[16] The struggle among male anatomists to claim scientific priority in this discovery assumed that women across the ages had known nothing of this intimate part of their own corporal geography.

Embedded in the story of the clitoris is one of Laqueur's major contributions to body studies: identifying a fundamental shift within the medical sciences from a 'one-sex' to a 'two-sex' model of human sexuality. Since Aristotle, woman was viewed by and large as built on the same architectural plan as man, but of lesser perfection, a 'monster', an 'error of nature'.[17] Within the one-sex model, no special terms existed for women's reproductive parts: the ovaries were female testicles, the clitoris was a female penis. As Laqueur shows, the shift to a 'two-sex' model underlay and promoted increasing impulses to identify sex as both a biological and a political category. 'To be a man or a woman [before the eighteenth century] was to hold a social rank, a place in society, to assume a cultural role, not to be organically one or the other of two incommensurable sexes.'[18] In the course of the Enlightenment, the sexes came to be seen as physically and morally distinct, each uniquely suited by nature for a separate social sphere: the robust male for the public sphere of science and government, industry and the military, and 'his better half' for the tender care of children and the home.

Nelly Oudshoorn's 'The Birth of Sex Hormones' takes us into the early twentieth century and uncovers the political imperatives fuelling the notion of females run wild by fluctuating hormones. Her 'archeology of sex hormones' reveals how cultural ideas about gender can lead researchers to attribute maleness or femaleness to things like hormones that, in fact, have no inherent sex or gender. With the discovery of sex hormones, scientists suggested that they had found the key to understanding what made a man a man and a woman a woman. This discovery fits well the two-sex model with testosterone parading as the male sex hormone and estrogen as the female sex hormone, when in fact both sexes need a full complement of both. Trouble threatened when Ernst Laqueur (Thomas's uncle) isolated the female hormone from the urine of stallions in the 1930s and raised the spectre of 'endocrinological androgyny'. The idea that the stallion—an emblem of robust masculinity since the eighteenth century—excretes high levels of estrogen in its urine was disconcerting to a society that sought neat and discrete divisions between the sexes.[19]

Any lingering notion that absolute distinctions can be made between sex and gender was exploded by both Anne Fausto-Sterling's argument for five sexes and Suzanne Kessler's demonstration that what we heretofore considered natural aspects of sexed bodies are often constructed either by biologists (in our understanding of sexual difference) or by medical doctors (in the flesh of the patient).[20] The first international meeting of the Intersex Society of North America held in the mid-1990s further detailed (in very personal terms) how the medical profession has tended to deal with intersexed children by fixing what are seen as ambiguous genitalia through many painful surgeries. In the most extreme cases, the subspeciality of the physician handling the case can determine the ultimate sex of the child.

Alice Dreger's history of the medical treatment of hermaphrodites in *fin de siècle* England and France highlights these issues through tragic tales of 'doubtful sex'. As Dreger tells, doubtful sex was not to be tolerated: too much was invested in the two-sex model of clear and distinct sexes. But what constituted 'true sex'? Dreger discusses the 'Age of Gonads' when true sex depended on possessing either ovaries or testicles, when the essence of femininity—'all that we admire and respect in a woman as womanly'—was thought to issue from the ovaries.[21] (Gonads were later unseated by sex hormones, the chemical messengers produced by the gonads, as discussed by Oudshoorn.) Are these methods of sex typing any more or less enlightened than ours in the late twentieth century, where we depend on penile function in males and reproductive function in females?

As feminist theorists in the 1980s and 1990s proliferated differences in order to better represent the contours of twentieth-century life, science kept pace. In the spring of 1998 researchers announced 'the lesbian ear'; scientists claimed to have found a biological marker for female homosexuality in the receptor cells in the cochlea, which in homosexual and bisexual women were found to be more like men's than like heterosexual women's.[22] This finding paralleled the search for 'gay genes' and brain structure which historically focused on males.[23] All of these attempts to identify and accentuate sexual difference—which might be labelled 'sexual science'—seek biological structures that will justify important social and political divides in contemporary society.[24]

THE BODY POLITIC

This section investigates not scientific representation of the body but the place of bodies in political orders, especially in the shift from early modern, monarchical states to modern, democratic polities. In early modern society, social place was by and large established by birth. There was little notion of individualism or social mobility dependent on individual merit. As head of state, the monarch's body took on great significance in defining a metaphorical 'body politic'. Kings in France and England were thought to have two bodies: a visible, corporeal, mortal body and an immaterial, symbolic, and sacred body that consisted of all the king's subjects, united harmoniously in the legal fiction of a corporate body. For some, the king was styled the 'head' of the body politic; his subjects were his 'members'.[25] In Thomas Hobbes's *Leviathan*, the relationship between absolute monarch and subject was that of a male sovereign whose body encompassed within it the bodies of his subjects. As historians Sara Melzer and Kathryn Norberg have emphasized, important rituals of the absolute state centred on the king's body: the royal body was anointed with holy chrism (oil) at his coronation; the royal touch was celebrated for curing scrofula and other virulent disease; royal potency was celebrated in public weddings and ceremonies for its abilities to secure a clear line of royal succession.[26]

The first essay in this section addresses these issues through the body of Queen Elizabeth I of England. Absolute monarchs in early modern Europe consolidated power through art, ceremony, and ritual. Portraiture aided audiences—many of whom lived in an oral rather than a written culture—in understanding the 'official' meaning of a sovereign's body. Philosopher Andrew Belsey and literary critic

Catherine Belsey discuss how portraits of Queen Elizabeth record her authority, wealth, and greatness, qualities that require absolute obedience. Yet, even in England where a woman could be queen (which was not the case in France), a female body threatened to disrupt the divine order of things. Consequently, the iconographic vocabulary for female sovereignty differed significantly from that of male sovereignty. The body and dress of King Henry VIII of England, Elizabeth's father, served as icons of masculinity and power, his extravagant codpieces denoting his virility, his wide, forceful stance and forthright gaze exuding power and kingship. The portraits of Elizabeth, by contrast, subdued the sexuality of this self-styled 'Virgin Queen' in order to proclaim her power, and in the process placed her outside the realm of nature.[27]

With a new political order towards the end of the eighteenth century, corporate society waned and the mythic individual with hypothetical inalienable rights became the foundation of the state. In the process the sacred, transcendental order visible in the body of the king dissolved, as historian Lynn Hunt has written, into the collectivity of bodies in the nation.[28] This new order was grounded not in divine right but in Nature and natural law—often as defined by anatomists and medical men. The natural qualities of individual bodies were now seen to place the individual within an economic, psychological, moral, and political order.

A major project of body studies has been to show how bodies came to mark and be marked with (in)equalities, the topic of the next three essays in this section. Lynn Hunt has brilliantly uncovered the cultural politics of bodily coverings: clothing. In the ancient regimes of Europe, strict sumptuary laws had regulated apparel to display social rank for both men and women. Even if wealthy merchant families could afford ermine, brocade, and certain laces, these were reserved for their social betters. The transition to modernity ushered in a bourgeois order, where male dress tended to blur these class distinctions. In what Hunt highlights as 'The Great Masculine Renunciation', men gave up their wigs and powdered hair, makeup, brightly coloured silks, knee-breeches, comely stockings, and high heels for the homogenizing effect of sombre colours, trousers, the business suit, and other occupational uniforms. In Western culture, masculinity—'the real man'—eschews fashion (and the vast sums females spend on fleeting fineries). Women, by contrast, have taken up the burden of ritualistic display of a family's social standing; in this transition female dress became more insistently tied up with consumerism and class distinctions. As a result, Hunt argues, class distinctions in male dress were blurred towards the end of the eighteenth century, while gender distinctions between men's and women's dress were heightened. Until the rapid masculinization of female clothing in the twentieth century, the simple dress of males contrasted sharply with the fashionable dress of females.[29]

But it was especially the body inside the clothing that came to place the individual in the new political order. With the rise of what Foucault has called 'political anatomy', the body, stripped as clean of history and culture as it was of clothes and sometimes skin, grounded political rights and social privileges.[30] Political anatomy—the intense search for sexual and racial differences characteristic of modern science—responded to the Enlightenment challenge that 'all men are by nature equal'. The expansive mood of the Enlightenment gave middle- and lower-class men, women, Jews, Africans, and other outsiders living in Europe or its colonies reason to believe that they, too, might begin to share the privileges heretofore reserved for European elites. Optimism rested in part on the ambiguities inherent in the word 'man' as used in revolutionary documents of the period. The 1789 *Declaration of the Rights of Man and Citizen* said nothing about race or sex, leading many to assume that the liberties it proclaimed held universally. The future president of the French National Assembly, Honoré-Gabriel Riqueti, comte de Mirabeau, declared that no one could claim that 'white men only are born and remain free [while] black men are born and remain slaves'.[31] Nor did the universal and celebrated 'man' seem to exclude women. Addressing the Convention in 1793, an anonymous woman declared: 'Citizen legislators, you have given men a constitution . . . as the constitution is based on the rights of man, we now demand the full exercise of these rights for ourselves.'[32]

Within this revolutionary republican framework, an appeal to natural rights could be countered only by proof of natural inequalities. In other words, if social inequalities—the slavery of Africans in American colonies and the continued disenfranchisement of women—were to be justified, scientific evidence would have to show that human nature is not uniform but differs according to age, race, and sex.

The eighteenth century saw the rise of scientific sexism, on the one hand, and scientific racism, on the other. These two movements shared many key features: both regarded women and non-European men as deviations from the European male norm; both deployed scientific methods to measure and discuss difference; both sought natural foundations to justify social inequalities between the sexes and races. I have argued elsewhere that these two movements also suffered from certain fundamental asymmetries, namely, that scientific sexism drove distinctions between European bodies (male and female), while scientific racism focused primarily on male bodies (of different races). As a result, neither the dominant theory of race nor that of sex in this period applied to women of non-European descent. Like other females, women of African origins did not fit comfortably in the great chain of racial being; like other Africans, they did not fit European gender ideals.[33]

In her essay, 'Gender, Race, and Nation: The Comparative Anatomy of "Hottentot" Women in Europe, 1815–1817', biologist Anne Fausto-Sterling investigates the scientific study of the Khoikhoi women, known in nineteenth-century Europe as 'Hottentots', and highlights the most infamous of these cases, the dissection in 1815 of Sarah Bartmann by France's leading figure in comparative anatomy, Georges Cuvier.[34] The abundant nineteenth-century ideology idealizing woman as the angel of the home did not apply to women like Sarah Bartmann. The very name given her—Cuvier always referred to her as *Vénus Hottentotte*—emphasized her sexuality. (Passionate tendencies found in warm climates were often attributed to the planetary influence of Venus.) His interest in her body focused on her sexual parts; he devoted many pages to recording the dissection of her genitalia, breasts, buttocks, and pelvis, but only one short paragraph to her brain. In his memoir on the Hottentot Venus, Cuvier took up the issue of whether science had African origins: 'No race of Negro', he declared, 'produced that celebrated people who gave birth to the civilization of ancient Egypt, and from whom we may say that the whole world has inherited the principles of its laws, sciences, and perhaps also religion.' Without exception, the 'cruel law' of nature, he concluded, had 'condemned to eternal inferiority those races with a depressed and compressed cranium'.[35] Such was the fate of Sarah Bartmann.

Europeans also produced what might be called 'colonial bodies'. These developments, taking place outside Europe, held important implications for regimes of body inside Europe. In *Torrid Zones*, literary critic Felicity Nussbaum has described the mutually reinforcing dualities of the domestic and the exotic, the civil and the savage, the passionless virtue of European women and the relentless sexualization of colonized females (and feminizing of colonial males).[36] This literature, like the others discussed here, is immense. I have chosen to draw attention to the colonial bodies of Caribbean slave women under English rule. Slavery, it need not be said, savagely attacked bodies. Slave bodies were mangled by the whip, broken on the wheel, and disfigured by the surgeon's knife (run-a-ways' Achilles tendons were sliced at the first offence, a leg amputated at the third). To find release from their extreme suffering, some slaves swallowed their own tongues, ate dirt, and even leapt into cauldrons of boiling sugar, 'thus at one blow depriving . . . [the master] of his crop and his servant'.[37] On their side, slaves who managed to escape attacked the bodies and property of their oppressors, slicing open the bellies of their former mistresses large with child, poisoning entire plantations with clandestine substances carried under a single finger nail, and burning plantations.

All too often study of the bodies of non-Europeans mires in victimology. Barbara Bush's work, by contrast, reveals how Caribbean slave women used their bodies as a political weapon against the brutalities of slavery. Far from passive victims, these women developed a politics of resistance by quietly killing the infants in their wombs by means of well-known plant poisons to spite masters keen to have more 'manpower' to fuel colonial economies. For many reasons (including self-induced abortion), West Indian slave populations did not reproduce themselves; plantation owners were forced continually to purchase new slaves from Africa.[38]

....

EMBODIED IDEALS

....

Bodies have not only been prodded and poked by scientists and defined by legislatures as rights or non-rights bearing individuals, they have also come to embody—at different times and in different ways—cherished religious, scientific, and cultural ideals.

'Liberty', 'Equality', 'Fraternity', and other grandly abstract ideals underpinning democratic orders have taken concrete form in bodies, many of them imaged as female. In her chapter on visual representations of 'Liberty', Marina Warner decodes the icons of modern political orders that have become so familiar to us that we rarely think about their meanings. Warner addresses the question of why modern societies perpetuate massive allegories (blind Justice and her scale, the Statue of Liberty standing majestically in the New York Harbour) when revolution sought to topple such seemingly outmoded ways of thinking.[39] Absolutist regimes enjoyed extended allegory in the bodies of monarchs, as discussed above, but Elizabeth I was a real historical woman whose body was given universal meaning above and beyond its material form. Lady Liberty, by contrast, is a mythic female approximating a goddess—at one and the same time everywoman and no woman—who has come to embody cardinal virtues of the state.

Science, too, has had its prized images and modes of thought. Feminist theory cut its teeth on attempting to displace the dualistic notion that 'nature is to female as culture is to male'.[40] The association of female and body served as a catalyst to feminist body studies, as mentioned above. Carolyn Merchant's classic 'Nature as Female' explored the historical origins of another gender ideal: the persistent sexing of nature as female—alternatively mother, virgin, whore.[41] My own work on the iconography of early modern science revealed that *Scientia* was also imaged female—much like Delacroix's looming Liberty—until the end of the eighteenth century when scientists (in efforts to secure the clean-cut line of objectivity) self-consciously stripped science of all metaphysics, poetry, and rhetorical ornament.[42] While *Natura* and *Scientia* were imaged female, they were foils to a historically real protagonist, the scientist himself who for most of history has, in fact, been male.

Body studies have not given sufficient attention to the gendering of non-human nature. If we shift the focus from human to non-human bodies, new topics, perspectives, and concerns come into view. Nature's body has been sexed and gendered in various ways, as when in the late seventeenth century human (hetero)sexuality was overlaid onto innocent and unsuspecting plants.[43] Or, as Donna Haraway in her path-breaking *Primate Visions* has shown, in the twentieth century human notions of proper bourgeois behaviour infused understandings of animal bodies and characters, especially those of anthropoid apes. Visions of female apes especially have paralleled those of female humans in the past several decades: female apes in the 1950s and 1960s were seen as docile creatures (stay-at-home moms) who traded sex and reproduction for protection and food; in the 1970s they were seen as 'liberated females' who could be the equal of any male; and in the 1980s through to the 1990s, females—both human and ape—were often celebrated as unique creatures who expressed themselves 'in a different voice'.[44] Because the argument from nature is still strong in Western culture, these images of our evolutionary ancestors often prescribe and reinforce ideals for human bodies and behaviours in culture and society.[45]

Another way that our understandings of nature and nature's body have been gendered is in the association between animals and men. In human evolution, the virile 'man-the-hunter' has long been hypothesized as a key force driving early hominids into the human state, characterized by bi-pedalism and extensive brain development.[46] The gendering of nature has led, on the other hand, to a long-standing association between plants and women. Many years ago, the German philosopher Georg Wilhelm Friedrich Hegel compared the female mind to a plant because, in his view, both were essentially placid. The 'woman-the-gatherer' hypothesis, that emerged in the 1970s as a counterweight to the dominant 'man-the-hunter', perpetuated this association. Patty Jo Watson and Mary Kennedy's essay on the origins of horticulture shows that the associations man/hunt/animals/active vs. women/gather/plants/passive continue to drive archaeology today. According to Watson and Kennedy, when evidence suggests that women were primarily responsible for important cultural innovations such as the domestication of plants, the association of women with nature has led archaeologists to assume that 'plants virtually domesticate themselves'.[47] Women are not seen as active creators or inventors of culture, but as passive tenders of nature.

Finally, scholars are just beginning to turn attention to how humans have made and remade nature. European global mercantile and colonial expansion beginning in the sixteenth century changed the face of nature; naturalists not only studied plants and animals but moved them from place to place (sometimes intentionally, sometimes not). Historian Richard Grove has cited the problem of goats on St Helena, an island in the South Atlantic, introduced in 1533 as a ready meat supply to provision passing European ships. By 1582 the animals had completely overrun the island—'thousands of goats were seen up to 200 together and sometimes in a flock almost a mile long'—so that within a few years the island was completely deforested.[48] Similar episodes took place throughout the world—massive monocultural agriculture undercut biodiversity in many areas, while extensive seed exchange between botanical gardens and horticultural stations around the world promoted new mixes of flora in other areas. Human populations, too, were moved about through migration and diaspora; in some cases entire human populations were made extinct through warfare and disease.[49] These areas of study—for which the gender implications remain largely unexplored—provide new vistas for expanding body studies beyond the merely human.

MASCULINITIES

It may seem that this volume has focused inordinate attention on the female body. The same forces in Western culture that have identified female with body have forged the Cartesian transubstantiation of masculine body into mind.[50] Great men of science are often celebrated for ignoring bodily appetites: otherwise occupied, Sir Isaac Newton is said to have forgotten to eat the finely roasted chicken served to him in his study, and William Hamilton left half empty plates accumulating for days as he worked.[51] By leaving male bodies unscrutinized, feminists have tended to reinforce the notion of the male as the unmarked sex, the human standard of perfection from which the female can only deviate.[52]

Bodies, even philosophers' bodies, do, however, have bearing on the life of the mind. Recent scholarship has called into question the notion that there exists a 'free play of imagination', a transcendent and disembodied truth, guaranteed precisely because it stands apart from the body. Janet Browne's masterful essay on Charles Darwin explores how Darwin's ill health (his painful retching and antisocial flatulence) influenced his scientific career. Darwin's poor health eventually became the rationale for the premature publication of his *On the Origin of Species* (his preface mentions health considerations, not the fact that A. R. Wallace threatened to scoop his life-time of work). Furthermore, Browne argues, Darwin used his illness to shield himself from scientific controversy; in his later years he rarely went out, citing ill health as his reason for not engaging in public debates concerning human evolution. In the second part of her essay, Browne analyzes the photographic portraits made of Darwin over his lifetime. This study of the cultural presentations of the 'great man of science' compares nicely to the portrayals of royal power discussed in other sections of this volume.[53]

Continuing the analysis of masculinities, Sander Gilman's essay reveals the interplay between anti-Semitism and dominant forms of masculinity in his perusal of the cultural meanings of the Jewish male foot. During the Middle Ages, Christians associated the Jew's foot with the cloven hooves of the devil. Gilman tells how during the nineteenth century the supposed failings of the Jewish foot were modernized and reformulated. Now Jewish males' 'weak' feet were seen as making them unfit for German military service, thus stripping them of the robust masculinity prescribed by bourgeois culture and relegating them to second-class citizenship within the nation state. Gilman remarks how the epistemological foundations shifted from the rhetoric of religious anti-Judaism to the rhetoric of scientific anti-Semitism. As with women's bodies, pre-modern prejudices are reinscribed into the modern body politic, now in the language of science. The deeply-rooted belief that moral character is rooted in the body raised Jewish flat-footedness to an emblematic racial characteristic under Nazism.[54]

The German ethnologist Sabine Gieske in her essay, 'The Ideal Couple: A Question of Size?', draws our attention to the cultural imperative that romantic couples consist of a man who stands significantly taller than his female mate. Gieske notes that in Western cultures a short man escorting a tall woman can evoke the same kind of looks and unconscious reprimands as a couple of mixed race. The notion of a large, robust male and a petite, delicate female arose towards the end of the eighteenth century as part of the same shift in politics and culture from early modern to modern regimes that is discussed in a number of the essays in this volume. Gieske points out that considerations of height played an insignificant role for aristocratic couples in the *ancien régime*. Within this class, marriages were arranged primarily to enhance property holdings, political power, or social standing. A couple might be engaged or even married at such an early age that their relative mature heights simply were not known. Gieske argues further that in this period upper-class fashion tended to accentuate rather than diminish a woman's size. Women could easily overpower many a man visually with their towering powdered wigs, high-heeled shoes, and wide skirts.

The rise of the propertied middle-classes brought with it the ideal of the big man and little woman. The ideal man, as citizen in the state and actor in the competitive marketplace, was supposed to be large, forceful, and protective of his dependent wife. Gieske notes that fashion helped where nature failed: the top hat, invented in the nineteenth century, served to add a good six inches to a male so that if he were the same height or slightly shorter than his wife, in public at least he appeared to be of appropriate stature.

RESTRAINED BODIES

Not only have cultural ideals been embodied in various allegorical images, they have also been etched into the flesh and blood of very real bodies. Jacqueline Urla and Alan Swedlund's essay on Barbie provides a fine example. An emblem of hyperfemininity, this platinum bombshell with the hourglass figure and feet permanently imprinted for her impossibly high heels remained in character when, in 1992, she uttered her first words and told the most recent of her 800-million owners that 'math class is tough'. Barbie has not only been a huge commercial success for over forty years, her unreal proportions have also dangerously disciplined the bodies of young American girls. As the vast scholarship on eating disorders and anorexia in the United States has revealed, Barbie's impossibly slender waist plummets girls as young as 9 years old into depression and even death as they try to make their own bodies conform to the ideal.[55] Barbie was originally fashioned for white American girls. The makers of Barbie, however, not to miss a potential market, attempted to broaden the appeal of their product by introducing 'Black Barbie' in 1980, followed by Spanish Barbie, Jamaican Barbie, Malaysian Barbie, and so forth.[56]

In this wonderfully playful essay, Urla and Swedlund undertake a physical anthropology of Barbie, analysing her dimensions in an effort to understand why the 'big-hair girl' represents fond notions of femininity in many contemporary Western societies. While the authors emphasize the conjunction between consumption and femininity, we should note that Western men are increasingly being brought into the commercial nexus that promises youth and good looks; each year more men succumb to plastic surgery, facials, and treatments for baldness and impotence.

Other cultures, too, have etched cultural ideals of femininity directly into the body. Foot-binding in China was prohibited in 1644, but commonly practised until 1958. In his essay, C. Fred Blake calls attention to the 'mindful body' (a notion taken from medical anthropologists Nancy Scheper-Hughes and Margaret Lock) in order to understand this deformation of the body perpetrated by loving mothers upon young daughters. As Blake explains, foot-binding represents a body mindful of its fate (a mother's love, a daughter's duty) and a body mindful of the ever-present need to exercise self-control. Blake further argues that foot-binding placed women in the 'body politic' of Chinese society, recruiting women's bodies into labour-intensive production of economic goods and equally labour-intensive reproduction of patriarchal families. In addition to thinness and foot-binding, other examples of female-controlled body sculpting and decimation include clitorectomy and sati, the Hindi custom of a widow burning herself on the funeral pyre of her dead husband.[57]

The final essay in this volume takes up the issue of veiling within Muslim communities. Clothing signifies different things in different cultures. The body often derives meaning and mystery to the extent that it is revealed or concealed by clothing. Veiling—the wearing of headcoverings or long gowns that cover sometimes the entire body and face, which is practised by some Muslims—appears to Western eyes as an extreme form of negating the body by hiding it or at least the sexually charged aspects of the body, such as a woman's hair, from sight. While Westerners often flaunt a scantly clothed body, Easterners tend to conceal especially the female body.

Reactions to veiling have differed over the centuries. Lady Mary Wortley Montagu, wife of the British ambassador to Adrianople in the early eighteenth century, felt veiling enhanced women's social freedoms, since it allowed them to walk out in the streets without fear of being molested or recognized. She even suggested that this form of dress permitted women to carry out secret assignations without detection and wrote, 'the Turkish ladies . . . have more liberty than we have.'[58] Nilüfer Göle, in 'The Forbidden Modern', evaluates veiling quite differently. The veil, she urges, is the symbol *sine qua non* of the 'otherness' of Islam to the West; Islamic veiling is a political issue in both Muslim and Western European countries. Rather than seeing this as a practice restricting or enhancing women's freedoms and opportunities, she argues that veiling is a living tradition tied up in the politics of Islamic self-identity and its adherence to 'modernism' both inside Turkey and in relation to the West.[59]

The essays in this volume represent only the tip of the scholarly iceberg that comprises feminist body studies. Considerations of space have in some cases required me to shorten endnotes, remove textual notes, and delete illustrations. For a fuller treatment of each topic, readers are referred to the authors' books and articles from which these selections have been reprinted. A word of thanks to my many students whose good humour and comments on this volume are warmly appreciated. Special thanks also to Robert Proctor, Joan Landes, Mary Pickering, Lynne Fallwell, Sarah Goodfellow, Debra Hawhee, and Julie Vedder who contributed in various ways to the success of this project. Thanks, too, to Teresa Brennan and Susan James, the series editors, for their interest in this topic, and to Tim Barton, Angela Griffin, and Lesley Wilson at Oxford University Press for seeing the project successfully through the press.

Notes

1. Bryan Turner, 'Recent Developments in the Theory of the Body', and Arthur Frank, 'For a Sociology of the Body: An Analytical Review', in Mike Featherstone, Mike Hepworth, and Bryan Turner (eds.), *The Body: Social Process and Cultural Theory* (London: Sage Publications, 1991), pp. 1–35; Susan Bordo, *Unbearable Weight: Feminism, Western Culture, and the Body* (Berkeley: University of California Press, 1993), intro.; Carolyn Bynum, 'Why All the Fuss about the Body: A Medievalist's Perspective', *Critical Inquiry*, 22 (1995).
2. Evelyn Fox Keller and Helen Longino (eds.), *Feminism and Science* (Oxford: Oxford University Press, 1996).
3. Edward O. Wilson, *Sociobiology: The New Synthesis* (Cambridge, Mass.: Harvard University Press, 1975), p. 553.
4. For a critique of evolutionary psychology, see Natalie Angiers, *Woman: An Intimate Geography* (Boston: Houghton Mifflin, 1999).
5. François Poullain de la Barre, *De l'égalité des deux sexes: Discours physique et moral* (Paris, 1673); also the classic statement by Simone de Beauvoir: 'women are made, not born' in *The Second Sex*, trans. H. M. Parshley (1949; New York: Vintage Books, 1974).
6. As the first use of 'gender' in this sense, the *Oxford English Dictionary* cites A. Comfort, *Sex in Society* (1963), ii. 42; see also Ann Oakley, *Sex, Gender, and Society* (London: Temple Smith, 1972).
7. Bordo, *Unbearable Weight*, p. 17.
8. Londa Schiebinger, *Has Feminism Changed Science?* (Cambridge, Mass.: Harvard University Press, 1999), ch. 6.
9. Martha Minnow, 'Learning to Live with the Dilemma of Difference: Bilingual and Special Education', *Law and Contemporary Problems*, 48 (1984), 157–211.
10. For current feminist political theory, see Joan Landes (ed.), *Feminism, the Public and the Private* (Oxford: Oxford University Press, 1998).
11. Robert Nye (ed.), *Sexuality* (Oxford: Oxford University Press, 1999).
12. See e.g. Emily Martin, *The Woman in the Body: A Cultural Analysis of Reproduction*, 2nd edn. (Boston: Beacon, 1992); Margaret Lock, *Encounters with Aging: Mythologies of Menopause in Japan and North America* (Berkeley: University of California Press, 1993); Elizabeth Fee and Nancy Krieger (eds.), *Women's Health, Politics, and Power: Essays on Sex/Gender, Medicine, and Public Health* (Amityville, NY: Baywood Publishing, 1994); Sheryl Ruzek, Adele Clarke, and Virginia Olesen (eds.), *Women's Health: Complexities and Differences* (Columbus: Ohio State University Press, 1997).
13. Carolyn Bynum, *Fragmentation and Redemption: Essays on Gender and the Human Body in Medieval Religion* (New York: Zone Books, 1991); Sarah Coakley, *Religion and the Body* (Cambridge: Cambridge University Press, 1997).
14. Londa Schiebinger, *The Mind Has No Sex? Women in the Origins of Modern Science* (Cambridge, Mass.: Harvard University Press, 1989), ch. 7; Elizabeth Fee, 'Nineteenth-Century Craniology: The Study of the Female Skull', *Bulletin of the History of Medicine*, 53 (1979), 415–33; Stephen Jay Gould, *The Panda's Thumb: More Reflections in Natural History* (New York: Norton, 1980), ch. 14.
15. Boston Women's Health Book Collective, *Our Bodies, Ourselves: A Book by and for Women* (New York: Simon and Schuster, 1973).

16. Thomas Laqueur, 'Amor Veneris, vel Dulcedo Appeletur', in Michel Feher (ed.), *Fragments for a History of the Human Body* (New York: Zone, 1989), pp. 90–131. For a recent history of the formerly ineffable, see Rachel Maines, *The Technology of Orgasm: 'Hysteria,' the Vibrator, and Women's Sexual Satisfaction* (Baltimore: Johns Hopkins University Press, 1999).

17. Kathleen Mendelsohn, Linda Nieman, Krista Isaacs, Sophia Lee, and Sandra Levison, 'Sex and Gender Bias in Anatomy and Physical Diagnosis Text Illustrations', *Journal of the American Medical Association*, 272 (26 October 1994), 1267–70.

18. Thomas Laqueur, *Making Sex: Body and Gender from the Greeks to Freud* (Cambridge, Mass.: Harvard University Press, 1990).

19. Nelly Oudshoorn, *Beyond the Natural Body: An Archeology of Sex Hormones* (London: Routledge, 1994).

20. Anne Fausto-Sterling, 'The Five Sexes', *The Sciences* (March/April 1993), 20–5; Suzanne Kessler, 'The Medical Construction of Gender: Case Management of Intersexed Infants', *Signs: Journal of Women in Culture and Society*, 16 (1990), 3–26.

21. Alice Dreger, *Hermaphrodites and the Medical Invention of Sex* (Cambridge, Mass.: Harvard University Press, 1998). For the Intersex Society of North America, see <http://www.isna.org>.

22. Constance Holden, 'A Marker for Female Homosexuality?' *Science*, 279 (13 March 1998), 1639.

23. Simon LeVay, *Queer Science: The Use and Abuse of Research into Homosexuality* (Cambridge, Mass.: MIT Press, 1996).

24. Cynthia Russett, *Sexual Science: The Victorian Construction of Womanhood* (Cambridge, Mass.: Harvard University Press, 1989).

25. Ernst Kantorowicz, *The King's Two Bodies: A Study in Medieval Political Theology* (Princeton: Princeton University Press, 1957).

26. Sara Melzer and Kathryn Norberg, *From the Royal to the Republican Body: Incorporating the Political in Seventeenth- and Eighteenth-Century France* (Berkeley and Los Angeles: University of California Press, 1998), intro., esp. 1.

27. Andrew Belsey and Catherine Belsey, 'Icons of Divinity: Portraits of Elizabeth I', in Lucy Gent and Nigel Llewellyn (eds.), *Renaissance Bodies: The Human Figure in English Culture* (London: Reaktion Books, 1990), pp. 11–35.

28. Lynn Hunt, 'Freedom of Dress in Revolutionary France', in Melzer and Norberg (eds.), *From the Royal to the Republican Body*, pp. 224–49.

29. Ibid. Hunt notes that the phrase 'The Great Masculine Renunciation' is J. C. Flügel's. See also Philippe Perrot, *Fashioning the Bourgeoisie: A History of Clothing in the Nineteenth Century*, trans. Richard Bienvenu (Princeton: Princeton University Press, 1994).

30. Michael Foucault, *Discipline and Punish: The Birth of the Prison*, trans. Alan Sheridan (New York: Pantheon, 1977), p. 193.

31. Cited in Charles Hardy, *The Negro Question in the French Revolution* (Menasha, Wis.: George Banta Publishing Co., 1919), p. 15. Condorcet expressed similar sentiments in his 'Lettres d'un bourgeois de New Haven à un citoyen de Virginie' (1787), *Oeuvres de Condorcet*, ed. A. O'Connor and M. F. Arago (Paris, 1847), ix. 15–19.

32. Cited in Jane Abray, 'Feminism in the French Revolution', *American Historical Review*, 80 (1975), 48.

33. Londa Schiebinger, *Nature's Body: Gender in the Making of Modern Science* (Boston: Beacon Press, 1993).

34. Anne Fausto-Sterling, 'Gender Race, and Nation', in Jennifer Terry and Jacqueline Urla (eds.), *Deviant Bodies* (Bloomington: Indiana University Press, 1995), pp. 19–48.

35. Georges Cuvier, 'Extrait d'observations faits sur le cadavre d'une femme connue à Paris et à Londres sous le nom de Vénus Hottentotte', *Mémoires du muséum d'histoire naturelle*, 3 (1817), 259–74, esp. 272–3.

36. Felicity Nussbaum, *Torrid Zones: Maternity, Sexuality, and Empire in Eighteenth-Century English Narratives* (Baltimore: Johns Hopkins University Press, 1995); Mrinalini Sinha, *Colonial Masculinity: The 'Manly Englishman' and the 'Effeminate Bengali' in the Late Nineteenth Century* (Manchester: Manchester University Press, 1995); and David Arnold, *Colonizing the Body: State Medicine and Epidemic Disease in Nineteenth-Century India* (Berkeley: University of California Press, 1993).

37. John Stedman, *Narrative of a Five Years Expedition against the Revolted Negroes of Surinam* (1796; repr. Baltimore: Johns Hopkins University Press, 1988), p. 272.

38. Barbara Bush, 'Hard Labor: Women, Childbirth, and Resistance in British Caribbean Slave Societies', *History Workshop Journal*, 36 (1993); see also Bush, *Slave Women in Caribbean Society, 1650–1838* (Bloomington: Indiana University Press, 1990).

39. Marina Warner, *Monuments and Maidens: The Allegory of the Female Form* (New York: Atheneum, 1985). For recent work on this topic, see Sumathi Ramaswamy, 'Body Language: The Somatics of Nationalism in Tamil India', *Gender & History*, 10 (1998), 78–109.

40. Sherry Ortner, 'Is Female to Male as Nature is to Culture?' *Feminist Studies*, 1 (1972), 5–31.

41. Carolyn Merchant, *The Death of Nature: Women, Ecology, and the Scientific Revolution* (San Francisco: Harper & Row, 1980).

42. Schiebinger, *The Mind Has No Sex?*, ch. 5.

43. Schiebinger, *Nature's Body*, chs. 1–3.

44. Carol Gilligan, *In a Different Voice: Psychological Theory and Women's Development* (Cambridge, Mass.: Harvard University Press, 1982).

45. Donna Haraway, *Primate Visions: Gender, Race, and Nature in the World of Modern Science* (New York: Routledge, 1989).

46. Adrienne Zihlman, 'The Paleolithic Glass Ceiling: Women in Human Evolution', in Lori Hager (ed.), *Women in Human Evolution* (New York: Routledge, 1997), pp. 91–113.

47. Patty Jo Watson and Mary Kennedy, 'The Development of Horticulture in the Eastern Woodlands of North America: Women's Role', in Joan Gero and Margaret Conkey (eds.), *Engendering Archaeology: Women and Prehistory* (Oxford: Blackwell, 1991), pp. 255–75.

48. Richard Grove, *Green Imperialism: Colonial Expansion, Tropical Island Edens and the Origins of Environmentalism, 1600–1860* (Cambridge: Cambridge University Press, 1995), p. 96.

49. Alfred Crosby, *Ecological Imperialism: The Biological Expansion of Europe, 900–1900* (Cambridge: Cambridge University Press, 1986).

50. Genevieve Lloyd, *The Man of Reason: 'Male' and 'Female' in Western Phil-

osophy (Minneapolis: University of Minnesota Press, 1984); David Noble, *A World Without Women: The Christian Clerical Culture of Western Science* (New York: Knopf, 1992); Nancy Tuana, *The Less Noble Sex: Scientific, Religious, and Philosophical Conceptions of Woman's Nature* (Bloomington: Indiana University Press, 1993).

51. David Brewster, *The Life of Sir Isaac Newton* (London: John Murray, 1831), p. 341.

52. On this point, see Nelly Oudshoorn, 'On Bodies, Technologies, and Feminisms', in Angela Creager, Elizabeth Lunbeck, and Londa Schiebinger (eds.), *Science, Technology, and Medicine: What Difference has Feminism Made?* (Chicago: Chicago University Press, forthcoming).

53. Janet Browne, 'I Could Have Retched All Night: Charles Darwin and His Body', in Christopher Lawrence and Steven Shapin (eds.), *Science Incarnate: Historical Embodiments of Natural Knowledge* (Chicago: University of Chicago Press, 1998), pp. 240–87.

54. Sander Gilman, *The Jew's Body* (New York: Routledge, 1991).

55. Bordo, *Unbearable Weight*; Joan Brumberg, *The Body Project: An Intimate History of American Girls* (New York: Random House, 1997).

56. Jacqueline Urla and Alan Swedlund, 'The Anthropometry of Barbie: Unsettling Ideals of the Feminine Body in Popular Culture', in Terry and Urla (eds.), *Deviant Bodies*, pp. 277–313.

57. C. Fred Blake, 'Foot-binding in Neo-Confucian China and the Appropriation of Female Labor', *Signs*, 19 (1994), 676–712.

58. Christopher Pick (ed.), *Embassy to Constantinople: The Letters of Lady Mary Wortley Montagu* (New York: New Amsterdam, 1988), 'to Lady Mar—Adrianople April 1, 1717', p. 111.

59. Nilüfer Göle, *The Forbidden Modern: Civilization and Veiling* (Ann Arbor: University of Michigan, 1996); see also Lila Abu-Lughod, *Veiled Sentiments: Honor and Poetry in a Bedouin Society* (Berkeley: University of California Press, 1986).

Part I. Scientific (Mis)representations

1 Skeletons in the Closet: The First Illustrations of the Female Skeleton in Eighteenth-Century Anatomy

Londa Schiebinger

....

INTRODUCTION
....

In 1796 the German anatomist Samuel Thomas von Soemmerring published in a separate folio what he claimed to be the first illustration of a female skeleton.[1] This was a remarkable claim since Andreas Vesalius and modern anatomists had drawn human skeletons from observation and dissection since the sixteenth century. Though Soemmerring's claim to originality was a bit exaggerated, he was, indeed, among the first illustrators of a distinctively female skeleton. But the importance of his illustration goes beyond the fact that it was one of the first. With his drawing of a female skeleton, Soemmerring became part of the eighteenth-century movement to define and redefine sex differences in every part of the human body. It was in the eighteenth century that the doctrine of humors, which had long identified women as having a unique physical and moral character, was overturned by modern medicine. Beginning in the 1750s, doctors in France and Germany called for a finer delineation of sex differences; discovering, describing, and defining sex differences in every bone, muscle, nerve, and vein of the human body became a research priority in anatomical science. It was as part of this broader search for sex differences that the drawings of the first female skeletons appeared in England, France, and Germany between 1730 and 1790.

What sparked interest in the female skeleton in the eighteenth century? Did portrayals of female bones follow the 'clean cut' of scientific objectivity? Or was interest in female anatomy molded by broader

From Londa Schiebinger, 'Skeletons in the Closet: The First Illustrations of the Female Skeleton in Eighteenth Century Anatomy' in *Representations*, 14 (1986): 42–82. Reprinted with permission.

social movements? Was there a connection between eighteenth-century movements for women's equality and attempts on the part of anatomists to discover a physiological basis for female 'inequality'?

I argue here that it was in the context of the attempt to define the position of women in European society that the first representations of the female skeleton appeared in European science. The interests of the scientific community were not arbitrary: anatomists focused attention on those parts of the body that were to become politically significant. When the French anatomist Marie-Geneviève-Charlotte Thiroux d'Arconville published drawings of the female skeleton in 1759, she portrayed the female skull as smaller than the male skull, and the female pelvis as larger than the male pelvis.[2] This was not, however, simply the product of the growth of realism in anatomy. The depiction of a smaller female skull was used to prove that women's intellectual capabilities were inferior to men's. This scientific measure of women's lesser 'natural reason' was used to buttress arguments against women's participation in the public spheres of government and commerce, science and scholarship. The larger female pelvis was used in parallel fashion to prove that women were naturally destined for motherhood, the confined sphere of hearth and home.

'Nature' played a pivotal role in the rise of liberal political thought. In the seventeenth and eighteenth centuries, natural-law philosophers such as Locke and Kant sought to found social convention on a natural basis.[3] Appeals to 'the natural reason and dignity of man' provided important philosophical underpinnings for arguments in favor of individual freedom and equality. Yet, when women asked for equality, they were denied it.

It is important to understand how the eighteenth-century denial of civil rights to women could be justified within the framework of liberal thought. To the mind of the natural-law theorist, an appeal to natural rights could be countered only by proof of natural inequalities. In his *Lettres Persanes*, Montesquieu, for example, framed the crucial question: 'Does natural law [*loi naturelle*] submit women to men?'[4] Nature and its law were considered above human politics. According to this view, there exists an order in nature that underlies the well-ordered *polis*. If social inequalities were to be justified within the framework of liberal thought, scientific evidence would have to show that human nature is not uniform but differs according to age, race, and sex.

In the course of the eighteenth and nineteenth centuries, the study of the 'nature' of woman became a priority of scientific research. The increasing tendency to look to science as an arbiter of social questions depended on the promise that hotly debated social issues—such as women's rights and abilities—could be resolved in the cool sanctuaries of science. When modern science first turned an 'impartial' eye to the study of women, however, there emerged a paradox that still plagues women's relationship to science. The 'nature' and capacities of women were vigorously investigated by a scientific community from which women (and the feminine) were almost entirely absent. As a consequence, women had little opportunity to employ the methods of science in order to revise or refute the emerging claims about the nature of women. As science gained social prestige in the course of the nineteenth century, those who could not base their arguments on scientific evidence were put at a severe disadvantage in social debate. Thus emerged a paradox central to the history of modern science: women (and often what they value) have been largely excluded from science, and the results of science often have been used to justify their continued exclusion.

I want to stress from the outset that it is not my purpose to explain away physical differences between men and women but to analyze social and political circumstances surrounding the eighteenth-century search for sex differences. This study of the first representations of the female skeleton is intended to serve as a case study of more general problems. Why does the search for sex differences become a priority of scientific research at particular times, and what political consequences have been drawn from the fact of difference? As we will see, the fact of difference was used in the eighteenth century to prescribe very different roles for men and women in the social hierarchy. In the course of the eighteenth century, some anatomists were even moved to believe that women held a lowly rank in the natural hierarchy; it became fashionable to find in women the qualities of children and 'primitives'. By locating woman's social worth in her physical nature, anatomists hoped to provide a sure and easy solution to the 'woman' problem.

SEX DIFFERENCES AND THE RISE OF MODERN ANATOMY, 1600–1750

> *L'esprit n'a point de sexe.*
> (François Poullain de la Barre[5])

The identification of sex differences in the human body is not unique to modern times. In the ancient world, Hippocrates, Aristotle, and Galen drew a picture of the nature of woman that provided a thoroughgoing justification of women's inferior social status: Aristotle argued that women are colder and weaker than men, and that women do not have sufficient heat to cook the blood and thus purify the soul. Galen, following the Hippocratic doctrine of the four humors, believed that women are cold and moist while men are warm and dry; men are active, women are indolent. The medical assumptions of these ancients were incorporated into medieval thinking with few revisions and dominated much of Western medical literature until well into the seventeenth century.[6]

. . .

Andreas Vesalius, widely recognized as the founder of modern anatomy, typifies a meshing of old and new attitudes. For Vesalius, sex differences were only skin deep. Vesalius did not believe that sex differences derived from the humors (as had the ancients); nor did he believe that sex differences penetrated the skeleton (as would the moderns). In the *Epitome* of his great work on the fabric of the human body, Vesalius drew a male and female nude where he pointed out differences in the curves and lines of the two bodies and the two sets of reproductive organs.[7] To accompany his male and female nudes, Vesalius drew a single skeleton that he labeled a 'human' skeleton. By believing that one 'human' skeleton gives shape to both the male and the female body, Vesalius did not sexualize the bones of the 'human' body. Though he made clear in textual notes that the skeleton was drawn from a 17- or 18-year-old male, Vesalius did not give a sex to his skeleton.

In Vesalius' view, sex differences between male and female bodies are limited to differences in the outline of the body and the organs of reproduction. In his *Epitome*, Vesalius drew two manikins or paper dolls that were to be cut out by medical students and 'dressed' with their organs; this was an exercise designed to teach medical students the position and relation of the various viscera. One manikin represented a female form and displayed the system of nerves; the other represented a male figure and showed the muscles. Vesalius presented both the male and female manikins in order to demonstrate the position and nature of the organs of generation. Apart from the reproductive organs, Vesalius considered all other organs interchangeable between the two figures. In his instructions for construction of the manikins Vesalius made this explicit: 'The sheet [of organs to be attached to the male manikin] differs in no way from that containing the figures to be joined to the last page [the drawing of the female manikin] except for the organs of generation.'[8]

Though Vesalius rejected the ancient view that sex differences pervade the body, he did accept ancient views on the inferior nature of women's reproductive organs. In his *De corporis humani fabrica*, Vesalius adopted Galen's view of female sex organs as analogous to men's but imperfect because they were inverted and internal.[9] Vesalius not only accepted Galen's view of women's reproductive organs, he also provided the best visual rendering of Galen's conception (Fig. 5).

The relative indifference of early modern anatomists to the question of sex differences did not derive from an ignorance of the female body. A look at anatomical illustrations from the fourteenth through the seventeenth centuries shows that women were, in fact, dissected. The Montpellier Codex of 1363 includes an illustration showing the dissection of a female body.[10] A statute enacted in France in 1560 required midwives to attend the dissection of female bodies so that they would know enough about female anatomy to be able to testify in abortion cases.[11] The frontispiece of Vesalius' 1543 *De corporis humani fabrica* depicts a public dissection in a theater teeming with men, dogs, and one lone monkey; on the table, under the knife, is a woman.[12] Vesalius based his drawings of female organs of reproduction on dissections of at least nine female bodies. Vesalius did not procure these bodies without difficulties, however; at least one was stolen. Hearing that a woman who had been the mistress to a certain monk had died, Vesalius and his helpers snatched her body from the tomb.[13] This remained a common practice for quite some time: William Cheselden, an English physician, reported in 1713 that he procured female bodies for dissection from 'executed bodies and . . . a common whore that died suddenly'.[14]

The indifference of early modern anatomists to the question of sex differences did not, however, lead them to 'desexualize' the bodies they studied. On the contrary, until the nineteenth century the sex of the bodies used for dissection was explicitly portrayed either by genitalia or breasts, or by a wisp of hair falling over the shoulder in the case of a woman or a prominent beard in the case of a man. The Dutch anatomist Godfried Bidloo produced a set of plates in the late seventeenth century unique for their explicit portrayal of the sex of the body dissected. Bidloo's 'true to life' drawings always portrayed the sex of the cadavers being dissected. Male and female bodies were used indiscriminately to illustrate various parts of the body. In William Cowper's 1698 publication of Bidloo's plates, a woman model appears in a series of plates describing the muscles in the upper half of the human body.[15]

Bidloo and Cowper, like Vesalius, focused on two major differences between men and women: external bodily form and reproductive organs. In 1697, Cowper reproduced the Bidloo drawings of the Apollo-Pithius and Medici Venus, in order to portray differences in symmetry and proportion between man and woman. Cowper found that

most remarkably the shoulders of the woman are narrower; the man having Two Length or Faces in the Breadth of his Shoulders, and one and a Half in his Hips; whereas Woman, on the contrary, has but one Face and a Half in her Shoulders, and Two in her Hips. Secondly, the Clavicule or Channel-bones, and Muscles in general do not appear in Women as in Men; whence it is, the out Line of the one, as Painters call it, differs very much from the other.[16]

These three Bidloo figures were drawn not from life but from classical statues; Bidloo claimed these figures exhibited 'the most beautiful proportions of a man and woman as they were fixed by the ancients'. Venus' distinctive parts are the breasts and genitalia. Cowper did not attribute differences in the male and female outline to any deep structural differences between men and women 'either in their whole frame, or in the intimate Structure of their Parts'. Rather, the differences in appearance between men and women emanate from 'the great quantity of Fat placed under the skins of women'.[17]

It is true of course that throughout the sixteenth and seventeenth centuries anatomists in the vanguard of science—Vesalius, Bidloo, Cowper—made passing remarks about differences in male and female bodies apart from differences in exterior form or sex organs. Cowper, for example, noted that the skin is softer in women and the thyroid gland is larger. Yet few attempts were made to delineate sex differences throughout the body. Drawings of skeletons, although most often done from male bodies, were thought to represent the bones of the human body.

SEX IS MORE THAN SKIN DEEP

> Sexual differences are not restricted merely to the organs of reproduction but penetrate the entire organism. The entire life takes on a feminine or masculine character.
>
> (J. J. Sachs[18])

A fundamental shift in the definition of sex differences emerged in the course of the eighteenth and early nineteenth century. Beginning in the 1750s, a body of literature appeared in France and Germany calling for a finer delineation of sex differences. In 1750, Edmond Thomas Moreau published a slim book in Paris entitled *A Medical Question: Whether Apart from Genitalia There Is a Difference Between the Sexes?*[19] In 1775, the French physician Pierre Roussel reproached his colleagues for considering woman similar to man except in sexual organs. 'The essence of sex,' he explained, 'is not confined to a single organ but extends, through more or less perceptible nuances, into every part.'[20] In 1788, German anatomist Jakob Ackermann stated that the present definition of sex differences was inadequate. The great physiologists, he complained, have neglected the description of the female body. 'Indeed, sex differences,' he emphasized, 'have always been observed, but their description has been arbitrary.'[21] In his two-hundred-page book on sex differences in bones, hair, mouths, eyes, voices, blood vessels, sweat, and brains, Ackermann called for a 'more essential' description of sex differences and encouraged anatomists to research the most basic parts of the body to discover 'the essential sex difference from which all others flow'.[22]

These anatomists' interest in the female body was shaped, in part, by changes in the broader culture. Mercantilist interests in population growth played a role in the rise of the eighteenth-century ideal of motherhood.[23] The ideal of motherhood, in turn, profoundly changed medical views of the uterus.[24] Prior to the eighteenth century, the uterus was much maligned in natural philosophy. Plato thought it an animal with independent powers of movement.[25] Democritus cited the uterus as the cause of a thousand sicknesses. Baglio thought that women suffer each sickness twice, because of the uterus.[26] Galen and even (for a time) Vesalius reported that horns bud from the sides of the womb.[27] As the ideal of motherhood gained acceptance, however, anatomists rejected the view that women are 'imperfect men' or monsters of nature.[28] Rather anatomists such as Jacques-Louis Moreau found women uniquely equipped to contribute to the propagation of the human race—even more so, in fact, than men.[29] Reviewing the history of women's medicine in 1829, Carl Ludwig Klose rejected the comparison of men's and women's sex organs that had, he maintained, occupied natural scientists from Aristotle to Albrecht von Haller. Klose argued that the uterus, woman's most important sex organ, has no analogue in man; hence the comparison with men's organs is worthless.[30]

The eighteenth-century ideal of motherhood encouraged doctors to view women as sexually perfect. Yet, if the uterus was to be viewed as perfect, did this mean that women were physiologically perfect beings? Certainly this hypothesis left the door open for those who, like Poullain de la Barre or Astell in the seventeenth century, argued that the physically perfect woman could be considered the social equal of man. Yet the debate did not end here. If the uterus was to be considered a unique and perfect organ in its own right, perhaps there were other sex differences that revealed a natural inferiority of women to men.

It was as part of this broader investigation of sex differences that drawings of the first female skeletons appeared in England, France, and Germany between 1730 and 1790. The skeleton, as the hardest part of the body, was thought to provide a 'ground plan' upon which muscles, veins, and nerves were to be drawn. In 1749, anatomist Bernhard Siegfried Albinus wrote:

I must pitch upon something . . . as the base or foundation to build my figures upon. And this is the skeleton: which being part of the body, and lying below the muscles, the figures of it ought first to be taken off as certain and natural direction for the others.[31]

If sex differences could be found in the skeleton, then sexual identity would no longer be a matter of sex organs appended to a neutral human body, as Vesalius had thought, but would penetrate every muscle, vein, and organ attached to and molded by the skeleton.

In 1734, Bernhard Albinus produced the definitive illustration of the human skeleton, which remained unsurpassed for at least three-quarters of a century (Fig. 1). The work was laborious, taking three months to complete. Albinus drew the skeleton from three different perspectives—front, side, and back—'not free hand as is customary, but from actual measure . . . and by collecting data from one body after another, and making a composite according to rules so that actual truth will be displayed . . . All [drawings] have been measured, brought down to scale . . . as architects do.'[32] Having produced the

FIG. 1.   Bernhard Albinus, bones of the human body, in *Tabulae sceleti et musculorum corporis humani* (Leyden, 1747), plate 1. By permission of the Francis A. Countway Library of Medicine.

most perfect possible drawing of the human skeleton (which Albinus made clear was drawn from a male body), Albinus lamented, 'We lack a female skeleton.'[33]

Albinus had good grounds for complaining that the study of female anatomy was inadequate before 1740. The standard studies of the human skeleton by Vesalius and Bidloo had been of the male. Only one 'crude' illustration of a female skeleton published by Gaspard Bauhin in 1605 had appeared before the eighteenth century.[34] Within fifty years of Albinus' plea, however, basic anatomical descriptions of the female body had been established. One of the first drawings of a female skeleton appeared in England in 1733 (William Cheselden). Two appeared in France: one in 1753 (Pierre Tarin),[35] and another in 1759 (d'Arconville). One appeared in Germany in 1796 (Soemmerring). Even though each of these drawings purported to represent *the* female skeleton, they varied greatly from one another.

. . .

One of the earliest drawings of a female skeleton was done by William Cheselden in 1733. This was a new interest of his in 1733; the 1713 edition of his *Anatomy* did not include an illustration of a female skeleton. Cheselden's matched set of male and female skeletons followed the Bidloo tradition of comparing idealized male and female figures drawn from art. His female skeleton is drawn in the 'same proportion as the Venus of Medici'; his male skeleton is drawn in the same proportion and attitude as the Belvedere Apollo.[36]

Text and image came together in the French rendering of a female skeleton that made its debut in 1759, capturing the imagination of medical doctors for more than half a century. The skeleton appears to be one of the very few drawn by a woman anatomist. Marie-Geneviève-Charlotte Thiroux d'Arconville, who studied anatomy at the Jardin du Roi, directed the drawing of illustrations from dissections for her French translation of Monro's *Anatomy*. D'Arconville's plates were published under the protection of Jean-J. Sue, member of the Académie Royale de Chirurgie. D'Arconville's name does not appear in the volume, and the illustrations were generally attributed to Sue.[37]

In 1796, the German anatomist Samuel Thomas von Soemmerring produced a rival female skeleton (Fig. 2).[38] Although d'Arconville's (Sue's) work was known in Germany, Soemmerring's reviewers praised his female skeleton for 'filling a gap which until now remained

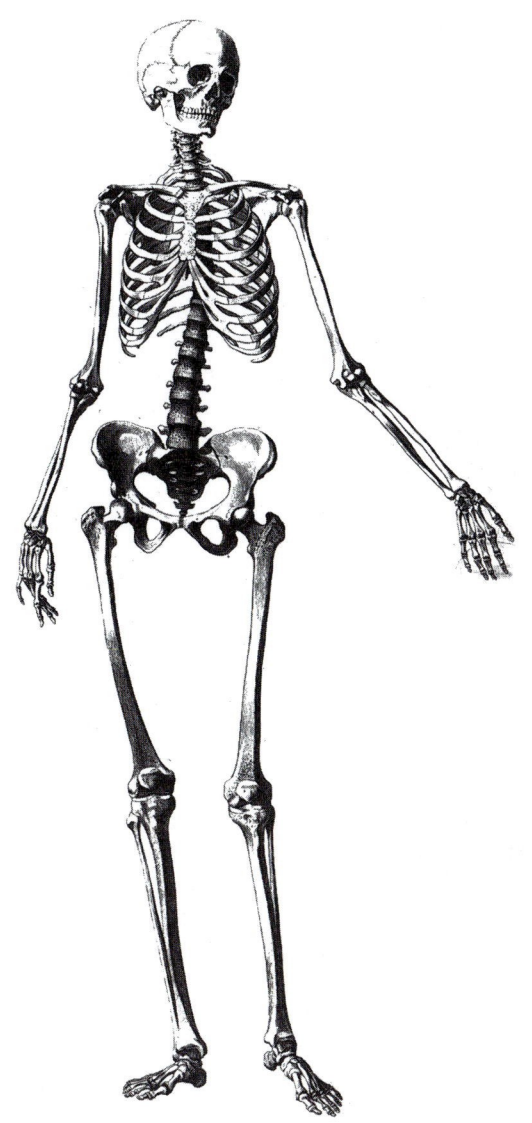

FIG. 2. Samuel Thomas von Soemmerring, female skeleton, in *Tabula sceleti feminini* (Frankfurt, 1796). By permission of the Francis A. Countway Library of Medicine.

in all anatomy'.[39] Directly answering Albinus' plea, Soemmerring spent years perfecting his portrayal of the female skeleton, and he considered his female to be of such 'completeness and exactitude' that it made a perfect mate for the great Albinus male. As a model, he selected the skeleton of a 20-year-old woman from Mayence who had borne a child.[40] Soemmerring also checked his drawing against the classical statues of the Venus di Medici and Venus of Dresden to achieve a universal representation of woman. Soemmerring intended that his skeleton represent not an individual woman but (as Ludwig Choulant put it) 'the most beautiful norm as it was imagined to exist in life, with all the carefully observed minutiae of the differential sexual characters of the entire bony structure of woman'.[41]

Although d'Arconville and Soemmerring applied the same criteria of exactitude in drawing the female skeleton from nature, the skeletons differ greatly from one another. D'Arconville/Sue depicted the skull (incorrectly) as smaller in proportion to the body than a man's, the hips as much broader than men's, and the ribs as extremely narrow and confining. In the commentary to her plate, d'Arconville described the chest of the female as narrower, the spine more curved, and the haunches and pelvis larger in women than in men.[42] Soemmerring, by contrast, portrayed his female skeleton with the ribs as smaller in proportion to the hips, but not remarkably so. Ackermann argued that women's hips appear larger than men's because their upper bodies are narrower; which by comparison makes the hips seem to protrude on both sides.

The d'Arconville skeleton is, in fact, remarkable for its proportions. The skull is drawn extremely small, the ribs extremely narrow, making the pelvis appear excessively large. D'Arconville apparently either intended to emphasize the cultural perception that narrow ribs are a mark of femininity, or she chose as the model for her drawing a woman who had worn a corset throughout her life.

Great debate erupted over the exact character of the female skeleton. Despite its exaggerations, the d'Arconville/Sue skeleton became the favored drawing in Britain.[43] Soemmerring's skeleton, by contrast, was attacked for its 'inaccuracies'. John Barclay, the Edinburgh physician, wrote, 'although it be more graceful and elegant [than the Sue skeleton] and suggested by men of eminence in modeling, sculpture, and painting, it contributes nothing to the comparison [between male and female skeletons] which is intended.'[44] Soemmerring was attacked, in particular, for showing the incorrect proportion of the ribs to the hips:

Women's rib cage is much smaller than that shown by Soemmerring, because it is well known that women's restricted life style requires that they breathe less vigorously ... The pelvis, and it is here alone that we perceive the strongly-marked and peculiar characters of the female skeleton, is shown by Soemmerring as improperly small.[45]

Anatomists concluded that Soemmerring was an artist, but no anatomist. The French doctor Jacques-Louis Moreau, however, found Soemmerring's portrayal of the female skeleton the most sublime and, in his two-volume work on the natural history of women, modeled his skeleton on Soemmerring's.[46]

What are we to make of this controversy? Did even the most exact illustrations of the female skeleton represent the female body accurately? The ideal of anatomical representation in the eighteenth century was exactitude. In his 'Account of the Work', Albinus recounted how he prepared his male skeleton carefully with water and vinegar so that it would not lose moisture and change appearance over the three months of drawings. Though anatomists attempted to represent nature with precision, they also intended to represent the body in its most beautiful and universal form. Albinus quite consciously strived to capture the details not of a particular body but of a universal and ideal type. 'I am of the opinion,' he stated, 'that what Nature, the arch workman ... has fashioned must be sifted with care and judgment, and that from the endless variety of Nature the best elements must be selected.'[47]

The supposedly 'universal' representations of the human body in eighteenth-century anatomical illustrations were, in fact, laden with cultural values. The male and female skeletons shown in Figures 1 and 2 represent not merely the bones of the male and the female body; they also serve to produce and reproduce contemporary ideals of masculinity and femininity. For his illustrations, Albinus collected data from 'one body after another'. He then selected one 'perfect' skeleton to serve as his model. Albinus also revealed the criteria by which he 'sifted' nature in the drawing of his male skeleton:

As skeletons differ from one another, not only as to the age, sex, stature and perfection of the bones, but likewise in the marks of strength, beauty and make of the whole; I made choice of one that might discover signs both of strength and agility: the whole of it elegant, and at the same time not too delicate; so as neither to shew a juvenile or feminine roundness and slenderness, nor on the contrary an unpolished roughness and clumsiness.[48]

In his preface to Ackermann's book on sex differences, Joseph Wenzel discussed the difficulties anatomists have in choosing models for their work. He noted that a sharp physiological delineation between the sexes is impossible because the great variation among individual men and women produces continuity between the sexes. In fact, he wrote, one can find skulls, brains, and breast bones of the 'feminine' type in men. Wenzel then defined a standard of femininity that he used as the basis of his own work:

> I have always observed that the female body which is the most beautiful and womanly in all its parts, is one in which the pelvis is the largest in relation to the rest of the body.[49]

Soemmerring strived, like Albinus, for exactitude and universality in his illustrations. He made every possible effort to 'approach nature as nearly as possible'. Yet, he stated, the physiologists should always select the most perfect and most beautiful specimen for their models.[50] In identifying and selecting the 'most beautiful specimen', Soemmerring intended to establish norms of beauty. According to Soemmerring, without having established a norm by means of frequent investigations and abstractions, one is not able to decide which cases deviate from the perfect norm. Soemmerring chose the 'ideal' model for his illustration of the female skeleton with great care:

> Above all I was anxious to provide for myself the body of a woman that was suitable not only because of her youth and aptitude for procreation, but also because of the harmony of her limbs, beauty, and elegance, of the kind that the ancients used to ascribe to Venus.[51]

In their illustrations of the female body, anatomists followed the example of those painters who 'draw a handsome face, and if there happens to be any blemish in it, they mend it in the picture'.[52] Anatomists of the eighteenth century 'mended' nature to fit emerging ideals of masculinity and femininity.

In the nineteenth century, the bones of the human body took on more overtones of masculinity and femininity. In 1829, John Barclay, the Edinburgh anatomist, brought together the finest illustrations from the European tradition for the sake of comparison. As the finest example of a male skeleton Barclay chose the Albinus drawing. Then looking to the animal kingdom, Barclay sought an animal skeleton, one that would highlight the distinctive features of the male skeleton. The animal he found as most appropriate for comparison to the male

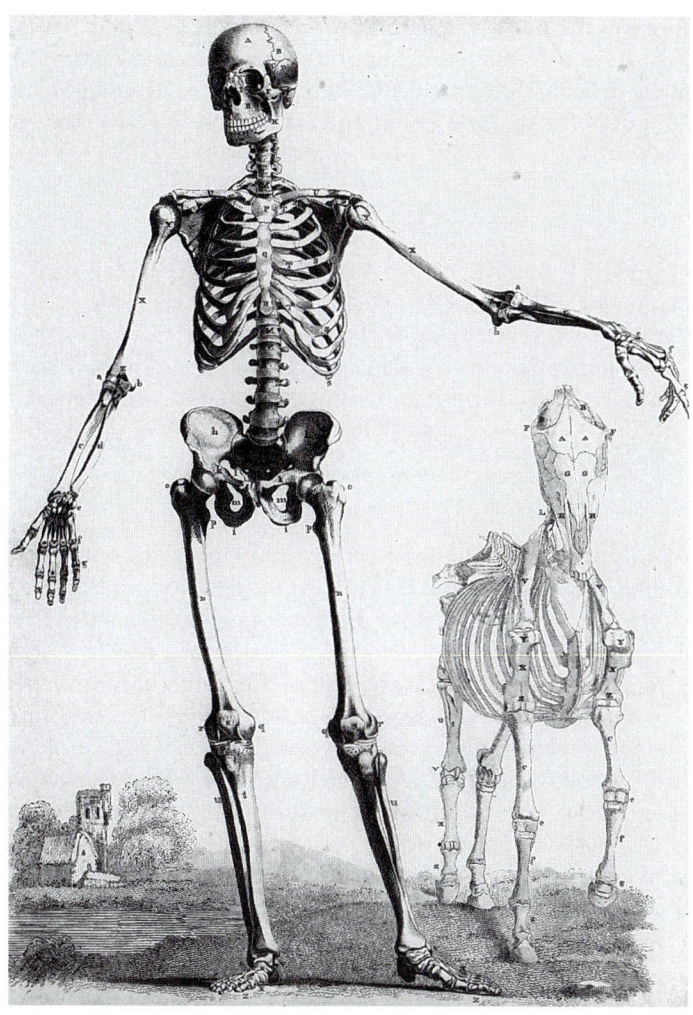

FIG. 3. John Barclay, 'Male Skeleton Compared to the Horse', in *The Anatomy of the Bones of the Human Body* (Edinburgh, 1829), plate 1. By permission of the Francis A. Countway Library of Medicine.

skeleton was the horse, remarkable for its marks of strength and agility (Fig. 3). As the finest representation of the female skeleton Barclay chose the delicate d'Arconville/Sue rendition. This he compared to an animal noted for its large pelvis and long, narrow neck—the ostrich (Fig. 4).

MAN (WHITE AND EUROPEAN) THE MEASURE OF ALL THINGS: WOMEN AS CHILDREN AND PRIMITIVES

> In approaching puberty, woman seems to distance herself less than man from her primitive constitution. Delicate and tender, she always conserves something of the temperament characteristic of children.
>
> (Pierre Roussel[53])

The flood of medical literature on sex differences did not subside in the course of the nineteenth century. As the century progressed, some anatomists came to believe that differences between male and female bodies were so vast that women's development had been arrested at a lower stage of evolution. Measurement of the distinguishing characteristics of the skeleton—the skull and pelvis—led some anatomists to conclude that white women ranked below European men in the scales of both ontogeny and phylogeny. Neither in the development of the species nor in the development of the individual were women thought to attain the full 'human' maturity exemplified by the white male.[54] In terms of both physical and social development, these anatomists classified women with children and 'primitive' peoples.

In the course of the late eighteenth and nineteenth centuries, categories of sex and race increasingly came to define standards of social worth. At the same time, these standards came to reflect the structure of the scientific community. Those who possessed the tools of science took themselves as the standard of excellence. In the absence of women, the largely male scientific community studied women using male anatomy as the norm against which to measure female anatomy. In similar fashion, in the absence of blacks, the white scientific community studied blacks using the white male as the standard of excellence. Excluded from the practice of science, women and blacks (not to mention other groups) had little opportunity to dispute the findings of scientists.

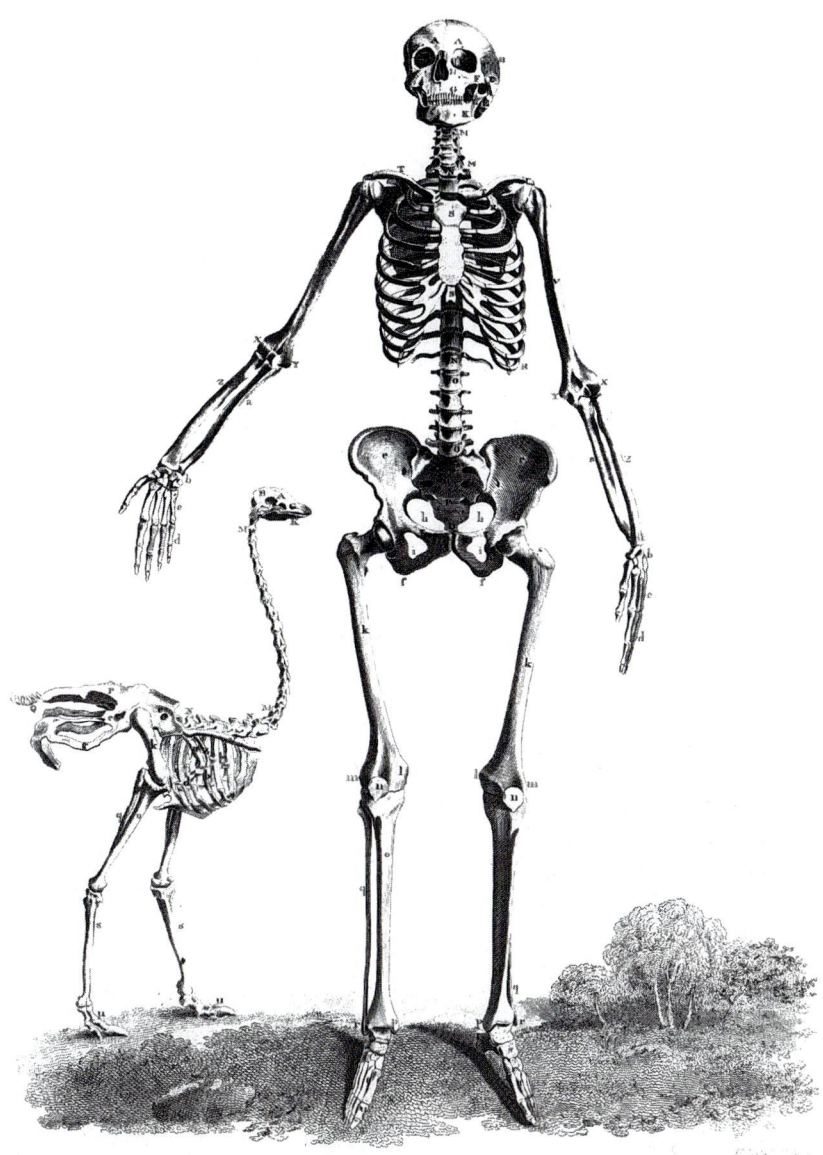

FIG. 4.   John Barclay, 'Female Skeleton Compared to the Ostrich', in *The Anatomy of the Bones of the Human Body* (Edinburgh, 1829), plate 4. By permission of the Francis A. Countway Library of Medicine.

The illustrations of female skeletons by d'Arconville/Sue and Soemmerring sparked a protracted debate about the size of the female skull. The question of intelligence was becoming more and more important; 'natural reason' was increasingly conceived to be a prerequisite for many political rights and social opportunities. In this context, the skull became an important focus for providing an objective measure of intelligence. Hegel, following F. J. Gall, held that the brain 'molded the skull—here pressing it out around, there widening or flattening it'.[55] Craniologists analyzed the size and shape of the skulls of men and women, whites and blacks, hoping to answer the much debated question: whether or not the intellectual capacities of women and 'primitive peoples' were equivalent to those of white men.

The assumption fueling the debate over skull size was that intelligence, like sexual identity, is innate and not dependent on educational opportunities. Soemmerring repudiated d'Arconville's findings and portrayed the female skull as larger in proportion to the body than the male skull. Soemmerring found that women's skulls are actually *heavier* than men's in the sense that the female skull occupies greater proportion of total body weight (1/6 for women; 1/8 or 1/10 for men).[56] From this Ackermann (Soemmerring's student) concluded that women's brains are larger than men's:

> Since women lead a sedentary life, they don't develop large bones, muscles, blood vessels and nerves as do men. Since brain size increases as muscle size decreases, it is no surprise that women are more adept in intellectual pursuits than men.[57]

The debate among anatomists about the correctness of d'Arconville/ Sue's or Soemmerring's representation of women's skull size found one resolution in Barclay's work in the 1820s. Women's skulls were larger in relation to their body size than men's—but then, so were children's. According to Barclay, one needn't conclude that women's larger skulls are loaded with heavy and high-powered brains. Rather than a mark of intelligence, large skulls signal women's incomplete growth. In 1829 Barclay, using d'Arconville/Sue's plates, presented a skeleton family. Although anatomical drawings of children and fetuses had been published since the early eighteenth century, skeletons of man, woman, and child were brought together for the first time by Barclay 'for sake of comparison'. In his commentary, Barclay upbraided Soemmerring for drawing comparisons between only the male and female skeleton; Barclay introduced the child's skeleton 'to shew that many of those characteristics, which he [Soemmerring] has described as peculiar to the female, are more obviously discernible in the fetal skeleton.'[58] Barclay pointed out that along with equivalent skull sizes, both women and children have the fissure separating the frontal skull bone; both have smaller bones compared to men; the rib cage, jaw shape, and feet size of women are more similar to those of children than to men. Woman's one distinguishing characteristic, Barclay argued, is her pelvis:

It is there [in the pelvis] that we cease to trace the analogies between its [the female skeleton's] proportions and those of the foetus: or in other words, it is there that, in deviating from those characters which at one time were common to both [male and female], we regularly find it [the pelvis] deviating farther than that of the male—the pelvis of the foetus being always proportionally the smallest of the three, and that of the female proportionally the largest.

By mid-century, the image of the childlike woman had become common. In *Das Weib und das Kind* (1847) the German doctor E. W. Posner gave one of the clearest statements of the physical similarities between women and children:

Women's limbs are short and delicate ... and the shortness of the limbs determines the smaller size of the female body and proves the similarity of the female body with the child's body. . . . As is true in the child, the woman has a larger and rounder abdomen in relation to her breast. . . . The entire trunk, which in man forms a pyramid with the base turned upward, is reversed in woman with the point of the pyramid at the shoulders which are smaller and narrower, while the stomach is the broad base from which the even broader hips and strong thighs proceed. . . . Women's heads also tend toward the childish type. The finer bone structure, the tender, less sharply developed facial features, the smaller nose, the larger childish roundness of the face clearly show this similarity. . . . The nerves and blood vessels of women are also as delicate and fine as those of children . . . and the skin with its layer of rich fat is childish.[59]

Posner accounted for these similarities between women's and children's bodies by arguing that since 'the female sex ends its growth earlier than men before reaching full individual maturity [at age 14 rather than 18] . . . the woman retains her childish roundness.' While the comparison of women to children is not in itself negative (children in the nineteenth century also represent innocence, freshness, and youth), Posner's explanation implied that women have failed to reach full human maturity. Like Charles Darwin, Posner assumed that there is one unified, natural hierarchy of physical, intellectual, and cultural development within which every category of human being has its place. Posner made it clear that the European male body type was the norm of maturity against which women, blacks, and children were to be measured.

It should be noted that the comparison of women with children was by no means new to the nineteenth century. By drawing parallels between the anatomy of women and children, anatomists restated in the language of modern science an ancient prejudice against women. The ancients—Zenocrates, Galen, and Hermagoras—commonly held that a woman can never be more than a child.[60] Galen thought that both women and children suffered from cold and moist humors that accounted for their lack of self-control. Aristotle grouped women, children, and slaves together in the three states of *consilium*.[61] In the biblical tradition the man was the head of the household, since God gave Adam dominion over women and children. The supposed evidence emerging in the early nineteenth century that women were physiologically linked to children served to translate these traditional prejudices into the language of modern science. The physical image of the childlike woman also reflected certain aspects of European custom. In the late eighteenth century, middle-class wives were on average ten years younger than their husbands;[62] it is not surprising that middle-class women should have appeared 'childish' in comparison to their husbands.

I want to mention again that it has not been my purpose here to minimize physical differences between men and women but rather to analyze the social and political circumstances surrounding the search for sex differences. While identifiable differences between the sexes do exist, they have often been exaggerated. It is important to remember what Wenzel stressed in 1789, namely that 'individual members of each sex differ significantly from one another; one can find male bodies with a feminine build, just as one can find female bodies with a masculine build.'[63] Moreover, many physical differences among human beings are not absolute but relative to social conditions.[64] Eighteenth-century anatomists repeatedly pointed to the fact that women are shorter and smaller than men. Women's smallness, they thought, was dependent on bone size, which provided an absolute index of women's weakness and delicacy. As current studies show, however, the difference in average height between European men and women has decreased over the last century. Height is, in part, determined by nourishment. In many cultures, it is customary for the woman to eat what remains after the man has eaten, or to give the man the best portion.[65]

Even if we do away with the exaggerations that have plagued studies of sex differences in Western medicine, however, we cannot do away with the fact of difference. Yet it is unclear what difference difference makes. Why should a set of physical differences be used to underwrite a system of social inequality? Why has the argument that physical difference implies intellectual and moral difference been so persuasive to so many for so long?

'SOCIAL INEQUALITY AS NATURAL LAW'

> The laws of divine and natural order reveal the female sex to be incapable of cultivating knowledge, and this is especially true in the fields of natural sciences and medicine.
>
> (Theodor L. W. von Bischoff[66])

Why did the comparative anatomy of men and women become a research project for the medical community in the late eighteenth century?[67] I want to argue here that it was the attempt to define the position of women in bourgeois European society at large and in science in particular that spawned the first representations of the female skeleton. One powerful assumption underlying much of nineteenth-century social theory was that physical evidence—nature—provided a point of certainty from which social theory could depart. A look at how the findings of anatomists were used will illuminate the role that the study of anatomical sex differences played in underwriting the increasing polarization of gender roles in European society.[68]

The growth of democratic tendencies brought about a reshuffling of the social order. In the eighteenth century it was not yet clear what women's role in the new social order was to be. Certainly, throughout the seventeenth and eighteenth centuries there was a struggle between those advocating full social equality for women and those advocating the continued subordination of women.[69] Some philosophers, such as Rousseau, hoped to lay a solid foundation for the theory of natural rights by disentangling the natural from the social in human nature. In his *Emile*, Rousseau appealed to a hypothetical state of nature, and to the nature of the human body as defined by comparative anatomy in his search for the natural relation between the sexes.[70]

Rousseau's pedagogical writings opened the flood gates of modern prescriptive literature on the proper character and education of the sexes.[71] Rousseau abhorred the public influence he saw in French women and the stirrings of feminism in the years leading up to the French Revolution. He took on proponents of women's equality by arguing that woman is not man's equal but his complement. Rousseau was instrumental in initiating the view that the inherent physical, moral, and intellectual differences of women suited them for roles in society vastly different from those of men.[72] To Rousseau's mind, it was the purpose of natural philosophy to read in the book of nature 'everything which suits the constitution of [woman's] species and her sex in order to fulfill her place in the physical and moral order'.[73]

There were also those who opposed Rousseau's argument of women's different nature and argued instead for the social equality of women. Among French philosophes, Helvétius, d'Alembert, and Condorcet traced the inferior intellectual achievement of women to their inferior education.[74] This line of argument was also promoted by Mary Wollstonecraft in England, Olympe de Gouges in France, and Theodor von Hippel in Germany, to name a few.[75] It was in the eighteenth century that debates about women's character were framed by the modern opposition of 'nurture' versus 'nature'. Those who found women's weakness a matter of 'nurture' envisioned social and educational reform as the brightest path toward social equality of the sexes. Those who traced women's weakness to women's 'nature' assumed that 'whatever is, is right'.

Many, however, were to apply the new findings of anatomy to the 'woman question'. In French medical camps in particular, the physical became tied to the moral in such a way that medical arguments from 'nature' increasingly penetrated social theory. The *Encyclopédie* article of 1765 on the 'skeleton' devoted half its text to a comparison of the male and female skeleton. In great detail the differences are laid out between the male and female skull, spine, clavicle, sternum, coccyx, and pelvis. The article ends with one prescriptive phrase: 'All of these facts prove that the destiny of women is to have children and to nourish them.'[76]

In 1775, the medical doctor Pierre Roussel expanded upon this message. He claimed that differences in the male and female skeleton were of great importance. 'Nature,' he wrote, 'has revealed through that special form given to the bones of woman that the differentiation of the sexes holds not only for a few superficial differences, but is the result perhaps of as many differences as there are organs in the human body.[77] Spirit or mind were among the organs Roussel listed. To Roussel's mind, moral and intellectual qualities were as innate and as enduring as the bones of the body. Roussel argued against 'writers' who insisted that differences between men and women resulted from custom, education, or climate.[78] It was, he believed, rather the unerring findings of medicine that provided a certain ground for ethics. Philosophy, he wrote, cannot determine the moral powers of human beings without taking into consideration the influence of the bodily organization.[79]

In Germany, Soemmerring also believed that gender differences were to be traced to 'nature', not to 'nurture'. In his book on the comparative anatomy of the 'Negro' and European, Soemmerring reported his observations on the physical and moral character of woman:

A boy will always dominate a girl, without knowing that he dominates, and knowing even less that he dominates because of his solid, strong body. He will dominate even when he has received the same nourishment, love and clothing as a girl. I have had the rare opportunity of seeing definite proof of this fact. From his earliest youth, Prince D . . . G was raised alongside his sister. Their training in all moral and physical matters was equivalent in every way. And yet, differences of masculinity and femininity in physical and moral character were always conspicuous. This is a fact of experience.[80]

Johann Ziegenbien propagated Soemmerring's ideas in German schools. Reading to the parents of a girls' school, Ziegenbien opened his lecture on 'Female Nature, Character, and Education' with Soemmerring's findings that 'already in the earliest stages of the embryo one finds sex differences. That boys will seize a stick while girls will take up a doll . . . that men rule the affairs of state while women govern the affairs of the home reflects nothing other than what is already in the seed of the embryo.'[81]

Medical manuals for women that appeared in France and Germany between the 1780s and 1830s also made use of the new findings on sex differences. These health manuals emphasized that physical differences between men and women must be taken into account for the proper treatment of their illnesses. These manuals also emphasized that the well-being of each sex depended on establishing a lifestyle appropriate to its particular physiology. The authors of these manuals spoke of the physiological and moral character of woman in one and the same breath.[82]

Those who thought that social life in 'harmony' with nature ensured individual well-being also thought that such a life ensured social stability. In his 1806 book on the character of the female sex, Carl Friedrich Pockels asserted that differences in bodily strength are 'designed by nature as a necessary basis to ensure the social order between man, woman, and the family'.[83] Drawing from Rousseau, Roussel, and Georges Cuvier, Pockels found man and woman perfect but complementary beings. In 1830, medical doctor J. J. Sachs explained how physical complementarity led to complementary social roles for men and women:

The male body expresses positive strength, sharpening male understanding and independence, and equipping men for life in the State, in the arts and sciences. The female body expresses womanly softness and feeling. The roomy pelvis determines women for motherhood. The weak, soft members and delicate skin are witness of woman's narrower sphere of activity, of home-bodiness, and peaceful family life.[84]

The French father of positivism, Auguste Comte, also considered the proper relation between the sexes the foundation of solid social order. Comte believed that the strength of the family rested on the natural subordination of the female to the male. Thus in his epic *Cours de philosophie positive*, Comte reasserted the growing belief that the proper social role of women is a question not of politics but of biology:

The sound philosophy of biology, especially the important theories of Gall, begins to offer a scientific resolution to the much acclaimed equality of the sexes. The study of anatomy and physiology demonstrates that radical differences, at once physical and moral . . . profoundly separate the one [sex] from the other.[85]

J. S. Mill's moral arguments for women's equality were unable to sway Comte from this view. Comte persisted in his belief that biology 'has established the hierarchy of the sexes'.[86]

One important aspect of the definition of women's nature emphasized women's inability to do intellectual work or science of any kind. This view, expressed by Wilhelm von Humboldt but stretching back to Rousseau, held that women's physical weakness results in intellectual weakness. According to Humboldt, women are incapable of analytical or abstract thought. Women, he wrote, are given to subjective impressions; they reveal their inner life, they think in wholes, but don't ground their information in empirical data.[87] Hegel strengthened the growing association of masculinity with reason and science, and of femininity with feeling and the moral sphere of the home. Woman's ethical and intellectual character, he argued, made her inherently incapable of philosophy or science:

Women are capable of education, but they are not made for the more advanced sciences, philosophy and certain forms of artistic production, all of which require a universal faculty. Women may have quick wit, taste, and elegance, but they cannot attain the ideal. . . . Women regulate their actions not by the demands of universality, but by arbitrary inclinations and opinions . . . they follow the dictates of subjectivity, not objectivity.[88]

It is important to look at these texts on women's nature—medical and philosophical—in the context of women's participation in medicine. The search for sex differences on the part of anatomists also coincided with changes in the structure of seventeenth- and eighteenth-century medical care. Anatomists' interest in the distinctive character of women's body build came at a time when the professionalization of medical sciences was taking women's health care out of the hands of female midwives.[89] On the one hand, doctors stressed their concern for distinctively female problems in health care. Ackermann, for example, appended a chapter on women's health to his book on sex differences and argued that traditional medicine did not take into account differences in body build that might influence the cause and course of a sickness.[90] In order to treat an illness properly a physician was to take into account women's distinctive anatomy. The implication here was, of course, that midwives did not have the proper training to care for women's health. On the other hand, women were not thought capable of getting that training. They were not to work in the increasingly public realm of medicine but to remain in the home.[91] Women were defined as incapable of doing science by a medical community that was itself attempting to become more scientific.

At the same time that the idea of the 'unscientific' woman drove a final nail in the coffin of midwifery, scientists presented their findings as 'objective' or 'value neutral'. In the course of the nineteenth century, there emerged an increasing tendency to look to science as an arbiter of social questions. The promise of objectivity inspired hope that the 'woman question' could be resolved by science. Anatomists stressed the 'impartial' nature of their work. In his work on the comparative anatomy of the 'Negro' and European, for example, Soemmerring concluded that the Negro was 'somewhat nearer the ape, particularly in respect to the brain, than the European'.[92] Yet Soemmerring argued that his work reflected what he called the 'cold-blooded' findings of science and did not take a 'moral' stand.[93] In a similar manner, Pockels insisted that his four-volume study of the character of woman was 'impartial', 'purely empirical', and following the 'principles of reason'.[94]

If the scientific argument against women in science played a role in removing women from the medical profession, it was also invoked repeatedly to keep them out. In 1872, Professor Theodor Bischoff argued against the admission of women to the medical profession. Using, as he said, the 'impartial and certain' methods of science, he promised to prove that the 'pure and unadulterated feminine nature' of woman is not a scientific one.[95] Bischoff's central argument against women's equal participation in medicine rested on scientific definitions of sex differences. Anyone who is familiar with the physical and mental differences between man and woman as discovered in anatomy and physiology, he wrote, could never support the equal participation of men and women in medicine. Equality, he stated, can only be achieved where initial conditions are equal. He then spun off the (by then) familiar list of sex differences—in bones, muscles, nerves, and skulls—relying heavily on the work of Soemmerring and citing the supporting passages from Kant's *Anthropologie*.[96]

Opposition to these and similar views on women came largely from outside the medical community. Already in 1744, Eliza Haywood reported in her journal, *The Female Spectator*, that anatomists were finding women to be physiologically incapable of deep thought. She believed this to be incorrect, and argued that the supposed delicacy of the female brain does not necessarily render it less 'strong' than the male brain:

> The Delicacy of those numerous Filaments which contain, and separate from each other what are call'd the Seats of Invention, Memory, and Judgment, may not, for anything they [the anatomists] can prove to the contrary, render them [women's brains] less strong.

Haywood was not a natural philosopher, however, and qualified her authority to speak of such matters; she wrote that 'as I am not Anatomist enough to know whether there is really any such Difference or not between the Male and Female Brain, I will not pretend to reason on this Point.'[97] Like others of her time, Haywood bowed to claims of science to have a privileged access to truth.

Asymmetries in social power forced adversaries on either side of the 'woman question' to use markedly different forms of arguments to state their case. Barred from science, women used what were essentially *moral* arguments to demand education and voting rights; scientists (by and large male) used the findings of anatomy and other sciences to argue that women were incapable of scientific endeavor. As science gained in social prestige, those who could not base their arguments on scientific evidence were put at a disadvantage in social debate. Nineteenth-century feminist Hedwig Dohm, for example, was not reticent in her critique of science. Yet even she qualified her rebuttal of anatomist Theodor Bischoff with the words 'I am no physiologist.'

Because Dohm wrote from outside the academy and not under the guise of science, her work was commonly viewed as merely polemical.[98] In 1875, an Austrian minister of health, Friedrich W. H. Ravoth, made explicit a common conception of the relationship between feminism of any kind and science. In a speech to the Assembly of German Scientists and Doctors, Ravoth stated:

Competent and qualified scientific research can and must face the dialectical, or rather, sophistical talk about the so-called woman question with a categorical imperative, and uphold unchangeable laws.[99]

Thus in the absence of women (one might say in opposition to women) nineteenth-century science defined feminine nature as essentially incommensurate with masculine nature. The 'natural' inequalities between men and women seemed to justify social inequalities between the two sexes. Many believed that the social order parallels the natural order. Cultural historian Wilhelm Riehl, for example, held that the incommensurate nature of men and women shows that inequalities of society are based in natural law.[100] Scientific definitions of human 'nature' were thus used to justify the channeling of men and women (as well as whites and blacks) into vastly different social roles. It was thought 'natural' that men, by virtue of their 'natural reason' should dominate public spheres of government and commerce, science and scholarship, while women, as creatures of feeling, fulfilled their natural destiny as mothers, conservators of custom in the confined sphere of the home.

The consequences of this story, however, go beyond the exclusion of various groups—women or ethnic minorities—from the scientific community. Along with women, scientists excluded from science a specific set of moral and intellectual qualities defined as feminine. It was woman's distinctive moral qualities—feeling and instinct—that were thought to dull her abilities to practice science. The true scientist was to be a man of 'reason and truth'.

The femininization of feeling and the masculinization of reason was produced and reproduced by specific divisions of labor and power in European society. At the same time, it should be recognized that the 'nature' of science is no more innate than the moral 'nature' of man or woman. Science too has been shaped by social forces. One of those forces has been the persistent desire to distance science from the feminine, and to identify it with the masculine. The irony in the case of the female skeleton is that as modern science plunged headlong into the study of sex differences in the eighteenth century, it helped construct its own gender.

Notes

1. Samuel Thomas von Soemmerring, *Tabula sceleti feminini juncta descriptione* ([Utrecht], 1796).
2. D'Arconville was one of a small number of women working in anatomy in eighteenth-century France. She published her work under the name and protection of Jean-J. Sue.
3. See Ernst Cassirer, *The Philosophy of the Enlightenment*, trans. F. C. A. Koellen and J. P. Pettegrove (Princeton, 1951), 234–52. See also Maurice Bloch and Jean Bloch, 'Women and the Dialectics of Nature in Eighteenth-Century French Thought', in *Nature, Culture and Gender*, ed. Carol P. MacCormack and Marilyn Strathern (Cambridge, 1980), 25–41.
4. Charles Louis de Secondat, Baron de Montesquieu, *Lettres Persanes* (Paris, 1721), letter 38.
5. François Poullain de la Barre, *De l'égalité des deux sexes: Discours physique et moral* (Paris, 1673), 59.
6. See M. C. Horowitz, 'Aristotle and Woman', *Journal of the History of Biology*, 9 (1976), 183–213. On the continuity in medical views from the ancient to the medieval worlds, see Julia O'Faolain and Lauro Martines, *Not in God's Image: Women in History* (London, 1973), 130–9; Ian Maclean, *The Renaissance Notion of Woman: A Study in the Fortunes of Scholasticism and Medical Science in European Intellectual Life* (Cambridge, 1980).
7. Andreas Vesalius, *De corporis humani fabrica* (Basel, 1543).
8. J. B. Saunders and C. D. O'Malley, *The Illustrations from the Works of Andreas Vesalius* (New York, 1950), 222–3.
9. In Galen's view, women's sex organs were analogous to men's but 'imperfect and, as it were, mutilated'. Women lacked sufficient heat to propel their sex organs outward. *On the Usefulness of the Parts of the Body*, trans. Margaret May, vol. 2 (Ithaca, NY, 1968), 628–30. See also Esther Fischer-Homberger, *Krankheit Frau und andere Arbeiten zur medizingeschichte der Frau* (Bern, 1979).
10. See Fritz Weindler, *Geschichte der gynäkologisch-anatomischen Abbildung* (Dresden, 1908), fig. 37, p. 41.
11. Kate Campbell Hurd-Mead, *A History of Women in Medicine* (Haddam, Conn., 1938), 358–9.

12. Vesalius, *De corporis humani fabrica*, frontispiece.

13. Saunders and O'Malley, *Illustrations from the Works of Vesalius*, 170.

14. William Cheselden, *Anatomy of the Bones* (1713; Boston, 1795), 276.

15. William Cowper, *The Anatomy of Humane Bodies* (London, 1698).

16. Cowper, *Anatomy of Humane Bodies*, commentary to plate 2.

17. Ibid.

18. J. J. Sachs, *Ärztliches Gemälde des weiblichen Lebens im gesunden und krankhaften Zustände aus physiologischem, intellektuellem und moralischem Standpunkte: Ein Lehrbuch für Deutschlands Frauen* (Berlin, 1830), 1.

19. Edmond Thomas Moreau, *Quaestio medica: An praeter genitalia sexus inter se discrepent?* (Paris, 1750).

20. Pierre Roussel, *Système physique et moral de la femme, ou tableau philosophique de la constitution, de l'état organique, du tempérament, des moeurs, & des fonctions propres au sexe* (Paris, 1775), 2. The German physician Carl Ludwig Klose also argued that it is not the uterus that makes woman what she is. Even women from whom the uterus has been removed, he stressed, retain feminine characteristics; *Über den Einfluss des Geschlechts-Unterschiedes auf Ausbildung und Heilung von Krankheiten* (Stendal, 1829), 28–30.

21. Jakob Ackermann, *De discrimine sexuum praeter genitalia* (Mainz, 1788); *Über die körperliche Verschiedenheit des Mannes vom Weibe ausser Geschlechtstheilen*, trans. Joseph Wenzel (Koblenz, 1788), 2–5. The German edition is used throughout.

22. Ibid. See Johann Christian Jörg, *Handbuch der Krankheiten des Weibes, nebst einer Einleitung in die Physiologie und Psychologie des weiblichen Organismus* (Leipzig, 1831), 6ff.

23. Yvonne Knibiehler and Catherine Fouquet, *La Femme et les médecins* (Paris, 1983), 124.

24. On the rise of the ideal of motherhood, see Marlene LeGates, 'The Cult of Womanhood in Eighteenth-Century Thought'. *Eighteenth-Century Studies*, 10:1 (1976):21–39; Margaret Darrow, 'French Noblewomen and the New Domesticity, 1750–1850', *Feminist Studies*, 5: 1 (1979), 41–65.

25. Plato *Timaeus* 91c, in *The Collected Dialogues of Plato*, ed. Edith Hamilton and Huntington Cairns (Princeton, 1961). Throughout the ancient period the uterus was portrayed in a variety of forms—a tortoise, a newt, a crocodile; Harold Speert, *Iconographia Gyniatrica: A Pictorial History of Gynecology and Obstetrics* (Philadelphia, 1973), 8.

26. Ackermann, *Über die körperliche Verschiedenheit des Mannes vom Weibe ausser Geschlechtstheilen*, appendix.

27. For Vesalius, see the six plates originally printed without title, now known as *Tabulae anatomicae sex* (1538), plate 87, figs. 2–4.

28. Maclean dates the reevaluation of women's sex organs much earlier. He found that after 1600, one sex is no longer thought to be an imperfect and incomplete version of the other, and that by the end of the sixteenth century, most anatomists abandon Galenic parallelism; *Renaissance Notion of Women*, 33. Nonetheless the Galenic view persisted well into the eighteenth century in the works of writers such as Diderot in France and Rosenmüller in Germany. Important naturalists such as Georges-Louis Buffon held that women's ovaries were testicles as late as 1749; *Histoire naturelle*, vol. 3 (Paris, 1749), 264.

29. Jacques-Louis Moreau, *Histoire naturelle de la femme, suivie d'un traité d'hygiène appliquée à son régime physique et moral aux différentes époques de sa vie*, vol. 1 (Paris 1803), 68–9.

30. Klose, *Über den Einfluss des Geschlechts-Unterschiedes*, 28–33.

31. Bernhard Albinus, 'Account of the Work', in *Table of the Skeleton and Muscles of the Human Body* (London, 1749).

32. Bernhard Albinus, *Annotationes academicae* (1754–68), 7, 11–14, 30–50; quoted in Ludwig Choulant, *History and Bibliography of Anatomic Illustration*, trans. Mortimer Frank (Chicago, 1920), 277.

33. Albinus gave a description of a female skeleton in his *De sceleto humano* but does not provide an illustration; B. S. Albini, *De sceleto humano* (Leiden, 1762), ch. 126.

34. Gaspard Bauhin, *Theatrum anatomicum* (Frankfurt, 1605), plate 4, p. 247.

35. Pierre Tarin, *Ostéo-graphie, ou description des os de l'adulte, du foetus* (Paris, 1753), plate 23.

36. William Cheselden, *Osteographia or the Anatomy of the Bones* (London, 1733), plates 34 and 35.

37. D'Arconville's role in the publication of the *Ostéologie* remains unclear. The title page attributes the work to Sue: *Traité d'Ostéologie, traduit de l'Anglois de M. Monro, Où l'on ajouté des Planches en Taille-douce, qui représentent au naturel tous les Os de l'Adulte & du Foetus, avec leurs explications, Par M. Sue, Professeur & Démonstrateur d'Anatomie aux Ecoles Royales de Chirurgie, de l'Académie Royale de Peinture & de Sculptur* (Paris, 1759). Sue's son, Jean-J. Sue, reproduced the plates under his own name in 1788; *Elémens d'anatomie, à l'usage des peintres, des sculpteurs, et des amateurs* (Paris, 1788). In Britain, Andrew Bell and John B. Barclay also attributed the plates to Sue. See Andrew Bell, *Anatomica Britannica: A System of Anatomy; Illustrated by Upwards of Three Hundred Copperplates, from the Most Celebrated Authors in Europe* (Edinburgh, 1798), and John Barclay, *The Anatomy of the Bones of the Human Body, Represented in a Series of Engravings, Copied from the Elegant Tables of Sue and Albinus* (Edinburgh, 1829). Current literature, such as Paule Dumaitre's *Histoire de la medicine et du livre medical* (Paris, 1978), 285, credits d'Arconville with the translation but Sue with the illustrations. The confusion about d'Arconville's role arises from the fact that, as Pierre-Henri-Hippolyte Bodard tells us, she was careful always to guard her anonymity; *Cours de botanique médicale comparée*, vol. 1 (Paris, 1810), xxvi–xxx. D'Arconville was a sharp critic of society women of her day, in particular those who (in her view) vainly paraded their literary achievement; see her essay 'Sur les Femmes', in *Mêlanges de littérature, de morale et de physique*, 7 vols. (Amsterdam, 1775), 1: 368–83.

38. Soemmerring, *Tabula sceleti feminini*.

39. *Journal der Empfindungen: Theorien und Widersprüche in der Natur- und Artzneiwiwissenschaft*, 6: 18 (1797), 17–18.

40. It was thought that a woman did not reach maturity with the onset of menstruation but only with age 18 or 20, after the birth of her first child; see Jörg, *Handbuch der Krankheiten des Weibes*, 6ff.

41. Choulant, *History and Bibliography of Anatomic Illustration*, 306–7.

42. Sue, *Ostéologie*, text to plate 4.

43. Bell, *Anatomica Britannica*; Barclay, *Anatomy of the Bones of the Human Body*.

44. Ibid., commentary to plate 32.
45. Ibid.
46. Moreau, *Histoire naturelle de la femme*, 1: 95.
47. Quoted in Choulant, *History and Bibliography of Anatomic Illustration*, 277.
48. Albinus, 'Account of the Work'.
49. Quoted in Ackermann, *Über die Verschiedenheit*, 5–7.
50. Choulant, *History and Bibliography of Anatomic Illustration*, 302.
51. Soemmerring, *Tabula sceleti feminini*, commentary to plate.
52. Albinus, 'Account of the Work'.
53. Roussel, *Système physique et moral de la femme*, 6.
54. See Stephen Jay Gould, *Ontogeny and Phylogeny* (New York, 1977).
55. Georg Wilhelm Friedrich Hegel, *Phänomenologie des Geistes* (1807), in *Werke*, ed. Eva Moldenhauer and Karl Michel, 20 vols. (Frankfurt, 1969–71), 3: 248.
56. Samuel Thomas von Soemmerring, *Vom Baue des menschlichen Körpers*, 5 vols. (1796; Frankfurt, 1800), vol. 1, section 61, p. 82.
57. Ackermann, *Über die Verschiedenheit*, 146.
58. Barclay, *Anatomy of the Bones of the Human Body*, text to plate 32.
59. E. W. Posner, *Das Weib und das Kind* (Glogau, 1847), 9–10.
60. Ruth Kelso, *The Doctrine of the Renaissance Lady* (Chicago, 1978), 213.
61. Aristotle, *The Politics*, trans. H. Rackham (London, 1932), 63.
62. Heidi Rosenbaum, *Formen der Familien* (Frankfurt, 1982), 288–9.
63. Quoted in Ackermann, *Über die Verschiedenheit*, 5.
64. On this point, see Marian Lowe, 'The Dialectic of Biology and Culture', in *Woman's Nature: Rationalizations of Inequality*, ed. Marian Lowe and Ruth Hubbard (New York, 1983), 39–62.
65. In the nineteenth century, poor women consistently had more deficient diets than their husbands, because they allowed their men and children more and better food; Mary Chamberlain, *Old Wives' Tales: Their History, Remedies and Spells* (London, 1981), 107.
66. Theodor L. W. von Bischoff, *Das Studium und die Ausübung der Medicin durch Frauen* (Munich, 1872), 45.
67. On 'social inequality as natural law', see Wilhelm Riehl, *Die Naturgeschichte des Volkes als Grundlage einer deutschen Socialpolitik*, vol. 3 (1856; Stuttgart, 1956), 3.
68. On the polarization of gender roles, see Karin Hausen, 'Die Polarisierung der Geschlechtscharaktere', *Sozialgeschichte der Familie in der Neuzeit*, ed. Werner Conze (Stuttgart, 1976), 363–93.
69. See Joan Kelly, 'Early Feminist Theory and the Querelle des Femmes, 1400–1789', *Signs*, 8: 1 (1982), 4–28; Susan Bell and Karen Offen (eds.), *Women, the Family and Freedom: The Debate in Documents, vol. 1, 1750–1880* (Stanford, Calif., 1983).
70. Jean-Jacques Rousseau, *Emile* (1762), in *Oeuvres complètes*, ed. Bernard Gagnebin and Marcel Raymond, 4 vols. (Paris, 1959–69), 4:693.
71. Rousseau's views are so well known that I have chosen to focus here on less well-known followers of Rousseau. On Rousseau, see Susan Moller Okin, *Women in Western Political Thought* (Princeton, 1979); Jean Elshtain, *Public Man, Private Woman: Women in Social and Political Thought* (Princeton, 1981); Joel Schwartz, *The Sexual Politics of Jean-Jacques Rousseau* (Chicago, 1984).
72. Bloch and Bloch, 'Women and the Dialectics of Nature', 35.

73. Rousseau, *Emile*, 692.
74. C. Helvétius, *De l'esprit* (Paris, 1758); Jean le Rond d'Alembert, *Lettre de M. d'Alembert à M. J.-J. Rousseau* (Amsterdam, 1759); M.-J. Condorcet, *Esquisse d'un tableau historique des progres de l'esprit humain* (Paris, 1794).
75. Mary Wollstonecraft, *Vindication of the Rights of Woman* (London, 1792); Olympe de Gouges, 'Declaration of the Rights of Woman and Citizen in 1791', in Bell and Offen (eds.), *Women, the Family and Freedom*, 1: 105–9; [Theodor von Hippel], *Über die bürgerliche Verbesserung der Weiber* (Berlin, 1792).
76. *Encyclopédie ou Dictionnaire Raisonné des Sciences, des Arts, et des Métriers* (Neuchatel, 1765), s.v. 'squelette'.
77. Roussel, *Système physique et moral de la femme*, 12.
78. Ibid. 22–3.
79. Ibid., xvi.
80. Samuel Thomas von Soemmerring, *Über die körperliche Verschiedenheit des Negers vom Europäer* (Frankfurt, 1785), ix. That Soemmerring wrote about women in a book on race is not surprising, for Soemmerring viewed race in the same way he viewed sex: as penetrating the entire life of the organism. He wrote: 'If skin is the only difference then the Negro might be considered a black European. The Negro is, however, so noticeably different from the European that one must look beyond skin color' (2).
81. Johann Wilhelm Heinrich Ziegenbien, *Aussprüche über weibliche Natur, weibliche Bestimmung, Erziehung und Bildung* (Blankenburg, 1808), 1.
82. On the relation between the moral and physical characteristics of woman, see Knibiehler and Fouquet, *La Femme et les médecins*, esp. ch. 4.
83. Carl Friedrich Pockels, *Versuch einer Charakteristik des weiblichen Geschlechts*, vol. 1 (Hanover, 1806), 6 and 8.
84. Sachs, *Ärztliches Gemälde des weiblichen Lebens*, 25, 47.
85. Auguste Comte, *Cours de philosophie positive*, vol. 4 (Paris, 1839), 569–70.
86. Auguste Comte to J. S. Mill, 16 July 1843, in *Lettres inédites de J. S. Mill à A. Comte avec les résponses de Comte*, ed. L. Lévy-Bruhl (Paris, 1899), 231.
87. Wilhelm von Humboldt, 'Über den Geschlechtsunterschied und dessen Einfluss und die organische Natur' and 'Über die männliche und weibliche Form', in *Neudrücke zur Psychologie*, ed. Fritz Giese, vol. 1 (1917), 110.
88. Georg Wilhelm Friedrich Hegel, *Grundlinien der philosophie des Rechts* (1821), in *Werke*, 7:319–20.
89. Midwives held a traditional monopoly on the certification of virginity and on birthing until the seventeenth century. Midwives received certification from other midwives, or they merely went into practice after serving an apprenticeship with an experienced midwife. On midwifery, see Barbara Ehrenreich and Deirdre English, *For Her Own Good: 150 Years of the Experts' Advice to Women* (New York, 1978); Jean Donnison, *Midwives and Medical Men* (New York, 1977); Toby Gelfand, *Professionalizing Modern Medicine: Paris Surgeons and Medical Science and Institutions in the Eighteenth Century* (Westport, Conn. 1980).
90. Ackermann, 'Krankheitslehre der Frauenzimmer', appendix to *Über die Verschiedenheit*.
91. Hilda Smith 'Gynecology and Ideology in Seventeenth-Century England', in *Liberating Women's History*, ed. Bernice A. Carrol (Chicago, 1976), 98.

92. Soemmerring, *Über die körperliche Verschiedenheit des Negers vom Europäer*, xiv.

93. Ibid., xix.

94. Pockels, *Versuch einer Charakteristik des weiblichen Geschlechts*, 1: viii–xviii.

95. Bischoff, *Das Studium und die Ausübung der Medicin durch Frauen*, 47.

96. Ibid. 14, 48, 15, and 20.

97. Eliza Haywood, *The Female Spectator*, 2 (1744), 240–1.

98. Hedwig Dohm, *Die wissenschaftliche Emancipation der Frauen* (Berlin, 1874). On Dohm, see Renate Duelli-Klein, 'Hedwig Dohm: Passionate Theorist', in *Feminist Theorists: Three Centuries of Women's Intellectual Traditions*, ed. Dale Spender (London, 1983), 165–83.

99. Friedrich W. T. Ravoth, 'Über die Ziele und Aufgaben der Krankenpflege', quoted in Helga Rehse, 'Die Rolle der Frau auf den Naturforscherver- sammlungen des 19. Jahrhundert', *Die Versammlung deutscher Naturforscher und Ärzte im 19. Jahrhundert*, ed. Heinrich Schipperges in *Schriftenreihe der Bezirksärztehammer Nordwürttemberg*, 12 (1968), 126.

100. Riehl, *Die Naturgeschichte des Volkes*, 3:3–5.

2 'Amor Veneris, vel Dulcedo Appeletur'

Thomas W. Laqueur

'If it is permissible to give names to things discovered by me, it should be called the love or sweetness of Venus,' urged one of the great and immodest anatomical explorers of the Renaissance. Like Adam, he claimed the privilege of naming what he had been the first to see: that which was 'preeminently the seat of women's delight.'[1] This Columbus, not Christopher but Renaldus, announced with much fanfare in 1559 that he had discovered the clitoris. ('O my America, my new found land!')

In 1905 Sigmund Freud rediscovered the clitoris, or in any case the clitoral orgasm, by inventing its vaginal counterpart. After four hundred, perhaps even two thousand years, there was all of a sudden a second place postulated from which women derived sexual pleasure. In 1905, for the first time, a doctor claimed that there were two kinds of orgasm and that the vaginal sort was the expected norm among adult women.

Both discoveries were and are controversial. Columbus's colleagues disputed his claim to precedence, arguing that the organ about which he made such a fuss either had been discovered by someone else or, had been common knowledge since Antiquity. Freud's discovery generated an immense polemical and clinical literature. More ink has been spilled, I suspect, about the clitoris than about any other organ, or at least about any organ its size.[2]

I shall not enter directly into these controversies. Instead I want to sketch the history of the clitoris in Western, predominantly medical, literature in order to make two points. In the first place, prior to 1905 no one thought that there was any other kind of female orgasm than

From Thomas Laqueur, 'Amor Veneris, vel Dulcedo Appeletur', in *Frangments for a History of the Human Body*, ed. Michel Feher (Zone, New York, 1989), pp. 90–131. Reprinted with permission.

the clitoral sort. It is well and accurately described in hundreds of learned and more popular medical texts as well as in a burgeoning pornographic literature. Thus, it simply is not true that, as Robert Scholes has argued, there has been 'a semiotic coding that operates to purge both texts and language of things [the clitoris as the primary organ of woman's sexual delight] that are unwelcome to men.' The clitoris, like the penis, was for two millennia both 'precious jewel' and sexual organ, a connection not 'lost or mislaid' through the ages, as Scholes would have it, but only—if then—since Freud.[3] To put it differently, Masters and Johnson's revelation that female orgasm is almost entirely clitoral would have been a commonplace to every seventeenth-century midwife and had been anticipated in consider-able detail by nineteenth-century investigators. For some reason, a great amnesia in this matter descended on scientific circles around 1900 so that hoary truths could be hailed as earth-shatteringly new in the second half of the twentieth century.

My second point is that there is nothing natural about how the clitoris is construed. It is not self-evidently the female penis nor is it self-evidently opposed to the vagina. Nor have men always regarded clitoral orgasm as absent, threatening or unspeakable because of some primordial male fear of, or fascination with, female sexual pleasure. The history of the clitoris is part of the history of sexual difference generally and of the socialization of the body's pleasures. Like the history of masturbation, it is a story as much about sociability as about sex.

FREUD AND HISTORY'S CLITORIS

'If we are to understand how a little girl turns into a woman,' Freud writes in the third of his epochal *Three Essays on the Theory of Sexuality*, 'we must follow the further vicissitudes of [the] excitability of the clitoris.' During puberty, so the story goes, there occurs in boys 'an accession of libido', while in girls there is 'a fresh wave of repression in which it is precisely clitoroidal sexuality that is effected'. The development of women as cultural beings is thus marked by what seems to be a physiological process: 'what is overtaken by repression is a piece of masculine machinery.'[4]

Like a Bahktiari tribesman in search of fresh pastures, female sexuality is said to migrate from one place to another, from the 'male-like' clitoris to the unmistakably female vagina. The clitoris, according to

Freud, does not, however, entirely lose its functions as a result of pleasure's short but significant journey. Rather, it becomes the organ *through which* excitement is transmitted to the 'adjacent female sexual parts', to its permanent home, to the true locus of a woman's erotic life, to the vagina. The clitoris, in Freud's less than illuminating simile, becomes 'like pine shavings', used 'in order to set a log of harder wood on fire'.[5]

This unlikely and strangely inappropriate identification of the cavity of the vagina with a hot log is not my concern here. Stranger still is the tension between the biological and the cultural claims of Freud's famous essay. A little girl's realization that she does not have a penis and that her sexuality resides in its supposed opposite, in the cavity of the vagina, elevates a 'biological fact' into the cornerstone of culture. He writes as if he has discovered the basis in anatomy for the entire nineteenth-century world of gender. In an age obsessed with being able to justify and distinguish the social roles of women and men, science seems to have found in the radical difference of the penis and vagina not just a sign of sexual difference but its very foundation. 'When erotogenic susceptibility to stimulation has been successfully transferred by a woman from the clitoris to the vaginal orifice, it implies that she has adopted a new leading zone for the purposes of her later sexual activity.'[6]

In fact, Freud goes even further by suggesting that the repression of female sexuality in puberty, marked by its abandonment of the clitoris, heightens male desire and thus tightens the web of heterosexual union on which reproduction, the family and, indeed, civilization itself appear to rest: 'The intensification of the brake upon sexuality brought about by pubertal repression in women serves as a stimulus to the libido of men and causes an increase in its activity.'[7] When everything has settled down, the 'masculine machinery' of the clitoris is abandoned, the vagina is erotically charged, and the body is set for reproductive intercourse. Freud seems to be taking a stab at universal historical bio-anthropology, to be making the claim that female modesty incites male desire while female acquiescence, in eventually allowing it to be gratified, lies at the very foundation of humanity's journey from the savage's cave.

Perhaps this is pushing one paragraph too hard, but in the passages I have quoted Freud is very much in the imaginative footsteps of Diderot, for example, who argues that civilization, and with it the family, began when women began to discriminate, to limit their availability. Possessing them, being the preferred one, loomed up as supreme happiness; monogamy and marriage became the price paid by men to once again gain access:

Then, when the veils that modesty cast over the charms of women allowed an inflamed imagination the power to dispose of them at will, the most delicate illusions competed with the most exquisite of senses to exaggerate the happiness of the moment; the soul was possessed with an almost divine enthusiasm; two hearts lost in love vowed themselves to each other forever, and heaven heard the first indiscreet oaths. [8]

Freud is not quite so explicit in the *Three Essays* but he does appear to be arguing that femininity, and thus the place of women in society, is grounded in the developmental neurology of the female genitals.

He cannot, however, really have meant this. In the first place, history would have shown him that the vagina fails miserably as a 'natural symbol' of interior sexuality, of passivity, of the private against the public, of a critical stage in the ontogeny of woman. In the one-sex model that dominated anatomical thinking for two millennia, woman was understood as man inverted. The uterus was the female scrotum, the ovaries were testicles, the vulva a foreskin, and *the vagina was a penis*. This account of sexual difference, though as phallocentric as Freud's, offered no real female interior, only the displacement inward to a more sheltered space of the male organs, as if the scrotum and penis in the form of uterus and vagina had taken cover from the cold.[9]

Freud may not have been aware of this arcane history. But he must have known that there was absolutely no anatomical or physiological evidence for the claim that 'erotogenic susceptibility to stimulation' is successfully transferred during the maturation of women 'from the clitoris to the vaginal orifice'. The abundance of specialized nerve endings in the clitoris and the relative impoverishment of the vagina had been demonstrated half a century before Freud wrote and had been known in outline for hundreds of years. Common medical knowledge available in any nineteenth-century handbook thus makes a farce of Freud's story if it is construed as a narrative of biology. And, finally, if the advent of the vaginal orgasm was the consequence of neurological processes, then Freud's question of 'how a woman develops out of a child with bisexual dispositions' could be resolved by physiology without any help from psychoanalysis.

Freud's answer, therefore, must be a narrative of culture in anatomical disguise. The story of the clitoris is a parable of culture, of how the body is forged into a shape usable by civilization despite, not because of, itself. The language of biology gives this tale its rhetorical authority but does not describe a deeper reality in nerves and flesh.

. . .

FIG. 5. From Vesalius, *De humani corporis fabrica* (Bâle, 1543). By permission of the Francis A. Countway Library of Medicine.

The history of the clitoris can be pushed back. In 1612, three years before 'clitoris' was first used in English, a French doctor discusses the organ and the word in great detail. 'Cleitoris', he says, is known by many Latin names, including Columbus's neologism. As for the vernaculars,

In French it is called temptation, the spur to sensual pleasure, the female rod and the scorner of men: and women who will admit their lewdness call it their great joy [En François elle dite tentation, aiguillon de volupté, verge feminine, le mespris des hommes: et les femmes qui font profession d'impudicité la nomment leur *gaude mihi*].[10]

Duval echoes the certainties and tensions of both later and also earlier accounts. On the one hand, the clitoris is the organ of sexual pleasure in women. On the other, its easy responsiveness to touch makes it difficult to domesticate for reproductive, heterosexual intercourse. This was Freud's problem, and I will return to it in the last section of this essay.

Freud may not have been aware of the detailed history of genital anatomy I have just recounted but it is impossible that he would not have known what was in the standard reference books of his day. He was, after all, especially interested in zoology during his medical student days and was an expert neurologist. Furthermore, one did not have to be a scientist to know about clitoral sexuality. Walter, the protagonist of *My Secret Life*, notes in his review of the copulative organs that the clitoris is an erectile organ which is 'the chief seat of pleasure in a woman'.[11] Nicholas Venette's *Tableau de l'amour conjugal*, first published in the late seventeenth century, translated and republished in many languages and in innumerable editions and still popular in the nineteenth century (the pharmacist's son in *Madame Bovary* consults it), remarks that in the clitoris 'nature has placed the seat of pleasure and lust, as it has on the other hand in the glans of a man . . . there is lechery and lasciviousness established.'[12] And, of course, Freud himself points out that biology has been 'obliged to recognize the female clitoris as a true substitute for the penis', though it does not follow from this, as Freud seems to think, that children recognize that 'all human beings have the same (male) form of genita', or that little girls therefore suffer penis envy because their genital is so small.[13]

Freud, in short, must have known that what he wrote in the language of biology regarding the shift of erotogenic sensibility from the clitoris to the vagina had no basis in the facts of anatomy or physiology. Both the migration of female sexuality and the opposition between the vagina and penis must therefore be understood as representations, namely, something made to present a social ideal in another form. On a formal level the opposition of the vagina and penis represents an ideal of parity. The social thuggery which takes a polymorphously perverse infant and bullies it into a heterosexual man or woman finds an organic correlative in the body, in the opposition of the sexes and their organs. Perhaps because Freud is the great theorist of sexual ambiguity he is also the inventor of a dramatic sexual antithesis: that between the embarrassing clitoris that girls abandon and the vagina whose erotogenic powers they embrace as the mark of the mature woman.[14]

More generally, what might loosely be called patriarchy may have appeared to Freud as the only possible way to organize the relations between the sexes, leading him to write as if its signs in the body (external active penis versus internal passive vagina) were 'natural'. But in Freud's question of how it is that 'a woman develops out of a child with a bisexual disposition', the word 'woman' clearly refers not to 'natural' sex, but to 'theatrical' gender, to socially defined roles.[15] As Gayle Rubin puts it, gender is 'a socially imposed division of the sexes', the 'product of the social relations of sexuality'. The supposed opposition of men and women, 'exclusive gender identity', in Rubin's terms, 'far from being an expression of natural differences . . . is the suppression of natural similarities'.[16]

Indeed, if structuralism has taught us anything, it is that humans impose their sense of opposition on a world of continuous shades of difference and similarity. In nearly all of North America, to use Lévi-Strauss's example, sage brush (Artemisia) plays 'a major part in the most diverse rituals as the opposite of other plants: "Solidaga", "Chrysothamnus", "Gutierrezia".' It stands for the feminine in Navaho ritual while 'Chrysothamnus' stands for the masculine, which does not follow from features of these plants readily detected by the outsider.[17] Their differences or similarities, like Freud's, are in the most immediate sense 'man-made'.

This is not the place to give and defend a structuralist reading of Freud generally. Yet *Civilization and Its Discontents* bears the most poignant reminders of the processes Lévi-Strauss describes. Civilization, like a conquering people, subjects 'another one to its exploitation', proscribes 'manifestations of sexual life in children', makes 'heterosexual genital love' the only permitted sort, and in so doing takes the infant, 'an animal organism with (like others) an unmistakably bisexual disposition', and forces it into the mold of *either* a man or a woman.[18] The power of culture thus represents itself in bodies, and forges them, as on an anvil, into the required shape.

Freud's myth of sexuality's migration, his invention of the vaginal orgasm that supplants a clitoral orgasm appropriate to an earlier developmental stage, is so powerful precisely because it is not a story of biology. What Rosalind Coward has called in another context 'ideologies of appropriate desires and orientations' must struggle— one hopes unsuccessfully—to find their signs in the flesh.[19] Freud's argument, flying as it does in the face of three centuries of anatomical knowledge, is a testament to the freedom with which the authority of nature can be rhetorically appropriated to legitimize the creations of culture.

···

DID COLUMBUS DISCOVER THE CLITORIS?

···

Even if he was not the first anatomist to do so, there seems little question that in 1559 Columbus saw and described the clitoris. But then the other Columbus is said to have discovered America sixty-seven years earlier and his claim rests upon visiting a continent which he took to be Japan, which had been visited by other Europeans centuries before and which was populated by the decendants of much earlier explorers from Asia. Renaldus Columbus's claim is just as complicated.

'This, most gentle reader, is that: preeminently the seat of women's delight,' he announced. Is what? His initial mapping is unfamiliar: 'These protuberances, emerging from the uterus near that opening which is called the mouth of the womb [*Processus igitur ab utero exorti id foramen, quod os matricis vocatur*]. . .' But the clitoris we know today does not emerge from the mouth of the *matrix*, whatever that might be. What follows—I quote it in full including passages I have already mentioned—is even more curious:

. . . [that] preeminently [is] the seat of women's delight, while they engage in sexual activity [*venerem*], not only if you rub it vigorously with a penis, but touch it even with a little finger, semen swifter than air flows this way and that on account of the pleasure even with them unwilling. If you touch that part of the uterus while women are eager for sex and very excited as if in a frenzy and aroused to lust [and] they are eager for a man, you will find it a little harder and oblong to such a degree that it shows itself a sort of male member. Therefore, since no one else has discerned these projections and their working, if it is permissible to give names to things discovered by me, it should be called the love or sweetness of Venus [. . . *illa praecipue sedes est delectionis mulierum, dum venerem exercent, quam non modo si mentula confricabis, vel minimo digito attrectabis, ocyus aura semen hac atque illac pre voluptate vel illis invitis profluet. Hanc eadem uteri partem dum venerem appetunt mulieres, [et] tanquam oestro percitae virum appetunt ad libidinem concitatae: si attingues, duriusculam et oblongam redditam esse comperies: adeo ut nescio quam virilis mentulae speciem prae se ferat. Hos igitur processus atque eorundum usum cum nemo hactenus animadverterit, si nomina rebus a me inventis imponere licet, amor Veneris, vel dulcedo appelletur*].[20]

He again calls the organ he is describing here 'that part of the uterus' and a third time refers to 'those projections' or 'processes' which he had earlier described as 'coming out of the mouth of the uterus'. The reference seems indeed to be specifically the vagina's 'prepuce', the outer skin of the female penis long recognized and erotically linked to pudendal pleasure. Columbus does not seem to be describing a newly discovered female penis. In any case, his nomenclature is decidedly odd and imprecise. His organ may be the clitoris but it is not the clitoris of modern anatomy.

Because Columbus claims to be discovering the female penis, his account is also unabashedly homoerotic. If the male penis or even the male finger is rubbed against the female, penis semen—presumably hers—flies this way, even though one would think that his would, under the circumstances, also fly. In any case, the ancient homologies are maintained. In the classical one-sex model some other parts of the female anatomy—parts that we would call the cervix, vagina and vulva—are the female penis against which the male's rubs. In Columbus's version a new external female penis is the partner object of the male.

Woman is submerged in the next sentence as well. In one of the few instances in which *mulieres* (women) are the grammatical subject, in the temporal clause they are literally surrounded by desire. *Appetunt* (from *appetere*, to be eager for) is repeated to flank *mulieres* (women); the redundant predicate adjectives, *percitae* (from *percire*, to arouse) and *concitae* (from *concitare*, to stir up or excite), attest further to her sexual arousal. But then the sentence takes an unexpected turn and the scientifically objective, presumably male, reader is told that the part of the female anatomy in question will become hard and oblong if touched. (A finger or a penis will do to make her 'semen [flow] swifter than air.')[21]

Columbus is clearly not briefing a case for clitoral as opposed to vaginal sexuality. The interest in this disjuncture is entirely Freud's. Rather, he and his contemporaries imagine a cascade of pleasure from the inside out. The cervix, he says, has circular folds so that during intercourse it 'may embrace and suck out the virile member'; these folds cause the friction 'from which lovers experience wonderful pleasure'. Further along, 'at the end of the cervix, approaching the vulva', there are several little pieces of flesh and by these as well 'pleasure or delight in intercourse is not a little increased.' Then finally comes the clitoris.[22]

The simultaneous presence of both an interior and an exterior female penis, both subject to erection, pleasure and ejaculation did not disturb Columbus or other sixteenth- and seventeenth-century writers. To the contrary, it provided a second register on which to play the old tune of hierarchical ordering of two genders in the one flesh. Columbus himself and his contemporaries remained committed to the old notions that women had the same organs as men, and that these organs functioned roughly in the same way.

A more popular literature echoes these views. Jane Sharp's self-consciously commonsensical midwifery guide asserts on one page that the vagina, 'which is the passage for the yard, resembleth it turned inward', while two pages later, with no apparent embarrassment, she reports that the clitoris is the female penis. 'It will stand and fall as the yard doth and makes women lustful and take delight in copulation.' And, with the usual credit given pleasure: 'were it not for this they would have no desire nor delight, nor would they conceive.'[23] Perhaps Sharp would resolve the dilemma of the two female penises by saying that the vagina only *resembles* a penis while the clitoris actually is one. But she is unworried by the problem.

Sharp was, moreover, as undisturbed by ambiguity as by contradiction. Thus, the fact that the labia fit nicely into both systems of homologies and functioned as the foreskin for both the vagina and the clitoris unasked the question of which one was the true female penis. Indeed, both could be regarded as erectile organs. The reader of another midwifery manual was told that 'the action of the clitoris is like that of the yard, which is erection' and that 'the action of the neck of the womb [the vagina and cervix] is the same with that of the yard; that is to say, erection.'[24]

There seemed to be no difficulty in holding that women had a topological inversion of the male penis within them, which embraced and sucked on the male member, and a morphological homologue without, which worked throughout reproductive adulthood in precisely the same way as the man's. The tip of the clitoris, argues Thomas Gibson, is called the *amoris dulceda* or *oestrum veneris* because, like the glans of the penis, it is where the sweetness of love, the venereal frenzy, is most intensely felt.[25] Sixteenth- and seventeenth-century writers would have found very curious Helene Deutsch's notion that 'the competition of the clitoris which intercepts the excitations unable to reach the vagina ... Create[s] the dispositional basis of permanent sexual inhibition,' or Marie Bonaparte's contention that 'clitoroidal women' suffer from one of the stages of frigidity or protohomosexuality.[26] Rather, as Nicholas Culpepper put it in 1675, 'it is agreeable both to reason and authority, that the bigger the clitoris in a woman, the more lustful they are.'[27] And lust, of course, was thought to be immediately relevant to the ultimate purpose of sexual intercourse: reproduction. It is 'by stirring the clitoris [that] the imagination causeth the vessels to cast out that seed that lyeth deep in the body,' explained Jane Sharp.[28] The erection of both penis and vagina, said another authority is 'for motion, and attraction of the seed'.[29] Presumably the 'spermatical vessels', the 'handmaidens to the stone', as Philip Moore called them a century earlier, 'carried the excitement from the external organs to the testes [what we would call ovaries] within.'[30]

Columbus claims to have made a very precise discovery in the female reproductive anatomy while at the same time he remained rooted in a conceptual universe that contained no women, only a colder version of man. The controversy over precedence must be understood in this context. It was fraught with language inadequate to genital specificity, a language of the one-sex body in which corporeal difference threatened always to collapse into sameness. Barriers of gender, precariously imposed on one flesh, were consequently all the more important.

The absence of a standard anatomical nomenclature for the female genitals, and reproductive system generally, testifies, however, not to the haplessness of Renaissance anatomists but to the absence of an imperative to create incommensurable categories of male and female through words. The issue here is not the vast elaboration in most languages of terms for organs and functions that are risqué or shameful. The point is rather that until the late seventeenth century it is often impossible to determine, in medical texts, to which part of the female reproductive anatomy a particular term applies.[31] Language constrains the seeing of differences and sustains the male body as the canonical human form. And, conversely, in the one-sex model even the words for female parts ultimately refer to male organs. The post-eighteenth-century words—vagina, uterus, vulva, labia, clitoris—do not have their Renaissance equivalents.[32]

'It does not matter,' says Columbus with more insight than he was perhaps aware, 'whether you call [the womb] matrix, uterus or vulva.'[33] And it does not seem to matter where one part stops and the other starts. He does not want to distinguish the true cervix—the 'mouth of the womb' (*os matricis*) which from the outside 'offers to your eyes . . . the image of a tenchfish or a dog newly brought to light,' which in intercourse is 'dilated with extreme pleasure,' and which is 'open during that time in which the woman emits seed'—from what we would call the vagina 'that part into which the penis [*mentula*] is inserted, *as it were*, into a sheath [*vagina*].'[34] (Note the metaphoric use of 'vagina', the standard Latin word for scabbard, which was otherwise never used in the medical literature for the part to which it applies today.) But Columbus offers no other term for 'our' vagina: he describes the labia minora as 'protuberances [*processus*], emerging from the uterus near that opening which is called the mouth of the womb . . .' (the same location as the clitoris).[35] The precision Columbus sought to introduce by naming the true 'mouth of the womb' *cervix* vanishes as the vaginal opening becomes the mouth of the womb and the clitoris one of its parts. The language simply does not exist, nor need exist, for distinguishing male from female organs clearly. The haze of one sex still hangs over genital differentiation.

This same sort of tension is evident in other anatomists. Fallopius is anxious to differentiate the cervix proper from the vagina but has no more specific name for it other than 'female pudenda', a part of a general 'hollow' (*sinus*). The Fallopian tubes, as he describes them, are not patent tubes that convey eggs from the ovaries to the womb but rather hollow twin protuberances of sinews (*neruei*) that penetrate the peritoneum, without seeming to have an opening into the uterus. Fallopius remains committed to a male-centered system and, despite his revolutionary rhetoric, assumes the commonplace that 'all parts that are in men are present in women'.[36]

Caspar Bauhin (1560–1624), professor of anatomy and botany in Basel, sought to clear up the nomenclature but with an equal lack of success because one sex was too deeply embedded in language. He begins, from a modern perspective, promisingly, 'Everything pertaining to the female genitalia is comprehended in the term "of nature" [*phuseos*], and the obscene term cunt [*cunnus*].' Clear enough. But then Bauhin informs his readers that some ancient writers called the male genitalia *phuseos* as well. Among the terms he offers for the labia is the Greek *mutocheila*, meaning snout, with its obvious phallic connection, or more explicitly translated, 'penile lips'. (The latter is the more likely sense in an age whose greatest anatomist could regard the cervix as an erectile organ.)[37]

The conflation of labia into foreskin, of female into male, is however much older. A tenth-century Arabic writer, for example, points out that the interior of the vagina 'possesses prolongations of skin called the lips' which are 'the analogue of the prepuce in men and has as its function to protect the matrix against cold air.'[38] For Mondinus, author of the major medieval anatomy text, the labia become a bifurcated foreskin:

They [the labia] hinder entrance of air and external matter into the neck of the womb or bladder as the skin of the prepuce guardeth the penis. Therefore . . . Haly Abbas calleth them *praputia matricis* [prepuce of the uterus, of the vagina?].[39]

Likewise in Berengario's text, a major Renaissance illustrated expansion of Mondino, *nymphae* refers to both the foreskin of the penis and the foreskin of the vagina, i.e., the labia minora.[40] John Pechy, a popular English writer during the Restoration, revises this trope to accommodate the new female penis so that the 'wrinkeled membranous production cloath the clitoris [not the vagina] like a foreskin.'[41]

Much of the controversy around who 'discovered' the clitoris arose out of precisely this sort of blurring of metaphorical and linguistic boundaries that, in turn, were the consequence of a model of sexual difference in which unambiguous names for the female genitals did not matter. A web of words, already pregnant with a theory of sexual difference or sameness, thus limited how genital organs would be seen and discussed. There were no terms in the Renaissance for what, since the eighteenth century, have been construed as essential signs in the body of incommensurable difference. The boundaries of organs themselves and what could be known about them constrained and were constrained by the openness of the language in which they could be thought.

Who then discovered the 'clitoris'? Fallopius says he was the first to see the organ and makes much of it. After trying to straighten out the terminological confusion between vagina and cervix he announces 'but these are trivial matters; weightier things to follow':

This genital [clitoris], because it is concealed among the fatter parts of the pubis, it therefore is hidden even from the anatomists and thus I was the first who detected that same thing in previous years and any others who speak or write about this, *be assured that they learned of the thing itself either from me or from my followers.* You will find the outermost gland of this penis ['*penis*,' not the more common word '*mentula*'] straightaway in the upper part of the outer genitals itself where the 'wings' [i.e., the labia minora] are joined or where they begin.[42] (Emphasis added.)

This seems straightforward enough: Fallopius says he saw it first. But even though Fallopius advertises his skills and his precedence, he elsewhere shifts to the rhetoric of a Renaissance humanist tracing a word or idea back to classical Antiquity. Like Copernicus he wants to proclaim both his originality *and* his rootedness in tradition:

> ... in book 3, section 21 [of the *Canon*], Avicenna mentions something situated in a certain part of the female genitals that he calls a penis [*virga*] or *albathara*. Albucasis in book 2, section 1, terms this part the penis [*tentigenem*, literally tenseness] which is accustomed sometimes to come to such an increase that women having this can copulate with others as if they were men.[43]

Fallopius then goes on to say that the Greeks called the structure in question *kleitoris*, from which the 'lewd verb, "to touch the clitoris",' is derived. 'Our anatomists,' he laments, have neglected it. Such neglect might justify the detailed description which I quoted earlier but scarcely the aggressive claim of priority. Two professional imperatives seem to be at war here: the anatomist's desire for personal glory through besting his contemporaries, immediate predecessors and the ancients, and the humanist's concern for recovering the classical past. In any case, it is difficult to say whether Fallopius discovered the clitoris because by his own account he both did and did not.

Bartholinus, a major anatomist of the early seventeenth century, is similarly puzzled. How is it possible, he seems to be asking, that 'Fallopius arrogates to himself the Invention or first Observation of this Part; and Columbus gloriously, as in other things he is wont, attributes it to himself,' while the organ in question has been known to everyone from the second-century physicians Rufus of Ephesus and Julius Pollux to the Arabic anatomists Albucasis and Avicenna.[44]

A half century later Regnier De Graaf argues that Columbus had no right to name something that had been sighted, named and renamed by generations of anatomists stretching back to the Greeks. Hippocrates, he says, had called it *columnella*, little column, from the Greek *kios;* Avicenna named it *albatra* or *virga*, the rod; to Albucasis it was the *amoris dulceda*, the sweetness of love; and to others still it was the *sedes libidinis*, seat of lust, the *irritamentum libidinis*, the goad to lust, the *oestrus Veneris*, the frenzy of Venus. It was known to still other ancient anatomists, he reports, as *nympha*, the water goddess.[45]

De Graaf missed some names. Julius Pollux, the second-century compiler of medical word lists, says that in the cleft of the female pudenda, 'the throbbing bit of flesh in the middle is the *nymphē* or *myrton* [because it resembled a myrtle berry?] or *epidermis* or *kleitoris*.'[46] Other terms in the literature include *tentigo*, tension; *cauda muliebris, coles muliebris*, or *mentula muliebris*, all versions of 'woman's penis' since *cauda* means literally 'tail' as did penis in its earliest usage, while *mentula* was the standard classical obscene term for the male organ. *Crista* [as in cockscomb] and *landica* occurred in nonmedical texts.[47]

But perhaps Columbus was not being quite as self-aggrandizing and wrongly dismissive of his predecessors as Bartholinus and De Graaf suggest. Since Antiquity there had obviously been a metaphorical association between the penis and what we would call the clitoris. But the latter's erotogenic power, which Columbus regards as its essence, had been far less explicitly recognized. And because of this, the various words listed above did not quite mean what 'clitoris' meant after the sixteenth century.

I will at least try to brief a case for Columbus. In the first place, De Graaf was mistaken in attributing to classical or Arabic writers those terms—'sweetness of love', 'venereal frenzy', 'seat of lust', 'spur of lust'—which referred, not to the structure, but to the function of the clitoris. With the possible exception of 'spur of lust', all these terms were either first used by Columbus or derived from his description. Furthermore, although all modern translators and some ancient authorities tell us that terms like *nympha* and *tentigo* mean 'clitoris', they were used in such a way as to obscure precisely *what* they meant.[48]

To be sure, some early uses seem wholly transparent. 'Kleitoris' first appears in the verb form *kleitorizein* in Rufus of Ephesus (second century A.D.), who tells us that it means 'to touch [the *nymphē*] lasciviously'. Moreover, Hyrtl, philologist and professor of anatomy at the University of Vienna while Freud was there, explicitly notes that Rufus was not confusing the singular form of the noun with the plural *nymphae*, i.e., the labia minora: *nymphē* = the muscular piece of flesh in the midst, which some call the 'hypodermis' and others the 'kleitoris', Hyrtl says, citing Rufus.[49]

But when we consider how, for example, Galen uses the word *nymphē* it is apparent that he is not interested in the specificity of anatomical and erotic signification that would be commonplace after Columbus. The part called *nympha* (or *nymphē*), he writes, 'gives the same sort of protection to the uteri that the uvula gives to the pharynx; for it covers the orifice of their neck by coming down into the female pudendum and keeps it from being chilled.'[50] Perhaps the clitoris does resemble the uvula with which it is also linguistically related. In addition, one can certainly make a case for the metaphorical connection in ancient texts between the vagina = *valvē*, i.e., an organ that breathes in and out, and the throat. But Galen's anatomy remains difficult to interpret. The clitoris does not come down into the pudendum to cover the mouth of the vagina. Moreover, Galen assigns the same function to the 'outgrowths of skin at the ends of the two pudenda', that is, to the labia minora that also act as covers to keep the uterus warm.

Soranus, the great second-century gynecologist, has no more interest in the clitoris as an erectile, erotogenic organ: it, the *nymphē*, 'is the origin of the two labia and by its nature it is a small piece of flesh almost like a muscle; and it has been called 'nymphē' because this piece of flesh hides like a bride.'[51] But then so do the labia minora, which he does not distinguish by name and which are associated with nymphs because of their alleged function of directing the urine.

The same sort of muddle arises regarding the word *tentigo* as a putative synonym for clitoris in the translations of the Arab physician Albucasis. It comes from the verb *tendere*, meaning 'to stretch'; as a noun it means 'tenseness' or 'lust' and was used in Antiquity to refer to an erect penis and even to an erect *landica*, in other words, clitoris.[52] Therefore, when Thomas Vicary, surgeon to Henry VIII, wrote in 1548 (before Columbus published his work) that the vulva 'hath in the middest a Lazartus pannicle, which is called in Latin *tentigo*,' the reference would seem to be unambiguous. *Tentigo* in early seventeenth-century English still bore the word's Latin meaning: 'a tenseness or lust; an attack of priapism; an erection'. The structure's location as Vicary describes it allows even less doubt that it is the clitoris. But when he reports on the functions of this part, its 'two utilities', he seems to be discussing an entirely different organ. There is no mention of pleasure or sexual arousal: 'The first [utility] is that by it goeth forth the urine, or else it should be shed through out all the Vulva: The seconde is, that when a woman does set hir thies abrode, it altereth the ayre that commeth to the Matrix for to temper the heate.' What the name led us to expect, a female penis, turns out to be a pair of workaday flaps, a dual purpose female foreskin.[53] But whatever Vicary means, it is impossible to translate across the chasm that divides his world from ours.

The nomenclature is finally straightened out in the generation after Columbus when *nymphae*, i.e., the plural of *nympha*, was first used by Adrianus Spigelius (1578–1625; the last of the great Vesalian line in the chair of anatomy at Padua) to mean what *nymphae* means today, the labia minora. Thus, right after a precise new erotic center is defined in women's bodies, language comes to articulate that definition. *Nymphae* and *clitoris*, 'wings' and 'rod', went their separate linguistic ways.

By the early seventeenth century the term 'clitoris' came to have its modern meaning: a female homologue of the penis which is the primary locus of sexual pleasure.[54]

Ironically, the first use of 'clitoris' in English denies this. In 1615 Helkiah Crooke, an English anatomist of some note, argued against all sorts of male/female homologies—the Galenic assignment of the vagina as the female penis as well as the novel description of the clitoris in that role. The clitoris, he says, is not a penis because it does not have 'a passage for the seede.' It may seem that Crooke is here striking a blow against the epistemological basis of the entire one-sex model until one looks more carefully at his argument against the vagina being a penis. 'Howsoever the necke of the wombe shall be inverted, yet it will never make the virile member.' Why? Because 'the yard consisteth of three hollow bodies,' the 'two hollow nerves' and a 'common passage for seed and urine,' while 'the necke of the womb hath but one cavity.' Clearly, 'three hollow bodies cannot be made of one.' Moreover 'neither is the cavity of a man's yard so large and ample as that of the necke of the wombe.' In short, the penis is not a vagina because it is thrice hollow or because it is not hollow enough![55]

Claude Duval in 1612 used the difference between the clitoris and the penis to establish the sex of a man mistaken for a woman and the great seventeenth-century specialist in legal medicine Paulo Zachias likewise distinguishes between the penis and the enlarged clitoris of some women on the basis of the latter's lack of a duct.[56] In 1779 the English surgeon John Hunter summed up the wisdom of two centuries by pointing out that precisely 'the part common to both' was the sign of a person's true condition.[57]

But these sorts of arguments do not take away from Columbus's claim. Perhaps he did discover for medical science the organic source of the 'sweetness of Venus'. Or perhaps he did not. It does not matter—or it is difficult to say—because the dominant medical paradigm of his day held that there was only one sex anyway, differing only in the arrangement of a common set of organs. The problem in Colombus's day and well into the seventeenth century was not finding the organic signs of sexual opposition but understanding heterosexual desire in the world of one sex. Being sure that 'jackdaw did not seek jackdaw,' that 'like did not seek like' as Aristotle had put it, took tremendous cultural resources. It was by no means certain that each time nature would 'be to her bias drawn', that, as in the end of *Twelfth Night*, nature would swerve from a straight path and make sure that male and female would couple. The Renaissance shared this concern with Freud. But the clitoris was only a very small part of the problem, if a problem at all, when the entire female genitalia were construed as a version of the male's.

In much Renaissance writing about what might be called the social problem of the clitoris, the issue is making sure that women engage in sexual intercourse as befits their station and not as befits men. By the eighteenth century when the notion of two opposite sexes made hetero-sexual coupling natural—opposites attract—the problem of the clitoris became the same as that of the solitary penis: masturbation. The threat to society was solitude, not collapse into homoeroticism, which was, by now, more safely 'against nature' in a reductionistic sort of way.

THE CLITORIS AS SOCIAL PROBLEM

In Western Europe the most dramatic attack on the clitoris—clitorectomy or excision—was for practical purposes never undertaken until the notorious and quickly condemned antimasturbatory oper-ations of the 1870s. The procedure was widely known, but only in the context of what *other* people did.

From Antiquity to at least the late nineteenth century it was thought that the clitoris of Egyptian women, and more generally of women in hot climates, grew preternaturally large. These unfortunate creatures suffered from what seems to have been conceived of as a racially and geographically specific clitoral hypertrophy that required surgical treatment.[58]

Not all classical reports of clitorectomy concern anatomical deformation. The Greek geographer Strabo observes in his account of travels through North Africa and the eastern Mediterranean that Creophagi males 'have their sexual glands mutilated and the women are excised in the Jewish fashion'. He elsewhere noted that the Jews circumcised males and excised females, though he does not offer clit-oral enlargement as the reason for these practices. Presumably, cul-tural factors instead were at work.[59]

The sixth-century Byzantine physician Aetius of Amida and Paulus Aegineta were more straightforwardly clinical in their allusions. Aetius discusses the *nymphae* in their pathological condition:

This organ in some women attains such a size that it may eventually consti-tute a deformity and lead to a feeling of shame. And further, it is greatly irritated by constant contact with the clothing and stimulates venery and coitus. On that account it appeared feasible to the Egyptians to amputate the nympha before it became too large, especially before the marriageable virgins were to be assigned. The amputation is done in the following manner. . . .

He then describes how the woman is to be seated, presumably leaning backward on a stool or small table. A young man supports her from behind with his arms under her knees thereby keeping her legs raised and apart. The surgeon takes hold of the offending flesh—which is presumably the covering of the clitoris and the labia minora—and excises them, being careful not to cut too deeply.[60]

Paulus, in describing the same operation, suggests that perhaps clitorectomy is not quite so routine nor so much an attack on excess female sexuality generally but, rather, that it is a measure to be used against women who cross gender boundaries in the sexual act. 'In some women,' he reports,

the nympha is excessively large . . . insomuch that, as has been related, some women have had erections of this part like men, and also venereal desires of a like kind. Wherefore, having placed the women in a supine position, and seizing the redundant portion of the nympha in a forceps. . . .[61]

He describes essentially the same procedure as does Aetius. However we interpret Paulus's clinical rationale, it is clear that clitorectomy was rejected in the West until the nineteenth century. Robert James, the compiler of a standard eighteenth-century medical encyclopedia, quotes the accounts I have cited, reports that the operation is practised among Arabians and Egyptians to remove 'from newborn girls whatever was indecently prominent in that part' and assures his readers that 'such an operation is indeed rarely practised among Europeans'.[62] In a remarkable seventeenth-century compilation of the anthropological lore of bodily transformation, John Bulwer similarly associates clitorectomy with the barbarity of 'the other'. In Arabia, he notes, a people called Creophagi circumcise ('Judaically') not only the men, but women also; 'the women of the Cape of Good Hope also excise themselves, not from a notion of Religion, but as an ornament.' And most worthy of scorn,

[in] Ethiopia, especially in the Dominions of Prester John, they Circumcise women. These Abassines have added errour upon errour, and sin upon sin, for they cause their Females to be circumcised. . . . A thing which was never practiced in Moses Law, neither was there ever found any expresse Commandment to do it; I know not where the Noselesse Moores learned it. . . .[63]

The only account of clitorectomy in Western Europe that I have encountered arises not from the victim's excessive sexual lust, nor even from her erotic encounters with another woman, but from her having pretended to be what she was not. Like those men and women accused of having lain with their own sex under the pretence of being the opposite, Henrica Shuria stands accused of having violated some supposedly self-evident sumptuary law of the body; she was charged as a woman who duplicitously played the part of a man, as if a peasant had dressed up as a lord with intent to deceive.

Her accusers claimed that Shuria was 'a woman of masculine demeanor who had grown weary of her sex'. She dressed as a man, enlisted in the army and served as a soldier at the siege of Sylva Ducis under Frederick Henry, stadholder (1625–47). She seems to have kept out of trouble until after the war when, as a civilian, it was presumably more difficult for her to sustain a male persona. It was then that she was accused of 'immoral lust',

for sometimes even exposing her clitoris outside the vulva and trying not only licentious sport with other women . . . but even stroking and rubbing them . . . so that a certain widow, who burned with immoderate lusts, found her depraved longings so well satisfied that she would gladly—except for the legal prohibition—have married her.[64] Her clitoris, it was said, 'equalled the length of half a finger and in its stiffness was not unlike a boy's member.' Shuria was convicted and sentenced to death, though it is not clear whether her primary offense was 'sodomy'—essentially placing some body part in the wrong orifice—or very nearly getting away with being a female bridegroom. But, in any case, instead of being put to death as various great jurists had recommended, 'this "tribas" finding a merciful judge, was, as it were, nipped in the bud, and sent into exile.' The widow, though punished in an unspecified manner, was allowed to remain in the city.[65]

The great French surgeon Ambroise Paré gives a complex account of the same sort that discusses female genital mutilation in the context of exorcism and the exotic East. But it also speaks to a more general cultural concern about the 'naturalness' of heterosexual union, of the mating of opposites. The 'nimphes' of women, he informs his lay audience, 'two excrescences of muscular flesh' which cover the urinary meatus and 'which hang and, even in some women, fall outside the neck of the womb; lengthen and shorten as does the comb of a turkey, principally when they desire coitus; and when their husbands want to approach them, they grow erect like the male rod. . . .'[66] So far this seems to be much like the other accounts of clitoral excitation that arises from the rubbing of penis on 'penis', or from the focusing of diffuse lust. But Paré then unexpectedly turns to a discussion of what he takes to be a more serious problem, i.e., the performance by females alone of what would otherwise occur between men and women. In the course of heterosexual loveplay, the 'nimphes' swell 'so much so that [women] can disport themselves with them, with other women.' This externally visible erection renders them 'very shameful and deformed being seen naked' and a surgeon must be called to cut out what is superfluous and subject to abuse. By this time it is not clear whether Paré is warning men about women who during marital sex might prefer to engage in masturbation or homoerotic practices, or whether he is alluding to the 'problem' of clitoral hypertrophy among Eastern women, i.e., only to that class of women in whom the 'nimphes' 'fall outside the neck of the womb', and who shamefully act like men. The question of who sees their 'shameful' condition when naked is also left unspoken.[67]

Moreover, the case that Paré gives to assure his readers that his account is 'as true as it is monstrous and difficult to believe' is so fraught with crosscurrents of meaning that its specific illustrative power is greatly confused. The venue is North Africa. In Fez, we are told, there lives a group of female 'prophets', exorcists, who give people 'to understand that they have familiarity with demons'. These women change their voices so that one thinks that the spirits are speaking through them; they live from gifts people leave for the demons *and* they 'rub one another for pleasure, and in truth they are afflicted of that wicked vice of using one another carnally'. More sinister still than speaking in the voices of men and having sex like men, they use the spirit's voice to ask beautiful women who consult them to pay by 'carnal copulation'. And to further complicate matters their patients apparently come to enjoy this form of remuneration, feign sickness and send their husbands to fetch the healers. The poor dupe, whose permission is required for his wife to visit the prophetesses, is thereby made the instrument of his own cuckolding. Indeed, the benighted husband might even prepare a feast for the whole lecherous band. Some husbands, however, get wise to the ruse and drive the putative spirits out of their wives' bodies with a good clubbing, or 'deceive them by the same means as the prophetesses have done' (presumably by having sex with themselves?). There are in Africa—so the story ominously ends—'castrators' who 'make a trade of cutting off such caruncles, as we have shown elsewhere under *Surgical Operatians*.'[68]

Once again the offense—in addition to trafficking with demons—is deception and inversion, not perversion. The lowly would be high, women would play the part of men. The narrative slipperiness of

Paré's account belies its diffuse anxiety about a sexual world turned upside down. This story is also a curious turn on the quite common trope of exorcism and sexual exploitation. Men who cast out demons, charges Samuel Harsnett in his condemnation of exorcists and exorcism in sixteenth-century England, seem to prefer young women to old because they writhe more actively when the spirits are wrenched from their bodies. These charlatans further exploit their position by touching the cross to the genitals of their patients. Harsnett's condemnation is Protestant propaganda but it was a commonplace of the anticlerical literature of the Middle Ages and the Renaissance that priests and monks hid their sexual exploits behind priestly or monastic garb. The Fez prophetesses seem therefore in good male company.[69]

And finally, Pare's account might be regarded as a kind of medical pornography that plays on the erotic power of the one-sex model. He was asked by the medical faculty of Paris, and he agreed, to cut the passages I have quoted from his popular book on monsters in which they first appeared. (They ended up in his *On Anatomy*, a book less appealing to, and presumably less read by, the laity.) It is difficult to believe that in either context the passages in question could have been written primarily for their prurient interest since the vast bulk of the surrounding material is decidedly unerotic. But by the late seventeenth century similar accounts in popular medical books have an unambiguously naughty quality. 'The Lesbian Sappho would never have acquired so wicked a reputation [*une si méchante réputation*] if this part [the clitoris] had been smaller,' Nicholas Venette writes in his best-selling sex guide regarding the organ which 'is often abused by lascivious women'. Clitoral sexuality—masturbation, stimulation by men, or 'rubbing' with another woman—finds its way into eighteenth-century pornography and is a commonplace by the nineteenth century.[70]

This, however, is not the occasion to analyze male erotic interests in particular practices. Parés tale and Venette's coy aside are very much part of that genre of titillating literature that purports to allow men a view into the secret lives of women.[71] When among themselves, they seem to be saying, women rub *their* penises together. But since sex between men and women is metaphorically construed in a wide range of contexts as the friction of two like parts—whether these be the male penis with the vagina or the clitoris—tribadism, women 'rubbing' one another, would become merely another variation on an old theme. Indeed, all sex becomes homoerotic. Clitoral arousal—and male homoerotic acts—as signs of their practitioners' intrinsic sexual deviance, of their being 'in-between sexes', would have to await the eighteenth century's redefinition of man and woman as being essentially different sorts of creatures. Lesbianism and homosexuality as categories were not possible before the creation of men and women as opposites, thereby leaving room in-between for those whose bodies (and/or psyches) made a choice impossible. Thus, the elaboration in medical literature, as well as in pornography of a 'new' female penis and specifically clitoral eroticism, was a re-presentation of the older homology of the vagina and the penis, not its antithesis.

By the nineteenth century the problem of the clitoris was like that of the penis: the solitary vice. 'Clitorisme,' according to the *Dictionnaire encyclopédique des sciences médicales*, is not found in ordinary dictionaries but finds a place here. It is the act in which women substitute by 'a kind of artifice' the pleasures reserved for love between the two sexes. 'It is, for women, the same thing that masturbation is for men.' These two definitions are, of course, not versions of one another. The first refers primarily to love between women, to what the 'dames de Lesbos' do. But this concerns the author less than young women who learn, from bad company or bad literature, to touch their clitorides and who then become addicted to the practice. Eventually, however, the two definitions merge. Women who masturbate do not do so alone; they make converts, and by exciting their clitorides enough develop a sort of penis themselves (perhaps through some unspoken Lamarckian mechanism). They even act as men, and sink further and further into drink and debauchery. 'See further "Masturbation"' the entry concludes.[72]

Clitorisme leads to the collapse of the social order, the *Dictionnaire* argues. But abuse is not quite like male masturbation in that it leads not just to self-destruction but to homosexuality as well. The solitary vice has a social outcome even if it is a perverted one. Clearly the Renaissance male's fear of being left out is still alive but is secondary to the massive new concern with masturbation as an offence not so much against chastity as against society.

By the late nineteenth century the clitoris takes on still another signification. No longer an encouragement to lesbian practices, it becomes in the work of forensic anthropologists like Lombroso a mark of excessive heterosexuality, the mark of the prostitute. (Parent du Chatellet, the major early nineteenth-century expert on the subject, had declared on the basis of considerable observation that prostitutes did not have unusually large or in any way remarkable clitorides.)

But Freud is not part of this history. His concern is much more that of the Renaissance, of getting bodies whose anatomies do not guarantee the dominance of heterosexual procreative sex to dedicate themselves to their assigned roles. However, he is at the same time a product of nineteenth-century biologism that postulates two sexes, each with their distinctive organs and physiologies, and of an evolutionism which guarantees that even if the genital parts are not born to make heterosexual intercourse natural, they will somehow adapt. In the end, the cultural myth of the vaginal orgasm is told in the language of science. And thus, not because of but despite neurology, a girl becomes the Viennese bourgeoise ideal of a woman.

POSTSCRIPT

When I finished the foregoing essay I remained uneasy that I had not offered a coherent version of Freud's argument once it could no longer be regarded as grounded in neurology. In other words, if Freud knew —as I show—that the migration of erotic sensibility from the clitoris to the vagina could not be interpreted as the result of prior anatomy or of anatomical development (for example, the supposed accession of pain receptors in neonates) then how did he think it might be interpreted? I offer the following reconstruction.

In the first place Freud remained a Lamarckian all his life. He believed in the inheritance of acquired characteristics which he generalized to include traits of the psyche—aggressions and needs, for example. 'Need' in a Lamarckian sense, he wrote to his colleague Karl Abraham, is nothing other than the 'power of unconscious ideas over one's own body, of which we see remnants in hysteria, in short, "the omnipotence of thought."'[73]

Hysteria is the model for mind over matter. The hysteric, like the patient who feels pains or itches in a missing limb, has physical symptoms that defy neurology. The hysteric's seizures, twitches, coughs and squints are not the result of lesions but of neurotic cathexes, of the pathological attachment of libidinal energies to body parts. In other words, parts of the body in hysterics become occupied, taken possession of, filled with energies that manifest themselves organically. (The German noun *Bezetsung* is translated by the English neologism 'cathexis'. The verb *besetzen* also has the sense of 'charged as is a furnace, or tamped down, as is a blasting charge, or put in place as is a paving stone or a jewel'.

Freud knew that the natural locus of woman's erotic pleasure was the clitoris and that it remained competitive with the necessary cultural locus of her pleasure, the vagina. Marie Bonaparte reports that her mentor gave her Felix Bryk's *Neger Eros* to read in which the author argues that the Nandi engage in clitoral excision on nubile seventeen and eighteen year old girls so as to encourage the transfer of orgiastic sensitivity from its 'infantile' zone to the vagina, where it must necessarily come to rest. The Nandi were purportedly not interested in suppressing female pleasure but merely in facilitating its redirection to social ends. Freud drew Bonaparte's attention to the fact that Bryk must have been familiar with his views and that the hypothesis regarding Nandi orgasmic transfer was worth investigating.

Bonaparte's efforts to discover the fortunes of 'clitoroidal' versus 'vaginal' sexuality in women whose clitorides had been excised proved inconclusive but she did offer a theoretical formulation of the transfer of erotic sensibility which fits my understanding of Freud's theory not only in the *Three Essays* but also in *Civilization and Its Discontents* and in his 1931 papers on female sexuality. 'I believe,' writes Bonaparte,

that the ritual sexual mutilations imposed on African women since time immemorial . . . constitute the exact physical counterpart of the psychical intimidations imposed in childhood on the sexuality of European little girls.[74]

Society takes the bisexual body of a little girl and forces its erotic energies out of their infantile phallic place where nerves ensure pleasure and into the vagina where they do not. 'Civilized' people no longer seek to destroy the old home of sensibility—an ironic observation for Bonaparte, since she collected cases of European excision and herself underwent painful and unsuccessful surgery to move her clitoris nearer her vaginal opening so that she might be 'normally orgasmic' but enforce the occupation, i.e. cathexis, of a new organ by less violent means.

If we put all of this together, Freud's argument might work as follows. Whatever polymorphous perverse practices might have obtained in the distant past, or today among children and among animals, the continuity of the species and the development of civilization depend upon the adoption by women of their correct, that is nonphallic, vaginal, sexuality. For a woman to make the switch from clitoris to vagina is to accept the feminine social role that only she can fill. Each woman must adapt anew to a redistribution of sensibility which furthers this end, must reinscribe on her body the racial history of bisexuality. But neurology is no help. On the contrary. Thus, the move is hysterical, a recathexis that works against the organic structures of the body. Like the missing limb phenomenon, it involves feeling what is not there. Becoming a sexually mature woman is therefore living an oxymoron, becoming a lifelong 'normal hysteric', for whom a conversion neurosis is termed 'acceptive'.

Notes

1. Renaldus Columbus, *De re anatomica* (Venice, 1559), bk. 11, ch. 16, pp. 447–8.
2. For a review of this literature up to 1968, see the *Journal of the American Psychoanalytic Association* 16: 3 (July 1968), pp 405–612.
3. Robert Scholes, 'Uncoding Mama: The Female Body as Text', in *Semiotics and Interpretation* (New Haven: Yale University Press, 1982), pp.130–1 and *passim.*
4. Sigmund Freud, *Three Essays on the Theory of Sexuality* (1905), trans. and edited by James Strachey (Avon: New York, 1962), p. 123.
5. Ibid.
6. Ibid., p. 124.
7. Ibid., p. 123
8. Denis Diderot, *L'Encyclopédie*, 'La jouissance' (Neuchâtel, 1765), vol. 5, p. 889.
9. On the one-sex model, see *Making Sex: Body and Gender from the Greeks to Freud.* (Cambridge, Mass.: Harvard University Press, 1990), especially chs. 2 and 3.
10. Jacques Duval, *Traité des hermaphrodites* (Rouen, 1612; repr. Paris, 1880), p. 68.
11. Anonymous, *My Secret Life*, ed. G. Legman (New York: Grove Press, 1966), p. 357.
12. *Tableau de l'amour conjugal* (London: 'Nouvelle édition', 1779), pp. 20–1.
13. 'Infantile Sexuality', in *Three Essays*, p. 93.
14. I am indebted in my account of the 'aporia of anatomy' in Freud's essay on feminity in the *New Introductory Lectures* to Sarah Kofman, *The Enigma of Woman* (Ithaca and London: Cornell University Press, 1985), esp. pp. 109–14.
15. I argue more generally against the tenability of the sex/gender distinction in ch. 1 of *Making Sex.*
16. Gayle Rubin, 'The Traffic in Women: Notes on the "Political Economy" of Sex', in *Toward an Anthropology of Women*, ed. Rayna R. Reiter (New York and London: Monthly Review Press, 1975), pp. 179–80 and 187.
17. Claude Lévi-Strauss, *The Savage Mind* (Chicago: University of Chicago Press, 1962), pp. 46ff. and ch. 2 generally.
18. *Civilization and Its Discontents*, trans. James Strachey (New York: W.W. Norton, 1962).
19. Rosalind Coward, *Patriarchal Precedents: Sexuality and Social Relations* (London: Routledge and Kegan Paul, 1983), p. 286.
20. Columbus, *De re anatomica*, 11.16.447–8.
21. Ibid., 11.16.446–47. I owe this novel and important grammatical analysis of Columbus entirely to my research assistant, Mary McCary of the Comparative Literature Department, Berkeley.
22. Columbus, *De re anatomica*, 'Concerning the womb or uterus', 119.16. 445.
23. Jane Sharp, *The Midwives Book, or the Whole Art of Midwifery Discovered,*

Directing Childbearing Women How to Behave Themselves in Their Conception, Breeding, Bearing and Nursing Children (London, 1671), pp. 40, 42.

24. Ibid., pp. 45–6; Mayern et al., *Complete Midwives Practice*, 4th edn. (1680), p. 67.
25. Thomas Gibson, *The Anatomy of Humane Bodies Epitomized*, 4th edn. (London, 1694), p. 199; *oestrus* in Latin means a gadfly or horsefly and figuratively a frenzy.
26. Helene Deutsch, *The Psychology of Women: A Psychoanalytic Interpretation* (New York: Grune and Stratton, 1944), pp. 229–38, 319–24 *passim*.
27. Ibid., p.22.
28. Sharp, *The Midwives Book*, p. 45.
29. John Pechey, *The Compleat Midwives Practice Enlarged*, 5th edn. (London, 1698), p. 49.
30. Philip Moore, *The Hope of Health* (London, 1565), p. 6.
31. I have not studied the nomenclature for the male reproductive anatomy thoroughly and I know of no general study of the subject. There are to be sure many different words for penis, testicle or scrotum but in my reading the referents of these terms are unambiguous. Perhaps this is the linguistic correlative of the corporeal telos generally: the male body is stable, the female body more open and labile.
32. Ovary, or its modern equivalents in other languages, was simply not used, in any form, until the late seventeenth century. The standard terms for the organ that we now understand to be an ovary were 'female testicle' and its variants, like 'woman's stones'.
33. Columbus, *De re anatomica*, 11.16.443.
34. Ibid. 11.16.445, emphasis added.
35. Ibid., pp.447–8.
36. Fallopius, *Observationes*, pp. 193, 195–96.
37. Caspar Bauhin, *Anatomes*, 1. 12. 101–2.
38. Jacquart and Thomasset, *Sexualité et savoir médical au Moyen Age* (Paris, 1985), p. 34, quoting Al-Kunna al-maliki.
39. *The Anatomy of Mondinus*, in *Fasciculo de medicine*, ed. Charles Singer, 2 vols. (Florence, 1925), vol. 1, p. 76 and n. 64.
40. Jacopo Berengario da Carpi, *Isagogae Brevis*, ed. and trans. by L. R. Lind as *A Short Introduction to Anatomy* (Chicago, 1959), p. 78.
41. Pechy, *The Complete Midwives Practice Enlarged*, p. 49 generally, and the actual quote is from *A General Treatise of the Diseases of Maids, Bigbellied Women* (1696), p. 60.
42. Gabrielis Fallopius, *Observationes anatomicae* (Venice, 1561), p. 194.
43. *Observationes anatomicae*, p. 193.
44. [Caspar] Bartholinus, *Anatomy Made from the Precepts of His Father and from Observations of All Modern Anatomists, Together with His Own* (1668), p. 75.
45. Regnier De Graaf, *New Treatise Concerning the Generative Organs of Women*, trans. and ed. H. D. Jocelyn and B. P. Setchell, *Journal of Reproduction and Fertility*, supp. 17 (Dec. 1972), p. 89.
46. Julius Pollux, *Onomasticon* (Leipzig: Teubner, 1900), p. 174 . I owe the translations from the Greek to Mary McCary.
47. Joseph Hyrtl, *Onomatalogia anatomica*; 'Clitoris-Kitzler'; and James N. Adams, *The Latin Sexual Vocabulary* (Baltimore: Johns Hopkins University Press, 1983), pp. 7–98 ; this whole semantic field needs exploration.

48. Jocelyn and Setchell, the editors of De Graaf's *New Treatise*, point out these errors without drawing attention to their significance; Albucasis in fact used the term *tentigo*.

49. Rufus of Ephesus, *Du nom des parties des corps*, in *Oeuvres de Rufus d'Ephèse*, ed. Charles Daremberg (Paris, 1879, 1963), p. 147, i.e., lines 110–12 of the accompanying Greek; Hyrtl, *Onomatologia*, entry 248, 'nymphae und myrtiformis'.

50. Galen, *On the Usefulness of the Parts of the Body* (Ithaca: Cornell University Press, 1968), vol. 2, p. 661.

51. Soranus, *Gynaecology*, ed. and trans. Owsei Temkin (Baltimore: Johns Hopkins University Press, 1956), p.16.

52. Albucasis uses the term in his *Chirurgia* 2.71: regarding *tentigo, see* Hyrtl, 'clitoris'; Adams, *The Latin Sexual Vocabulary*, 103–4 and the *Oxford English Dictionary*.

53. Thomas Vicary, *The Anatomy of the Bodie of Man*, ed. F. J. and P. Furnivall (Oxford: Early English Text Society, 1888), p. 77.

54. The claim for Spigelius's primacy is made in Hyrtl, *Onomatologia anatomica*, 'nympha'; his *De humani corporis fabrica* (Venice, 1627) was published posthumously; see Charles Singer, *A Short History of Anatomy and Physiology from the Greeks to Harvey* (New York: Dover, 1957), p. 163.

55. Crooke, *Microcosmographia: A Description of the Body of Man* (London, 1615), p. 250 .

56. Zachias, *Quaestionem medico-legalium* (1661), vol. 2, p. 502, par. 16.

57. John Hunter, 'Account of the Free Martin', in *Philosophical Transactions of the Royal Society of London* 69 (1779), p. 281.

58. It is not clear from ancient accounts whether the clitoris or labia minora are at issue.

59. Strabo, *Geography*, ed. H. L. Jones (London: Loeb Classical Library, 1966–70), 16.2.37 and 16.4.9.

60. Aetius of Amida, *Tetrabiblion*, translated from the Latin edition of 1542 by James Ricci (Philadelphia, 1950), bk. 16, ch. 103, p. 107, and n.l. p. 163.

61. *The Seven Books of Paulus Aegineta*, translated from the Greek in 3 volumes by Francis Adams, (London: Sydenham Society, 1856), vol. 2, p. 381, sec. 70.

62. Robert James, *Medicinal Dictionary* (London, 1745), vol. 3, 'Clitoris'; Dr. Johnson helped James plan this work and wrote the biographical entry for Paulus Aegenita.

63. J. B. (John Bulwer), *Anthropometamorphoses: Man Tranformed: or, the Artificial Changeling* (London, 1653) p. 380.

64. I have taken this account from Nicolaas Tulp (1593–1674), *Observationum medicarum libri tres* (Amsterdam, 1641), pp. 3, 35, 54, 244 as quoted in Latin by L. S. A. von Romer, '*Der Uranismus in den Niederlanden bis zum 19. Jahrhundert' mit besonderer Berucksichtigung der grosen Uranierverfolgung in Jahre 1730,' Jahrbuch für Sexuelle Zwischenstufen*, vol. 8, pp. 378–79. I am grateful for this reference to Kent Girard.

65. Tulp, ibid.

66. I have taken this account from Ambroise Paré, *On Monsters and Marvels*, trans. with an introduction by Janis L. Pallister (Chicago: University of Chicago Press, 1982), p. 188, n.35.

THOMAS W. LAQUEUR

67. Ibid.
68. Ibid.; Paré gives as his source the early sixteenth-century Arab geographer Leo Africanus's *History of Africa*.
69. On Harsnett, see Stephen J. Greenblatt, 'Shakespeare and the Exorcists', in *Shakespeare and the Question of Theory*, ed. Pat Parker and Geoffrey Hartman (New York and London: Methuen, 1985).
70. Paré's paragraph on the so-called *nymphae* was shortened or lengthened again in the various 1573, 1575 and 1579 editions of the book on monsters before being relegated to the section on the womb in *De l'anatomie de tout le corps humain* (1585).
71. See, for example, the seventeenth-century pornographic tale *The School for Venus*, ed. with introduction by Donald Thomas (London, 1972); and more generally Roger Thompson, *Unfit for Human Ears* (Totowa, NJ: Rowman and Littlefield, 1979), pp. 30, 90, 158–75, and *passim*.
72. *Dictionnaire* (1813), vol. 5, pp. 376–8.
73. Freud to Abraham, Nov. 11, 1917, cited in Peter Gay, *Freud: A Life for Our Times* (New York: W. W. Norton, 1988), p. 368 .
74. Marie Bonaparte, *Female Sexuality* (New York: International Press, 1956), p. 203.

The Birth of Sex Hormones

Nelly Oudshoorn

Nowadays, we can hardly imagine a world without hormones. We have to travel in time to find other worlds that are not yet inhabited by them. Imagine a scene on a lazy Sunday afternoon in the late nineteenth century. Ladies are chattering about the exciting events of the past days. If we could eavesdrop on these conversations, we would hear detailed, intimate accounts of how these women try to cope with daily life. Maybe we are lucky and we can overhear them exchanging experiences about pregnancy and delivery. We will never know precisely which words women used in those days to express themselves, but we know one thing for sure: women did not refer to hormones to explain their lives. Simply because the very word hormones did not exist in the nineteenth century.

Where does the concept of hormones, particularly sex hormones, come from? How did it become included in the medical discourses about the body? What inspired scientists to develop a totally new model for understanding bodies? Obviously, scientists who introduce new concepts into science do not start from scratch. Or to paraphrase Nelson Goodman: 'scientific development always starts from worlds already on hand' (Goodman 1978: 6). I would like to know which worlds inspired scientists to introduce the concept of sex hormones. Was the introduction of the concept of sex hormones linked to any previously existing beliefs about women and men; and if so, how did these prescientific ideas then become integrated into the emerging field of sex endocrinology? Using Ludwig Fleck's (1979) notion of prescientific ideas, I shall trace which cultural ideas became embodied in the concept of sex hormones and how and to what extent scientists actively transformed these ideas, once they were incorporated in

From Nelly Oudshoorn, 'The Natural Body: An Archeology of Sex Hormones', *Journal of the History of Biology* (1990), pp. 163–86. Reprinted with permission.

research practice. My strategy is to focus on the different disciplines that became involved in hormone research. Sex endocrinology, like other fields in the life sciences at the turn of the twentieth century, was characterized by two different approaches: a biological approach and a chemical approach (Clarke 1985: 390; Kohler 1982). In the early years the study of sex hormones was dominated by scientists who adopted a biological style: physiologists, gynecologists, anatomists and zoologists. In the 1920s, chemical approaches came to dominate the field. I shall examine the extent to which differences in disciplinary styles may account for the transformation of prescientific ideas.

Finally, I shall evaluate the impact of the introduction of the concept of sex hormones on the conceptualization of sex differences. What were the consequences of this new approach in the study of the human body? I describe how sex endocrinology caused a revolutionary change in the study of sex differences and led to a conceptualization of sex that meant a definitive break with prescientific ideas about the female and the male body.

HOW THE CONCEPT OF SEX HORMONES ORIGINATED

The first use of the term 'hormone' can be traced back to Britain. In 1905, Ernest H. Starling, Professor in physiology at University College in London, introduced the concept of hormones:

> These chemical messengers . . . or 'hormones' as we may call them, have to be carried from the organ where they are produced to the organ which they affect, by means of the blood stream, and the continually recurring physiological needs of the organism must determine their production and circulation through the body. (Starling 1905)

The concept of hormones as potent substances regulating physical processes in organisms implied a drastic change in the paradigm of physiology. Edward Schaefer, Professor at University College in London, evaluated this shift as follows:

> The Old Physiology was based, as we have seen, on nervous regulation; the New Physiology is based on chemical regulation. . . . The changes of physiology which have resulted from this knowledge constitute not merely an advance in degree but an alteration in character. . . . We must in future explain physiological changes in terms of chemical regulation as well as nervous regulation. (Edward Schaefer as cited in Medvei 1983: 339)

The 'New Physiology' enabled scientists to conceptualize the development of organisms in terms of chemical agencies, rather than just nervous stimuli. The chemical messengers believed to originate from the gonads (sex glands) were designated sex hormones, with male sex hormone designating the secretion of the testis and female sex hormone designating ovarian secretion. With the introduction of the concept of sex hormones scientists suggested that they had found the key to understanding what made a man a man and a woman a woman. In the General Biological Introduction to the first textbook of sex endocrinology, *Sex and Internal Secretions*, the French-Canadian zoologist Frank R. Lillie evaluated this rapidly expanding research field:

One of the most interesting and promising lines of experimental biological investigations of the present century has been in the biology of sex. It has been discovered that sex characteristics in general are subject to certain simple mechanisms of control that operate throughout the life history, and which determine whether male or female characters shall develop in the individual. . . . The mechanisms of control are exceedingly simple compared with the sex machinery itself. . . . This book deals predominantly with a method of control of sex characters which is especially characteristic of vertebrates including man, mediated by hormones circulating in the blood. Of these, the specific internal secretions of the testis, or male sex hormone, and the specific internal secretion of the cortex of the ovary, or female sex hormone, are the most important, and probably occur in all vertebrates. (Lillie 1939: 5–6)

In the same textbook Lillie described the function of sex hormones:

As there are two sets of sex characters, so there are two sex hormones, the male hormone controlling the 'dependent' male characters, and the female determining the 'dependent' female characters. (Lillie 1939: 11)

Sex hormones were thus conceptualized as the chemical messengers of masculinity and femininity.

To what extent were these developments linked to any prescientific ideas? The idea of testes and ovaries as agents of masculinity and femininity can be traced back to several periods in which these ideas emerged, prior to the hormonal era.

The idea that the ovaries are in one way or another related to female sexual development can be traced back as far as Aristotle. In his *History of Animals* Aristotle wrote that 'the ovaries of sows are excised with the view of quenching in them sexual appetites and of stimulating growth in size and fatness' (Aristotle as cited in Corner 1965: 3). Aristotle was referring here to the custom of removing ovaries in domestic animals, a widespread practice among European farmers in the Middle Ages, that seems to have been kept in use till the late nineteenth century. The idea that the ovaries are somehow linked to female sexuality remained, however, for a long time restricted to the domain of agricultural practices (Corner 1965: 4).

Let us first focus our attention on the prescientific precursors of the concept of male sex hormones, before tracing how the idea of ovaries as agents of femininity became incorporated in the life sciences. The idea that the male gonads are the seat of masculinity is a very old one. From the earliest times, the testis has been linked with male sexuality, longevity and bravery. Greeks and Romans used preparations made from goat or wolf testes as sexual stimulants. The seventeenth century brought a revival of these ancient ideas of virility. In this period, the prescientific idea of testes as agents of masculinity became for the first time incorporated into medical science. European reformers, like the physician Paracelsus, used testes extracts in the treatment of 'imbecility of the instruments of generation'. The official pharmacopoeia of the London College of Physicians of 1676 gave directions for the extraction of animal reproductive organs as a treatment for numerous illnesses and as sexual stimulants. In the eighteenth century, belief in the testis as the controller of virility was abandoned to the realm of folk-wisdom and quackery. By 1800, testicular extracts had disappeared from the official pharmacopoeias in Europe. The belief in the testis as agent of masculinity remained, however, very much alive in popular culture. Although the medical profession strongly disapproved of testicular therapy, potions made from testis extracts were among the wares of eighteenth-century quacks and were quite popular all over Europe.

How did the pre-ideas of testes and ovaries as seats of masculinity and femininity become integrated into the modern life sciences? The most conspicious actor in advocating the doctrine of the gonads was the French physiologist Charles-Edouard Brown-Séquard. In 1889, Brown-Séquard addressed his colleagues at the Société de Biologie in Paris, reporting the results of self-medication in which he had treated himself with injections prepared by crushing guinea-pigs' and dogs' testicles, resulting in 'a marked renewal of vigour and mental clarity'. He also reported the practices of a midwife in Paris, who treated women with the filtered juice of guinea-pigs' ovaries for 'hysteria, various uterine affections, and debility due to age' (Corner 1965: 5). On this occasion Brown-Séquard suggested that the testes produced a secretion that controlled the development of the male organism. These 'internal secretions' might be discovered by using extracts in treatment for certain diseases. Brown-Séquard's advocacy gave rise to a renewed interest in the 1890s in what was now called 'organotherapy': the use of extracts of animal organs as therapeutic agents.

The scientific community reacted for the most part with hostility to Brown-Séquard's claims. In their eyes the clock was being put back to the dark ages of quackery. On the other hand, Brown-Séquard, as the successor to Claude Bernard at the College de France, was considered a distinguished neurophysiologist (Hamilton 1986: 12, 15). Moreover, Brown-Séquard's ideas harmonized with Victorian notions of masculinity. He suggested that the secretions of the testis were present in the seminal fluid containing sperm. In this manner, Brown-Séquard linked his ideas to the then popular notion that loss of semen, through sexual intercourse or, even worse, masturbation, was weakening the male. Brown-Séquard's claims thus not only reflected prescientific ideas about the power of the testis that can be traced back to early civilization, but also reflected the sexual assumptions of Victorian days (Hamilton 1986: 16).

If we want to trace how the prescientific idea of the ovaries as the seat of femininity became integrated into the center of the life sciences, we have to direct our attention to another place: the gynecological clinic. It was through changes in medical practice that the prescientific idea of ovaries as agents of femininity became incorporated into scientific theory and practice. Until the mid-nineteenth century, medical scientists studying the female body focused their attention primarily on the womb. The uterus was known several millennia before the ovary (the 'female testicle') was described as an anatomic unit. Scientists located the essence of femininity in the uterus. Beginning in the middle of the nineteenth century, medical attention gradually shifted from the uterus to the ovaries. The ovaries came to be regarded as the essence of femininity itself: the study of these organs would lead to an understanding of woman's whole being, including all women's diseases. (Gallagher and Laqueur 1987: 27). In gynecological textbooks the ovary was described as 'the organ of crisis which is missing in the male body.' This shift from the uterus to the ovaries provided the gynecological profession with their own 'paradigm-specific' organ which enabled them to delineate the boundaries between gynecology and obstetrics the profession that focused primarily on the uterus (Honegger 1991: 82–3).

In this period the role of the ovaries was not yet described in terms of chemical substances, but rather in the then popular terms of regulation by the nervous system. Gynecologists were the first to introduce the idea that ovaries secreted chemical substances that regulate the development of the female body (Medvei 1983: 215). They were already familiar with the changes in the body that followed the removal of ovaries, due to the widespread medical practice of surgical operations for the removal of the ovaries in the late nineteenth century (Corner 1965: 4; T. Laqueur 1990: 176). Two Viennese gynecologists, Emil Knauer and Josef Halban, described the secretion of chemical substances by the ovaries as early as 1896 and 1900. Gynecologists were thus the first to recognize the relevance of Brown-Séquard's theory of 'internal secretions' to the female sex glands. This branch of the medical profession came under the spell of the glands because of their therapeutic promises. Gynecologists were particularly attracted to the concept of female sex hormones because it promised a better understanding and therefore greater medical control over the complex of disorders in their female patients frequently associated with the ovaries, such as disturbances in menstruation and various diseases described as 'nervous' in medical literature. Moreover, by linking female disorders to female sex hormones, 'women's problems' remained inside the domain of the gynecologists.

THE EMERGENCE OF SEX ENDOCRINOLOGY

In the first decade of the twentieth century, the study of sex hormones developed into a major field of research that became known as sex endocrinology. The concept of sex hormones as agents of masculinity and femininity functioned as a paradigm, focusing previously scattered research around a generally accepted theory. Compared with gynecologists, physiologists were relatively slow to recognize the relevance of the theory of internal secretions for the sex glands. One of the main reasons for this was the association of the sex glands with human sexuality and reproduction, an area that previously had been taboo in biomedical research. This negative association was reinforced by the therapeutic claims of Brown-Séquard about the effects of testis extracts on the sexual activity of men, claims that caused a controversy among clinicians and laboratory scientists. Physiologists who took up the study of ovary and testes preparations did so cautiously, avoiding association with these therapeutic claims. Obviously, the subject was more legitimate for gynecologists (Borell 1985: 2).

After the turn of the twentieth century, physiologists also gradually came under the spell of the gonads. Schaefer's laboratory in London was one of the first physiological laboratories that took up the study of the ovaries (Borell 1985: 13). The physiologists were particularly interested in the study of the glands because the concept of hormones provided a new model for understanding the physiology of the body. In the first decade of the twentieth century, physiologists included the study of the ovaries and testes as a branch of general biology (Corner 1965: 7). Hereby the traditional borders between two different groups of actors—the physiologists and the gynecologists—changed drastically. Before the turn of the twentieth century the study of ovaries, particularly in relation to female disorders, had been the exclusive field of gynecologists. With the introduction of the concept of sex hormones, laboratory scientists explicitly linked female disorders with laboratory practice, thus entering a domain that had traditionally been the reserve of gynecologists. Whereas gynecologists were particularly interested in the functions of the ovaries in order to control all kinds of disorders ascribed to ovarian malfunction, physiologists had a broader interest in the role of the ovaries and testes in the development of the body.

The concept of hormones triggered a new experimental approach in laboratory science. At the turn of the twentieth century scientists began to search actively for the chemical substances in the sex glands using the techniques of castration and transplantation. In this surgical approach, scientists removed ovaries and testes from animals like rabbits and guinea-pigs, cut them into fragments, and reimplanted them into the same individuals at locations other than their normal positions in the body. With these experiments scientists tested the concept of hormones as agents having control over physical processes without the mediation of nervous tissue. In transplantation the nervous tissue of the glands was dissected, so the effects of the reimplanted glands on the development of the organism had to take place through another medium, such as the blood (Borell 1985).

The acceptance of the hormonal theory in the biological sciences was facilitated by the fact that it fitted into a major debate among biologists about the sexual development of organisms. In the 1910s, the topic of sexual development was most controversial, particularly between physiologists and geneticists. Physiologists at that time suggested that the determination of sexual characteristics is affected by environmental and physiological conditions during the development of the embryo. Geneticists suggested however that sex is irrevocably fixed at conception by nuclear elements: the sex chromosomes.

With the introduction of the concept of sex hormones, sex endocrinologists claimed they had found the missing link between the genetic and the physiological models of sex determination. In 1916, Frank Lillie provided arguments for the role of sex hormones as well as sex chromosomes in the sexual development of higher animals by studying intersexes in cows. Lillie looked at the anatomical characteristics of the freemartin, the sexually abnormal co-twin of a male calf, usually possessing female as well as male external genitalia. Lillie suggested that the freemartin, a 'natural experiment', is genetically female but that 'a powerful blood-born chemical produced in the male had altered the sex that the genes intended for the freemartin'. Lillie thus suggested that 'the intentions of the genes must always be carried through by appropriate hormones developed in the gonad.'

This suggestion provided geneticists and sex endocrinologists with arguments to demarcate the fields of the two young sciences with respect to the study of sex, a demarcation coined with the concepts of sex determination and sexual differentiation. Geneticists focused on the study of sex determination, defined in *Sex and Internal Secretions* as 'the establishment of internal conditions leading to the development of one or the other set of sex characters'. Sex endocrinologists restricted their research to the study of sexual differentiation: 'the development of sexual characteristics in the course of the individual's life history.' Lillie described this demarcation of the domains of genetics and sex endocrinology as follows:

> It is clear that we must make a radical distinction between sex determination and sex differentiation. In most cases the factors of sex determination are chromosomal, and subject to the usual laws of Mendelian inheritance. . . . In the higher vertebrates, the mechanism of sex differentiation is taken over by extracellular agents, the male and the female sex hormones. (Lillie 1939: 7–8)

In the early decades of the twentieth century sexual development came thus to be defined as the result of two processes: sex determination regulated by genetic factors, and sexual differentiation influenced by hormonal factors.

In the 1920s, biochemists became involved in the study of sex hormones. Following advances in organic chemistry in the late 1910s, the surgical approach of transplanting gonads was replaced by chemical extraction of the gonads. The introduction of the chemical approach into the study of sex hormones was—compared with other fields—somewhat belated. This delay was partly due to technical problems. In the 1910s, biochemists were preoccupied with proteins. In this period the biochemists had 'neither the incentive nor the information' to enter the field of sex hormone research. This situation changed in the 1920s when lipid chemistry emerged as a new line of inquiry in biochemistry. In the 1920s, sex hormones were classified as steroids, a class of substances that could be extracted with the same solvents applied in extracting lipids, thus providing biochemists with both the information and the tools to enter the study' of sex hormones (Long Hall 1975: 83).

In addition to clinicians and laboratory scientists, the emerging field of sex endocrinology also attracted a third group to the scene: the pharmaceutical industry. The manufacturing of extracts from animal organs offered a new and promising line of production. Pharmaceutical companies started producing ovary and testes preparations, and not without success. At the turn of the century the advertising pages of medical journals were full of recommendations for the prescription of these preparations under a wide variety of trade names, indicating a flourishing trade in 'biologicals' (Corner 1965: 6). Many researchers involved in the study of sex hormones worked in close cooperation with pharmaceutical companies.

SEX HORMONES AS DUALISTIC AGENTS OF SEX

By 1910, the prescientific idea of the gonads as agents of sex differences had been transformed into the concept of sex hormones as chemical messengers of masculinity and femininity. With this conceptualization, sex endocrinologists reformulated the cultural notion of gonads as the seat of masculinity and femininity. Sex endocrinologists focussed their attention on the secretions of the gonads, rather than on the gonads themselves. In other respects, the scientific conceptualization of sex remained very close to common-sense opinions about masculinity and femininity. In the early period of sex endocrinology, the concept of sex hormones was straightforward and simple: there existed just two sex hormones, one per sex. In the period between 1905 and 1920, scientists defined sex hormones as exclusively sex specific in origin and function. With this conceptualization sex endocrinologists suggested that sex had to be considered as a strictly dualistic concept.

This conceptualization not only harmonized with the prescientific idea of a sexual duality located in the gonads, but also found a ready acceptance given the cultural notions of masculinity and femininity of the day. In this period the dominant cultural idea of sex was determined by the Doctrine of the Two Sexes, a concept of sex developed in Victorian times but still prevalent in the opening decades of the twentieth century. According to this doctrine, women's activities were in most respects the opposite of those of men. Therefore female and male were understood as opposite categories, not as two independent or complementary dimensions (Lewin 1984: 169–170).

The idea of female and male as opposite categories was further reinforced by those sex endocrinologists who advocated the idea of sex antagonism. Some scientists—anticipating the women's liberation movement at the turn of the century—had advocated the idea of sex antagonism. Among them was the British physiologist Walter Heape, who was the first to study the menstrual cycle of women in relation to the menstrual cycle in animals. In 1913, Heape published a book entitled *Sex Antagonism*, an anthropological study claiming that women's biological destiny was the opposite of men's (Heape 1913). Heape refuted the claims of feminists to equal rights for women and argued that biology restricted women's destiny to motherhood. Although it was criticized for its unfounded biological determinism, Heape's colleagues subscribed to his view of the relations between the sexes. In his review of *Sex Antagonism* the British evolutionary biologist J. Arthur Thomson emphasized that it was good biology to emphasize that woman's usefulness depends on her dissimilarity to man (Thomson 1914: 346 as quoted in Long Hall 1975: 85).

In the 1910s, the Viennese gynecologist Eugen Steinach attributed the idea of sex antagonism to the concept of sex hormones. With Heape and scientists like the Dutch sexologist Van de Velde, Steinach shared a conservative reaffirmation of the traditional distinction between the sexes, emphasizing that the appropriate social roles for women were rooted in biology and opposite to men's roles (Long Hall 1976). Steinach conceptualized the organism as a system of competing forces and persuaded his colleagues that male and female gonads secreted opposite, antagonistic hormones (Long Hall 1975: 88).

In itself, the idea of sex antagonism was new to the field of sex hormones. Earlier workers like Brown-Séquard had restricted the function of sex hormones to stimulating the development of 'homologous' sexual characteristics, suggesting that female sex hormones controlled female characteristics and male sex hormones male sexual characteristics. Steinach, however, attributed a double potentiality to sex hormones by suggesting that 'sex hormones simultaneously stimulated homologous sexual characteristics and depressed heterologous sexual characteristic'. Besides stimulating the development of female sexual characteristics, female sex hormones were thought to suppress the development of male sexual characteristics.

In more popular writings on sex hormones, this supposed antagonism between both type of hormones was compared with the relationship between men and women: 'the chemical war between the male and the female hormones is, as it were, a chemical miniature of the well-known eternal war between men and women' (Kruif undated: 167). The conceptualization of sex hormones as antagonists thus fitted seamlessly with Victorian notions of the proper relationship between the sexes, and was consistent with the dualistic idea that each sex had its own specific sex hormone.

The emerging field of sex endocrinology shows a striking unanimity in interpretations of the concept of sex hormones. The groups involved in research on sex hormones, during this period mainly gynecologists and some biologists (physiologists, anatomists and zoologists), basically agreed about the conceptualization of sex hormones. This unanimity changed drastically when the field became more specialized and new groups entered the arena of hormonal research. In the 1920s, there emerged a lively dispute in the scientific community about the dualistic assumption that sex hormones are strictly sex-specific in origin and function. A growing number of publications appeared contradicting the prescientific idea of a sexual duality located in the gonads and underlying the original concept of the sexual specifity of sex hormones. The next sections describe how the different disciplines involved in the study of sex hormones gradually transformed the prescientific idea of sexual duality into a new meaning of sex.

SEX-SPECIFIC ORIGIN

The first challenges to the prescientific idea of a sexual duality located in the gonads appeared in the early 1920s. In 1921, the Viennese gynecologist Otfried Fellner published an article in *Pflueger's Archiv* describing experiments with rabbits in which extracts of the testis produced effects on the growth of the uterus similar to those produced by ovarian extracts. Fellner suggested that the testis of the rabbit obviously contained female sex hormones (Fellner 1921: 189). It took several years before this report evoked reactions from his colleagues. Remarkably, this response came from the 'newcomers' in the field of sex endocrinology: the biochemists. The biochemical focus on the chemical identification and isolation of sex hormones generated a new set of needs for raw materials, amongst others urine (Clarke 1987: 331). In this biochemical search for new resources to obtain female sex hormone (to replace the expensive ovaries), scientists reported the presence of female sex hormones in men. Ernst Laqueur's research group at the Pharmaco-Therapeutic Laboratory of the University of Amsterdam—the Amsterdam School, as they were known by other scientists—reported in 1927 that female sex hormone was present not only in the testis but also in the urine of 'normal, healthy' men. (E. Laqueur et al. 1927: 1,859).

What really made an impact was an article which appeared in 1934 in *Nature* by the German gynecologist Bernhard Zondek, then working at the Biochemical Institute of the University of Stockholm. In this article, entitled 'Mass excretion of œstrogenic hormones in urine of the stallion', Zondek described his observations as follows:

Curiously enough, as a result of further investigations, it appears that in the urine of the stallion also, very large quantities of œstrogenic hormone are eliminated. . . . I found this mass excretion of hormone only in the male and not in the female horse. The determination of the hormone content, therefore, makes hormonic recognition of sex possible in the urine of a horse. In this connexion we find the paradox that the male sex is recognized by a high œstrogenic hormone content. (Zondek 1934*a*)

In the same publication Zondek reported that 'the testis of the horse is the richest tissue known to contain œstrogenic hormone'. One month later he published a second article in *Nature* describing the identity and the origin of female sex hormones in males. Zondek's observations were startling at the time. Who could have expected that the gonads of a male animal would turn out to be the richest source of female sex hormone ever observed?

The female of the species did not escape this confusion: in the same period, reports were published of the presence of male sex hormones in female organisms—but it is interesting that this phenomenon received far less attention. The articles indexed in the *Quarterly Cumulative Index Medicus* show that the number of articles written on females is considerably less than on males. The first publication reporting the presence of male sex hormone in female organisms was published in 1931 by the German gynecologist Siebke. In 1932, this observation was confirmed by Elisabeth Dingemanse, a biochemist from the Amsterdam School (Jongh 1934). In the years to follow, the British physiologist Hill published a series of contributions under the flagrant title, *Ovaries secrete male hormone* (Evans 1939: 597).

The above observations contradicted the original concept of the sexual specificity of sex hormones. What label should be attached to substances isolated from male organisms possessing properties classified as being specific to female sex hormones, and vice versa? Scientists decided to name these substances female sex hormones and male sex hormones, thus abandoning the criteria of exclusively sex-specific origin. Female sex hormones were no longer conceptualized as restricted to female organisms, and male sex hormones were no longer thought to be present only in males. Here we see how scientists gradually moved away from the prescientific idea that the essence of femininity is located in the ovaries, while the testes are the seat of masculinity. This shift in conceptualization led to a drastic break with the dualistic cultural notion of masculinity and femininity that had existed for centuries.

Because the prescientific idea of a sexual duality located in the gonads had dominated research for years, scientists were rather taken aback by the idea that female sex hormone could also be found in male bodies. In the first reports these results were evaluated as 'one of the most surprising observations in the sex hormone field' and 'a strange and apparently anomalous discovery'; the reports contained phrases like 'curiously enough', 'unexpected observation' and 'paradoxical finding' (Frank 1929: 292; Jongh 1934: 1,209; Parkes 1938; Zondek 1934a: 209). Many scientists found their own results so surprising that they felt obliged to emphasize the fact that they had used the urine and blood of 'normal, healthy men and women'. In *The Female Sex Hormone*, Robert Frank, a gynecologist at Mount Sinai Hospital in New York, legitimized the identity of his test subjects by the statement that he had observed female sex hormones in the bodies of males 'whose masculine character and ability to impregnate females' were unquestioned (Frank 1929: 120). Other scientists, however, concluded that the tested subjects, though apparently normal were 'latent hermaphrodites' (Parkes 1966: 26).

SOURCE AND IDENTITY OF 'HETEROSEXUAL HORMONES'

Confronted with these unexpected data, scientists started looking for a plausible theory to explain the source and identity of these 'heterosexual hormones' (as female sex hormones in male organisms, and vice versa, were named) (Jongh 1934: 1,209). In the 1930s, different hypotheses were proposed to explain the presence of female sex hormones in male organisms. In some of these hypotheses, scientists tried hard to maintain the dualistic conceptualization of sex according to which male and female were defined as mutually exclusive categories.

In 1929, Robert Frank suggested that female sex hormones were not produced by the male body itself, but that they originated from food (Frank 1929: 293). In the late 1930s, this hypothesis was criticized as implausible by, among others, the Amsterdam School (Dingemanse et al. 1937). Despite this criticism, the food hypothesis remained popular. In *Sex and Internal Secretions*, the food hypothesis is advanced without paying attention to the critical notes of the author of the cited paper:

Following the demonstration of the wide occurrence of estrogens in foods, it became apparent that estrogens in male urine need not necessarily imply their secretion in the male body. Eng (1934) reported that when a young man was placed on an estrogen-free diet, the excretion of estrogen in both urine and faeces dropped to 3 mouse units per day. On a standard diet, between 13 and 44 units per day could be recovered from the urine of the same man. (Allen et al. 1939: 561)

The food hypothesis seems to have been postulated only for the presence of female sex hormone in male organisms: no reports were published attempting to explain the presence of male sex hormones in females by the intake of male hormones from food.

In contrast with Frank, British scientists like the biologists Nancy and Robert Callow and Alan Parkes from the National Institute for Medical Research in Hampstead, London, suggested that male bodies should be considered capable of producing female sex hormones and proposed the adrenals as one of the sites of the production of sex hormones (Callow and Callow 1938; Parkes 1937). The adrenal hypothesis was still consistent with the prescientific idea of a sexual duality, in which it was impossible to consider the gonads capable of excretion of both sex hormones. In 1951, Samuel de Jongh, one of the laboratory scientists of the Amsterdam School, evaluated the proposal of the adrenal hypothesis as follows:

By proposing the hypothesis of an extra-gonadal source to explain the presence of female sex hormones in male bodies, scientists could avoid the necessity to attribute the secretion of male sex hormones to the ovary. (Jongh 1951: 20)

It was the gonadal hypothesis, suggested by Zondek in 1934, that first broke with the dualistic concept of sex hormones. In this hypothesis it was proposed that the regular occurrence of female sex hormones in male organisms was the result of the conversion from male sex hormones into female ones (Zondek 1934*b*). The idea of conversion was strongly advocated by biochemists, who posited a close interrelationship between male and female sex hormones and brought about the general acceptance of the idea that the gonads produce both type of sex hormones. After 1937, the adrenals and the gonads of both sexes were considered as the sites of production of male as well as female sex hormones (Kochakian 1938; Parkes 1938).

This succession of hypotheses illustrates how the prescientific idea of sexual duality located in the gonads was gradually reshaped into a new conceptualization of sex differences. In this period, scientists definitely broke with the cultural notion that the essence of femininity and masculinity was located only in the gonads. They suggested that the chemical messengers of masculinity and femininity were present in the adrenals as well as in the gonads of all organisms, rather than being restricted to the gonads of one sex. In the 1930s, scientists reshaped the original dualistic assumption of the sex-specific origin of sex hormones into a conceptualization in which the categories male and female were no longer considered mutually exclusive. By the end of the 1930s, scientists supported the idea that male bodies could possess female sex hormones and vice versa, thus for the first time combining the categories of male and female into on sex.

The debate over the sex-specific origin of sex hormones shows how the different disciplines played different roles in this controversy. Although the presence of female sex hormones in male bodies was reported by gynecologists and biologists, the biochemists took a key position in the debate on the sex-specific origin of sex hormones, as the following citation from the Amsterdam School illustrates:

In the second place I want to reflect on the question of whether the oestrus-producing substance that is present in the organs and body fluids of men, and apparently has a function there, is really female sex hormone. In fact we know nothing more than that it can cause the same effects as female sex hormone. An identification can only be possible after the chemical isolation of this oestrus-producing substance from male products, which has not yet been done. (Jongh 1934: 1,213)

Prior to the 1930s, the question of the identity of sex hormones could be addressed only by using the techniques of biological assays. Substances isolated from male urine which passed the biological assays specific for female sex hormones were defined as female sex hormones. It was only after 1929 that scientists could assess the identity of sex hormones with chemical methods, thanks to developments in organic chemistry in the area of steroid and lipoid compounds. Sex hormones—classified as steroids—could now be chemically identified and isolated (Long Hall 1975). Female sex hormones were first chemically isolated from the urine of horses and pregnant women in 1929. In 1932, English and German chemists classified female sex hormones as steroid substances, and a colorimetric test was developed to detect the presence of female sex hormone in organisms (Walsh 1985). Male sex hormones were first isolated from men's urine in 1931 and were classified two years later in the same group of chemical substances as female sex hormones: the steroids.

In 1938, the Amsterdam School finally reported the isolation of female sex hormone from male urine. The delay had been mainly caused by the limited availability of raw material, a general obstacle to the chemical isolation of sex hormones, which could be surmounted only by scientists working in close cooperation with pharmaceutical companies. Elisabeth Dingemanse explained the problem in an article in *Nature* as follows:

The necessity arose shortly afterwards of identifying these active substances chemically. The small amounts in which these active substances occur in adult male urine has so far made this impossible. . . . Through the intervention of the N.V. Organon-Oss, for which we take this opportunity of expressing our thanks, it has been possible for us to process 17,000 litres of male urine. . . . In this way we succeeded in obtaining 6 milligrams of a single crystalline substance. This proved to be identical with oestrone (female sex hormone). (Dingemanse et al. 1928: 927)

Thus, the biochemists claimed to possess a definitive answer to the question of the identity of female sex hormone in male organisms. They defined female and male sex hormones as closely related chemical compounds, differing in just one hydroxyl group, which could be detected by chemical methods in both sexes. In this manner they broke with the dualistic concept of male and female as mutually exclusive categories.

THE FUNCTION OF HORMONES

The dispute over the sexual specificity of sex hormones did not only address the origin of sex hormones. In the 1920s, the function of 'heterosexual' hormones was also frequently discussed. If female sex hormones were present in males, should the concept of an exclusively sex-specific function of sex hormones then be reconsidered as well? Scientists questioned whether female sex hormones had any function in the development of male organisms and vice versa. This section analyzes the positions the different disciplines involved in the study of sex hormones took in the debate about the sex-specific function of sex hormones.

The debate about the sex-specific function of sex hormones developed along lines similar to the discussion about the origin of sex hormones. Initially, scientists adhered to the prescientific notion of sexual duality. This assumption made it difficult to conceive of any function for heterosexual hormones at all, and therefore different hypotheses were proposed suggesting a functionless presence of the hormones. In 1929, the Amsterdam School suggested that female sex hormones probably had no function in the male body, because the concentration of female sex hormone in males was too small (Jongh et al. 1929: 772); in this period the amount of female sex hormone was thought to be considerably less in males than in females.

The assumption that female sex hormones had no function in male bodies directed research throughout the 1920s. In 1934, the Amsterdam School described how in 1928 they had observed the growth of the seminal vesicles in castrated rats after treatment with female sex hormones, but they had simply overlooked this function of female sex hormone in male animals. They described this observation as follows:

Menformon (female sex hormone)—also the completely pure preparation —enlarges the seminal vesicles of animals castrated when they were young. Although this enlargement does not develop into the adult size of the seminal vesicles, this enlargement is beyond dispute. We had observed this enlargement already for years. However, we had neglected this observation because we thought it fell within the margin of error. Freud and de Jongh made this observation in several different experiments. Also outside our laboratory [here the authors referred to the German scientist Loewe] the same observation was made, but was attributed to the presence of small amounts of male sex hormones in the ovarian extracts. From histological analysis I learned (and I had the pleasure of convincing Professor Loewe during a visit to our laboratory) that this conception is not right. Female sex hormone does enlarge the seminal vesicle, but in its own specific way. Male sex hormones stimulate the growth of epithelial parts and female sex hormones stimulate the growth of non-epithelial parts of the seminal vesicles. (Jongh 1934: 1,209)

The concept of sexual specificity also structured the debate about the function of 'heterosexual' hormones in human bodies. Clinicians suggested that female sex hormones had no function in the normal development of male bodies. Instead, they conceptualized female sex hormones as agents that caused diseases, in particular sexual and psychological disorders. Others suggested that female sex hormones affected a specifically female development, thus focusing research on homosexuality.

By the end of the 1930s, scientists had largely accepted the idea that 'heterosexual hormones' have a function in the normal development of male organisms, thus abandoning the dualistic concept of an exclusive sex-specific hormonal function. This led, in turn, to the reconsideration of another assumption in the original conceptualization of sex hormones. At the moment that the idea of the sex-specific function of sex hormones was overthrown, scientists questioned the concept of sex antagonism as well. If female sex hormones were present in male bodies, it was hard to maintain the idea of an antagonistic effect on the development of male sexual characteristics. In the early 1930s, the idea of an antagonism between male and female sex hormones was widely disputed. In this debate we see again how the disciplinary background of scientists structured their claims. From a chemical perspective, Ernst Laqueur, professor at the Pharmaco-Therapeutic Laboratory at the University of Amsterdam, rejected the idea of sex antagonism in 1935 as follows: 'Our chemical knowledge makes the original exaggerated assumption of the antagonism between male and female substances rather unlikely' (E. Laqueur 1935).

We saw above how biochemists had defined sex hormones as chemically related compounds. In this definition it is not necessary to assume an antagonistic relationship between sex hormones; other relationships are possible as well. This enabled the Amsterdam School to emphasize a different relationship between female and male sex hormones: instead of an antagonism they reported on the cooperative actions of sex hormones in the development of male secondary sexual organs such as the seminal vesicles, the ductus deferens and the prostate gland. Other scientists reported synergistic actions of male and female sex hormones in female rats, in such processes as stimulation of the growth of the uterus, the first opening of the vagina, and changes in the uterus similar to those seen during pregnancy—processes 'typical of the most female sexual function' (Korenchevsky et al. 1937).

In this debate about sex antagonism, we see how sex endocrinologists actively transformed the prescientific idea that femininity and masculinity reside in the gonads. The American researchers Carl Moore and Dorothy Price, both experimental biologists from the Department of Zoology of the University of Chicago, extended the conceptualization of sex from the gonads to the brain. In 1932, they postulated the idea of a feedback system between the gonads and the hypophysis—that is, they suggested that the inhibiting effects of female sex hormones on male sexual characteristics could not be understood in terms of a direct antagonistic effect on the male gonads, but rather in terms of a depressing effect of female sex hormones on the hypophysis, thus diminishing the production of male sex hormones by the gonads (Moore and Price 1932). In 1939, Frank Lillie evaluated these developments in *Sex and Internal Secretions* as follows:

> If both sex hormones were present simultaneously . . . there should be an 'antagonism' of the two hormones, each striving, so to speak, to control the development of the sex character in question. Such an antagonism was in fact postulated by earlier workers in this field, e.g., Steinach, but more recent work, especially that of Moore, seems to remove the necessity of assuming any antagonism in the simultaneous action of the two hormones, by showing that each operates independently within its own field. (Lillie 1939: 12)

In the late 1930s, the hypothesis of an endocrine feedback system was gradually accepted as the theory to explain interrelations between male and female sex hormones. Scientists thus transformed the prescientific idea that the essence of femininity and masculinity resides in the gonads into a conceptualization of sex that included the brain as the organ that controls sexual development. The extension of the conceptualization from the gonads to the brain also included the postulation of a new type of hormone: the gonadotropic hormones. In 1930, Bernhard Zondek, then working at the Department of Obstetrics and Gynecology at the City Hospital in Berlin, suggested that 'the motor of sexual function' is located in a specific part of the hypophysis, the pituitary gland, which produces two separate chemical substances. The function scientists ascribed to these 'master hormones' was to induce the gonads to produce sex hormones (Zondek 1930: 245; Zondek and Finkelstein 1966).

Additionally, scientists reshaped the prescientific idea of a sexual duality located in the gonads. Initially, scientists had translated this cultural notion into the idea that there existed just two sex hormones, one per sex. In the 1930s, scientists dropped the claim of the existence of just one single female sex hormone. Scientists now suggested that the ovaries are capable of producing two distinct types of female sex hormone. The 'one hormone' doctrine was gradually replaced by the theory that different parts of the ovaries secreted two separate chemical substances. Between 1929 and 1930, three research groups in Europe and in the United States reported the isolation of chemical substances originating from the follicular fluid of the ovaries, which they named estrogenic hormones. By 1934, research groups reported the isolation of 'a second female sex hormone' that, as they claimed, originated from another part of the ovaries: the corpus luteum. This hormone became known as progesterone. From this time onwards, the idea of a single female sex hormone finally disappeared (Parkes 1966: 21). Some scientists depicted progesterone as 'the most female type' of the female sex hormones, because this hormone was considered as sexual-specific in origin (Jongh 1936: 5,370). Other scientists suggested that this 'second' female sex hormone was the more masculine of the two hormones, emphasizing its similarity in function with male sex hormones (Klein and Parkes 1937).

The 1930s were tumultuous years. Scientists reconsidered earlier assumptions about the function of sex hormones as well. In the original concept, sex hormones were understood as substances that affected only those anatomical features which were related to sexual characteristics. In the 1930s, however, reports were published suggesting that the hormones were not so restricted in their function: experiments were described in which they affected the weight of the hypophysis and liver, nitrogen metabolism, and total body weight (Korenchevsky and Hall 1938: 998). As had happened before, scientists adopted a new perspective on the function of sex hormones. After 1935, sex hormones were no longer considered as exclusively sex-specific in function, nor as merely sex hormones or antagonists; instead, they were seen as substances that could generate manifold synergistic actions in both the male and the female body. In this manner, sex endocrinologists thoroughly transformed the prescientific idea of a sexual duality located in the gonads.

TERMINOLOGY AND CLASSIFICATION

In the previous sections we have seen how in the decade from 1920 to 1930 the original concept of sex hormones was transformed with regard to the basic assumptions underlying the early conceptualization of sex hormones. The question that emerges from this is whether scientists still adhered to the concept of sex hormones. Did the drastic changes in conceptualization affect in any way the very naming of sex hormones as male and female? A perusal of the publications of the 1930s suggests that the debate about the sexual specificity of sex hormones was also extended to the terminology and classification of sex hormones.

In the 1930s, many scientists expressed discontent with the terminology and classification of sex hormones, referring to these substances as 'so-called' female sex hormones or simply putting the labels between parentheses. In this debate the Amsterdam School seems to have played an important role. They repeatedly criticized the use of the names 'male' and 'female' sex hormones. In the Dutch journal *Het Chemisch Weekblad* John Freud concluded in 1936:

On purpose we avoid classification in terms of male and female hormones. Maybe our laboratory has contributed the most to overthrowing this classification, since it has been proven experimentally that the oestrogenic substances of the male and the comb-growth stimulating substances of the female have certain functions and are found in the urine of both sexes. (J. Freud 1936)

The Amsterdam School even expressed their doubts about whether these substances should be classified as sex hormones:

These substances are historically named sex hormones not because they are very important for sexual development, but merely because the changes these substances bring about in the organism can be observed with rather crude techniques of observation. (J. Freud 1936)

In addition to the Dutch scientists, the British physiologist Vladimir Korenchevsky was also rather averse to the classification of male and female sex hormones; and both groups proposed other classifications (Korenchevsky and Hall 1938: 998). Korenchevsky and his colleagues proposed a classification of sex hormones into three groups: purely male and female hormones (active only in male or female organisms), partially bisexual hormones (hormones with chiefly male or female properties) and true bisexual hormones (active in both sexes) (Korenchevsky et al. 1937).

107

Laqueur's group suggested that these substances might better be classified as catalysts, thus accommodating the wide variety of their function (J. Freud 1936: 12). From a chemical perspective, they even proposed abandoning the entire concept of sex hormones:

If we understand the hormones as catalysts for certain chemical conversions in cells, it would be easier to imagine the manifold activities of each hormonal substance . . . maybe the greatest discovery in the area of sex hormones will be the detection of the chemical conversions in the cells which are caused by steroids with certain structural qualities. Then the empirical concept of sex hormones will disappear and a part of biology will definitely pass into the *property of biochemistry.* (J. Freud 1936: 12–14)

As against this proposal, the zoologist Frank Lillie intended to adhere to the old names:

The great advances that have been made and consolidated especially in the chemistry and chemical relationships of the male and female sex hormones and in the study of the relations between gonads and hypophysis and of the gonadotropic hormones have served to complicate rather than to simplify our conceptions of the mechanisms of control of sexual characteristics. It seems inadvisable to include in a biological introduction the newer chemical terminology. The old terms male and female sex hormones carry the implication of control of sexual characteristics and represent conceptions that would be valid whatever the outcome of further chemical and physiological analysis. (Lillie 1939: 6)

The debate about the terminology and classification of sex hormones makes it clear how the different professional backgrounds of the disciplines involved in hormonal research led to a different conceptualization of sex hormones. Biochemists assigned meanings to their objects of study that were different from those of biologists. The hormone of the biochemist is in many respects quite different from the hormone of the biologists. From the chemical perspective, hormones were conceptualized as catalysts: chemical substances, sexually unspecific in origin and function, exerting manifold activities in the organism, instead of being primarily sex agents. From the biological perspective, hormones were conceptualized as sexually specific agents, controlling sexual characteristics.

What exactly happened to these different interpretations? Which interpretation of hormones became accepted as the dominant conceptualization of sex hormones? Although the chemical interpretation —emphasizing the resemblance of male and female sex hormones and the possibility of conversion from one to the other—did provoke confusion in the field of sex endocrinology, the prediction of the Amsterdam School was not fulfilled: the biological concept of sex hormones did not disappear. In the 1930s, we see the frequent use of a more specialized, technical terminology for sex hormones. Female sex hormones became known as estrin and estrogen (as a collective noun). For the male sex hormone the names androsterone and testosterone as specialized terms, and androgens as a collective term, became more frequently used. This new terminology did not, however, replace the old terms of female and male sex hormones. Although scientists abandoned the concept of sexual specificity, the terminology was not adjusted to this change in conceptualization. The concept of sex hormones thus showed its robustness under major changes in theory, allowing talk of sex hormones to continue unabated, even though new properties were being ascribed to the hormones. From the 1930s until recently, the names male and female sex hormones have been kept in current use, both inside and outside the scientific community. In this respect the biological perspective overruled the chemical perspective.

This outcome illustrates the strength of the tradition of biologists in the young field of sex endocrinology. The biologists had established a much longer tradition in the field than the biochemists, who were after all newcomers in the field. This does not mean that biologists can be portrayed as the 'winners' of the debate. Both biochemists and biologists adjusted their interpretations of the concept of sex hormones: biologists adjusted their original interpretation of sex hormones as substances sexually specific in origin and function; while biochemists dropped their interpretation of hormones as catalysts. The interpretation that finally came to dominate the field may thus be considered as the result of a compromise between biologists and biochemists.

ON MASCULINE WOMEN AND FEMININE MEN

In the previous sections we have seen how sex endocrinologists gradually transformed the prescientific idea of a sexual duality located in the gonads into a conceptualization of sex that became more and more remote from common-sense opinions about sex and the body. In the paradigm of sex endocrinology, the essence of femininity and masculinity was no longer located primarily in the gonads, but extended to the adrenals. Moreover, the control of sexual characteristics was not restricted to the gonads but was conceptualized as a complex feedback system between the gonads and the brain. This conceptualization meant a definitive break with common-sense opinions that the essence of femininity and masculinity was located in the gonads, a cultural notion that could be traced back to early civilization. That sex endocrinologists were aware of this break is shown in the following quotation from *The Sex Complex: A Study of the Relationship of the Internal Secretions to the Female Characteristics and Functions in Health and Disease*:

It used to be thought that a woman is a woman because of her ovaries alone. As we shall see later, there are many individuals with ovaries who are not women in the strict sense of the word and many with testes who are really feminine in many other respects. (Bell 1916: 5)

The major question that emerges now is: what were the consequences of this new approach to the study of the human body? What was the impact of the introduction of the concept of sex hormones on the conceptualization of sex? A comparison with other fields in the life sciences reveals that the introduction of sex hormones generated a revolutionary change in the study of sex. For the first time in the history of the life sciences, sex was formulated in terms of chemical substances in addition to bodily structures such as organs or cells. Prior to the emergence of sex endocrinology, the study of sex differences had been traditionally the domain of anatomists and physiologists. In the sixteenth century, only the organs directly related to sexuality and reproduction were sexualized. In the course of the eighteenth century, the study of sex differences became a priority in scientific research. Since then, the sexualization of the body has been extended to anatomical structures not related to sexuality and reproduction, such as the skeleton, the blood and the brain.

The introduction of the concept of sex hormones as chemical messengers controlling masculinity and femininity meant a shift in the conceptualization of sex from an anatomical entity to a chemical agency. Instead of identifying which organ was considered as the seat of femininity and masculinity, sex endocrinologists looked for the causal mechanism which regulates the development of the organism either into a male or a female. With the emergence of sex endocrinology and genetics the study of sex differences focused on the causality rather than on the identification of sex. Sex endocrinologists claimed to provide the basic mechanism of sexual differentiation, a knowledge considered far more fundamental to a thorough understanding of sex than that which anatomists had provided.

The introduction of the concept of sex hormones not only meant a shift in the study of sex away from an anatomical identification of the body to a causal explanation of sexual differentiation, but also entailed another major change in the study of sex. Instead of locating the essence of femininity or masculinity in specific organs, as the anatomists had done, sex endocrinologists introduced a quantitative theory of sex and the body. The idea that each sex could be characterized by its own sex hormone was transformed into the idea of relative sexual specificity. Sex endocrinologists suggested that, although female sex hormones were more important for women (especially during pregnancy) than for men, their potency was the same in both sexes. Male sex hormones were thought to be of greater importance for the internal and external phenotype of men, but this was regarded as only a difference of degree. In *The Annual Review of Physiology* published in 1939, the American physiologist Herbert Evans postulated this theory as follows:

It would appear that maleness or femaleness cannot be looked upon as implying the presence of one hormone and the absence of the other, but that differences in the absolute and especially relative amounts of these two kinds of substances may be expected to characterize each sex and, though much has been learned, it is only fair to state that these differences are still incompletely known. (Evans 1939: 578)

Or as Robert Frank suggested:

The explanation naturally suggesting itself is in favor of the theory of a quantitative and fluid transition from male to female. (Frank 1929: 115)

With this quantitative theory, the endocrinologists introduced a new conceptualization of sex. In the earlier anatomical definition of sex individuals could be classified into two, or actually three, categories: on the basis of the type of sexual organs an individual was categorized as male or female, and in cases where an individual possessed the sexual organs of both sexes, as a hermaphrodite. However, with the new concept of relative sexual specificity endocrinologists constructed a biological foundation for a definition of sex in which an individual could be classified in many categories varying from 'a virile to effeminate male' or from 'a masculine to feminine female,' as Robert Frank described it. Laqueur's group described this conceptualization of sex in 1928 as follows:

The occurrence of the female hormone in the male body gives rise to many fantastic reflections. . . . It is now proved that in each man there is something present that is inherent in the female sex. Whether we will succeed in determining the individual ratio of each man, in terms of a given percentage femininity, we don't know. (Borchardt et al. 1928 1,028)

Other scientists joked, privately, that the new biochemistry meant the end of sex differences: 'there but for one hydroxyl group go I' (Long Hall 1976: 20).

The new model of sex in which sex differences are ascribed to hormones as chemical messengers of masculinity and femininity, agents that are present in female as well as male bodies, made possible a revolutionary change in the biological definition of sex. The model suggested that, chemically speaking, all organisms are both male and female. Sex could now be conceptualized in terms of male/masculine and female/feminine, with the elements of these two pairs no longer considered *a priori* as exclusive. In this model, an anatomical male could possess feminine characteristics controlled by female sex hormones, while an anatomical female could have masculine characteristics regulated by male sex hormones.

This hormonal model of sex provided the life sciences with a model to explain the 'masculine characteristics' in the female body and vice versa. In obstetric science, for example, the hormonal model was used in the 1930s to substantiate the classification of female pelves into 'masculine' and 'feminine' types. Physicians and clinicians used the quantitative model of sex to account for the purportedly feminine character of homosexual men. Sex endocrinologists introduced diagnostic tests to measure the degree of femininity and masculinity in the human body.

This shift in thinking was found in other fields of science as well. After 1925, the concepts of masculinity and femininity became the central focus of research in psychology, next to the study of sex differences (Lewin 1984: 158). And similarly in anthropology: in 1935, Margaret Mead wrote her classic work *Sex and Temperament* in which she postulated the idea that masculinity and femininity are randomly distributed between the sexes by nature but are assigned to only one sex by society (Lewin 1984: 166).

CONCLUSIONS

Evaluating this case study, we may conclude that prescientific ideas are a major factor in structuring scientific development. The early history of sex endocrinology illustrates in the first place how the prescientific idea of a sexual duality located in the gonads functioned as a major guideline structuring the development of endocrinological research. At the beginning of each new line of research, scientists proposed hypotheses corresponding to the cultural notion of sexual duality. In research on the origin as well as on the function of sex hormones, hypotheses were directed by this assumption, thus producing friction between expectations and experimental data.

How science gives meaning to sex differences is thus partly shaped by cultural notions of masculinity and femininity. Scientists use cultural notions as one of their cognitive resources. This conclusion is in line with feminists' claims that the development of knowledge is shaped by cultural norms about women and men. This does not imply that science leaves these cultural notions unchanged, as is often assumed in feminist studies. Scientists not only use cultural ideas as cognitive resources, but also actively modify these notions. This study showed how sex endocrinologists incorporated the cultural notion of a sexual duality located in the gonads, and subsequently transformed this notion into a model of sex differences that implied a drastic change in the conceptualization of sex differences.

In this respect, the story of sex hormones deviates from Fleck's story of syphilis. Fleck concluded that, despite the different cognitive approaches, scientists eventually maintained the cultural notions about syphilis. He described the development of the Wasserman test in terms of the ultimate realization of the prescientific idea of syphilis as bad blood (Belt and Gremmen 1988). The concept of sex hormones, however, is not simply a realization of prescientific ideas. In the case of sex hormones, differences in disciplinary styles contributed to a drastic change in the cultural notion of sexual duality. I described how biochemists approached the study of sex hormones from a perspective different from that of biologists and gynecologists. The hormone of the biochemist differed from the hormone of the biologist and the gynecologist. Gynecologists (studying dysfunctions of organs) and biologists (studying the physiological development of the organism) shared a common interest in functions. Biochemists (focusing mainly on chemical structures) were devoted to the study of structures rather than functions. Consequently, the disciplines gave different interpretations to the function of sex hormones, and they opted for different forms of classification and naming. Biochemists saw the object of their research in terms of catalysts: chemical substances, sexually unspecific in origin and function, inducing chemical conversions in the cells, thus emphasizing the manifold activities of the substances as major characteristics of sex hormones. Gynecologists and biologists interpreted their object of study as sex hormones: substances sexually specific in origin and function, thus highlighting as a major characteristic the function of these substances in the sexual development of the organism. Sex hormones thus embody the ideas and interests of different disciplinary traditions. This multidisciplinary context accounts for the changes that took place in the conceptualization of sex differences, a process in which the prescientific idea of a sexual duality located in the gonads became transformed into a chemical model of sex that enabled the construction of new meanings and practices that became attached to the human body.

References

Aberle, S. D., and Corner, G. W. (1953). *Twenty-Five Years of Sex Research: History of the National Research Council Committee for Research in Problems of Sex 1922–1947*. Philadelphia: W. B. Saunders.

Allen, E., Hisaw, F. L., and Gardner, W. U. (1939). 'The Endocrine Function of the Ovaries', in E. Allen (ed.), *Sex and Internal Secretions*, 2nd edn. Baltimore: William and Wilkins.

Bell, B. (1916). *The Sex Complex: A Study of the Relationship of the Internal Secretions to the Female Characteristics and Functions in Health and Disease*. London: Bailliere, Tindall & Cox.

Belt, H. v. d., and Gremmen, B. (1988). 'Het specificiteitsbegrig in de tijd van Koch en Ehrlich: Een nieuwe interpretatie van de 'serologische denkstijl' van Ludwik Fleck'. *Kennis en Methode*, 12: 334–50.

Borchardt, E., Dingemanse, E., Jongh, S. E. de, and Laqueur, E. (1928). 'Over het vrouwelijk geslachtshormoon Menformon, in het bijzonder over de anti-masculine werkin'. *Nederlands Tijschrift Geneeskunde*, 72: 1,028.

Borell, M. (1976*a*) 'Brown-Séquard's Organotherapy and its Appearance in America at the End of the Nineteenth Century'. *Bulletin of the History of Medicine*, 50: 309–20.

—— (1976*b*) 'Organotherapy, British Physiology, and the Discovery of the Internal Secretions'. *Journal of the History of Biology*, 9: 235–68.

—— (1985) 'Organotherapy and the Emergence of Reproductive Endocrinology'. *Journal of the History of Biology*, 18: 1–30.

Caldwell, W. E., Moloy, H. C., and D'Esopo D. A. (1934). 'Further Studies on the Pelvic Architecture'. *American Journal of Obstetrics and Gynecology*, 28: 482–97.

Callow, N. H. , and Callow, R. K. (1938). 'The Isolation of Androsterone and Transhydroandrosterone from the Urine of Normal Women'. *Biochemical Journal*, 32: 1,759–1,762.

Clarke, A. K (1985). 'Emergence of the Reproductive Research Enterprise: A Sociology of Biological, Medical and Agricultural Science in the United States, 1910–1940'. Dissertation, University of California, San Francisco.

—— (1987). 'Research Materials and Reproductive Science in the United States, 1910–1940', in L. Gerald Geison (ed.) *Physiology in the American Context 1850–1940.* New York: American Physiological Society.

Corner, G. W. (1965). 'The Early History of Oestrogenic Hormones'. *Proceedings of the Society of Endocrinology*, 33: 3–18.

Dingemanse, E., Laqueur, E., and Muhlbock, O. (1928). 'Chemical Identification of Estrone in Human Male Urine'. *Nature*, 141: 927.

—— Borchardt, A., and Laqueur, E. (1937). 'Capon Comb Growth-Promoting Substances ('Male Hormones') in Human Urine of Males and Females of Varying Ages'. *Biomedical Journal*, 31: 500–7.

Eng, H. (1934). 'Resorption und Ausscheiding des Follikulins im menschlichen Organismus II. Mitteilung: Zur Kenntnis der Follikelausscheiding in Harn und Fezes normaler Maenner'. *Biochemische Zeitschrift*, 274: 208–11.

Evans, H. M. (1939). 'Endocrine Glands: Gonads, Pituitary and Adrenals'. *Annual Review of Physiology.* The Hague: Martinus Nijhoff.

Fellner, O. (1921). *Pflueger's Archiv,* 189.

Fleck, L. (1979). Genesis and Development of a Scientific Fact. Chicago: University of Chicago Press.

Frank, R. T. (1929). *The Female Sex Hormone.* Springfield: Charles C. Thomas.

Freud, J. (1936). 'Over Geslachtshormonen'. *Chemisch Weekblad,* 33 (4) 3: 1–14.

Freud, S. (1905). 'Three Essays on the Theory of Sexuality', in *Sigmund Freud: The Standard Edition of the Complete Psychological Works of Sigmund Freud,* vol. 7, ed. and trans. James Strachey, in collaboration with Anna Freud, assisted by Alix Strachey and Alan Tyson. London: Hogarth Press and the Institute of Psychoanalysis.

Gallagher, C., and Laqueur, T., (eds.) (1987). *The Making of the Modern Body: Sexuality and Society in the Nineteenth Century.* Berkeley: University of California Press.

Goodman, N. (1978). *Ways of Worldmaking.* Indianapolis: Hackett.

Gruhn, J. G., and Kazer, R. R. (1989). *Hormonal Regulation of the Menstrual Cycle: The Evolution of Concepts.* New York: Plenum Medical.

Hamilton, D. (1986). *The Monkey Gland Affair.* London: Chatto & Windus.

Heape, W. (1913). *Sex Antagonlsm.* London: Constable. (ed.), *De moderne aspecten der endocrinologie: speciaal der praktische hormonologie.* Amsterdam: Scheltema en Holkema's Boekhandel un Uitgevrijmaatschappij.

Hiddiga, A. (1990). 'Geordende bekkens: classificatie-systemen in de verloskunde'. *Tildschrift voor Vrouwenstudies,* 11 (2): 158–75.

Honegger, C. (1991). *Die Ordnung der Geslechter: Die Wissenschaften vom Menschen und das Weib.* Frankfurt: Campus Verlag.

Jongh, S. E. de (1934). 'De betekenis van vrouwelijk hormoon, menformon, voor mannelijke individuen'. *Nederlands Tijdschrift voor Geneeskunde* 78: 1,208–1,216.

Jongh, S. E. de (1936). 'Vrouwelijk' en 'mannelijk' geslachtshormoon. *Nederlands Tijdschrift voor Geneeskunde*, 80: 5,366–5,375.

—— (1951). 'In hoeverre verdient testosteron de naam mannelijke hormoon?', in Faculteit der Geneeskunde der Rijsuniversiteit Utrecht.

—— Laqueur, E., and Dingemanse, E. (1929). 'Over vrouwelijk hormoon (menformon) in mannelijke organismen; iets over hetbegrip specificiteit'. *Nederlands Tidschrift Geneeskunde*, 73: 771–5.

Klein, M., and Parkes, A. (1937). 'The Progesteron-like Action of Testosterone and Certain Related Compounds'. *Proceedings of the Royal Society London*, 3: 574–9.

Kochakian, Ch. D. (1938). 'Excretion and Fate of Androgens: Conversion of Androgens to Estrogens'. *Endocrinology*, 23: 463–7.

Kohler, R. E. (1982). From Medical Chemistry to Biochemistry: The Making of a Biomedical Discipline. Cambridge: Cambridge University Press.

Korenchevsky, Vl., and Hall, K. (1938). 'Manifold Effects of Male and Female Sex Hormones in Both Sexes'. *Nature*, 142: 998.

—— Dennison, M., and Hall, K. (1937). 'The Action of Testosterone Propionate on Normal Adult Female Rats'. *Biochemical Journal*, 31: 780–85.

Kruif, P. de (undated). *Het Mannelijk Hormoon*, Nederlandse bewerking door C.C. Bender, Uitgeverij Keesing, Amsterdam.

Kuhn, T. (1970). *The Structure of Scientific Revolutions*, 2nd edn. Chicago: University of Chicago Press.

Laqueur, E. (1935). 'Sexualhormone und Prostratahypertrophie'. Unpublished lecture. Organon's Archive on the correspondence between Ernst Laqueur and Organon. Lecture dated 9 November.

—— Dingmanse, E. Hart, P. C., and Jongh, S. E. de (1927). 'Female Sex Hormone in Urine of Men'. *Klinische Wochenschrift*, 6: 1,859.

Laqueur, T. (1990). *Making Sex: Body and Gender from the Greeks to Freud.* Cambridge, Mass.: Harvard University Press.

Lewin, M. (1984). 'Rather Worse than Folly? Psychology Measures Femininity and Masculinity 1', in M. Lewin (ed.), *In the Shadow of the Past: Psychology Portrays the Sexes.* New York: Cambridge University Press.

Lillie, F. R. (1939). 'Biological Introduction', in E. Allen (ed.), *Sex and Internal Secretions*, 2nd edn. Baltimore: Williams & Wilkins.

Long, D. (1987). 'The Physiological Identity of American Sex Researchers' in G. L. Geison (ed.), *Physiology in the American Context 1850–1940.* Bethesda, Md. American Physiological Society.

Long Hall, D. (1975). 'Biology, Sexism and Sex Hormones in the 1920s', in M. Wartofsky and C. Could (eds.) *Women and Philosophy.* New York: Putnam.

—— (1976). 'The Social Implications of the Scientific Study of Sex', in *The*

Scholar and the Feminist, iv. New York: Women's Center of Barnard College, 11–21.

Medvei, V. C. (1983). *A History of Endocrinology*, The Hague: MTP Press.

Meyer-Bahlburg, H. F. L. (1977). 'Sex Hormones and Male Homosexuality in Comparative Perspective'. *Archives Sexual Behavior*, 297–326.

—— (1984). 'Psychendrocrine Research on Sexual Orientation, Current Status and Future Options'. *Progress in Brain Research*, 61: 375–99.

Money, J. (1980). 'History of Concepts of Determination', in J. Money, *Love and Lovesickness: The Science of Sex Gender Difference and Pairbonding*. Baltimore: Johns Hopkins University Press.

Moore, C. R., and Price, D. (1932). 'Gonad Hormone Functions: Reciprocal Influence between Gonads and Hypophysis with its Bearing on Problems of Sex Hormone Antagonism'. *American Journal of Anatomy*, 50: 13–71.

Parkes, A. S. (1937). 'Androgenic Activity of Ovarian Extracts'. *Nature*, 139: 9.

—— (1938). 'Terminology of Sex Hormones'. *Nature*, 141: 12.

—— (1966). 'The Rise of Reproductive Endocrinology 1926–1940'. *Proceedings of the Society of Endocrinology*, 34 (3): 20–32.

Price, D. (1975). 'Feedback Control of Gonadal and Hypophyseal Hormones: Evolution of the Concept', in J. Meites and B. M. McCann (eds.), *Pioneers in Neuroendocrinology*. New York: Plenum.

Schiebinger, L. (1986). 'Skeletons in the Closet: The First Illustrations of the Female Skeleton in the Nineteenth-Century Anatomy'. *Representations*, 14: 42–83.

Starling, E. H. (1905). 'The Croonian Lectures on the Chemical Correlation of the Functions of the Body'. *Lancet*, ii: 339–41.

Steinach, E. (1926). 'Antagonische Wirkungen der Keimdrusen Hormone'. *Biologia Generalis*, 2: 815–34.

Thompson, J. A. (1914). 'Review of Sex Antagonism'. *Nature*, 93: 346.

Walsh, J. (1985). 'The Science Work of Guy Marrian (1904–1981) Mainly with Respect to Steroid Hormones'. Library-based Dissertation, University of Oxford.

Wijngaard, M. van den (1991). 'Reinvention the Sexes: Feminism and Bio-medical Construction of Masculinity and Femininity 1959–1985'. Thesis, University of Amsterdam.

Zondek, B. (1930). 'Ueber die Hormone des Hypophysenvorderlappens'. *Klinische Wochenschrift*, 9.

—— (1934a). 'Mass Excretion of Oestrogenic Hormone in the Urine of the Stallion'. *Nature*, 1933: 209–10.

—— (1934b). 'Oestrogenic Hormone in the Urine of the Stallion'. *Nature*, 133: 494.

—— and Finkelstein, M. (1966). 'Professor Bernard Zondek: An Interview'. *Journal of Reproductive Fertility*, 12: 3–19.

4 Doubtful Sex

Alice Domurat Dreger

In reality, is the subject a boy or a girl?

(Dr. Gaffe, reporting on a case of hermaphroditism in the *Journal de médecine et de chirurgie pratiques* [1885])

Reality is crushing me, is pursuing me. What is going to become of me?

(Abel (née Alexina) Barbin, autobiography [*c.* 1868])

I wrestled with reality for thirty-five years, Doctor, and I'm happy to state I finally won out over it.

(Jimmy Stewart as Elwood P. Dowd in *Harvey*)

The history of hermaphroditism is largely the history of struggles over the 'realities' of sex—the nature of 'true' sex, the proper roles of the sexes, the question of what sex can, should, or must mean. Much recent scholarship has been devoted to documenting the nineteenth-century medical and scientific construction of sex and gender, especially of women and femininity.[1] Such studies show how scientific and medical men used their minds, tools, and rhetoric to build up powerful ideas about the 'true natures' (and limits) of femininity and masculinity. My review of medical literature on hermaphrodites supports in part the findings of these studies in that it similarly demonstrates rampant sexing of—that is, attributing the designation of 'male' or 'female' to—various body parts, behaviours, and desires by medical and scientific men and lay people of the time. The literature on hermaphroditism also reveals, however, that there was not a single, unified medical opinion about which traits should count as essentially or significantly feminine or masculine. In France and Britain, the sexes

From Alice Domurat Dreger, *Hermaphrodites and the Medical Invention of Sex* (Harvard University Press, 1998), Chapter 1. Reprinted with permission.

were constructed in many different, sometimes conflicting ways in hermaphrodite theory and medical practice, as medical men struggled to come up with a system of sex difference that would hold. Ultimately it was not only the hermaphrodite's body that lay ensconced in ambiguity, but medical and scientific concepts of the male and the female as well. We see here not stagnant ideas about sex, but vibrant, growing, struggling theories. Sex itself was still open to doubt.

Witness now the stories of two nineteenth-century hermaphrodites, one the relatively well known Abel/Alexina Barbin of France, the other a 'living specimen of a hermaphrodite' who quietly surfaced in England twenty years after Barbin's death.

TALES OF TWO HERMAPHRODITES

In Paris in 1868, during the month of February, a local police commissioner and one Monsieur Régnier, a physician in the employ of the state registry office, were called to investigate the suicide of a young man who had worked as an administrative clerk for the Parisian railroads.[2] When the two officials mounted the stairs to the scene of the death, a squalid, sparsely furnished room on the fifth floor of a boarding house on the rue de l'École-de-Médecine, they found the corpse of Abel Barbin lying across the bed. Barbin had evidently used the small charcoal stove in his room to end his life by carbon monoxide poisoning. A discharge of dark, frothy blood spilled from his still lips.[3] Surveying the lamentable scene, Régnier decided to remove the few articles of clothing in which Barbin had died, and to examine the subject's genitalia. The doctor suspected the young man's melancholy had grown either directly or indirectly from a familiar cause: syphilis.[4] During a genital examination, however, much to his surprise, Régnier discovered not signs of syphilis, but a strange melange of sexual anatomy; a short, imperforate penis, curved slightly backward and pointed toward what Régnier could only call a vulva—labia minora and majora, and a vagina large enough to admit an index fnger.[5]

News of the discovery of a recently dead 'hermaphrodite' soon reached the Faculty of Medicine, including the *lauréat* Dr. E. Goujon, who feared that if some member of the faculty did not act quickly, 'this observation', this unusual body, 'would be lost for science'.[6] Goujon quickly mobilized, determined to see the anomaly for himself. He located Régnier and arranged to have the corpse turned over to the faculty for study. While performing the autopsy Goujon took careful notes, for he intended to publish his findings of this strange case.

Before long, however, Goujon discovered his publication would not be the literary debut of this hermaphrodite. Eight years earlier, a doctor in La Rochelle by the name of Chestnet had reported the history and anatomy of this very same 'man' in the pages of the *Annales d'hygiène publique* (Annals of Public Hygiene) on the occasion of the revision of the subject's civil status from female to male. Moreover, to the great excitement of all the medical men who would involve themselves in or with the story of Barbin, the subject himself had left behind extensive memoirs, dating from about 1864, or approximately three years after the legal sex revision.[7] In thick, dramatic prose, Barbin had recorded every detail of the trials, tribulations, loves, and losses of her/his tortured soul, as well as her/his persistent conviction that s/he could not go on, that death was imminent. In this auto-biography—a text that would be published in 1874 by one fascinated medical doctor as part of a treatise on the 'Medico-Legal Question of Identity'[8]—Barbin imagined the day of her/his death:

When that day comes a few doctors will make a little stir around my corpse; they will shatter all the extinct mechanisms of its impulses, will draw new information from it, will analyze all the mysterious sufferings that were heaped up on a single human being.[9]

It turned out to be an eerily prescient dream.

Twenty-nine years before his piteous death, in the town of Saint-Jean-d'Angély, Barbin was born and christened as a girl. It is not clear whether any questions were raised about the child's sex at that time. We do know that within a few days of the birth, as required by French law, the newborn was presented to the local mayor and civil status registrar so that they might record certain facts: the child's birthdate (November 8, 1838), the names of the parents (Jean Barbin and Adélaïde Destouches), the sex of the child (female), and her name (Adélaïde Herculine Barbin). From her childhood until her legal sex revision, Barbin's familiars knew her as Alexina.

Uncommon anatomy was not the only trouble with which Alexina Barbin was cursed. Her father died when she was a small girl, thus compelling her desperate mother to give up her only child to the care of an order of nuns, the Ursulines of Chavagnes. In the convent boarding school, Alexina enjoyed a stable life and a good education. She was a solid and obedient student, and eventually agreed to become a teacher of girls herself, though she confessed she 'was no more flattered by the prospect of being a *working woman*. I believed I deserved better than that.'[10] Trained at the normal school of Oléron in Le Château from 1856 to 1858, Alexina gained a position at a small girls' boarding school directed by an established teacher and her sister, the latter a young, unmarried woman by the name of Sara. Sara lived with her mother, Madame P., who provided the new teacher with lodging in their home.

Alexina—an individual of somewhat stocky build, who had, to her dismay, never menstruated—and Sara, whom Alexina saw as a model of feminine grace, grew intimate quickly. Indeed they became nearly inseparable, even going so far as to share a bed regularly. Witnessing their strong attachment, Madame P. scolded them, 'You are very fond of each other, and for my part I am very happy that you are; but there are proprieties that must be observed, even among *girls*.'[11] Still, Madame did not suspect the level of their intimacy. Not long after Alexina's arrival, she and Sara had begun having sexual relations. As Alexina remarked, Sara's mother 'saw me only as her daughter's *girlfriend*, while in fact I was her lover!'[12]

A combination of a weighty conscience and a painful abdomen finally led the tormented Alexina to a series of priest-confessors and medical men, the result of which was a consensus that Alexina was a man, a male who had been mistaken at birth for a female, and that therefore her legal and public identity ought to be 'rectified' to match her 'true sex'. In 1860 her case was brought before the tribunal of her home district, with the critical testimony entered by Dr. Chestnet of La Rochelle. On June 21, 1860, the register of Barbin's birth was amended. Her sex was changed to male, and her name to Abel. The record now showed she had always truly been a male.[13]

'According to my civil status, I was henceforth to belong to that half of the human race which is called the stronger sex,' Alexina—now Abel—noted in the memoirs.[14] Barbin hoped that this new public identity might enable him to marry Sara, a dream the two apparently shared. Such happiness was not to be; the scandal was too great. Instead a family benefactor aided Barbin by securing him a position of clerk to a Parisian railroad. So to Paris he went, where no one would know him and his past. Nevertheless, Barbin could not escape the painful incongruities of his life. Alexina had detested the thought of being a working woman, and now she found herself a working man. He bemoaned his fate: 'Reality is crushing me, is pursuing me. What is going to become of me?'[15]

Within a short time, before reaching the age of thirty years, the body of Abel Barbin lay on the dissecting table of Goujon. We do not know precisely what finally drove Barbin to suicide: the lost love? the sudden change in identity? something else instead? In any case, at the autopsy, the doctor saw to it that careful sketches of Barbin's unusual anatomy were made and recorded in the medical literature for any curious medical man to peruse. Over the next several decades, many did.[16] There is, however, no known picture of Alexina Barbin's or Abel Barbin's face.[17]

By contrast, the face of a hermaphrodite who surfaced in England in 1888 was recorded, though perhaps unintentionally, looming as it did in the background of a photograph focused on this subject's uncommon genital conformation. On Wednesday, April 25, 1888, at a regularly scheduled meeting of the British Gynaecological Society, the esteemed gynecologist Dr. Fancourt Barnes, physician to the Chelsea Hospital for Women, the British Lying-in Hospital, and the Royal Maternity Charity, announced that he had 'in the next room a living specimen of a hermaphrodite'. Barnes added that he used the term hermaphrodite 'on general principles, because it is the case of an individual who has been brought up to the age of nineteen as belonging to the female sex, when it is perfectly clear that he was a male.'[18] Or was it perfectly clear?

Although a few colleagues—among them the well-known gynecologist Lawson Tait, surgeon to the Birmingham and Midland Hospital for Women, a man like Barnes well versed in hermaphroditism—supported Barnes's diagnosis, several society members in attendance were not so sure. Pressed by the latter to defend his claim, Barnes explained that

his own reasons for believing the person to be a male were, (1) the appearance of the head, (2) the *timbre* of the voice, (3) the non-development of the breasts, (4) the undoubted existence of a well-formed prepuce and glans penis, (5) the imperfectly formed urethra running down from the tip of the glands [*sic*] and passing into the bladder, (6) the utter absence of anything like a uterus or ovaries, and (7) the appearance of the perineum. The thighs were covered with masculine hairs . . . Lastly, the patient never had the menstrual molimina.[19]

These seemed to Barnes significant and sufficient symptoms of malehood. But if Barnes had come to the meeting thinking that his diagnosis (that the patient was 'undoubtedly a man') would be uncontroversial, he was sorely mistaken.[20]

Criticism and controversy began with an objection from Dr. Charles Henry Felix Routh, consulting physician to the Samaritan Free

Hospital and founding member of the British Gynaecological Society. Routh insisted it 'was by no means clear that it was a man. Such a conclusion was mere guess work'.[21] Routh asked whether the rectal examination conducted by Barnes had been sufficient, specifically whether it was enough to have performed a digital exam, or if, rather, Barnes should have tried to 'pass his entire hand into her rectum' to search for evidence of a uterus. Furthermore, Routh objected, 'Even supposing there were no uterus, the mere fact was no argument against its being a woman.' This might simply be an instance of a woman deprived of that organ. Routh was unconvinced. Barnes offered the additional evidence for manhood that two or three years earlier there had appeared a moustache and beard, but Routh protested that 'this had absolutely no weight. Many Jewesses had quite a large quantity of both beard and moustache.' Beards and moustaches, like absent uteri, were not unheard of in women. In Routh's eyes these were not sure signs of sex.[22]

As consulting physician to the Chelsea Hospital for Women, Dr. James Aveling had to agree with Dr. Routh on the matter of facial hair, for he himself 'had often seen it in women just as well marked'. Aveling also noted of Barnes's subject that '[t]he face was feminine, [and] the throat was decidedly that of a woman, the *pomum Adami* not being at all prominent.'[23] He recommended putting the patient under an anesthetic and performing a bimanual examination 'per rectum and through the abdomen'.[24] He wanted, if at all possible, evidence of ovaries or testes. Without that, perhaps Barnes had been too quick to conclude.

For his part, Dr. G. Granville Bantock, surgeon to the Samaritan Free Hospital, concurred that '[t]hey all knew how unsafe facial characteristics were as a guide, the dress making such a very great difference.' Nonetheless he found the case 'clear as daylight'. To him the critical signs lay well below the face: 'The appearance of the sexual organ was that of a penis, and the remainder of the urethra was to his mind, incontrovertible evidence in favor of its being a male.'[25] Still this reasoning did not persuade everyone present. After all, deceptive genitalia were the root of the problem in such cases of 'doubtful sex'; it hardly seemed that in cases of ambiguous sex the offending organs should stand as trustworthy witnesses. So the fellows now argued about how much weight ought to be given to *these* signs, or to pubic hair and menstruation, or lack thereof, to the shape of the skull, to balding patterns. Dr. Routh pondered aloud the question of whether a woman without a uterus and ovaries would still be a woman. Piling doubt upon doubt, Dr. Bantock paused to criticize Dr. Aveling's handling of a similar case some years prior, with the result that a subdebate erupted between Bantock and Aveling over what should have been the proper diagnosis in that case.

Finally, in an effort to bring to a close this fractious exchange, the London physician Dr. Heywood Smith 'suggested that the Society should divide on the question of sex'.[26] Divide these men did, dramatically unable now to decide what they had seen and felt, incapable of agreeing on the nature of sex and its proper diagnosis. The 'living specimen' was twice photographed for the record, and apparently then departed with his/her mother, the doubt here left uncomfortably unresolved.

THE INTOLERANCE OF HERMAPHRODITES AND THE AGE OF GONADS

Many French and English readers of our time are familiar with the tragic story of Alexina/Abel Barbin, thanks mostly to the work of Michel Foucault who, about two decades ago, directed the republication in English and in the original French of Barbin's memoirs. It is an unfortunate fact, however, that Barbin's memoirs are the only known record of their type from nineteenth-century Europe. How much better off we would be in our understanding of sexual identity and hermaphroditism if we could, as Roy Porter suggested in 1985, do a good part of medical history 'from below', in this case from a study of the hermaphrodites' viewpoint.[27] How much more complex and rich the history would be if, for instance, Sophie V. and Barnes's 'living specimen' had left behind their impressions of their bodies and experiences, especially given that their 'true' sexes remained overtly in doubt, while Barbin's was publicly settled.

The absence of documents like Barbin's memoirs cannot justify the far too sweeping conclusion—about scientific and social concepts of sex, gender, and sexuality—that have been drawn by some recent scholars simply from the singular case of Barbin. As we can see from the cases of Sophie V. and Barnes's 'living specimen', the lack of documents like Barbin's memoirs does not mean that all knowledge of hermaphrodites and their experiences aside from Barbin's is lost to us. Important information about hermaphrodites' lives and views can be garnered from the sizable body of medical and scientific literature on the subject, keeping in mind that this information is filtered and shaped by contemporary biomedical observers and authors.

Indeed, we can see even from just the three cases discussed so far how variable the encounter of hermaphrodites and medical men could be in the nineteenth century. Sometimes the issue of the patient's 'true sex' was resolved, sometimes not. Sometimes the medical men's decisions were heeded, sometimes not. Sometimes there was agreement among the experts on what mattered in finding 'true sex', and sometimes not. These are important variations, and if we look at just one case like Barbin's, we lose sight of all the turmoil and negotiation that occurred over doubtful patients and the richness of experience of the hermaphrodites themselves. Some of this variation in case histories, as we shall see, is attributable to cultural differences in France and Britain. This study intentionally looks at hermaphroditism in two nations of the same period in order to explore how—even in fairly similar cultures—the understanding and treatment of hermaphrodites varies according to culture. But these encounters also varied because the individuals in them were of different motivations, convictions, educations, and experiences.

When I set out to do this work, I made a point of trying to locate as many documents as possible about cases of hermaphroditism in France and Britain during the period in which I was interested. I have been able to find roughly three hundred commentaries and accounts of human hermaphroditism published in the scientific and medical literature of France and Britain from 1860 to 1915. For comparison I have also read many others published before and after this period. These collectively refer to approximately two hundred different cases of hermaphroditism, roughly one-half to two-thirds of which seem to have surfaced in France and Britain after 1870 and before 1915. (It is not always possible to tell from the record where or when a case was uncovered or whether it is the same as one reported elsewhere.)

With the exception of Barbin's memoirs, all of the primary source texts I have located were written by medical and scientific men. The lack of documents written by hermaphrodites and the preponderance of men among biomedical authors means that we trace here the deveopment of a peculiarly masculine, biomedical view of masculinity, femininity, and hermpahroditism.[28] By the late nineteenth century, women were starting to make inroads into medical schools, but the fact was that throughout that century in Europe and America women were largely excluded from obtaining medical degrees and licenses.[29] Moreover, as Ornella Moscucci has demonstrated in her study of the nineteenth-century rise of gynecology in England, 'opposition to female practitioners was greatest among obstetricians and gynaecologists'—the types of medical professionals most likely to discover a hermaphroditic subject—because obstetricians and gynecologists 'had the most to fear from the competition of medical women, for the major argument in favour of lady doctors centered on women's need for medical care which did not violate women's modesty'.[30] A woman medical physician or surgeon was therefore uncommon; one who had the opportunity to discover a hermaphrodite even more so.

Of course, during the late nineteenth century many women were still serviced by a different kind of medical woman—a midwife—despite the fact that for over a century midwives had been suffering from the verbal and political attacks and from the professional competition of medical men. Midwives were likely to come upon cases of unusual genitalia with reasonable frequency. However, I have been unable to locate a body of literature on hermaphroditism by midwives of the late nineteenth century. Part of the reason for this may be that midwives did not share the same level of interest in publication that the professionalizing medical men did. Additionally, in searching for relevant documents I was obligated primarily to use indexes compiled by men often disinterested in the writings of midwives. When, in the literature on 'doubtful sex' by men physicians and surgeons, the labors of midwives are discussed, they are almost always mentioned in passing only to be condemned as contributing to disastrous cases of 'mistaken' sex assignment.

In the last few years of the nineteenth century, there occurred a virtual explosion of human hermaphroditism. Why did this happen? Some people have suggested to me that a significant increase in industrial pollution could have contributed to the apparent rise in numbers of cases of hermaphroditism, but it is very hard to know what, if any, significant material environmental changes might have given impetus to the rise. (The question of the general frequency of hermaphroditism is discussed later in this chapter.) Instead I think it reasonable to credit the steady rise to other sorts of important social changes.

First, because gynecology and access to medical care were on the rise, more and more people in France and Britain were subject to genital examinations. This meant an increase in the number of people considered to be of doubtful or mistaken sex; there was simply more opportunity for doubt to surface because more genitals fell under careful scrutiny. Second, a dramatic increase in the opportunities for reporting of medical findings—there were more and more outlets for medical publications—meant more opportunities for cases to be documented. Relatedly, there spread in consequence a conviction among medical practitioners that hermaphroditism was not all that rare and that practitioners should be on the look out for the phenomenon—and that, when they found it, they should report it.[31] Seek and ye are more likely to find, find and ye shall publish, publish and the door is opened.[32]

But the amount of ink and angst dedicated to hermaphrodites probably also increased among medical and scientific men in the late nineteenth century because that period saw a proliferation of people like feminists (including some women physicians) and homosexuals who vigorously challenged sexual boundaries. These border challenges resulted in a concomitant reaction on the part of many medical and scientific men to insist on tighter definitions of acceptable forms of malehood and femalehood, and more bodies therefore fell into the 'doubtful' range.

Hermaphrodites like Barbin and Barnes's unnamed 'living specimen' embodied serious practical problems for the medical and scientific men who grappled with them. This was not a time when most of these men were interested in seeing sexual boundaries blur, not a time when they found terribly amusing the sorts of obvious, profound disagreements about sex the British Gynecological Society had had to suffer in 1888. The late nineteenth century in France and Britain was a time (not unlike our own fin-de-siècle) in which the nature of the sexes stood as a politically charged issue. Many men of science, following in the footsteps of the great Charles Darwin, wrote with confidence and enthusiasm about the natural and profound differences of the male and the female types. The higher the organism on the ladder of life, they said, the more exquisitely differentiated the male and female of the species. But such lively, lofty, often poetic discussions of sex were carried on even while—and probably in part because—certain increasingly visible people challenged the borders that were said by most to divide the sexes naturally.

Homosexuals and feminists remained the most publicly problematic sorts. While a number of women sought university educations, medical degrees, and the vote—all supposedly the realm of a few, select men—some men openly loved other men, and some women other women. These 'behavioral hermaphrodites' were sometimes considered enough of a nuisance to warrant direct responses by their detractors in the medical and scientific literature. To make matters for scientific sex-difference theorists even more awkward, at the same time, adventurers reporting back from other lands told stories of cultures in which the behavior of men and women seemed not to follow what most Europeans took to be the natural sexual order.[33] Sex seemed to be changeable with conditions of climate or race—now challenged from within and without.

Still, in contrast (and no doubt partly in response) to these challenges, many medical and scientific men in France and Britain vigorously tried to argue and to evidence that the existing social sex boundaries in their cultures reflected and were therefore necessitated by 'natural' sex boundaries—that (most) women did what (most) women did because they were female, that (most) men were manly because of their malehood, that to do or be otherwise would not only be unusual, and perhaps immoral, but also unnatural. Never could the two sexes meet in ideas, talents, or roles, for they had parted so long in the past. Men like Patrick Geddes and J. Arthur Thomson, sold on the idea of human evolution, sometimes called on the entire history of life as witness to the necessity of the social sex order. In their popular 1889 tract, *The Evolution of Sex* Geddes and Thomson declared:

We have seen that a deep difference in constitution expresses itself in the distinctions between male and female, whether these be physical or mental. The differences may be exaggerated or lessened, but to obliterate them it would be necessary to have all the evolution [of life] over again on a new basis. What was decided among the prehistoric Protozoa cannot be annulled by Act of Parliament . . . We must insist upon the biological considerations underlying the relation of the sexes.[34]

The social sex order had to stand as it did because it was the natural order of things.

Yet in spite of such claims about sex order and profound sex distinctions, as the nineteenth century aged, more and more physicians and surgeons reported in the medical literature cases of anatomically unusual people they called hermaphrodites. Indeed, there seemed to be shocking numbers of anatomical hermaphrodites turning up in the general populace. One French medical man lamented that hermaphrodites seemed to 'literally run about on the street'.[35] Hermaphroditism was a sticky problem, one whose possible solution held important ramifications well beyond the life of the individual doubtful patient. So, while deeply fascinated by cases of human hermaphroditism, many medical and scientific men simultaneously expressed disgust at the very idea and resentment at the confusion hermaphrodites caused. In a 'Note on Hermaphrodites' in the 1896 volume of his *Archives of Surgery*, Jonathan Hutchinson lamented, 'So much of what is repulsive attaches to our ideas of the conditions of an hermaphrodite that we experience a reluctance even to use the word.'[36] What was one to do with a person who seemed to be neither or both male and female? What was one to do with the Woman Question, which concerned the proper roles and rights of women, if one could not exactly say what a woman was? How was one to distinguish 'normal' (heterosexual) from 'perverse' or 'inverted' (homosexual) relations if one could not clearly divide all parties into males and females?

Given what Cynthia Eagle Russett has termed 'an intense somatic bias' among nineteenth-century scientists—that is, given scientists' deep faith in materialism as the key to truth—the unusual bodies of hermaphrodites presented extremely powerful challenges to biomedical claims about the natural, inviolable distinctions between men and women.[37] Hermaphrodites did not consciously seek to crash sexual borders, but any body which does not clearly fit into the stereotypical categories of male or female necessarily raises questions about the integrity, nature, and limits of those categories. In *Gender Trouble*, a 1990 critical analysis of the relationships of sex, gender, desire, and identity, Judith Butler notes:

The presuppositions that we make about sexed bodies, about them being one or the other, about the meanings that are said to inhere in them or to follow from being sexed in such a way are suddenly and significantly upset by those examples that fail to comply with the categories that naturalize and stabilize that field of bodies for us within the terms of cultural conventions.[38]

As Butler suggests, anatomical hermaphrodites—however unintentionally—necessarily challenged what it meant to be female or male, woman or man. In doing so they forced observers to admit presuppositions and to make decisions about the categories of female and male—and they forced medical and scientific men to tighten up the borders.

This study takes as its basic starting point the year 1868, that is, the year of Barbin's suicide, and that date is chosen for two reasons. First, the publicity surrounding Barbin's memoirs, life, and death instilled in medical practitioners an appreciation of just how troublesome and urgent—and potentially common—the problem of hermaphroditism was. Medical men understood as never before the dangers of hermaphroditism and the importance of early and accurate diagnoses of an individual's 'true' sex. In many ways, Barbin shaped the biomedical treatment of human hermaphroditism for years to come. For example, as in many other instances, the medical man reporting the case of Sophie V. specifically referred to Barbin's suicide as reason to worry about forcing an unfamiliar new sex on a hermaphroditic patient. Meanwhile, Barbin's case, with its scandalous sexual exploits, also pushed many medical men to try to seek out and prevent or end the 'unnatural' sexual associations (like Sophie V.'s marriage) caused by hermaphroditism.

Second, and more significant, the year of Barbin's death makes a logical starting point for this history because it was roughly at that time that consensus began to solidify that the single reliable marker of 'true sex' in doubtful cases was the gonad, that is, the ovary or testicle.[39] This consensus managed to hold as the rule for assignment of sex until about 1915, that is, the time through which I follow this history. During this period, 1870–1915, what I call the Age of Gonads, scientific and medical men, faced with and frustrated by case after case of 'doubtful sex', came to an agreement that every body's 'true' sex was marked by one thing and one thing only: the anatomical nature of the gonadal tissue as either ovarian or testicular. Not coincidentally, such a definition virtually eliminated 'true' hermaphroditism in theory and practice, even if—probably because—people with challenging bodies kept popping up. Without the material and consequent social problems presented by the hermaphroditic body, this particular construction of 'true sex'—namely, that sex is ultimately determined by the gonad—might never have occurred.

Imagine saying that you are female only because you have ovaries, or male only because you have testicles—no matter what the rest of your body or experiences were like. Why pick such a narrow definition of 'true' sex? As we shall see, there were many reasons nineteenth-century medical and scientific men were so focused on the gonads: they were partly influenced by trends in contemporary scientific pathology, especially the sort of pathology that used histology as the key to understanding disease; they were partly influenced by trends in contemporary embryology which recognized the differentiation into ovarian or testicular as an embryonically early (and therefore, by the rules of the day, significant) phenomenon; they were partly influenced by trends in evolutionary theories which posited the key difference between males and females to be their reproductive capabilities and roles.[40] Nonetheless, the gonadal definition of sex was not a simple manifestation of trends in related sciences. Without a doubt, scientists and doctors who rallied around the gonadal definition of 'true' sex and 'true' hermaphroditism did so in part because it meant that nearly every body could officially be limited to one and only one sex. This was critical at a time when the social body itself seemed to be getting too blurry in terms of sexual distinctions. British and French medical practitioners agreed that, as two French physicians writing on the enigmatic problem noted in 1911, 'the possession of a [single] sex is a necessity of our social order, for hermaphrodites as well as for normal subjects.'[41] The gonadal definition of true sex seemed to preserve, in theory and in practice, a strict separation between males and females, a strict allotment of only one sex to each body—a way to enforce the one-body-one-sex rule.

THE ORDERING OF HERMAPHRODITES

I use the general term 'hermaphrodite' for all so-identified subjects of anatomically double, doubtful, and/or mistaken sex (that is, supposedly mislabeled sex). But I do this not because I think the category of 'hermaphrodite' is self-evident or because I think it forms a clearly bounded, ontological category that cannot be disputed. On the contrary, the histories of Barnes's 'living specimen', Barbin, and others demonstrate how problematic, diverse, and changeable the categories of hermaphrodite, female, and male are. I use 'hermaphrodite' to refer to my historical subjects first because it simplifies my narrative. A single so-called hermaphrodite could, in the scientific and medical literature of the late nineteenth and early twentieth centuries, go from being officially labeled male, to female, to hermaphrodite, to a subject of mistaken sex, to a subject of doubtful sex. So, while I try to note the changes and contradictions in the identities of any given subject, when speaking generally it is simplest to use the word 'hermaphrodite' for anyone whose 'true' sex fell into question among medical and scientific men. I also use 'hermaphrodite' because it was in fact the blanket term commonly used before and during the period of my study for persons suspected of being subjects of double, doubtful, or mistaken sex. The label 'hermaphrodite' was sometimes also given to people we would now call homosexuals, transvestites, feminists, and so on, but it was, by the nineteenth century, most commonly reserved for the anatomically 'ambiguous' bodies on which I focus. Editors and authors of medical and scientific literature of the nineteenth and early twentieth centuries nearly universally reserved the heading 'Hermaphrodite' for reports of cases of doubtful or mistaken sex.

In fact, in spite of repeated attempts on the part of many late nineteenth-century scientific and medical men to do away with the general, popular, and sometimes vague term 'hermaphrodite' in favor of a more rigorous terminology, only since the middle of the twentieth century has a different general term, 'intersexed', been regularly substituted for 'hermaphroditic' in medical literature. Although these two terms, 'intersexual' and 'hermaphroditic', are used to refer to the same sorts of anatomical conditions, they do signal different ways of thinking about the sexually 'ambiguous' body. 'Intersexed' literally means that an individual is *between* the sexes—that s/he slips between and blends maleness and femaleness. By contrast the term 'hermaphroditic' implies that a person has *both* male and female attributes, that s/he is not a third sex or a blended sex, but instead that s/he is a sort of double sex, that is, in possession of a body which juxtaposes essentially 'male' and essentially 'female' parts.

Richard Goldschmidt was apparently the first biomedical researcher to use the term 'intersexuality' to refer to a wide range of sexual ambiguities including what had previously been known as hermaphroditism. Goldschmidt discussed a variety of 'intersexuals' including hermaphrodites in a 1917 article on 'Intersexuality and the Endocrine Aspect of Sex,'[42] and the term 'intersexual' slowly gained popularity among medical professionals. Before Goldschmidt, some authors had used the term 'intersexuality' to refer to what we would call homosexuality and bisexuality, and even Goldschmidt himself suggested that human homosexuality might be thought of as one form of intersexuality.[43] (As we will see, the hermaphrodite and the homosexual share a surprising amount of medical history.) Today, however, 'intersexual' is used specifically among biomedical professionals to refer to anatomical sex variations considered ambiguous or misleading. Present-day historians sometimes use one term and sometimes the other, as do many of today's activist intersexuals.

Hermaphrodites were not new to the European scene in the nineteenth century. Indeed it was Ovid's myth of the Ur-Hermaphrodite that fixed in the Western imagination the long-standing image of the hermaphrodite as a tragicomic, double-sexed creature. According to the story Ovid told in Book IV of his *Metamorphoses*, the gods Hermes and Aphrodite, themselves 'the embodiments of ideal manhood and womanhood', together had a son named Hermaphroditos after his parents.[44] Hermaphroditos was, like his father, a beautifully formed male, and he might have remained so had not the nymph Salmacis one day chanced upon Hermaphroditos while he bathed in her fountain.

Having seen his potent form, his remarkable beauty, so badly did Salmacis wish to be united with Hermaphroditos that she begged the gods to join them together: 'Oh gods may you so order it, and let / no day take him away from me or me from him.'[45] The gods, with their strange sense of humor, took her literally. Hermaphroditos's and Salmacis's bodies were united permanently to become one—one Hermaphrodite—a being made both fully man and woman, 'mingled and joined . . . / just as, if someone puts branches through a tree's bark, / he sees them joined as they grow and [mature] together.'[46]

Both visions of the hermaphrodite—as an in-between sex and as a double sex—can actually be found as far back as the days of ancient Greece. Joan Cadden has traced the conception of hermaphrodites as in-between sexes back to the Hippocratic writers. The Hippocratic paradigm assumed that sex existed along a sort of continuum from the extreme male to the extreme female and that the hermaphrodite therefore was s/he who lay in the middle. By contrast, later thinkers formed from the writings of Aristotle a different, equally persistent tradition that imagined hermaphrodites to be doubly sexed beings. That tradition specifically held that hermaphrodites had extra sex (genital) parts added on to their single 'true' sexes via an excessive generative contribution of matter on the part of the mother. Hermaphroditism in this Aristotelian line of thinking was therefore like extra toes or nipples, in that it represented an overabundance of generative material.[47] Cadden shows that followers of both traditions can be found for centuries beyond ancient Greek times.

Indeed, stories of hermaphrodites are sprinkled throughout virtually every era of recorded history. Sometimes, as in Ovid's case, the hermaphrodite was employed in literature as an allegorical character. In the twelfth century, for instance, John of Salisbury used the image of the hermaphrodite to talk about the strange, double-natured position of a court philosopher, a position which was, according to John, plagued by inherently contradictory (hence 'hermaphroditic') loyalties.[48] But we also find references to nonallegorical—that is, living—hermaphrodites in practically every historical era. Some hermaphrodites in Roman and medieval times may have been put to death, as apparently were all kinds of 'monstrous' beings. The reasoning behind these murders held that the 'monster' was surely a supernatural portent, a messenger of evil, a demonstration (note the shared root with 'monster') of bad happenings, and that as such it deserved and even required prompt annihilation.[49] Of course, we must remember that later claims regarding the barbaric treatment of 'monsters' by earlier peoples often came during later people's claims to great progress and moral superiority, so they may well have been overestimated. Some hermaphrodites may have been tolerated, but probably only as long as they chose and stuck to a single male or female sexual identity.[50] Hermaphroditism, it seems, has always been a relatively risky identity.

Lorraine Daston and Katharine Park noted in a 1995 study that, in early modern France, scientific thinkers developed a pervasive and singular fascination with hermaphrodites.[51] Men of the sixteenth and seventeenth centuries recorded a dazzling array of opinions and inter-pretations of hermaphroditism, and Daston and Park suggest that this relatively abundant interest in hermaphrodites arose out of a 'sexual anxiety' fed by increasing concerns over transvestitism, sodomy (espe-cially between women), and the possible transgression of other social sex roles. (Interest in hermaphroditism seems almost always to wax with public challenges to sex roles.) In early modern France, herm-aphrodites counted chiefly as preternatural beings, that is, 'rare and unusual, outside the ordinary course of nature, but in principle fully explicable by natural causes'. No longer understood primarily as supernatural portents, hermaphrodites and other 'fantastic' creatures became 'marvels, not miracles'.[52] The fate of hermaphrodites in the cultural scene depended on the nature of others' understanding of the hermaphrodites' origins and actions. While the 'preternatural' ana-tomical hermaphrodites could be explained, understood, and given an ordered place in nature and perhaps as well in the social setting, 'arti-ficial' hermaphrodites (those created through human deceit or inven-tion) and 'unnatural' hermaphrodites (those whose sexual behavior seemed to go against the 'nature' of their 'true' sexes) were often persecuted.

By the early nineteenth century, an entire field dedicated to the scientific study of 'monstrosities' had emerged. Isidore Geoffroy Saint-Hilaire (1805–1861), a French anatomist, coined the name teratology for the new field. Geoffroy and his cohorts laid out an ambitious goal for the discipline, namely, the exploration of all known and theoret-ical anatomical 'anomalies' and the explication of those anomalies within a single 'anatomical philosophy' which would at once des-cribe, explain, and predict all normal and abnormal forms. 'Nature is one whole,' Geoffroy confidently declared, and all 'monsters', including the hermaphrodite, were therefore part of nature. Thus the hermaphrodite—like many other previously supernatural or preter-natural phenomena—came to be fully 'naturalized', so that by the early nineteenth century, hermaphroditism was understood by scientists and medical men as a phenomenon to be fully explained by the natural sciences, one existing within the realm of natural law.

More specifically, the development of the hermaphrodite and other 'anomalous' beings was now to be explained in terms of variations of normal development.[53] Doctors and others settled on the conviction that, in the case of hermaphroditism, the ultimate 'explanation is to be found in the embryology of the genitals'; the hermaphrodite was just a would-be male or female gone wrong in the womb.[54] It seemed one had only to understand normal embryological sex development to be able to come up with explanations for abnormal sex development. Nineteenth-century anatomists knew that future-female and future-male children began as embryos with the same basic parts, but that those parts developed differently in utero and after. By the 1870s all authorities in embryology agreed that, in the female fetus, the two proto-gonads took one path to become ovaries, and in the male fetus, they followed another and became testes.[55] In addition, they knew that ultimately-male and ultimately-female fetuses both began with Müllerian and Wolffian systems of proto-organs internally. In the female, however, the Wolffian system atrophied and the Müllerian system evolved to form 'female' internal organs, including the fallopian tubes, uterus, and vagina. In the male, the Müllerian system atrophied and the Wolffian system evolved to form 'male' internal organs, including the deferent canals and the prostate.[56] Knowledge of such common developmental pathways made it possible to explain, for instance, how a 'true male' could seem to have developed an otherwise inexplicable vagina or uterus.

With regard to the external genitalia, medical and scientific men of the nineteenth century saw the human male genitalia as a more elaborately developed version of the human female genitalia. The female seemed to represent simply a lower degree of external development in this sense. This assumption—that females were actually under-developed males—was based in part on a long-running anatomical and philosophical tradition that stretched at least back to the work of Aristotle. Aristotle had described women as monstrous, under-developed men, 'undercooked' for lack of an essential heat. The male heat, by contrast, cooked the male, pushing out the phallus— understood to be the vagina in the woman—and the gonads.[57] But in the nineteenth century, this theory of women as less-developed men was also based on the fact that the external conformation of young embryos seemed to observers to be nearer the ultimate feminine than the ultimate masculine state, and on the observation that masculine genitalia apparently continued to develop after the point at which the development of feminine genitalia stopped.[58] So, for instance, embryologists and anatomists noted that the clitoris in the female is relatively small, but its homologue (developmental counterpart) in the male, the penis, grows comparatively large.[59] Similarly, it seemed to nineteenth-century observers that proto-labia formed in both the male and the female fetus, but in the male, development continued such that the labia joined to form a scrotum.[60] It appeared, therefore, that if a female developed too much, she would look masculine or hermaphroditic, and that if a male developed too little, he would look feminine or hermaphroditic.[61]

Ultimately the work of Geoffroy Saint-Hilaire and other modernist teratologists resulted in what several historians of teratology have labeled the 'domestication of the monster'. This was the process by which the extraordinary body ceased to be truly extraordinary—that is, outside the realm of the natural—the process by which the teratological was stripped of its wonder and made simply pathological. 'Domesticated within the laboratory and the textbook,' writes Rosemarie Garland Thomson, 'what was once the prodigious monster, the fanciful freak, the strange and subtle curiosity of nature, . . . [became] the abnormal, the intolerable.'[62] Elizabeth Grosz joins with Thomson in remarking on the paradoxical fate of the modern hermaphrodite and other contemporary 'monster/freaks', namely, to finally be, in 'medical discourse and practice' made the subject of 'simultaneous normalization and pathologization'. Just when (and because) the hermaphrodite was normalized via the alliance of its development with the development of 'normal' creatures, simultaneously was it declared unacceptably pathological because of its variation from those very 'norms'.[63] So, while Geoffroy welcomed human 'monstrosities' into the natural realm of the living, at the same time he hoped the science of teratology would have a very useful application: the prevention of future human 'monstrosities'.[64]

CURRENT-DAY EXPLANATIONS AND TYPING

Embryological ideas about the parallels in male and female development are still used by biomedical researchers to explain human hermaphroditism, although today experts have added elaborate understandings of the role of genetics and endocrinology to their explanations. Philosophers of hermaphroditism like to recall that Ovid's Hermaphrodite was a sort of 'Siamese' twin, a double body at once completely male and completely female with all the attendant parts and traits. But gods and nymphs aside, the human body labeled 'hermaphroditic' presents not all the so-called female and male parts, but rather an unusual mix or blend of parts (depending how you think about and see things) as in the cases of Sophie V., Barbin, and Barnes's 'living specimen'. The organ located where the penis or clitoris is usually found might look like the 'wrong' organ (that is, the organ common to the sex 'opposite' to which the person is thought to belong), or like something in between the two, or not especially like either. The genitalia may appear to be of the female type, yet the labia might contain testicles. Or the genitalia may look mostly male, but include a seeming vagina. The appearance of the genitals might also change over the course of the lifetime. Change in sexual anatomy—for instance, relative growth of some parts, the growth or loss of hair on other parts—is the rule for most of us, but in some hermaphrodites, genitalia have also been observed to undergo unusual transformations from a more female-like to a more male-like conformation, and vice versa.

Today in the medical literature human hermaphroditism is divided into three basic theoretical types: male pseudohermaphroditism, female pseudohermaphroditism, and true hermaphroditism. Typing in this theoretical taxonomy dates back to the Age of Gonads and so is based primarily on the anatomical structure of the gonadal tissue as was done in the late nineteenth century. Thus if an 'ambiguous' individual has testicular tissue only, s/he is technically categorized as a male pseudohermaphrodite; if ovarian tissue only, s/he is categorized as a female pseudohermaphrodite; if s/he has one or more ovotestis, that is, an organ with both ovarian and testicular attributes, s/he is categorized as a true hermaphrodite. The tissue in question need not be functional in any sense. It is only the anatomy of the gonads that is used for this theoretical typing.

Given that medical theory still categorizes hermaphroditisms as pseudomale, pseudofemale, or true on the basis of the anatomy of the gonadal tissue, one might think that we are still in an Age of Gonads. But after about 1915—what I mark as the real end of the Age of Gonads—medical practitioners came to a general agreement that the gonads should not in practice be very important to sex assignment. Today in the United States, for example, sex assignment is based largely on a concern for penis functionality in males and for reproductive functionality in females. Gonads no longer have the kind of tremendous power they had from the time of Barbin's death to the time of World War I.

Today, 'true hermaphroditism'—cases in which an individual has one or more ovotestes—is considered 'extremely rare', but it is hard to know how many cases go undiagnosed.[65] Biomedical experts are also not sure how or why a proto-gonad would develop both ovarian and testicular attributes. The vast majority of 'true hermaphrodites' seem to have an XX chromosomal basis, although a small percentage exhibit XY chromosomes and a very few have some cells showing XX and others showing XY, a phenomenon known as 'chimerism' which probably occurs when two early embryos (one XX and one XY) fuse together to form one individual. The genitalia of true hermaphrodites sometimes look 'typically' male, sometimes 'typically' female, and sometimes something else.[66] This today is the least well understood type of hermaphroditism.

'Female pseudohermaphrodites' are by contrast much more common, perhaps 'accounting for about half of all cases of ambiguous external genitalia'.[67] In cases of female pseudohermaphroditism, the individuals in question by definition have ovaries, and they also exhibit an XX chromosomal basis. The external genitalia look 'masculinized' because the children are exposed in the womb to relatively high levels of androgens, a type of hormone important in sex-organ development. Sometimes the supposed-clitoris looks and acts more like a penis, sometimes the labia join to look like a scrotum, and so on. Meanwhile, upon examination the internal organs appear 'typically' feminine because the developmental anomaly does not profoundly affect internal development.

According to today's expert understanding of female pseudohermaphroditism, the 'masculinizing' phenomena of female pseudohermaphroditism can occur for at least three reasons. First, a tumor on the pregnant mother's suprarenal gland can produce excessive amounts of androgens that effect a more 'male' type development of the 'female' child's genitalia. This seems to be a rare cause. Second, administration of androgenic hormones to a woman pregnant with a 'female' fetus can cause 'masculinization'. Pregnant women have sometimes been given these hormones by their doctors to prevent miscarriage, and there may also be environmental toxins which can cause similar, externally induced 'masculinization'. Third, 'masculinization' in a 'female' fetus can be caused by a condition known as congenital adrenal hyperplasia, or CAH. This seems to be the most common cause of female pseudohermaphroditism today. In CAH, the fetus again has an XX chromosomal basis and ovaries, as in all cases of female pseudohermaphroditism, but the production of relatively large amounts of androgens by the adrenal glands of the fetus results in a rather 'male'-like development of the external genitalia. Female pseudohermaphroditism can range from very dramatic cases in which the child looks very much like the typical male (some children go years without being diagnosed) to relatively less dramatic cases in which the clitoris is a bit enlarged but most everyone assumes the child must be a girl.[68]

There are at least two major forms of male pseudohermaphroditism. In both of them, by definition, the individuals in question have testes and an XY chromosomal pattern. The first type is known today as testicular feminization syndrome or androgen insensitivity syndrome (AIS). In AIS, the testes produce the usual androgens effective in male development. The body lacks a key androgen receptor, however, and so the body cannot 'hear' or 'read' the androgen ('masculinizing') messages. Therefore, rather than developing along the typical masculine developmental pathway, the tissues develop along more 'feminine' lines. The genitals look quite feminine, with labia, a clitoris, and a vagina (usually relatively short). At puberty the body develops still more along the more 'typical' feminine pathway because, although the testes produce more androgens, the body still cannot 'hear' them.[69] Breasts fill out, the hips grow rounder. Generally AIS individuals do not develop very much noticeable body hair, and they grow tall with long arms and legs. Indeed, with these features—tall, smooth-skinned bodies with rounded hips and breasts and long limbs—they seem to fit the dominant feminine ideal in the United States today better than most medically 'true' females. (There is a persistent rumor that many 'female' high-fashion models are testically feminized males, but I do not know if this is true, and I have been unable to locate the ultimate source of the rumor.) Individuals with AIS often do not know about their condition until they go to a gynecologist when they fail to menstruate at puberty, and, even at that time, many are not told their true diagnosis, but are simply told they can only become mothers via adoption.[70] AIS patients are frequently told by their doctors that their ovaries have been removed, when in fact they had no ovaries and it is their testes that have been removed. (Their testes are removed because they have a relatively high chance of becoming cancerous, and, since the testes are often undescended, such cancers can easily go undetected until they have reached the point of being terminal).

The other major type of male pseudohermaphroditism is known as 5-alpha-reductase (5-AR) deficiency. This is one of the most striking forms of hermaphroditism because it results in an apparent female-to-male transformation at puberty. During fetal development the 'male' child's testes produce testosterone. But in order for the developmental 'message' of the testosterone to be 'heard' in the child, the tissues must have the enzyme 5-alpha-reductase, which converts the testosterone into 'readable' dihydrotestosterone. If it is lacking, as it is in cases of 5-AR deficiency, the fetus will develop female-like genitalia. Therefore 5-AR deficient individuals are born with feminine-looking genitalia, including generally a short vagina and apparent labia and clitoris. At puberty, however, the testes of these individuals produce more testosterone, and for the pubertal changes to occur the body doesn't need the converting work of the 5-AR enzyme. So now the testosterone messages *are* read, and 'masculinizing' puberty occurs. The body grows taller, stronger, more muscular, usually with the addition of significant body and facial hair but with no breast development, and the voice drops. Often at this time the testes descend into the assumed-labia, and the penis/clitoris grows to look and act more like a penis.[71]

These are the types of hermaphroditism generally discussed under the heading of 'intersexuality' or '(pseudo)hermaphroditism' in medical textbooks today. However, 'ambiguous' genitalia can result from other conditions besides those mentioned above. In Klinefelter's syndrome, for instance, the presence of several X chromosomes and one Y chromosome sometimes results in sexual 'ambiguity'.[72] We should also note that not all the causes of intersexuality are known or understood and that, more over, many babies are born with relatively unusually formed genitalia but are not categorized by their physicians as 'ambiguous' or 'intersexed'. Unquestioned females, for instance, are sometimes born with relatively large clitorises, and a large number of male babies—perhaps one in every one or two hundred—are born with hypospadic penises. Hypospadias is a condition in which, in a person with male-type genitalia, the urethra exits some place other than the tip of the glans. The penis may also be relatively small, and it may be bound-down, pointing backwards, to create a condition known as chordee.[73] Today operations are often performed very early on hypospadic penises, in part because it is considered very important that a boy be able to urinate standing up, and relatively large clitorises are often surgically reduced because it is generally considered inappropriate for girls to have large clitorises. While these cases may not officially be labeled cases of intersexuality, the genitalia are treated as cosmetically troublesome.

THE QUESTION OF FREQUENCY

Many people are understandably curious to know how common the phenomenon of 'intersexuality' or 'ambiguous genitalia' is. But for several reasons it is almost impossible to provide with any confidence an overall statistic for the frequency of sexually ambiguous births.

First, even the experts today cannot agree on how common various intersex conditions are. For instance, consider statistics given with regard to congenital adrenal hyperplasia. Three recently published, well-respected medical texts give the frequency of CAH alternately as 1 in 60,000 births, 1 in 20,000, and greater than 1 in 12,500.[74] Which statistic to trust? Similar variations are present in estimates with regard to 'intersexual' conditions besides CAH. The fact that, even in the most recent medical literature, there exists significant disagreement about the incidence of various intersex conditions makes it difficult to estimate the overall frequency.

Second, and relatedly, part of the problem in determining the frequency of intersexuality or hermaphroditism is a problem inherent in all statistics; there is always a question of whether or not a given sample from which one estimates the frequency of an event represents the larger population in which one is ultimately interested. For instance, when we try to estimate 'frequency of intersexual births,' are we seeking to estimate it for the entire human population? Since when? Ending when? Only for the United States today? Or for Victorian Britain? Obviously the numbers would differ depending on the parameters, again for several reasons.

Some forms of intersexuality, such as 5-AR deficiency, seem to have a strong genetic component. It seems that 5-AR deficiency can result from a 'spontaneous' mutation, but that it can also result from an inheritable genetic sequence. In fact, we know that several populations, through isolation and intermarriage, have come to manifest a much greater frequency of 5-AR deficiency than other populations. One such population exists in a rural area of the Dominican Republic and another among the Sambia people of Papua New Guinea.[75] Should we include or exclude such clusters of intersexuality in our 'overall' estimate of frequency of intersexuality—or even in our estimates of the frequency of 5-AR deficiency alone? Neither including nor excluding cluster cases in our count seems to be satisfactory.

Similarly, we know that certain kinds of 'environmental' factors can influence sexual development. For instance, in the 1960s, many pregnant women in the United States were given hormone treatments to prevent miscarriage, and these treatments resulted in a high number of CAH-like conditions. In other words, as in the case of 5-AR deficiency mentioned above, we can find 'clustering' of CAH-like cases in the late twentieth century United States, this time owing to local environmental factors. So even if we could come up with a reliable frequency count for CAH-like conditions in the United States in the 1960s–1990s, that frequency may not be anything like the frequency of CAH-like conditions in other periods and places. The frequency of various conditions is likely to vary quite significantly according to time and space because of local variations in the factors that influence sexual development.

Third, in trying to estimate the overall frequency of sexual ambiguity or sexual anomalies, we encounter the enormous problem of what exactly to count toward such a statistic. Take, for instance, the problem of hypospadias. Like CAH, hypospadias may but does not necessarily result in confusion and disagreement about to which sex an affected person should belong. Similarly, as mentioned above, Klinefelter's syndrome may result in so-called sexual ambiguity, but does not necessarily. Should the overall frequency of hypospadias, CAH, and Klinefelter's syndrome be counted into a general statistic of the frequency of sexual anatomical ambiguity? Or should we try to count only those cases which result in 'confusing' or 'mistaken' sex? Neither approach seems reasonable. The first would seem to result in too high a count for our purposes, because it would include cases in which the sex was never considered 'ambiguous'. The second would factor in unintended variables, like the frequency of accurate diagnosis and reporting of hypospadias, CAH, and Klinefelter's.

One might argue that we should simply not lump together all of these various conditions—that we should consider each condition or even each case separately. Yet the phenomenon variously called 'doubtful sex', 'ambiguous sex', 'hermaphroditism' and so on has been and continues to be treated as a singular sort of problem in medicine and culture, and so should be treated historically as such. Even if we can't find a for-all-time 'scientific' way to define intersexuality, that does not mean it does not exist as a very important cultural phenomenon.

In conclusion, it is not possible to provide with any great certainty a statistic of the frequency of births in which the child's sex falls into question. This problem highlights an important point of this book and especially this chapter, namely, that such a statistic is always necessarily culture specific. It varies with gene-pool isolation and environmental influences. It also varies according to what, in a given culture, counts as acceptable variations of malehood or femalehood as opposed to forms considered sexually ambiguous. And it varies according to what opportunities there are in a given culture for doubts to surface and be articulated on record. A culture in which genitalia are covered up, rarely examined, and not discussed would likely be a culture in which there were fewer cases of so-labeled ambiguous sex. A culture in which big clitorises or small penises are considered unacceptable would likely be one in which there were more cases of so-labeled ambiguous sex. Frequency is specific to particular cultural spaces.

It remains true, nevertheless, that the frequency of births in which the child exhibits a condition which today could count as 'intersexual' or 'sexually ambiguous' is significantly higher than most people outside the medical field (and many inside) assume it is. Curiously, few medical texts offer any estimate of the frequency of 'intersexuality' even though many treat its various forms under that single heading. One 1993 gynecology text does offer the claim that 'in approximately 1 in 500 births, the sex is doubtful because of the external genitalia', but unfortunately that author does not say how she reached that conclusion.[76] When I am pressed for a rough statistic, I suggest that today, in the United States, probably about one to three in every two thousand people are born with an anatomical conformation not common to the so-called typical male or female such that their unusual anatomies can result in confusion and disagreement about whether they should be considered female or male or something else. Anne Fausto-Sterling, through recent research, estimates the incidence of intersexed births to be in the range of 1 percent, although Fausto-Sterling warns that that figure 'should be taken as an order of magnitude estimate rather than a precise count'.[77] (In other words, the number might be closer to one in a thousand.) Fausto-Sterling's estimate comes from a careful study of the current medical literature, and after a review of her preliminary findings, I find her compilation of the data convincing, with all the provisos given above taken into account.[78]

It seems quite evident, then, that in the United States today the anatomical condition of intersexuality is about as common as the

143

relatively well known conditions of cystic fibrosis (roughly one in two thousand 'Caucasian' births is of a child with cystic fibrosis) and Down's syndrome (roughly one in eight hundred live births).[79] Given roughly four million births per year in the United States, that means there are now several thousand medically defined 'intersexuals' born in the United States each year, and there are now living in the United States tens of thousands of people who were born to be labeled 'intersexual,' although many of them probably don't know it. (This is largely because doctors avoid using the terms hermaphroditism or intersexuality around intersexuals and their parents.) Lay people do not hear about intersexuality much in the United States today because such conditions are often considered traumatic, scandalous, and shameful, so few people talk about such births when they happen. Our system of birth registry also covers up the phenomenon: every child, no matter how 'ambiguous', gets recorded as male or female very quickly. Nevertheless, intersexuality is becoming more well known in the United States as intersexuals make themselves known and as the phenomenon begins to show up in various pop-culture venues such as the television drama 'Chicago Hope,' in which a baby was born with AIS, and the feature film 'Flirting with Disaster,' in which one character (notably the only self-identified bisexual) mentions that he was born with a hypospadic penis.

BODY PAINTING

The latest opinions of scientific researchers are important, of course, to our understanding of intersexuality. But at the same time, human sex has never been and will probably never be a merely academic matter. Ideas about sex, however theoretical, carry with them tremendous ramifications. Consider our subject period: in the immediate sense, a medical practitioner's opinion with regard to the 'true sex' of a given patient could shape, in the late nineteenth and early twentieth centuries, at the very least his view of the patient and at most the patient's whole life, and perhaps even death, as in the suicides of Barbin and other hermaphrodites. Imagine what depended on the determination of one's true sex: name, dress, education, occupation, possibilities for enfranchisement and conscription, and status as wife or husband, maiden or bachelor, widow or widower, as sexual 'normal' or 'invert' (what we would call heterosexual or homosexual). Every body that slid through the divisions weakened those boundaries. If strict sex borders were to be maintained in the culture, sex had to be maintained and controlled in the surgical clinics and in the anatomy museums and in every body.

The demands put on the hermaphroditic body therefore are many, as many agendas—scientific, medical, personal, national, professional, moral, and political—meet. This is perhaps inevitable, for in any human culture, a body is never a body unto itself, and bodies that openly challenge significant boundaries are particularly prone to being caught in struggles over those boundaries. Rosemarie Garland Thomson has argued, 'By its very presence, the exceptional body seems to compel explanation, inspire representation, and incite regulation. The unexpected body fires rich, if anxious, narratives and practices that probe the contours and boundaries of what we take to be human.'[80] Mary Douglas suggested in her study of bodily pollution and taboo that these bodily divisions can in fact be made on virtually any body, extraordinary or not, for 'the body is a model which can stand for any bounded system. Its boundaries can represent any boundaries which are threatened or precarious. The body is a complex structure. The functions of its different parts and their relation afford a source of symbols for other complex structures.'[81]

So it was that, by the late nineteenth century in France and Britain, the hermaphroditic body became a symbol for many conflicts, as lines were repeatedly painted on it dividing the masculine from the feminine, the normal from the abnormal, the real from the apparent, those with authority from those without. More than any other conflict (and there were many involved), the battle over the 'nature' of males and females affected the biomedical treatment of hermaphrodites. Hermaphrodites were repeatedly construed by medical and scientific men so as to reinforce primarily what they threatened most: the idea that there was a single, knowable, male or female 'true' sex in every human body.

This idea—that the hermaphrodite's body gets caught up in cultural wars and has painted on it the fronts of those wars—is not novel to our time. In 1899 Xavier Delore published in the *Écho médicale de Lyon* a lengthy study of the cultural-historical, embryological, and evolutionary stages of hermaphroditism, and in his research, the Lyonnaise medical doctor was struck as we are by how, historically, the hermaphrodite had 'so strongly impressed public opinion that its history reflect[ed] the prominent traits of the various societies' in which it had been found. But Delore was convinced that by 1899 the hermaphrodite had finally ceased to be a mythological being, the kind constantly used for the justification of 'irrational' and 'immoral' beliefs and acts. The hermaphrodite, Delore insisted:

now no longer has adoring admirers who render it a cult; it is no longer the object of those gracious legends of Greek mythology; it is no longer sung of by the poets of Rome; philosophers no longer utilize its defective structure in order to palliate, by insidious sophisms, the licentiousness of their morals; . . . theologians will no longer use it in order to elucidate ridiculous heresies; but it also will no longer engender that sentiment of horror which unenlightened civilizations drew from it, [only] to exterminate it as a monster!

Now, Delore claimed, the hermaphrodite had become simply, perhaps pathetically, merely 'a scientific matter and a degraded organism'. As such it had become 'the domain of the medical doctors. It [was] incumbent upon them to reconcile its interests with those of society, in the midst of which they [would] mark for it its true place.'[82]

Ironically, even while Delore recognized the mapping of many belief systems and cultural agendas onto the body of the hermaphrodite throughout history, he assumed that this practice had essentially ended when the hermaphrodite became an object of modern medical and scientific study. Yet Delore himself readily admitted the exigencies of his own society and profession in matters of hermaphroditism, and the social needs to which Delore alluded necessarily shaped contemporary representations and treatment of hermaphrodites.

Of course the hermaphrodite did not cease to be a site of cultural conflict and cultural demands—of cultural body painting—when it became an object to be handled by science and medicine. It was only the type and status and media of the painters that changed.[83]

Notes

1. See, for example, Ornella Moscucci, *The Science of Woman: Gynaecology and Gender in England, 1800–1929* (Cambridge: Cambridge University Press, 1990); Elizabeth Fee, 'Nineteenth-Century Craniology: The Study of the Female Skull', *Bulletin of the History of Medicine*, 53 (1979), 415–33; Thomas Laqueur; *Making Sex: Body and Gender from the Greeks to Freud* (Cambridge Mass.: Harvard University Press, 1990), chap. 5; Robert A. Nye, *Masculinity and Male Codes of Honor in Modern France* (New York: Oxford University Press, 1993); Cynthia Eagle Russett, *Sexual Science: The Victorian Construction of Womanhood* (Cambridge Mass.: Harvard University Press, 1989).

2. See E. Goujon, 'Étude d'un cas d'hermaphrodisme bisexuel imparfait chez l'homme', *Journal de l'anatomie et de la physiologie normales et pathologiques de l'homme et des animaux*, 6 (1869), 599–616. For an English translation of parts of this article, see Herculine Barbin, *Herculine Barbin Being the Recently*

Discovered Memoirs of a Nineteenth Century Hermaphrodite, intro. Michel Foucault, trans. Richard McDougall (New York: Pantheon, 1980), pp. 128–44.

3. Goujon, 'Étude', p. 606.
4. Ibid., p. 599.
5. Ibid., pp. 607–8.
6. Ibid., p. 600.
7. Lutaud, for instance, reported that the suicide occurred three years after the change of civil status when in fact it occurred seven and a half years later. (See August Lutaud, 'De L'Hermaphrodisme au point de vue médico-légal. Nouvelle observation. Henriette Williams', *Journal de médecine de Paris* [1885], 386–96, p. 388).
8. See Ambroise Tardieu, *Question médico-légale de l'identité dans ses rapports avec les vices de conformation des organes sexuels*, 2nd edn. (Paris: J. B. Baillière et Fils, 1874).
9. Barbin, *Herculine Barbin*, p. 103.
10. Ibid., p. 20; original emphasis.
11. Ibid., p. 56; original emphasis.
12. Ibid., p. 52; original emphasis.
13. Ibid., pp. 150–1.
14. Ibid., p. 89.
15. Ibid., p. 104.
16. See, for instance, the commentaries on Barbin's case in Tardieu, *Question médico-légale de l'identité*; Lutaud, 'De L-Hermaphrodism', p. 388; Gérin-Roze, 'Un Cas d'hermaphrodisme faux,' *Bulletins et mémoires de la société médical' des hôpitaux*, 3rd ser., 1 (1884), 369–73, p. 372; Dandois, 'Un Exemple d'erreur de sexe par suite d'hermaphrodisme apparent', *Revue médicale*, 5 (1886), 49–52; Jean Tapie 'Un Cas d'erreur sur le sexe. Malformation des organes génitaux externes; pathogénie de ces vices de conformation', *Revue médicale de Toulouse*, 22 (1888), 301–13, p. 306; Xavier Delore, 'Des tapes de l-hermaphrodism', *L'Écho médicale de Lyon*, 4, no. 7 (1899), 193–205 and no. 8 (1899), 225–232, p. 230.
17. Franz Neugebaner's *Hermaphroditismus beim Menschen* (Leipzig: Werner Klinkhardt, 1908) purports to include a reproduction of a portrait of Barbin, but as Michel Foucault has noted, 'the printer by mistake put Alexina's name under a portrait that is obviously not her own' (Foucault in Barbin, *Herculine Barbin*, p. xiv; n. 1).
18. Fancourt Barnes, [report of a living specimen of a hermaphrodite, meeting of April 25, 1888], *British Gynaecological Journal*, 4 (1888), 205–12 p.205.
19. Ibid., p. 212.
20. Ibid., p. 206.
21. Ibid., p. 208.
22. Ibid.
23. Ibid.
24. Ibid., p. 210.
25. Ibid.
26. Ibid., p. 212.
27. See Roy Porter, 'The Patient's View: Doing Medical History from Below', *Theory and Society*, 14 (1985), 175–98.
28. For a review of how the preponderance of men in science and medicine

influenced visions of sex, gender, and sexuality from the eighteenth to the twentieth centuries, see Ludmilla Jordanova, *Sexual Visions: Images of Gender in Science and Medicine between the Eighteenth and Twentieth Centuries* (Madison: University of Wisconsin Press, 1989).

29. For discussions of women's attempts to be recognized as medical professionals, see Thomas Neville Bonnet, *Becoming a Physician: Medical Education in Britain, France, Germany, and the United States, 1750–1945* (New York: Oxford University Press, 1995), pp. 207–13, 312–15; see also W F Bynum, *Science and the Practice of Medicine in the Nineteenth Century* (Cambridge: Cambridge University Press, 1994), pp. 206–8, and Moscucci, *Science of Woman*, pp. 71–3.

30. Moscucci, *Science of Woman*, p. 73.

31. See Ian Hacking's *Rewriting the Soul: Multiple Personality and the Sciences of Memory* (Princeton: Princeton University Press, 1995).

32. Pierre Garnier, 'Du Pseudo-hermaphrodisme comme impédiment médico-légale à la déclaration du sexe dans l'acte de naissance', *Annales d'hygiene publique et de médecine légale*, 3rd ser., 14 (1885), 285–93, p. 290.

33. Russett, *Sexual Science*, p. 7.

34. Patrick Geddes and J. Arthur Thomson, *The Evolution of Sex* (London: Walter Scott, 1889), p. 267.

35. Garnier, 'Du Pseudo-hermaphrodisme', p. 287.

36. Jonathan Hutchinson, 'A Note on Hermaphrodites', *Archives of Surgery*, 7 (1896), 64–6, p. 64.

37. Russett, *Sexual Science*, p. 48.

38. Judith Butler, *Gender Trouble: Feminism and the Subversion of Identity* (New York: Routledge, 1990), p. 110.

39. I use the expression 'gonad' to stand for ovaries, testes, and ovotestes because it simplifies my narrative, especially when the nature of the sex gland had yet to be determined in a given case. I prefer not to use the expression 'sex gland', because it connotes an endocrinological understanding of the gonad.

40. On the 'scientizing' of medicine in the nineteenth century, particularly in Germany, see Bynum, *Science and the Practice of Medicine*, chaps. 4 and 5; see also Bonner, *Becoming a Physician*, chap. 10.

41. Th. Tuffier and A. Lapointe, 'L'Hermaphrodisme. Ses variétés et ses conséquences pour la pratique médicale (d'après un cas personnel)', *Revue de gynécologie et de chirurgie abdominale*, 17 (1911), 209–68, p. 256.

42. See Richard Goldschmidt, 'Intersexuality and the Endocrine Aspect of Endocrinology (Philadelphia), 1 (1917), 433–456. See also Richard Goldschmidt, *The Mechanism and Physiology of Sex Determination*, trans. William J. Dakin (London: Methuen & Co., Ltd., 1923), p. 77.

43. See Goldschmidt, *The Mechanism and Physiology of Sex Determination*, p. 249.

44. The quotation is from Joseph H. Keifer, 'The Hermaphrodite as Depicted in Art and Medical Illustration', *Urological Survey* 17 (April 1967), 65–70, p. 65.

45. Ovid, *Metamorphoses*, books 1–4, trans. D. E. Hill (Oak Park, Ill.: Bolchazy-Carducci, 1985), book 4, lines 371–2.

46. Ibid., lines 374–6.

47. See Joan Cadden, *Meanings of Sex Difference in the Middle Ages: Medicine, Science, and Culture* (New York: Cambridge University Press, 1993); see also Lorraine Daston and Katharine Park, 'The Hermaphrodite and the Orders of

Nature: Sexual Ambiguity in Early Modern France', GLQ (*A Journal of Lesbian and Gay Studies*), 1 (1995): 419–38, esp. pp. 420–3.

48. On this, see Cary J. Nederman and Jacqui True, 'The Third Sex: The Idea of the Hermaphrodite in Twelfth-Century Europe', *Journal of the History of Sexuality;* 6 (1996), 497–517, pp. 506–7.

49. For an overview of hermaphrodite history, see Julia Epstein, 'Either/Or—Neither/Both: Sexual Ambiguity and the Ideology of Gender' *Genders,* 7 (1990), 99–142.

50. Nederman and True ('The Idea of the Hermaphrodite', pp. 501–2) claim that 'there is much evidence to suggest that hermaphrodites were tolerated (at least so long as they conformed to one or another gender role)' in the twelfth century, but unfortunately they do not offer references to that evidence.

51. Daston and Park, 'The Hermaphrodite', p. 419.

52. Ibid., p. 429.

53. See Isidore Geoffroy Saint-Hilaire, *Histoire générale et particulière des anomalies de l'organization chez l'homme et les animaux . . . ou Traité de tératologie* (Paris: J. B. Baillière, 1832–6). For a summary review of this text, see Alice Domurat Dreger, '"Nature Is One Whole": Isidore Geoffroy Saint-Hilaire's *Traité de tératologie*', MA thesis (Indiana University, 1993).

54. J. W. Ballantyne, 'Hermaphroditism', in Chalmers Watson (ed.), *Encyclopaedia Medica*, vol. 4 (Edinburgh: William Green and Sons, 1900), p. 491.

55. See, for example, Oscar Hertwig, *Text-Book of the Embryology of Man and Mammals*, trans. from the third edition by Edward L. Mark (London: Swan Sonnenschein & Company, 1892), p. 374.

56. On this, see Delore, 'Des Étapes', pp. 201ff.; Xavier Delore, 'De L'Hermaphrodisme dans l'histoire ancienne et dans la chirurgie moderne', *Journal des sciences médicales de Lille*, 2 (1899), 63–70, p. 67; and Charles Sedgwick Minot, *Human Embryology* (New York: W. Wood, 1892), pp. 490–2.

57. See pp. 147–53 in Nancy Tuana, 'The Weaker Seed: The Sexist Bias of Reproductive Theory', in Nancy Tuana, (ed.), *Feminism and Science* (Bloomington: Indiana University Press, 1989), pp. 147–71.

58. See, for example, Tapie, 'Un Cas d'erreur', p. 309: 'the appearance of every embryo is rather feminine.' See also Arène, 'Un Cas de pseudohermaphrodisme', *Loire médical* 14 (1895), 187–95, esp. p, 192: 'the external genital organs have kept an embryonic disposition, and consequently [look] feminine.'

59. See, for example, Minot, *Human Embryology*, pp. 517–18. For a critique of the common way of thinking about male and female development, see Anne Fausto-Sterling, *Myths of Gender: Biological Theories about Women and Men*, rev. edn. (New York: Basic Books, 1992), pp. 77–85.

60. Minot, *Human Embryology*, p. 520.

61. This challenges Laqueur's contention that by the nineteenth-century a 'two-sex' model, according to which men and women were understood to be fundatmentally different, had taken root in Western science, medicine, and culture, completely displacing an older 'one-sex' model. (See Laqueur, *Making Sex.*) I do, however, agree wholeheartedly with Laqueur's major point that sex—anatomy and physiology—gets 'constructed' just as gender does.

62. Rosemarie Garland Thomson, 'Introduction: From Wonder to Error—A Genealogy of Freak Discourse in Modernity', in Rosemarie Garland Thomson

(ed.), *Freakery: Cultural Spectacles of the Extraordinary Body* (New York: New York University Press, 1996), pp.1–19, p. 4.

63. See Elizabeth Grosz, 'Intolerable Ambiguity: Freaks as/at the Limit', in Thomson, (ed.), *Freakery*, pp. 55–6, p. 58.

64. Geoffroy thought, for instance, that if teratological experiments showed physical trauma to be a cause of serious monstrosities in chick embryos, then that knowledge might be used to prevent monstrosities or at least be used to instigate searches for violent culprits (like wife-beaters) when a woman gave birth to a monstrous child. See Isidore Geoffroy Saint-Hilaire, *Histoire générale et particulière des anomalies*, vol. 3, pp. 582–3, and Dreger, 'Nature Is One Whole', pp. 156–7.

65. Keith L. Moore, *Before We Are Born: Basic Embryology and Birth Defects*, 3rd edn. (Philadelphia: W. B. Saunders Company, 1989), p. 195. See also Ethel Sloane, *Biology of Women*, 3rd edn. (Albany, NY: Delmar Publishers, 1993), p. 168.

66. For a discussion of the genetic aspects of true hermaphroditism, see Margaret W. Thompson, Roderick R. McInnes, and Huntington F. Willard, *Thompson & Thompson Genetics in Medicine*, 5th edn. (Philadelphia: W. B. Saunders Company 1991), pp. 242–3.

67. Moore, *Before We Are Born*, p. 195.

68. On female pseudohermaphroditism, see Thompson, McInnes, and Willard, *Thompson & Thompson Genetics in Medicine*, p. 244; Sloane, *Biology of Women*, pp. 169–70; 'Intersex States', in Robert Berkow (editor-in-chief), The *Merck Manual of Diagnosis and Therapy*, 15th edn. (Rahway, NJ: Merck Sharp and Dohme Research Laboratories, 1987), pp. 1962–3; Thomas E. Andreoli et al., *Cecil Essentials of Medicine*, 3rd edn. (Philadelphia: W. B. Saunders Company, 1993), p. 494.

69. Experts today believe that androgen insensitivity can range from partial to complete. My description here is of the more complete form of androgen insensitivity.

70. The phrase 'key androgen receptor' is from Berkow, 'Intersex States,' p. 1962. On AIS, see also James E. Griffin, 'Androgen Resistance: The Clinical and Molecular Spectrum', *New England Journal of Medicine*, 326 (February 27, 1992), 611–18.

71. On 5-AR deficiency, see Griffin, 'Androgen Resistance'; see also Berkow, 'Intersex States', p. 1962. On 5-AR deficiency and AIS, see also Jean D. Wilson, 'Syndromes of Androgen Resistance', *Biology of Reproduction*, 46 (1992), 168–73.

72. A discussion of Klinefelter's is provided in most medical reviews of intersex conditions. For a history of the discovery of the syndrome by Harry F. Klinefelter himself, see 'Klinefelter's Syndrome: Historical Background and Development', *Southern Medical Journal*, 79 (1986), 1089–93.

73. On hypospadias, see Moore, *Before We Are Born*, pp. 195–8; Berkow, 'Intersex States', p. 1962. For references to articles of historical interest with specific regard to hypospadias, see Alice Domurat Dreger, 'Doubtful Sex: Cases and Concepts of Hermaphroditism in France and Britain, 1868–1915', Ph.D. diss. (Indiana University, 1995), p. 44, n. 5.

74. For these statistics, see respectively Andreoli et al., *Cecil Essentials*, p. 494; Sloane, *Biology of Woman*, p. 169.

75. See Gilbert Herdt, 'Mistaken Sex: Culture, Biology and the Third Sex in New Guinea', in Gilbert Herdt (ed.), *Third Sex, Third Gender: Beyond Sexual Dimorphism in Culture and History* (New York: Zone Books, 1996), pp. 419–45.

76. Sloane, *Biology of Woman*, p. 168.

77. Anne Fausto-Sterling, pers. comm. and 'Building Bodies: Biology and the Social Construction of Sexuality' (forthcoming), chap. 2.

78. The highest modern-day estimate for frequency of sexually ambiguous births comes from the Johns Hopkins psychologist John Money, who has posited that as many as 4 percent of live births today are of 'intersexed' individuals (cited in Anne Fausto-Sterling, 'The Five Sexes', *The Sciences, 33* (March/April 1993), 20–5, p. 21). Money's categories tend to he exceptionally broad and poorly defined, and not representative of what most medical experts today would consider to be 'intersexuality'.

79. These estimates for the incidence of cystic fibrosis and Down's syndrome are taken from Thompson, McInnes, and Willard, *Thompson & Thompson Genetics;* for cystic fibrosis, see p. 67; for Down syndrome, see p. 224.

80. Thomson, 'Introduction: From Wonder to Error', p. 1.

81. Mary Douglas, *Purity and Danger: An Analysis of Concepts of Pollution and Taboo* (London: Ark, [1966] 1984), p. 115.

82. Delore, 'Des Étapes', pp. 231–2.

83. It has not escaped my attention that these problems of objectivity and representation inhere in my own work. I subscribe to the working model of 'objectivity as responsibility' presented in Lisa M. Heldke and Stephen H. Kellert, 'Objectivity as Responsibility', *Metaphilosophy*, 26 (1995), 360–78.

Part II. **The Body Politic**

5 Icons of Divinity: Portraits of Elizabeth I

Andrew Belsey and Catherine Belsey

....

I

....

The face is pale and perhaps slightly unearthly, with a high forehead sharply defined against red-gold hair. Brilliant lighting falls directly on the Queen, eliminating almost all shadow. The white lace ruff and the enormous pearls in her hair complete a circle, radiating outwards against a dark background from a face which seems in consequence to be the source of light rather than its object. Elizabeth gazes out to the left, beyond the frame of the painting, contemplating a destiny invisible to the spectator. She appears remote and a little austere. George Gower's 'Armada' portrait (1588?) presents the Queen as an emblem of majesty (Fig. 6).

Her richly jewelled dress, meanwhile, and the ropes of pearls round her neck, are palpable signifiers of magnificence. The skirt and the bodice, with its long sleeves hanging behind, are of black velvet, all bordered with a single row of pearls held between gold edging. Silk bows, matching the rose-coloured lining of the outer sleeves, are held in place by rubies and emeralds. The satin underskirt and sleeves are embroidered with pearls and with devices in gold thread which resemble the sun.

Only the face and hands are visible. The exaggerated sleeves and the gigantic skirt efface between them all other indications of a human body. The effect of the richly lined oversleeves is of a sumptuous casing which, without attributing to the Queen anything so specific as broad shoulders (she has, indeed, no *bones* at all), extends the

From Andrew Belsey and Catherine Belsey, 'Icons of Divinity: Portraits of Elizabeth I' in *Renaissance Bodies: The Human Figure in English Culture* (Reakton Books, 1990), pp. 11–35. Reprinted with permission.

FIG. 6.   George Gower, *Elizabeth I—The 'Armada' Portrait*, 1588. Woburn Abbey. By kind permission of the Marquess of Tavistock and the Trustees of the Bedford Estate.

dimensions of the figure to make of the arms an anatomically improbable, embracing, encompassing semi-circle. This image of Elizabeth invites comparison and contrast with the body of Henry VIII, famously portrayed by Holbein as an icon of masculinity and power, now in the National Portrait Gallery, London. Henry's equally improbable shoulders are made to seem flesh and blood by the King's stance: in the firmly modelled cartoon, which is all that survives of Holbein's original painting of the King and his family in 1537, Henry stands, legs astride, arms akimbo, the feet aligned with the shoulders in a posture that combines aggression with strength. The King meets the spectator's gaze, requiring submission.

Elizabeth must have been aware of the implications of the Holbein portrait, then hanging in the Privy Chamber at Whitehall. And she was, after all, happy for dynastic reasons to recognise her resemblance to her father[1]. But the same artistic-political strategies were not available to a woman. Henry's magnificence resides in his muscularity, in the picture's exaggeration of his physical attributes. However, in the sixteenth century as now, because of the cultural construction of gender, an emphasis on the physical character of the female body connotes something quite different: not authority but availability, not sovereignty but subjection. Where Henry's right to dominate is confirmed by his virility, represented in the extraordinarily prominent codpiece, Elizabeth's depends by contrast on sexuality subdued, on the self-containment and self-control of the Virgin-Queen. Louis Adrian Montrose, making an illuminating comparison between the two pictures, points out that in the place corresponding to Henry's codpiece, the 'Armada' portrait of Elizabeth displays a giant pearl. At the base of the triangle made by her stomacher, and foregrounded by a white ribbon tied in a bow, hangs an emblem of the Queen's chastity.[2] Indeed, the composition of the portrait insists on the pearl's emblematic significance. The near-vertical line which forms the fastening of the bodice links a succession of pearls, culminating in the vast pearl of the headdress, emphasising a symmetry between majesty and virginity. The juxtaposition of the two huge gems displays the Queen impregnable in both senses, powerful precisely to the degree that she is inviolable.

The point is allegorically reinforced. As so often in her portraits, the Queen stands before a chair of state, and in this instance the arm of the chair is supported by the carved figure of a mermaid. During the Middle Ages and the Renaissance mermaids represented the dangerous and destructive possibilities of uncontrolled female sexuality. According to legend, they offered sailors tempting glimpses of their beauty as they combed their long golden hair among the waves. Interchangeable with the sirens who threatened to destroy Odysseus and his crew, mermaids also sang with extraordinary sweetness and seductiveness. Sailors who failed to exercise the highest self-discipline were known to leap into the sea in pursuit of their charms, but cheated of satisfaction by the mermaids' scaly tails, they were drowned, and their ungoverned ships were lost among the rocks. In the painting the angle of the mermaid's eminently visible body, the direction of her gaze and even her surprisingly aquiline features suggest a parody of the image of the Queen, implying that she represents an alternative sexual identity to Elizabeth's. The Queen firmly turns her back on the form of femininity that the mermaid stands for.

Directly above the figure of the mermaid, however, the picture records a disaster at sea. A row of columns opens on to the loss of the Armada as Spanish galleons are lashed against huge rocks in the darkness. The shipwreck of the Spanish fleet was real enough, but its role in the portrait is as allegorical as the mermaid's. Elizabeth stands with her back to darkness and turmoil, confident in the power of the Protestant wind which God has shown himself willing to use to destroy the enemies of his Church and people. On the left of the picture, meanwhile, English fireships set out towards the approaching Armada. The episode is bathed in golden light, and the sea is calm. The weather, we are to understand, acknowledges the direction of Elizabeth's gaze and responds to the sun-like radiance of her face.

The imperialist allegory of the painting is explained in detail by Roy Strong in his invaluable *Gloriana: The Portraits of Queen Elizabeth I*. The Queen's hand caresses the globe, and her elegant fingers cover that portion of it which represents America. By 1588 the colonial project in Virginia had been established as the foundation of the British Empire in the New World. Above the globe, and echoing its shape, the crown links the Armada victory with colonial possession. The 'closed' Tudor crown resembled the crown of the Holy Roman Empire and reinforced the claim of the Tudor monarchs to equal status with the Emperor. In the 'Armada' portrait Elizabeth appears not only as the ruler of a victorious realm but also as aspiring Empress of the world.[3]

It is widely agreed that pictures of Elizabeth are supremely icons of magnificence. The 'Armada' portrait proclaims the sovereignty and the right to rule of this individual woman: the splendour of her appearance, her vision and her *self*-control are evidence of the majesty and the authority which inhere in the person of the Queen. But the painting also declares the magnificence of her realm: England's wealth and maritime prowess are evidence of its authority in the world, of national sovereignty divinely endorsed. In this sense the 'Armada' portrait is profoundly political, and it shares that character with many other representations of Elizabeth. Most of the specific claims it makes on her behalf recur elsewhere. The virgin-pearl, for example, reappears in exactly the same place in the 'Hardwick' portrait of 1599(?). Chastity is also the theme of the series of 'Sieve' portraits, painted between 1579 and about 1583. The reference is to Tuccia, a Vestal Virgin accused of breaking her vow of chastity, who refuted the slander by filling a sieve with water from the River Tiber and bearing it back to the Temple of Vesta without spilling a drop. When the Queen is portrayed holding what might otherwise be mistaken for a colander, she invokes the magic power of chastity, demonstrated by Tuccia, to seal the leaky orifices of the female body, to make it hard, impenetrable, invulnerable.[4] Elizabeth-Tuccia thus miraculously transcends the body of a weak and feeble woman, and makes of this transcendence a mystical sovereignty which has the effect of inverting the conventional hierarchy of gender. Her virginity shows her more than a woman, more, indeed, than human.

Similarly, the command of the weather she displays in the 'Armada' painting is also a theme of the 'Ditchley' portrait of *c.*1592, where Elizabeth stands with her back to a dark sky rent by lightning. In front of her, by contrast, is calm and sunshine as the clouds disperse (Fig. 7). And in the 'Rainbow' portrait, probably painted in about 1600 by Marcus Gheeræerts the Younger, the Queen wears a brilliant flame-coloured taffeta cloak and holds a tiny rainbow. The painting is inscribed 'Non sine sole iris' (No rainbow without the sun) (Fig. 8). The red-gold wig of Elizabeth's later years, assumed as her own hair lost its colour, was evidently more than a requirement of feminine vanity: it was an important element in the political meaning of her image.

The imperial theme, too, was a recurrent feature of royal portraiture in the 1580s and 1590s. The globe first appeared with the Queen in 1579 in the earliest of the 'Sieve' portraits by George Gower. This was two years after the publication of John Dee's *General and Rare Memorials Pertayning to the Perfect Arte of Navigation*, a plea for the establishment of a British Empire. The first 'Sieve' portrait shows part of a luminous globe behind the Queen's right shoulder. A much more detailed reworking of the 'Sieve' motif in 1580 includes a number of other icons of empire. Here the globe appears again, in this instance with a glowing Britain set like a precious stone in a sea of ships. The 'Armada' portrait reproduces the globe, and finally in 1592 Elizabeth appears in the 'Ditchley' portrait standing on an enlarged map of her realm, with the edges of the globe stretching out beyond it against the sky. Here the Queen is shown literally on top of the world.

Much of the story of the iconography of royal portraiture in this period is familiar from the work of Roy Strong and Frances Yates.[5] What has received rather less attention, however, is the remarkable geometry of the 'Armada' portrait. Of course, the style of painting is non-illusionist: it does not use either perspective or shadow to construct for the spectator the illusion of what the eye might actually see. The furnishings of the room are in different visual planes, and the chair is visible from two distinct angles at once. Moreover, the picture has no 'moment': two separate episodes in the history of the Armada appear to be taking place in the open air immediately behind an oblivious Queen. But in place of illusion the picture offers a complex

FIG. 7.   M. Gheerærts the Younger, *Elizabeth I—The 'Ditchley' Portrait*, c.1592. By courtesy of the National Portrait Gallery, London.

NON SINE SOLE
IRIS

FIG. 8. M. Gheeraerts the Younger, *Elizabeth I—The 'Rainbow' Portrait*. By permission of the Marquess of Salisbury, Hatfield House: photograph, Courtauld Institute of Art.

composition of circles and semi-circles, and this contributes to the impression that we are presented with a figure abstracted from any kind of immediacy or materiality. The globe echoes the shape made by the Queen's head with its surrounding ruff, like a planet in relation to the sun. Meanwhile, the beaded border of her skirt draws attention to the corresponding pearl edges of the crown, and the semi-circle is reproduced again in the Queen's forehead and in the black outline of the skirt. The figure of the Queen is thus composed of a series of arcs, bisected, like the crown, by the line of jewels on the front of the bodice, and contained by the encompassing semi-circle of the sleeves. The portrait is a design more than it is a representation of the human figure: pattern takes the place of substance to construct the image of a disembodied, extra-human Queen.

This is partly, of course, a matter of fashion. English painters had relatively little contact with Italy, and were decidedly not working in the Italian Renaissance tradition of perspective and *chiaroscuro*. As the 'Armada' portrait indicates, English painting of this period still had much in common with heraldry.[6] On the other hand, Spanish activity in the Netherlands produced an influx of immigrant Flemish painters throughout the period, and particularly in the 1570s, so that an alternative and much more illusionist tradition was available. Portraits of Elizabeth's subjects in the final decades of the sixteenth century were certainly stiff and formal by the standards of the next generation, but it would be hard to find among those which survive anything to equal the abstract geometry of the later portraits of the Queen. And as Lucy Gent points out, Elizabeth herself was certainly in a position to exercise a choice in the matter, since she had constant access to the work of Holbein.[7] In the early 1590s Elizabeth sat for Isaac Oliver, who was thoroughly familiar with the Renaissance strategies for producing the effect of depth and shadow. She did not, however, like the results at all. Her own expressed views, as far as we know them, seem on the whole to have been contradictory: she told Nicholas Hilliard that the Italians were the best painters; she also told him that she disapproved of shadows.[8]

Strong believes that Elizabeth's view of Italian painting was formed on the basis of her encounter with Federigo Zuccaro and that it was Zuccaro who painted the 'Darnley' portrait in 1575 (Fig. 9).[9] It is instructive to compare this with the 'Armada' portrait, painted some 13 years later. The strategies of the 'Darnley' painting, while duly respecting the Tudor preference for line and colour over perspective and modelling, are nevertheless relatively illusionist. Elizabeth's dress,

FIG. 9.   *Elizabeth I—the 'Darnley' Portrait,* unknown artist, *c.*1575.
By courtesy of the National Portrait Gallery, London.

of white and gold brocade, is much less elaborate than the 'Armada' costume, and the head-dress is comparatively unassuming. The pearls are there, but in nothing like the abundance of the later portrait. The 'Darnley' portrait seems to present a human figure rather than an icon.

At the same time, with hindsight it is possible to read in this picture indications of the signifying practices that were to come in royal portraiture. The crown and sceptre are visible on a table to the right. Both the austerity of the Queen's expression and her fashionably inflated shoulders perhaps in their different ways draw attention to her descent from Henry VIII. And the red and gold frogging on the bodice, in conjunction with the shape of the waist, hints at a softened, duly feminised, ornamental breastplate, and thus aligns Elizabeth with the warrior maidens of the psychomachia, Virtues defeating Vices in armed struggle against evil.[10]

Marina Warner traces a line of descent, which passes through Queen Elizabeth I, from Athena to Britannia and Margaret Thatcher. She might have added Spenser's Britomart, whose name indicates both her patriotism and her warlike character. These iron ladies are guardians of virtue, controlling and subduing masculine sexuality. Their function is to civilise and check impulses which are understood to be natural: they contain and discipline unruly male desire. But they do so without abandoning the virtues which are seen as specifically feminine. It is precisely as chaste maidens or model wives and mothers that they exercise the only form of power that patriarchy leaves to women: the right of prohibition. They are thus able to remain objects of desire without themselves being subject to it.[11]

This was precisely the role of Elizabeth in the many volumes of love poetry addressed to her. She became the virtuous focus of masculine desire, the unmoved mover who stirred her subjects to acts of gallantry and heroism.[12] It is as courtly lady, beautiful but aloof, and as armed maiden, austere and commanding, that the Queen appears in the 'Darnley' portrait. The allegory of Elizabeth is thus already beginning to be visible in this relatively illusionist portrait. Even in a painting which Strong attributes to an Italian painter, on the basis of its innovatory strategies and its formal debt to Titian, the Queen's iconic character is in the process of construction.[13]

A decade later, in a series of portraits attributed to John Bettes, she approaches pure geometry (Fig. 10). Golden rays extend downwards from her sun-like face, the more startling for their contrast with her black dress. Her body is more or less indecipherable, and

Fig. 10. *Queen Elizabeth I* (attributed to J. Bettes the Younger, c.1585–90). By courtesy of the National Portrait Gallery, London.

her arms are quite lost in the voluminous gauze oversleeves. Here only the jewels are illusionist, depicted with *trompel'œil* fidelity. The extraordinary shape shown here and in the 'Armada' portrait is, none the less, the image of Elizabeth which has become so familiar that it is readily adaptable to commercial purposes. In 1988 an advertisement proclaiming that 'Elizabeth I enjoyed her Burton ale' showed the Queen in an approximation to the 'Armada' costume, slightly humanised, but still displaying the remarkable semi-circle formed by the immense sleeves. To us now the geometry remains memorable, even if it is not quite possible to reproduce the repudiation of human attributes which characterises the sixteenth-century portraits.

II

Now the Queen has left the earth and appears in a position of dominance over the whole universe (Fig. 11). The image appears as the frontispiece to John Case's *Sphæra civitatis*, published in 1588, the year of the Armada.[14] Less an icon of colourful magnificence because of the different technical possibilities of the woodcut, the diagram nevertheless offers an image replete with mysterious, challenging significances. While it is true that the diagram is 'closely related in imagery to the "Armada" portrait'[15] what is immediately striking is that the latter places the Queen in relation to the earthly globe, whereas Case represents her with the *universe*. What is going on?

On first inspection the diagram appears to show what became known as the Ptolemaic system of the universe, a set of concentric spheres (shown in cross-section as circles).[16] At the centre is the globe of the Earth, labelled *Justitia Immobilis*. Then come the spheres of the seven celestial wanderers, the Moon, Mercury, Venus, the Sun, Mars, Jupiter and Saturn, each planet indicated by its astronomical symbol and identified with a corresponding regal virtue in a characteristic piece of Renaissance typology. The sphere beyond the wanderers is occupied by the fixed stars, accompanied by a rather obscure description. The outermost sphere is filled by an inscription of the official style and title of the Queen. Finally, above and beyond the spheres is the figure of the Queen herself, apparently in the act of embracing the universe. She is again an encompassing semi-circle, with a splendid cloak, an impossible ruff, and an imperial crown.

But this description of Case's diagram raises many questions of interpretation. How should it be understood? What are the meanings of the various labels and inscriptions? According to Charles B. Schmitt, the diagram offers a 'rather striking portrait of Queen Elizabeth in a sort of *regina universi* pose'.[17] The vague nature of this characterisation points to the fact that existing commentaries have not appreciated the full significance of Case's diagram. Frances Yates, for example, comments that the diagram shows the 'tendency of Virgo-Astraea-Elizabeth to expand until she fills the universe'.[18] But Elizabeth

FIG. 11.   Frontispiece to John Case, *Sphæra civitatis* (Oxford, 1588). By permission of the British Library.

does not fill the universe. She is outside it, beyond it. And that is a very remarkable place to find an earthly monarch.

The striking implications of the image arise from the fact that Case's diagram has a vital double aspect. Or perhaps it would be more accurate to say that it has more than one duality. On the one hand it is the universe. And on the other hand it is what it proclaims itself to be, *sphæra civitatis*, the state. On the one hand it is the world in the sixteenth-century sense, the cosmos. And on the other hand it is the world in the modern sense, the planet Earth. So on the one hand it shows the Queen ruling over the cosmic world. And on the other hand it shows her ruling over the mundane world. The claims of the 'Ditchley' portrait are here multiplied: Elizabeth is doubly on top of the world.

Frances Yates has interpreted the political aspect of these ambiguities by reference to yet another related duality, the analogy between unified rule in heaven and on Earth. 'As the heaven is regulated in all its parts. . . by the one first mover who is God, so the world of men is at its best when it is ruled by one prince.'[19] Understood in this way, Case's diagram clearly indicates the imperialist aspirations of the last two decades of the sixteenth century. Not only does Elizabeth claim the right to undisputed monarchy in her existing realms (no Pope, for instance), she also claims justification for extending and expanding those realms. It is as if 'the imperialist theme is given a universal application'.[20]

It is certainly true that the political aspect of Case's diagram can be illuminated by placing it within the appropriate tradition of medieval and Renaissance political theory. But the cosmological aspect equally needs the appropriate theoretical illumination, which is the iconographical tradition of medieval and Renaissance cosmology.[21] This, however, produces an even more political interpretation, for Case's diagram does not simply reproduce or conform to the conventional images of Renaissance cosmology. It transforms them and transcends them, sometimes in quite startling ways.

This iconographical tradition included representations of the human figure in relation to the universe: the macrocosm-microcosm relation. The idea goes back to antiquity (it was Aristotle who first defined the human being as a microcosm), but like other classical theories it was adapted to Christian use in the Middle Ages. Pictorial realisations of the idea range from that of Hildegard of Bingen in the twelfth century to Robert Fludd's in the seventeenth.[22]

Such diagrams make a number of claims about the nature of things. First, that since both macrocosm and microcosm were made by God, therefore there are important analogies between them. Next, that the human being, made in God's image, can to some extent understand the nature and laws of the universe, and even put them to the use and benefit of human life (the implication sometimes being the restoration, as far as is possible on Earth, of the prelapsarian state of perfect knowledge and control). So the human being enjoys some superiority over the non-human universe. And yet the human figure is always within the universe; it is enclosed by it, encompassed by it. If it rules, it rules from within.

Not so the Queen. The positions are audaciously reversed. She encloses the universe, encompasses it in her embrace. From without, she rules the cosmic world.

Elizabeth's cosmic world, Case's universe, was Ptolemaic. Copernicus had published his alternative heliocentric system in 1543, and it was a perfectly respectable and much-debated hypothesis among professional astronomers (some of whom, such as the Englishman Thomas Digges, were extremely strong adherents). But the earth-centred astronomy of Ptolemy was still the established system during the reign of Elizabeth. The medieval cosmological tradition which combined a Christianised Aristotelian physics with Ptolemaic mathematical astronomy was still strong. The *Tractatus de sphæra* (*On the Sphere*) by Johannes de Sacrobosco (John of Holywood),[23] delineated the rudiments of geocentric spherical astronomy and cosmology, and though written in the early thirteenth century, it was still a standard textbook in the sixteenth century, continuing in use even into the seventeenth.[24]

Sacrobosco described 'the heavens' in the language of medieval (i.e., Christianised) Aristotelian physics:

Around the elementary region revolves with continuous circular motion the etherial, which is lucid and immune from all variation in its immutable essence. And it is called 'Fifth Essence' by the philosophers. Of which there are nine spheres . . . namely, of the moon, Mercury, Venus, the sun, Mars, Jupiter, Saturn, the fixed stars, and the last heaven. Each of these spheres incloses its inferior spherically.[25]

Sacrobosco's 'last heaven' was the sphere of the *primum mobile*, which meant literally that it was the 'first moved', the origin of movement that mechanically transmitted regular motion to all the other spheres. But it was first only in the created universe. Something transcendent was required as the ultimate source of motion. So the *primum mobile* was thought of as the place where God, the First Mover, either directly through his intelligence or spirit or indirectly through his angels, applied the motive power that kept the cosmos turning.

This became the standard conception of the universe, but the details of the system were not beyond dispute: the medieval-Renaissance tradition of cosmology was not as ossified nor as unified as is sometimes thought.[26] There were disagreements about terminology and, more important, the number of spheres. Although Sacrobosco described all the spheres beyond the elementary region collectively as 'the heavens' (a usage which persisted, of course, and still persists in everyday speech today), other astronomers began to locate Heaven in the specifically Christian sense beyond the spheres as Sacrobosco describes them, placing it either in a tenth sphere or in the unbounded region beyond all spheres.[27] This clearly had an appeal for Christian cosmologists, as it demonstrated the role of God in ruling the universe and governing its motions.

The system of Sacrobosco, then, had nine spheres, the outermost being occupied by the activity of the First Mover who was, for a medieval Aristotelian like Sacrobosco, the Christian God. By the sixteenth century, however, variations had sprung up. Apian (1533), for example, favoured ten spheres, each labelled a 'heaven', and an unbounded Heaven beyond, described as 'Cœlum Empireum Habitaculum Dei et Omnium Electorum'.[28]

Even a switch to the Copernican system did not fundamentally change this aspect of the rival tradition. Thomas Digges's delightful and fabulous heliocentric diagram, 'A Perfit description of the Cælestiall Orbes', published in 1576, conflated Heaven with the infinite region of the fixed stars, and labelled it thus: 'This orbe of starres fixed infinitely up extendeth hit self in altitude sphericallye, and therefore immovable the pallace of foelicitye garnished with perpetuall glorious lightes innumerable farr excellinge our sonne both in quantitye and qualitye the very court of coelestiall angelles devoid of greefe and replenished with perfite endlesse joye the habitacle for the elect'.[29] This is a decidedly Anglicised version of Copernicanism, however, for Copernicus himself stuck strictly to astronomical tasks and did not concern himself with the physical location of Heaven.

Though Copernicanism was known in England during the reign of Elizabeth, it was not standard, and the Ptolemaic system was iconographically more significant for Case's project, since it identified the fixed earth with *Justitia Immobilis* at the centre, suitable as an image of global imperialism. The outer extremities of Case's universe, however, are even more interesting. The fixed stars of the eighth sphere are labelled *Camera Stellata Proceres Heroes Consiliarii,* which punningly identifies the Councillors of the Court of Star Chamber with the fixed stars.[30] This is a bold claim (with evident authoritarian political implications), but far more startling are the facts that the ninth sphere—that of the *primum mobile,* the place where God acts—is occupied by the description 'Elisabetha D. G. Angliæ Franciæ et Hiberniæ Regina Fidei Defensatrix', and that Elizabeth herself occupies the region beyond the spheres, the unbounded Heaven.

Whatever the differences of detail, the Ptolemaic astronomers all agreed that God and Heaven were located beyond the visible universe, and even the Copernican Digges put them in the outermost region. Yet in Case's diagram Elizabeth *twice* usurps the position and role of God, by appearing as both *primum mobile* and Sovereign of Heaven. In the latter role, the depiction of her embracing the universe from a superior position mimics the image of God as the creator of the universe (Fig. 12): both God and Queen are crowned and wear loose, ample cloaks fastened in front by jewelled clasps.

Other examples of Elizabeth claiming the place of God can be found in the iconography of the period. The depiction of the Sabbath in the *Nuremberg Chronicle* (1493) shows Heaven as an eccentric eleventh sphere, with God sitting enthroned above the created universe and surrounded by ranks of angels, his right hand raised in blessing and his left holding an orb (Fig. 13). The title-page of the *Bishops' Bible* (1569) shows Elizabeth enthroned in a similar position, but instead of God's creation it is the preaching of God's word that she surmounts—and authorises. The Queen's image guarantees God's truth. She too has her left hand round an orb. As in the Case woodcut, she is supported by regal virtues, two of whom, Justice and Mercy, crown her.[31]

From his position in Heaven above and beyond the universe, God sits not only as creator but also as judge. Plate 41 of Holbein's *Dance of Death* (1538) shows God surmounting the universe at the Last Judgment, and graphic representations of this event would have been familiar, as they were often found over entrances or chancel arches of churches. It is worth noting, therefore, that the Doom painting over the chancel arch of St Margaret's Church, Tivetshall, Norfolk, was overpainted with the royal arms of Elizabeth.[32] According to Strong, such 'manifestations of the royal governorship of the *Ecclesia Anglicana* became "portraits" of the Queen'.[33] What is more significant, however, is that this 'portrait' of Elizabeth would be replacing the image always found top-centre in a Doom painting: God-as-Christ-in-Majesty.

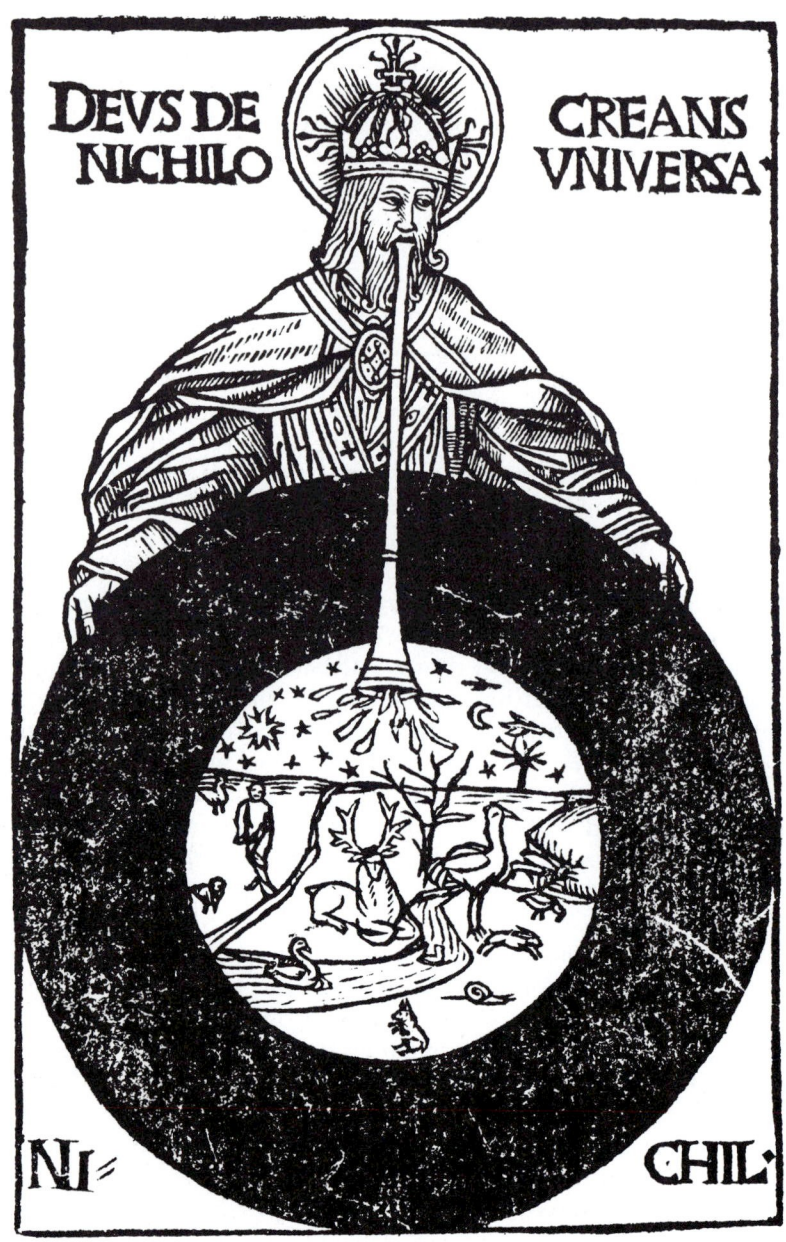

FIG. 12. Charles de Bouelles, *Libellus de nichilo* (Paris, 1510). By permission of the British Library.

De sanctificatione septime diei

Onsummato igitur mundo:per fabricam diuine solercie sex dierum. Creati em dispositi ꞇ ornati tande pfecti sunt celi ꞇ terra. Compleuit de gliosus opus suu: ꞇ requienit die septimo ab operib manuum suaru: postᵭ eictum mundu: ꞇ omnia que in eo sunt creasset: no quasi operando lassus: sed nouam creaturam facere cessauit: cuius materia vel similitudo non precesserit. Opus enim propaga tionis operari non desinit. Et dominus eidem diei benedixit: ꞇ sanctificauit illu: vocauitᵭ ipsum Sabatu quod nomen hebraica lingua requiem significat. Eo ᵱ in ipso cessauerat ab omi opere ᵭ patrarat. Un ꞇ Iudei eo die a laboribus propriis vacare dignoscitur. Quem ꞇ ante leges certe gentes celebrem obser uarunt. Iamᵭ ad calcem ventum est operum diuinozum. Illum ergo timeamus: amemus: ꞇ veneremur. In quo sunt omnia siue visibilia siue inuisibilia. Et a domino celi: domino bonoz omniu. Cui data omis potestas in celo ꞇ in terra. Et presentia bona: quatenus bona sint. Et veram eterne vite felicitatem queramus.

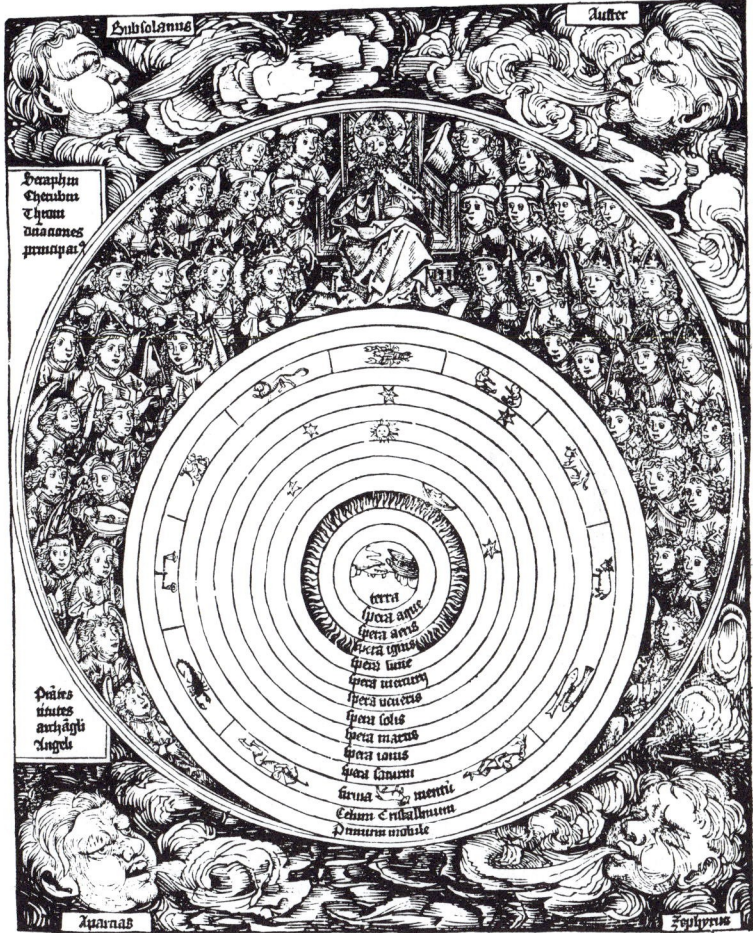

FIG. 13.   The Sabbath, from Hartmann Schedel, *Liber chronicarum* (Nuremberg, 1493). By permission of the British Library.

Not content with a mundane role, Elizabeth achieved lift-off to Heaven. To be portrayed as having imperialist ambitions, however grandiose, is nothing by comparison. To be turned into a Vestal Virgin, as in the 'Sieve' portraits, or to be 'soberly hailed as a second Virgin'—an extraordinary extension of the image of Mary as the second Eve—even these claims pale against the image of Elizabeth as transcendent supra-mundane ruler.[34] There are hints of divine attributes in the 'Armada' and 'Ditchley' portraits, where she commands not just the earthly globe but also the elements in the meteorological sense, but it is Case's wood-cut of the *Sphæra civitatis* that finally transforms Elizabeth into God, the extra-human Queen, the unmoved mover commanding the affairs not just of the realm, not just of the Earth, but of the cosmos.

III

It was in the Renaissance that personal portraiture first became an art form in its own right. In fifteenth-century Italian and Flemish painting, and during the course of the sixteenth century in England, individual subjects were increasingly identified, characterised and commemorated, their individuality fixed and recorded for the con-templation of themselves and others. Renaissance humanism put for-ward with mounting conviction the view that human beings, rather than supernatural forces, were the agents of human history. And in the cause of legitimating the existing inequalities of wealth and power, humanism was also in due course to promise each human being a unique subjectivity, which was in turn understood to be the origin of his or her personal history: what you were, justified—or limited—what you could become. Eventually, character was destiny. The Renaissance portrait initiated the process of recording that character in its uniqueness—as an example to others, as an affirmation of dynasty, or simply as a way of arresting the depredations of time.

The humanist subject of the Renaissance received an excess of confidence from the development in Venice in the early sixteenth century of flat glass mirrors, which were soon exported to the rest of Europe. Metal mirrors in the Middle Ages must have been both relatively rare and comparatively unsatisfactory.[35] In the medieval art of northern Europe they usually appear as emblems of (female) vanity: mermaids carry them; Idleness commonly looks in one in illustrations of *The Romance of the Rose*; the Whore of Babylon holds a looking-glass in the Apocalypse tapestries at Angers. Later Spenser's Lucifera looks admiringly at her face in a mirror.[36] But if mirrors retained something of this moralised meaning, the wider availability of accurate reflection in the sixteenth century must at the same time have contributed to the individual's sense of his or her own uniqueness.[37]

According to Jacques Lacan's rereading of Freud, mirrors are critical, mythologically at least, in the acquisition of subjectivity. Seeing in the mirror its own image, apparently unified and autonomous, the infant experiences a moment of jubilation as it assumes its own specular identity. This sense of triumph, in Lacan's account, is 'imaginary', that is, both imaged and illusory. The recognition of the self as a unity is at the same time a misrecognition, since the moment of singular identity is simultaneously one of division—between the I/eye that appears in the mirror image and the I/eye that perceives it. Renaissance mirrors must have reaffirmed the emergent humanist subject's misrecognition of its own unity and autonomy. Moreover, in the mirror the subject also becomes the object of its own knowledge. The knowing subject of humanism perceives, studies, reflects on, *knows* its identity in its image.[38]

Mirror images, however, are necessarily discontinuous: they move with the subject, turn away, change and grow old. The mirror reflects the loss of power and prowess, as well as its possession. Portraiture, by contrast, can be seen as offering to immortalise an ideal image of the subject; the portrait promises to fix and perpetuate the subject's autonomy and its knowingness, whether the I/eye meets the gaze of the spectator across time and space, or contemplates a distant vision which exceeds our knowledge. This fixing of the subject is also necessarily imaginary: the ideal image is always other than the subject itself. In any case, there is a sense in which the real mastery, the real knowingness, are understood ultimately to be a property of the artist, the person who knows, interprets and captures the true identity of the sitter. The Renaissance inaugurated the process, so familiar in the modern world that it seems inevitable, of naming and ranking artists in order of merit and corresponding (financial) value.[39] Like the Lacanian mirror, the portrait, too, betrays its promise in the moment of fulfilment.

Since Elizabethan portraiture increasingly recorded the identity of specific individuals, portraits and illusionist techniques developed alongside each other as humanism increased its hold. But while the bodies of her subjects were depicted with increasing verisimilitude, the image of the Queen became, we have suggested, progressively more geometrical, more heraldic, more remote from illusionism. Elizabeth I, after all, was not a subject but a sovereign, and as Charles I would later insist, from an absolutist point of view 'a subject and a sovereign are clean different things'.[40] Elizabeth had no need, therefore, of illusionist techniques, of resemblance, or even of identity in the obvious sense of the term. Portraits of the Queen are a record not of her subjectivity but of her authority, wealth and greatness, the qualities that require absolute obedience. These pictures do not present an image, not even an ideal image, of her body, since the available repertoire of likenesses of the human body offers no obvious image for a woman who is also a ruler in a patriarchal society. The portraits of the Queen subdue her sexuality in order to proclaim her power, and in the process they place her outside the realm of nature. In these images Elizabeth escapes the constraints of time and space; she represents a superhuman transcendence; and finally she takes the place of God.

But like the unity, autonomy and knowingness of the subject, the sovereign's divinity is also necessarily imaginary. Elizabeth's control, especially in the later years of her reign, was in practice increasingly precarious. The economy was in serious trouble. In the 1590s the cumulative effects of inflation combined with a succession of poor harvests from 1594 onwards, and as a result, by 1597 the purchasing power of wages had fallen to its lowest level since the thirteenth century. At the same time, the enclosures led to an increase in vagrancy which was almost beyond manageable proportions. There were serious food riots in 1595 and again in 1596 and 1597. The government, meanwhile, resorted increasingly to the sale of monopolies to raise funds. In 1601 Parliament delivered a direct challenge to the royal prerogative by initiating a bill against monopolies, and the Queen was obliged to yield on the substantive issue in order to save the prerogative. As a result, the parliamentary impediment to absolutist aspirations was momentarily visible for all to see.

The hounding out of heresy, whether religious or political, is always a symptom of instability in the state. Trials for witchcraft reached a peak in the 1590s. The execution of Mary Queen of Scots in 1587 failed to put a stop to the activity of the Catholics, and the decade between 1585 and 1594 witnessed a series of major treason trials. The Essex rebellion of 1601 was a devastating blow to Elizabeth's confidence. And though it was increasingly apparent that James VI of Scotland was the only possible heir to the English throne, Elizabeth resolutely refused to quell anxiety on this score by naming her successor.[41] The sphere of the state was by no means as unshakeable as Court propaganda implied.

The Case diagram awards Elizabeth the familiar title of Queen of England, France and Ireland. The Queen had no authority in France: the last English territory, Calais, fell to the French in 1558, before she came to the throne. The Irish problem, chronic from 1560 onwards, subsided in the 1580s, when 1,000 English settlers moved in to effect the plantation of Munster. But it became acute again when trouble broke out in 1594, and the Nine Years' War followed. By the time the news of the Irish surrender reached London in 1603, the Queen was dead.[42]

The presence of the globe as an icon of empire in the royal portraiture of the 1580s and 1590s was even more optimistic. In the last decade of the reign, despite the iconography of the royal portraits, the British empire was still no more than a devout wish, a story of courageous effort defeated by lack of state commitment, and especially by insufficient government funding. By this time the Virginia project had foundered, and it was not until the seventeenth century that British imperialism got seriously under way. At the end of Elizabeth's reign her subjects had attempted and failed to intervene in the Guinea trade; they had failed to break into the Caribbean slave trade; and they had failed to establish a lasting colony in North America.[43]

According to Case's diagram, the Elizabethan state in 1588 was blessed with Plenty. Within a few years this would come to seem a more doubtful claim. The Queen's Clemency was not much in evidence in the 1590s either, and if her Majesty remained in place, her Prudence was certainly severely tested. By the end of the century it was no longer obvious that Elizabeth-as-First-Mover was the only moving force in the realm, or that all the initiative any longer lay with the Queen.

In the deposition scene of Shakespeare's *Richard II* the King, who once resembled the sun (III. ii. 36–52; III. iii. 62–7), sends for a mirror, 'that it may show me what a face I have/Since it is bankrupt of his majesty' (*King Richard II*, IV. i. 266–7). Richard expects to find a difference between the image of a sovereign and the image of the subject he is about to become. In the event he is disappointed:

> No deeper wrinkles yet? Hath sorrow struck
> So many blows upon this face of mine
> And made no deeper wounds? O flatt'ring glass,
> Like to my followers in prosperity,
> Thou dost beguile me! Was this face the face

> That every day under his household roof
> Did keep ten thousand men? Was this the face
> That like the sun did make beholders wink?
> Is this the face which fac'd so many follies
> That was at last out-fac'd by Bolingbroke?

And in a theatrical gesture, Richard dashes the mirror to the ground:

> A brittle glory shineth in this face;
> As brittle as the glory is the face;
> For there it is, crack'd in a hundred shivers.
>
> (IV. i. 277–89)

In 1595, when the play was probably written, the glory of the monarchy was, in the Lacanian sense, imaginary. The awareness that the Queen's authority was precisely 'brittle' is registered in the refusal to allow the deposition scene to be printed during Elizabeth's lifetime. None the less, after the Essex rebellion, Elizabeth told the antiquary William Lambarde, with whatever ironic reference, 'I am Richard II, knowe ye not that?'[44] The comment sounds like an acknowledgement that the Queen's image has only narrowly avoided the fate of Richard's, 'crack'd in a hundred shivers'.[45]

Are the majestic affirmations of the royal portraits no more, then, than magnificent lies? How, in other words, should we read these images of a face 'That like the sun did make beholders wink'? The answer, we propose, is that the pictures are elements in a struggle at the level of representation for control of the state. The Tudors ruled as much by 'right' as by force. In the absence of a standing army, or anything like an effective system of policing, sixteenth-century monarchs owed their power in some degree to the cultural practices by which they enlisted the loyalty and the submission of their subjects. Castles gave way in this period to palaces, as property became evidence of the propriety of Tudor rule. Control of the Church permitted the preaching of obedience to the sovereign as God's will, reinforced by the appearance of the Royal Arms in churches. Progresses and pageants displayed the magnificence which legitimated monarchical power.

Images of the Queen affirm that power in every detail. They display the authority which demonstrates her right to rule in spite of her sex. And the more precarious the power, the more extravagant the claims of the royal portraits. They are in this sense weapons in an ideological and cultural struggle.

Viewed in this way, these dazzling paintings also betray to us now something of the nature of the absolutism they proclaim. From our own perspective, the image of the iron maiden, the proprietary hand on the globe, the 'immobile', inflexible justice of the Case diagram, and the repudiation of the human in its entirety, all tell, perhaps in spite of themselves, part of the ruthless but still familiar story of monarchy and of empire.

Notes

1. As she proceeded through the streets of London the day before her coronation, Elizabeth smiled when a voice in the crowd called out, 'Remember old King Henry VIII' (J. E. Neale, *Oueen Elizabeth* [London: Jonathan Cape, 1938], p. 68). And in her youth Elizabeth was prone to draw attention to the likeness by standing in front of the Holbein portrait (Leah S. Marcus, *Puzzling Shakespeare: Local Reading and its Discontents* [Berkeley: University of California Press, 1988], p. 55).
2. Louis Adrian Montrose, 'The Elizabethan Subject and the Spenserian Text', in Patricia Parker and David Quint (eds.), *Literary Theory/Renaissance Texts* (Baltimore: Johns Hopkins University Press, 1986), 303–40, esp. pp. 312–15.
3. Roy Strong, *Gloriana: The Portraits of Queen Elizabeth I* (London: Thames and Hudson, 1987), pp. 132–3.
4. Marina Warner, *Monuments and Maidens: The Allegory of the Female Form* (London: Pan, 1987), pp. 241–66.
5. See in particular Strong, *Gloriana*; also Roy Strong, *Portraits of Queen Elizabeth I* (Oxford: Oxford University Press, 1963) and Frances A. Yates, *Astraea: The Imperial Theme in the Sixteenth Century* (London: Routledge, 1975).
6. Lucy Gent, *Picture and Poetry 1560–1622* (Leamington Spa: James Hall, 1981), p. 20.
7. Ibid., p. 21.
8. Strong, *Gloriana*. p. 87; John Pope-Hennessy, *The Portrait in the Renaissance* (London: Phaidon, 1966), p. 204.
9. Strong, *Gloriana*, pp. 85–9.
10. In 1575, the year of the 'Darnley' portrait, the Queen was represented in the royal entertainment at Kenilworth as a questing knight (Philippa Berry, *Of Chastity and Power: Elizabethan Literature and the Unmarried Queen* [London: Routledge, 1989], pp. 95–100). Leah S. Marcus discusses the heterogeneity of Elizabeth's gender-identifications (*Puzzling Shakespeare*, pp. 53–66). See also Winfried Schleiner, '*Divina Virago*: Queen Elizabeth as Amazon', *Studies in Philology*, 75 (1978), 163–80.
11. Warner, *Monuments and Maidens*, pp. 38–60.
12. For examples, see E. C. Wilson, *England's Eliza* (Cambridge, Mass.: Harvard University Press, 1939). Philippa Berry points to the contradictory implications of this representation of the Queen, which was not, of course, necessarily of her own making. While on the one hand it secured the service of her 'knights', on the other it permitted them to escape in foreign exploits from the disturbing control of a monarch who was also a woman, and to show

themselves capable military leaders, which Elizabeth herself could never be (*Of Chastity and Power*, pp. 76–7).

13. Strong, *Gloriana*, pp. 85–9.

14. Case (?1540/6–1600) was in his time an influential Oxford philosopher who published a number of works on Aristotelian themes. See Charles B. Schmitt, *John Case and Aristotelianism in Renaissance England* (Montreal: McGill-Queen's University Press, 1983). The *Sphæra civitatis* was an exposition of political theory based on parts of Aristotle's *Politics*.

15. Strong, *Gloriana*. p. 133.

16. 'There is no trace in [Ptolemy's] *Almagest* of what the Middle Ages and the Renaissance called the Ptolemaic System, namely the description of the Universe as consisting of planetary spheres nested one inside the other so as to fill the space between the highest sublunary sphere (that of fire) and the sphere of the fixed stars' ('Ptolemaic Astronomy', *Dictionary of the History of Science* [London: Macmillan, 1981], p. 350). It is historically more accurate to call this cosmology Aristotelian.

17. Schmitt, *John Case*, p. 87.

18. Yates, *Astraea*, p. 64.

19. Ibid., quoting Dante's *De Monarchia*. See also Leonard Barkan, *Nature's Work of Art: The Human Body as Image of the World* (New Haven: Yale University Press, 1975), ch. 2.

20. Strong, *Gloriana*, p. 133.

21. See S. J. Heninger, Jr., *The Cosmographical Glass: Renaissance Diagrams of the Universe* (San Marino, C.A.: Huntingdon Library, 1977).

22. On the ancient tradition, see Barkan, *Nature's Work of Art*, ch. 1; on Hildegard, see Barkan, ch. 3; on Fludd, see Frances A. Yates, *The Rosicrucian Enlightenment* (London: Routledge, 1972), esp. plate 21, which reproduces the engraved title-page of Fludd's *Utriusque cosmi . . . historia* (Oppenheim, 1617).

23. Lynn Thorndyke, *The Sphere of Sacrobosco and its Commentators* (Chicago: University of Chicago Press, 1949).

24. Heninger, *Cosmographical Glass*, p. 37; Thorndyke, *Sphere*. p. 41. Case knew Sacrobosco's work (Schmitt, *John Case*, p. 159).

25. Thorndyke, *Sphere*, p. 119.

26. The variations are graphically demonstrated in Heninger, *Cosmographical Glass, passim*. See also Edward Grant, 'Cosmology', and Olaf Pedersen, 'Astronomy', in David C. Lindberg (ed.), *Science in the Middle Ages* (Chicago: University of Chicago Press, 1986), pp. 265–302 and 303–37, respectively.

27. A further complication is that some astronomers placed a crystalline sphere beyond the fixed stars, which made the *primum mobile* the tenth sphere, and the sphere of Heaven, if there was one, the eleventh.

28. Heninger, *Cosmographical Glass*, p. 38. Finé (1532) had only eight spheres, with the function of the *primum mobile* being amalgamated with that of an unbounded Heaven beyond. Cortes, in a book which appeared in England in 1561, had eleven spheres, the outermost being described as 'The Empyreal Heaven, the Abitation of the Blesed'. For both Finé and Cortes, see Heninger, *Cosmographical Glass*, p. 39.

29. Ibid., p. 50.

30. Yates, *Astraea*, p. 65.

31. Ibid., plate 10b.

32. Strong, *Gloriana*, p. 39.
33. Ibid., p. 40.
34. Ibid., p. 43. See also Yates, *Astraea*, pp. 78–9.
35. Northern European painting shows glass mirrors in the fifteenth century, but they are normally convex (Pope-Hennessy, *The Portrait in the Renaissance*, p. 129).
36. *The Faerie Queene*, I. iv. 10. For an account of Renaissance mirrors moralised, see Ernest B. Gilman, *The Curious Perspective: Literary and Pictorial Wit in the Seventeenth Century* (New Haven: Yale University Press, 1978), pp. 167–77.
37. Paul Delany, *British Autobiography in the Seventeenth Century* (London: Routledge, 1969), pp. 12–13.
38. See Jacques Lacan, 'The Mirror Stage as Formative of the Function of the I as Revealed in Psychoanalytic Experience', *Ecrits: A Selection* (London: Tavistock, 1969), pp. 1–7. We do not imply that any of this would or could have been acknowledged in the sixteenth century.
39. The text of *The Winter's Tale* credits its sculptor, Julio Romano ('that rare Italian master') with powers of illusionism which fall (just?) short of God's: 'had he himself eternity and could put breath into his work, [he] would beguile nature of her custom, so perfectly he is her ape' (V. ii. 94–7). Shakespeare references are to the one volume edition of the *Complete Works*, ed. Peter Alexander (London: Collins, 1951).
40. Roger Lockyer (ed.), *The Trial of Charles I* (London: Folio Society, 1959), p. 135.
41. For an account of the succession controversy, see Marie Axton, *The Queen's Two Bodies: Drama and the Elizabethan Succession* (London: Royal Historical Society, 1977).
42. David Beers Quinn, *The Elizabethans and the Irish* (Ithaca, NY: Cornell University Press, 1966).
43. Kenneth R. Andrews, *Trade Plunder and Settlement: Maritime Enterprise and the Genesis of the British Empire, 1480–1630* (Cambridge: Cambridge University Press, 1984), p. 2.
44. Peter Ure (ed.), *King Richard II* (London: Methuen, 1961), pp. lvii–lix.
45. Philippa Berry identifies a crisis in the literary representation of the Queen during the 1590s, *Of Chastity and Power*, pp. 134–65.

6 Freedom of Dress in Revolutionary France

Lynn Hunt

'Freedom of dress' is not usually counted as one of the inalienable human rights central to modern democratic politics. It might be related to Jefferson's 'pursuit of happiness' in the Declaration of Independence, but freedom of dress hardly seems central to life, liberty, or the protection of property. Even so, dress, with both its freedoms and constraints, turned out to be one of the most hotly contested arenas of revolutionary cultural politics. It provided the most visible marker of both adherence and resistance to new social and political conceptions, including notions of gender definition. Its powers of signification went far beyond the vagaries of *la mode*, beyond even the identification of political groupings; it touched on the very definition of the new body politic and especially on the definitions of the many different bodies, both male and female, that made up that new body politic.

As in many other questions of cultural politics, French revolutionaries veered between two extremes in their attitude toward dress: on the one hand, they wanted to erase all the supposed legal encumbrances of the Old Regime and make what one wore yet another arena for free choice (and market forces); on the other hand, they harbored a more repressive wish to insist on consensus by enforcing completely new regulations about personal adornment. The cockade is a prime example of the latter impulse. The red, white, and blue cockade appeared on men's hats as early as 16 July 1789, spontaneously signaling adherence to the new nation. Over the years, however, the appearance of the cockade, its size, and the material used in making it all became the subject of controversy. Large woolen ones soon seemed

From Lynn Hunt, 'Freedom of Dress in Revolutionary France', in *From the Royal to the Republican Body: Incorporating the Political in Seventeenth- and Eighteenth-Century France*, eds. Sara Melzer and Kathryn Norberg (University of California Press, 1998), pp. 224–49. Reprinted with permission.

more democratic than small silk ones. On July 1792 the Legislative Assembly ordered all men to wear it by law, a requirement repeated by the National Convention on 3 April 1793.[1]

These laws apparently did not apply to women. When the Society of Republican and Revolutionary Women was founded in May 1793, one of their first acts demanded that members wear the cockade. The club also sent an address to the forty-eight sections of Paris inviting all women to follow their example. In the summer of 1793, a 'cockade war' broke out pitting club militants against market women who made a practice of ripping off the cockade to show their disdain for the female Jacobines and their mimicry of men. At the beginning of September, a group of women from another club initiated a petition drive to force the municipal and national governments to take action. In response, the city government forbade women to appear in public without a cockade. The National Convention, meanwhile, temporized, and street battles erupted once again. On 21 September the Convention finally ordered all women to wear the cockade, promising a six-year prison term as punishment for tearing off someone's cockade.

Matters did not end there. In October, the struggle shifted to the red liberty cap, frequently worn by female club members during their meetings. On the twenty-eighth of that month a pitched battle erupted during one of the meetings of the Republican and Revolutionary Women. Some of the women who had come to watch shouted, 'Down with the red cap, down with the Jacobines, down with the Jacobines and the cockade.'[2] Faced with a full-scale melee, the club members fled.

The next two days their opponents took the floor of the Convention. A deputation of women appeared to present a petition complaining of the efforts of 'supposedly revolutionary women' to force them to wear the red liberty cap. The president assured them that the Committee on General Security was in the midst of preparing a report on the subject. Fabre d'Eglantine then took the floor in an angry mood:

There has already been trouble over the cockade. You decreed that all women must wear it. Now they want the red cap: they won't rest there, they'll soon ask for a belt with pistols, in such a way that this will coincide perfectly with the maneuver of raising crowds about bread, and you will see lines of women going for bread like men march to the trenches. It is very adroit on the part of our enemies to attack the strongest passion of women, that is, their attire [ajustement].

Fabre went on to attack women's clubs, claiming, to much applause, that they were composed of 'adventuresses, wandering female knights, emancipated girls, and amazons'. He called for a decree forbidding all efforts to force changes in dress. The Convention then ruled that 'no person of either sex may constrain any citizen or citizeness to dress in a particular manner, each individual being free to wear whatever clothing or attire of its sex that pleases him, under pain of being declared suspect.' At the same time, and in somewhat contradictory fashion, the Convention explicitly maintained its previous decrees requiring the cockade.[3]

The next day, 30 October, Amar delivered the eagerly awaited report of the Committee on General Security. He began with a description of recent troubles: 'Several women, so-called Jacobines, of a Society that pretends to be revolutionary, promenaded this morning at the markets . . . in pantaloons and red liberty caps; they tried to force other women to adopt the same costume. . . . A crowd of some 6,000 women formed. All the women agree in saying that violence and threats will not force them to wear a costume that they honor but which they believe to be reserved to men.'[4] He went on to develop at some length the rationale for the exclusion of women from the public, political sphere and for the suppression of women's clubs as dangerous to public order and to the natural division of sexual functions. The Convention thereupon officially suppressed all women's political clubs.

Thus the conflicts over women's fashions ended in the simultaneous declaration of freedom of dress and the suppression of women's political organizations. A curious progression in concerns had developed in the course of these disputes. The deputies to the Convention wanted to enforce wearing the cockade as a badge of revolutionary belonging, but they drew the line when it came to women forcing other women to wear more controversial tokens of revolutionary enthusiasm such as the red liberty cap. Such women supposedly played on 'the strongest passion of women', their concern for dress, and turned it into a form of masculinization, wearing 'a costume reserved to men'. Dress had become too public an issue for women, and it signaled their intrusion into a public sphere seen as masculine (thus requiring masculine dress). The answer, it seems, was to reaffirm freedom of dress for women as long as their choice of dress remained confined to the private sphere and to feminine sartorial options. It would soon become apparent that men did not enjoy the same freedom of dress.

The declaration of freedom of dress obviously grew out of motives much more complicated than a simple break with Old Regime sumptuary laws. As Daniel Roche has argued, sumptuary laws had long been disregarded and were in any case caught up as much in royal monetarist policy as in efforts to defend the distinctiveness of the nobility.[5] The revolutionaries, however, had been habituated by their reading of the philosophes, especially Voltaire, to see in sumptuary regulation an attack on liberty.[6] Lying behind this conception of liberty was some notion of the operation of the market, but this was never fully elaborated either in the discussion of sumptuary regulation or in the many revolutionary discussions of dress.[7]

In 1789 the crown seemed to revive sumptuary legislation when it insisted that the deputies of the three orders wear distinctive costumes for the opening session of the Estates General: the clergy wore clerical costumes reflective of rank; the nobility wore hats with white feathers and clothing adorned by lace and gold; and the deputies of the Third Estate wore the sober black and three-corner hats of the magistrature. These prescriptions revived the spirit if not the letter of sumptuary laws and instantly politicized the question of dress. On 10 May 1789 Mirabeau published a letter from Jean-Baptiste Salaville denouncing the distinction in costume decreed by His Majesty's master of ceremonies as 'absurd', 'ridiculous', and 'the height of despotism and debasement'. The prescription of different dress for the different orders, he charged, only reinforced 'that unfortunate distinction between orders that can be regarded as the original sin of our nation'. Yet even Salaville did not conclude from this that all notions of official costume should be discarded; instead he argued that the National Assembly should be responsible for adopting any such costume, operating on the principle that all deputies represented the 'universality of the kingdom' and hence should be dressed exactly the same.[8] This same tension between dressing for difference and dressing for equality would appear again and again. For the moment, however, Salaville's suggestion fell on deaf ears; the deputies soon opted for the elimination of signs of difference between the three orders, leaving to each deputy the choice of individual costume.[9]

Freedom of dress had a very ambiguous history during the revolutionary decade because the deputies followed competing agendas: they wanted to wrench French society away from past habits of social distinction and at the same time underline the legitimacy of the new regime, in part by introducing distinctive markings for officials and also, of course, by maintaining gender differentiations. New kinds of functional distinctions had to replace the old objectionable social ones. On 20 May 1790, for instance, the National Assembly voted to dress all mayors in a tricolor sash.[10] In April 1792, in contrast, the Legislative Assembly decreed that members of the suppressed religious orders could no longer wear their customary costume. In proposing this decree, Bishop Torné explicitly rejected the argument that this constituted an attack on freedom of dress (*la liberté des vetements*):

Would it be permitted to one sex to wear indistinctly the clothing of one or the other sex? Do not the police prohibit masks and cockades that might be a sign of a party opposed to the Revolution? Do not the police prohibit clothing that undermines morals? And if the simple clothing of a citizen is susceptible to a multitude of wise regulations, would religious costume, which can entail so many abuses, then not be submitted to any police rule?[11]

Freedom of dress could not mean complete absence of regulation, at least not for men.

Thus, it is not surprising that the Convention did not explicitly declare freedom of dress until confronted with the very particular problem of women's political insignia. Long after they rendered their declaration, moreover, dress continued to be a subject of potential regulation, especially insofar as official costume was concerned. In March 1794 a local Parisian deputation appeared on the floor of the Convention to demand the suppression of judge's costumes 'because they seem to us to recall monarchical, feudal, and chivalric ideas; because the robe, by its form and its color, retraces the memories of nobles and priests, which contrasts too violently with our republican sentiments.'[12] Decrees requiring the cockade were repeated all during the Directory government, but at the same time, the government also restated its insistence on freedom of dress in order to prevent street fights over the color of clothing, black collars in particular being associated with royalism.[13] Dress had to be regulated in its official modes, free in its absolutely private ones, and was therefore a subject of constant contention in the areas in between (in particular the privtate/public wearing of the cockade).

Despite the best efforts of revolutionary legislators, dress necessarily crossed over the line separating private and public, especially since the distinction between public and private itself came under intense pressure with the politicization of many aspects of daily, private life. As the case of the Society of Republican and Revolutionary Women shows, the question of dress intersected at least two different axes of concern: the gender axis of the appropriate differentiation between men and women, and the political axis of the necessary differentiation between friends and enemies, insiders and outcasts, in the new revolutionary order.

Gender and political differentiation operate in every society, but during the French Revolution they became especially charged because their very definition came into question. The system of gender differentiation by dress underwent a profound transformation, what J. C. Flügel called 'The Great Masculine Renunciation'. As he put it, 'Men gave up their right to all the brighter, gayer, more elaborate and more varied forms of ornamentation, leaving these entirely to the use of women. . . . Man abandoned his claim to be considered beautiful.'[14] Men began to dress alike, whereas women's dress took on more of the burden of class signification. I will return to Flügel's argument in greater detail below; for now I want simply to suggest that disputes about appropriate female dress involved more than a simple reassertion of gender distinctions. They took place against a backdrop of subtle but momentous realignments of the gender/dress system.

Similarly, the very foundations of the political system shifted. A body was no longer defined by its place in a cosmic order cemented by hierarchy, deference, and readily readable dress; each individual body now carried within itself all the social and political meanings of the new political order, and these meanings proved very difficult to discern. With sovereignty diffused from the king's body out into the multiple bodies of the nation, the old codes of readability broke down and new ones had to be elaborated. The uncertainty of this situation was reflected in the seesaw of regulatory practice concerning dress.

The body assumed such significance in the French Revolution because the long-term shift from a sacred to a secular framework of legitimacy made the workings of the social both more visible and more problematic. In an important theoretical work on the role of the French Revolution in the process of secularizing legitimacy, Brian Singer offers what he calls 'a history of social visibility'. Using concepts first developed by Cornelius Castoriadis, Singer argues that the Revolution fundamentally altered the relation of visible powers to the larger society. Under the Old Regime, schematically speaking, society seemed to be given from without, transcendentally, that is by the will of God, with the king—the only visible power—embodying the presence of the divine. Society had no immanent, self-given status; society was only appearance or façade. The separation of state and society did not become possible, Singer insists, until society was successfully designated as the source of its own power. The Revolution completed the auto-institution of society, and it was this transference that made the revolutionary process revolutionary (and, some would argue, violent, because auto-institution is inherently unstable). The Revolution was made possible by the claim that society provided its own legitimation, and it was driven forward by disputes over the exact meaning of this claim.[15]

The development of a secular, societal framework for legitimacy does not mean that all notions of the sacred disappeared. Sacrality was still present but now displaced from the king's single body into the collectivity of bodies in the nation. As a result, the social came more closely into focus, and the bodies of individuals were even more intensely invested with significance. In other words, the Revolution brought into consciousness a new awareness of the social as a category. Bodies had to be closely examined for their meanings, but the system of decodification was constantly in flux. The strangest incidents of the Revolution can only be explained in this way, from the concern over the greenish putrefaction of Marat's body during his funeral to the executioner's slap of the severed head of Charlotte Corday, causing it to blush. All bodies had to be examined closely because all bodies now made up the body politic; every body literally embodied a piece of sovereignty, or at least some connection to it. As a consequence, the whole subject of appearance was valorized in new and particularly significant ways.

This long-term shift explains, as well, why the body is so central to the disciplinary practices analyzed by Michel Foucault. Almost all of Foucault's investigations concerned the body in the crucial transition period toward modernity: the seventeenth, eighteenth, and early nineteenth centuries.[16] This is the period of the secularization of legitimacy, the challenge to divine-right theories and the rise of socially immanent ones as a replacement. Although he never says so, Foucault's image of the dispersion of power rests on the eruption of democracy in the eighteenth century, and much of the characteristic postmodern problem of differentiating between democracy and totalitarianism can be traced back to Foucault's own inability or unwillingness to distinguish between the two. For him, all forms of bodily discipline are essentially the same: democracy and totalitarianism are equally carceral. Although both democracy and totalitarianism are made possible by the shift of sovereignty from the king to the nation, their common sources of enablement do not make them the same phenomenon.

We do not yet have a history of the body that enables us to distinguish clearly between totalitarian and democratic forms of discipline.[17] The question of dress provides an especially significant window onto this set of problems because dress was the field in which both gender and political differentiation were played out, and at times played upon, as the case of the Society of Republican and Revolutionary Women shows. Foucault did not discuss dress, just as he did not discuss gender (except in the most abstract terms), yet it seems likely that dress—and—gender, one of whose most visible components is dress—are crucial to modern conceptions of the individual, and by extension to the modern conception of democracy. Anne Hollander makes the link in *Sex and Suits* when she argues that the modern aesthetic principles of fashion reflect the modern democratic 'ideal of self-perpetuating order, flexible and almost infinitely variable'.[18]

Democracy, especially in practice, creates enormous tensions about the various forms of social and gender differentiation.[19] In the most radical moments of the French Revolution, the idea of social differentiation itself came under attack. There is no need to belabor the obvious signs of the disintegration of a body politic based on deference and the clearly readable signs of social status: the killings of the king and queen, the elimination of titles of nobility, the drive to repress the *vous* formal form of address and the title *Monsieur* and replace them with a universal *tu* and *Citoyen*, the imitation of the dress of the sans-culottes by various political figures, and so on. Even the recommended style of writing letters changed. According to *Le Secrétaire des républicains*, a letter-writing manual of 1793, those who had been reduced by misfortune to serving 'their equals' should no longer be called lackeys or domestics but rather 'homme ou femme de confiance'. Letters should be simply signed 'salut' or 'salut et fraternité.[20] This attack on an ancien régime based on deference and evident social distinctions inevitably created anxiety about the legitimacy of any form of social difference (class especially, but also to some extent race and gender).

This is not to say that most revolutionaries dreamed of a society utterly without distinctions. They wanted to install a new system of signs, marking individuals by their utility and republican virtue (and, of course, maintaining gender differentiations as well). Thus festivals instituted new kinds of functional divisions. More sinister forms of marking were suggested but not taken up. One deputy proposed in October 1791 that all nonjuring priests should wear a badge on the left breast reading, 'Priest, suspected of sedition'. Others suggested that prostitutes be forced to wear special colors. Saint-Just proposed that injured soldiers wear gold stars on their clothing over the spots of their wounds.[21] These all reflected an obvious concern with transparency, inspired by Rousseau: clothes should reveal the inner person, not provide a means of dissimulation.

At the center of anxiety about social differentiation stood the body, the preeminent site for revolutionary signification and symbolization.[22] Although the body was central to the revolutionary reformulations of political and social meaning, its clothing always raised more questions than it could solve. Clothing inevitably underlined political and social differences, whereas republicans wanted to emphasize sameness and consensus (except in the arena of gender, where the republicans insisted on difference). As a consequence, contemporary clothing came to seem virtually incompatible with republican ideals. Rarely did either Liberty or Hercules, two of the most current republican icons, appear in contemporary garb in revolutionary allegories. Roman dress and nudeness both had the effect of erasing the reference to contemporary social distinctions and enforcing the sense of consensus that stood above social differentiation. Pornography, the underside of revolutionary idealization, operated in a similar fashion. The nude pornographic scene made the basic materialist point that all bodies are alike. Clothing in pornography—showing monks or nuns in their customary dress, for example—served to inject a social distinction into the action, but by implication, such figures were in the process of losing their clothing too, of becoming like everyone else.[23]

To control the signifying possibilities of clothed bodies in the revolutionary festivals, organizers tended to rely on either uniforms or special dress. As Mona Ozouf explains, the revolutionaries tried to reduce social man to biological man in the festivals. In the Festival of the Supreme Being of June 1794, everyone had an indispensable part to play, 'whether as father or husband, mother or daughter, rich or poor, young or old'. Yet when one club in Compiègne suggested that a simple village girl march alongside an elegant citizeness, the festival organizers rejected the proposal because it would call too much attention to social divisions. Even rich and poor were to be alike in some fashion in the festivals. The program for the Festival of the Supreme Being called for young ladies to use powder with restraint and to bunch up their skirts in the Roman style. In other festivals women most often appeared in white dresses crowned with oak wreaths.[24]

Given the heightened importance of the social and its signification by the body, it might be presumed that men's and women's public appearances would be equally freighted. Or alternatively, men's dress might matter more than women's because men mattered politically whereas women presumably did not (women could not vote or hold office, and after October 1793 they could not even form their own political clubs). The story is more paradoxical, however, than either of these formulations would suggest. Put most schematically, men's dress mattered more politically, but the result was to make it more uniform-like over the long run; in other words, because of its greater potential political signification, it came to be evacuated of some of the most obvious signs of difference.

The long-term shift in the coding of men's dress has recently attracted attention from film and literary critics and art historians much more than from historians. At issue is 'The Great Masculine Renunciation', the homogenizing of male dress at the beginning of the nineteenth century to eliminate the obvious sartorial display that characterized aristocratic society.[25] The sumptuary legislation that had once guaranteed the accurate display of social rank for both men and women under the Old Regime gave way to a more implicit code of masculine sameness and female difference. As Kaja Silverman has argued, this historical shift provides the grounds for the modern equation of spectacular display with female subjectivity (that women are objects to be looked at by men), and the very fact that it is historical undermines all the psychoanalytic explanations (insofar as they are ahistorical) of the controlling male 'gaze', so central to film theory. In the past, men were as much subject to being looked at as women. The difference between beholding and being beheld is therefore not part of an intrinsic gender identity, whether imagined as biological or cultural.

Silverman only sketches out her argument, but it is very suggestive. The history of fashion shows ornate dress to have been a class rather than gender prerogative during the fifteenth, sixteenth, and seventeenth centuries, a prerogative protected by law. (The eighteenth century represents a transitional period in this regard, one in which commercial fashion steadily replaced sumptuary legislation in the determination of dress.)[26] Under the ancien fashion régime, extravagance in clothing signified aristocratic power and privilege; it was, as Silverman insists, 'a mechanism for tyrannizing over rather than surrendering to the gaze of the (class) other.'[27] The elegance and finery of male dress at least equaled that of female dress during the classical period—and indeed, since the end of the Roman empire; hence visibility was a male as much as or even more than a female attribute. Most commentators seem to agree that a major change took place in the eighteenth and early nineteenth centuries, but no one has offered a compelling explanation of just why and how it occurred then. Why did men give up their hair, for example, their long locks, their wigs—or as one observer puts it, their century and a half of immunity to the crowning injustice of nature?[28]

In other words, class distinctions within male dress blurred toward the end of the eighteenth century, while gender distinctions between men's and women's dress became all the clearer. (Thus French republicans could insist on all men wearing the cockade even while insisting on freedom of dress for women.) Women's dress carried all the messages of social distinction, while men's dress began to shun them. Eventually all men wore trousers of somber colors, rather than expressing their class differences by the length and color of their pants (silk, brightly colored breeches for the rich; gray or brown woolen trousers for the working poor). Trousers, according to Philippe Perrot, constituted one of the rare examples of a fashion moving up rather than down the social scale.[29] Men also now wore their own and only their own hair rather than wigs or powdered hair for the upper classes. They stopped using makeup, except on the stage. In short, they cut off all resemblances to female finery.

The Revolution did not transform the dress codes overnight, and men's dress continued to carry some of the freight of social distinction. Yet the Revolution did put enormous stress on male dress and on the signification of the male body. Mona Ozouf has remarked, for example, that women and children in processions and festivals often appeared indiscriminately, without social categorization, while men were always carefully ordered and categorized.[30] The presence of women and children was taken to symbolize the community as a whole rather than any particular social station or function (other than motherhood or childhood, the biological categories). Men had more individual representativity, as it were.

As a consequence, it was perhaps inevitable that the republican government would consider the introduction of uniforms not only for all officials but for all male citizens. As early as 1792 David made sketches for a new national male costume. According to John Moore, his sketches at that moment resembled 'the old Spanish dress, consisting of a jacket with tight trowsers, a coat without sleeves above the jacket, a short cloak which may either hang loose from the left shoulder or be drawn over both; a belt to which two pistols and a sword may be attached, a round hat and a feather'.[31] In 1794, on official request, David designed a national male costume that combined Renaissance and antique motifs with a tight-fitting tunic, a cloak, and classical tights.[32] The pantaloon and tights obviated the choice between breeches and trousers.

The Popular and Republican Society of the Arts defended the idea of a national costume on the grounds that it fulfilled the goal of 'announcing or recalling everywhere and at every instant the fatherland, of distinguishing French citizens from those of nations still branded by the irons of servitude; it would offer easy means for designating both the age and the diverse public functions of the citizens, without altering the sacred bases of equality.'[33] In other words, a national costume would reinforce the break with the Old Regime and mold republican citizens for the new one. Here freedom of dress for men has clearly been much attenuated, if not effaced altogether.

Although the government never manufactured these civil uniforms, Denon produced an engraving of David's sketch, and discussion of appropriate dress continued even after the fall of Robespierre.[34] *La Décade philosophique*, soon to be one of the leading Directorial journals, carried lengthy discussions of dress throughout the summer of 1794. One of the contributors to the journal recommended a new style of men's dress that in many ways resembled a uniform: tunics with sleeves that did not go below the elbow (so that men's muscles would be evident—'that is the beauty of men'); unpowdered hair falling just below the neck; no hats except when traveling because they were not healthy; simple slippers or sandals that left the toes free and could be taken off at the doorway; and a formless coat, 'a kind of portable tent'. Similar but much briefer recommendations were offered for women's dress: they should also wear tunics and similar shoes and wear their own hair held back in a simple knot, washed and lightly perfumed with ribbons as decoration. *La Décade* went on at some length about the inconveniences of women's wigs.[35]

The Directorial republic eventually established official costumes for all branches of government, and the Consulate and Empire continued the practice.[36] When Grégoire presented the official report of the Committee of Public Instruction on the subject of official costumes on 14 September 1795, he specifically referred to the emotions caused by dress regulations at the opening of the Estates General, a long six years before:

The suppression of orders, which had presupposed a difference between civil and political modes of existence, brought in its train the suppression of costumes; but the Constituent Assembly was wrong not to substitute a costume common to its members. From that moment the dignity of its sessions steadily diminished. The harm worsened until the epoch when the tyrants who oppressed the National Convention did all but put cleanliness and decency on the list of counterrevolutionary crimes and prided themselves on wearing their contempt for propriety on their sleeves.[37]

The legitimacy of the government seemed to depend on being able to represent itself coherently in an immediately apprehendable fashion. But navigating between the excessive social distinctions of the ancien régime and the erasure of all social distinctions under the Terror proved nearly impossible.

The official costumes now seem quaint and sometimes even bizarre, yet they did announce an important break with Old Regime habits. In the royal court uniforms appeared only on exceptional occasions, such as the crowning of the king, and most often they had a religious significance; secular dress codes for the various professions conveyed social status and rank. Revolutionary and Napoleonic uniforms, in contrast, signaled the representativity of power: an official carried the marks of popular sovereignty rather than of his specific position within a social order defined by relative proximity to the body of the king. Even the uniforms introduced in Napoleon's imperial court to reestablish links to the monarchy signaled different purposes: they served as a kind of livery with a touch of militarism.[38]

The efforts to enforce uniformity of male dress predictably produced reactions, even among ardent revolutionaries. While some men donned the red liberty cap and 'licked the dress of the sans-culottes in 1793–94, others stuck by their middle-class forms of dandyism. Robespierre, for example, always wore a powdered wig, knee breeches, buckled shoes, and silk shirts. After the end of the Terror, sartorial opposition became the order of the day, the most spectacular examples of which were the *incroyables* and *muscadins* (a word used in the mid-eighteenth century for scented fops) who appeared in 1795 in the aftermath of the fall of Robespierre. They usually wore knee breeches in the old style, powdered hair or shaggy side locks falling like spaniel's ears, and fancy footwear. Their costumes seemed designed to emphasize individuality of choice rather than any consistent form of royalist dress. In Paris they roamed the streets looking for a fight, singing anti-Jacobin songs (including the taunting 'Remettez vos culottes'), effacing revolutionary inscriptions, and ripping the cockade off suspected Jacobins. 'Remettez vos culottes' played on the different possibilities of the term 'sans-culottes', which could mean (in addition to simply 'without culottes') not a wearer of knee breeches and hence someone who wore trousers instead. One verse ran: 'Don't trust the intriguer / Who praises the indecent costume of our false patriots. / Don't push liberty to the point of letting down your pants, / Put your pants back on! [*Remettez vos cutottes*].' The other side in these costume street battles would grab a *muscadin* and tear off his black or green collar (considered counterrevolutionary colors) and shear his hair 'à la Titus', that is, in the new short style associated with neoclassical austerity and republican virility.[39]

Still, the trend toward the Great Masculine Renunciation was clearly present, even if vehemently resisted by the *incroyables* and later by Napoleon himself. Napoleon tried to reinstitute much of ancien régime fashion after 1804, wearing the kind of coat worn at the court of Louis XVI and gradually replacing his republican pantaloons with aristocratic knee breeches (in part to hide his increasing portliness). Certain trends, however, were not reversible. More and more men wore English-inspired tailored costumes in dark colors with few adornments. Although returning aristocrats tended to favor powdered hair and tight-fitting knee breeches in the old style, most middle-class men wore trousers or pantaloons and kept their hair in a natural style, whether tousled or à la Titus. Even Napoleon kept the Titus cut of his youth. Whereas women wore clothes to show their sex, especially in the transparent whites of the Directorial period, men began to dress to look alike.[40]

The trend toward masculine sameness already attracted commentary during the decade of revolution. In February 1792, for instance, the *Journal de la mode* announced that 'for some time now, men's apparel has hardly been worthy of attention … coats, for the most part, are brown or black; frock coats of the worst taste; vests are almost all red.'[41] In contrast, women's fashion changed at a vertiginous pace, which the *Journal de la mode* attributed to the disappearance of the nobility: 'Since the abolition of titles, many women can only distingnish themselves by a continual variety in their attire.' Such comments were appearing as early as August 1790.[42]

The fashion journals that sprang up after the end of the Terror sounded similar notes. In 1800, *Le Mois* offered a little history of French fashion in its pages, claiming that 'in the first centuries of the monarchy, the clothing of men varied more in its major forms than that of women.'[43] This was no longer true. The post-Thermidor fashion journals showed continual variation in women's fashion and commented little if at all on men's fashion. *The Journal des modes et nouveautés*, for example, claimed in the spring of 1799 that the Titus cut so dominated among men that even those who had to wear wigs to cover their baldness wore wigs with a Titus cut. In 1797 the same journal insisted that women's fashion changed all the time: 'The fashion now is to follow none. In a circle of thirty women, you will not see two hairstyles, two dresses, two get-ups that resemble each other.'[44] In his *Encore un tableau de Paris* published in 1800 (Year VIII), Charles Henrion claimed that 'in the past, [a fashion] lasted three or four months, and sometimes a whole semester; now, it changes every fifteen days, and there is never one pronounced enough to enjoy preference over the others. This comes from the fact that there is no longer a court, and as a result, no longer a rallying point for a fashion. In the olden days, *Versailles* set the tone.'[45]

Men's fashion seemed to fall into oppositional categories rather than multiple variations, no doubt because of the politicization of the revolutionary decade. According to a 1799 issue of *Le Mois*, for instance, 'men are divided into two classes for hairstyles only, the *Titus* and the *Powdered*'—and powdered hair was definitely losing out. In 1800, the same journal described 'a great quarrel' that had arisen between the *pantalonistes* (trouser wearers) and the *culotistes* (*sic*, wearers of knee breeches). In its view, the *pantalonistes* were bound to win because the new consuls and ministers gave their audiences in trousers.[46] The trend in men's clothing was there to be seen, even if the final outcome remained in doubt in the 1790s and early 1800s.

The post-Terror fashion journals carried forward the same ambivalence about women's fashion that had characterized their eighteenth-century precursors.[47] The Rousseauian disdain for women's use of their adornment to get their way still appeared in the very journals that sold precisely because they provided women with indicators of the latest fashions. *Le Mois*, for instance, reproduced the Rousseau line in 1800: 'Women, this enchanting sex, born for the happiness of one part of our sex and for the torment of the rest, women, I say, discontent with the little that the laws have done for them in the distribution of direct power, have sought in all epochs to acquire by cunning what they cannot reasonably hope to obtain by open force.' Clothing was obviously central to this cunning. Yet the same journal concluded that 'we must be fair; women dress infinitely better today than in the past, with more taste and lightness.'[48]

Women's dress seemed to elude all facile categorizations such as the purported divisions between men who wore powdered hair and those who wore Titus cuts or those who wore trousers and those who wore knee breeches. Yet one stylistic innovation in women's wear does catch the eye and call out for further investigation: the white muslin dress worn in an antique manner. White gowns in neoclassical style had appeared in fashionable and artistic circles during the neoclassical revival of the 1780s. Vigée-Le Brun held a famous 'Greek supper' in 1788 in which she and her female guests wore white draped like a tunic. In September 1789 the painter David's wife and other wives and daughters of leading artists dressed in white with tricolor cockades in their hair when they went to publicly donate their jewelry to the National Assembly. David then chose the same style of dress for his festivals. Manon Roland and Lucille Desmoulins wore simple white dresses to their own executions, perhaps to make the point that they were the true republicans, not their executioners.

A similar style dominated Directorial fashions. Louis-Sébastien Mercier complained in 1798 that there is 'not a *petite maîtresse*, not a *grisette*, who does not decorate herself on Sunday with an Athenian muslin gown, and who does not draw up the pendant folds on the right arm, in order to drop into the form of some antique or at least equal Venus *aux belles fesses*.'[49] The *Tableau général du goût, des modes et costumes de Paris* that appeared in 1799 published twenty fashion plates; of the fifteen that depicted women's dresses, nine showed women in white dresses and six showed women in dresses of all other colors. The journal concluded that 'white is so advantageous to women that the use of it is constant, even though it is the most costly.'[50] The style was immortalized in David's 1800 portrait of Madame Juliette Récamier. Napoleon, however, explicitly discouraged the style, which depended on expensive English muslins, in favor of French made silks.[51]

But what are we to make of this fashion? Can it be termed a kind of uniform for women, since it figured so prominently in republican festivals and portraiture (or at least in David's imagination)? Viewed in this way, and contrasted with the wild variation of the costume of the *incroyables* and the *muscadins*, we might be tempted to conclude that the Revolution drove women toward uniformity and sameness and men toward variation and playfulness, thus inverting Flügel's proposed gender/dress system for a time. Like trousers for men, with their origins in the lower classes, the simple-looking muslin dress, often worn with natural-looking, even untidy hair, had filtered up as a fashion from milliners' assistants and farm girls, having been first adopted by Marie Antoinette and her circle.[52]

Many different interpretations might fit the evidence of the white muslin dress draped in antique fashion. Did the white suggest the effacement of difference implicitly required by the republican project of a 'new man' (or, in this case, woman)? Did the drapery reinforce the revival of neoclassical austerity and simplicity associated with republicanism? Was it as well an anti-aristocratic, antisilk, probourgeois statement, since the dress depended on fabric provided by the nation of shopkeepers, the English? Or was the naturalness of the line—often coupled with a high waistline that emphasized the maternal in women—a kind of preromantic gesture, a carrying through of Rousseauian ideals for women?

Faced with a plethora of possible meanings, the historian should probably hesitate. Dress can mean many things, perhaps because changes in fashion have no deep rationale beyond the desire to attract the attention of others. That desire takes undeniably historical and social forms, but there is no clear logic to their evolution or to the choices made at any particular moment. As Anne Hollander has argued, 'The art of dress has its own autonomous history, a self-perpetuating flow of images derived from other images.'[53] Desires for naturalness and simplicity, for purity and rejection of tradition, for sensibleness and practicality can and all do take a myriad of different forms, largely in reaction to the images that came before. Adam Gopnik insists, 'The truth is that reason really doesn't have much to do with the reasons of fashion.'[54] The white muslin dress, after all, was expensive, and for all their depictions of its multiple variations, fashion journals never offered much in the way of an explanation for its popularity.

Although it is difficult, if not impossible, to make much of specific changes in male or female fashion, it nonetheless remains true that questions of dress more broadly conceived went to the heart of the Revolution in both its democratic and totalitarian aspects. At some moments, republicans clearly hoped to break with the previous aristocratic domination of society by taking over the fashion system and literally redesigning it, that is, by providing a new code of signs for a new kind of society. But who should control these signs? The market with its drive toward constant novelty, constant expenditure, constant wasteful excess? Fashionable women with their reputed taste for luxury and control through bodily presentation? The women of the Society of Revolutionary and Republican Women with their penchant for cross-dressing and purported masculinization? Male fashion writers and publishers with their ambivalences toward the very world from which they profited? Or the government that invented those slightly ridiculous official uniforms, half antique, half Renaissance, which never aroused any contemporary enthusiasm (thus proving that fashion could not be legislated)? There was no single answer to these questions, which in itself shows that dress could not be controlled by any one group or agency This was so because dress was the social itself, the arena in which all social and gender distinction came into being and came under fire. As such it embodied the effervescence, as Durkheim called it, of all social interaction. We can search for its rules, but those rules constantly escape from our grasp, just as social life itself changes form just when we think we know what it is.

Notes

1. Lynn Hunt, *Politics, Culture, and Class in the French Revolution* (Berkeley: University of California Press, 1984), 57–9.
2. For the most complete account, see Dominique Godineau, *The Women of Paris and Their French Revolution* (Berkeley: University of California Press, 1998); originally published as *Citoyennes tricoteuses: Les Femmes du peuple à Paris pendant la Révolution française* (Aix-en-Provence: Alinéa, 1988), 163–77.
3. *Gazette nationale ou le Moniteur universel*, no. 39 (9 Brumaire Year II/30 October 1793), recounting the session of 8 Brumaire Year II. All translations from the French, unless otherwise noted, are mine.
4. *Moniteur universel*, no. 40 (10 Brumaire Year II/31 October 1793), recounting the session of the preceding day.
5. Daniel Roche, *La Culture des apparences: Une Histoire du vetêment (XVIIe–XVIIIe siècle)* (Paris: Fayard, 1989), 54.
6. Old Regime sumptuary laws have apparently attracted only the attention of legal historians interested in the long-term development of the practice. See, for example, Etienne Giraudias, *Etude historique sur les lois somptuaires* (Poitiers, 1910). I have not been able to locate any source that discusses eighteenth-century sumptuary laws in detail.
7. In the eighteenth century the *marchandes de modes* suddenly increased in number; they were first attached to the *merciers* and then, after 1776, to the *plumassiers-fleuristes*. See Madeleine Delpierre, 'Rose Bertin, les marchandes de modes et la Révolution', in Madeleine Delpierre et al., *Modes et révolutions, 1780–1804* [Catalogue of an exposition at Musée de la mode et du costume, Palais Galliera, 8 February–7 May 1989] (Paris: Delpierre Editions Paris-Musées, 1989), 21–5.
8. *Lettre du comte de Mirabeau à ses commettans*, as published in the *Moniteur's* account of the early days of the Estates General, *Réimpression de l'Ancien Moniteur* (Paris, 1847), 1:27 (6–14 May 1789).
9. Jean-Marc Devocelle, 'D'un costume politique à une politique du costume: Approches théoriques et idéologiques du costume pendant la Révolution français', in Delpierre et al., *Modes*, 83–103.
10. Ibid., 85.
11. *Réimpression de l'Ancien Moniteur*, 12:62 (7 April 1792).
12. Ibid., 20:64 (8 Germinal Year II/28 March 1794).
13. See, for example, ibid., 28:490 (26 Brumaire Year V/18 November 1796;

requirement to wear the cockade); 28:770, 771 (30 Thermidor Year V/17 August 1797; reaffirmation of freedom of dress and troubles over black collars); 28:797 (21 Fructidor Year V/7 September 1797; troubles over a black costume).

14. J. C. Flügel, *The Psychology of Clothes* (New York: Hogarth Press, 1969), III.

15. Brian C. J. Singer; *Society, Theory, and the French Revolution: Studies in the Revolutionary Imaginary* (New York: Macmillan, 1986), esp. 5.

16. It might be argued that Foucault's chronological focus steadily shifted from the seventeenth century toward the nineteenth century from the time of his writing *Madness and Civilization* (first published 1961) to *The History of Sexuality: An Introduction* (1976). Only in his last two posthumous volumes did he shift away from this central period.

17. Dorinda Outram certainly does not provide it, since her work is fundamentally confused on just this issue; see *The Body and the French Revolution: Sex, Class, and Political Culture* (New Haven: Yale University Press, 1989).

18. Anne Hollander; *Sex and Suits* (New York: Alfred Knopf, 1994), 9. She does not, however, recognize that this same ideal may be more closely tied to the operation of the capitalist market than to democracy.

19. I discuss the tensions surrounding gender differentiation in Lynn Hunt, *The Family Romance of the French Revolution* (Berkeley: University of California Press, 1992).

20. As quoted in Janet Gurkin Altmam, 'Teaching the "People" to Write: The Formation of a Popular Civic Identity in the French Letter Manual', in *Studies in Eighteenth-Century Culture*, vol. 22, ed. Patricia B. Craddock and Carla H. Hay (East Lansing: Michigan State University Press 1992), 159–60.

21. Nicole Pellegrin, *Les Vêtements de la liberté: Abécédaire des pratiques vestimentaires en France de 1780 à 1800* (Aix-en-Provence: Alinea, 1989), 123–4.

22. On the meanings of the French word *corps* and its allegorical and symbolic development during the Revolution, see the remarkably rich work of Antoine de Baecque, *Le Corps de l'histoire: Metaphores et politique, 1770–1800* (Paris: Calmann-Lévy, 1993).

23. For more on the relevance of pornography to these questions, see Lynn Hunt (ed.), *The Invention of Pornography: Obscenity and the Origins of Modernity, 1500–1800* (New York: Zone Books, 1993).

24. Mona Ozouf, *Festivals and the French Revolution*, trans. Alan Sheridan (Cambridge, Mass.: Harvard University Press, 1988), 114–15.

25. Flügel relates this shift to the French Revolution and the idea of fraternity, but his main center of interest lies elsewhere than in a historical account. Hollander explicitly dismisses Flügel's thesis only to reincorporate it into her own analysis: speaking of the eighteenth century, she claims, 'men gradually came to look similar; and to desire to look similar' (*Sex and Suits*, 97; cf 22).

26. Jennifer Jones of Rutgers University is preparing a book on this transitional period. I have benefited from the observations she offers in her introduction.

27. Kaja Silverman, 'Fragments of a Fashionable Discourse', in *Studies in Entertainment: Critical Approaches to Mass Culture*, ed. Tania Modleski (Bloomington: Indiana University Press, 1986), 139–52, quote 139.

28. Quentin Bell as cited in ibid., 140.

29. Philippe Perrot, *Fashioning the Bourgeoisie: A History of Clothing in the Nineteenth Century*, trans. Richard Bienvenu (Princeton: Princeton University Press, 1994), 31.

30. Ozouf, *Festivals*, 41.

31. Aileen Ribeiro, *Fashion in the French Revolution* (London: B. T. Batsford, 1988), 101.

32. For more on David's uniform, see Jennifer Harris, 'The Red Cap of Liberty: A Study of Dress Worn by French Revolutionary Partisans, 1789–1794', *Eighteenth-Century Studies*, 14 (1981), 283–312.

33. As quoted in *La Décade philosophique, littéraire et positique* I (10 Floréal Year II), 62.

34. On previous projects for civil uniforms, including some for children's dress, see Devocelle, 'D'un costume politique', 89–90.

35. *La Décade philosophique, littéraire et politique*, 2 (30 Thermidor Year II) 136–43 (Lettre de Polyscope au Rédacteur de la Décade sur les costumes); esp. 2 (20 Fructidor Year II), 279–86 (Troisième lettre de Polyscope); and 3 (30 Vendémiaire Year II), 147 (on wigs). I am grateful to Elizabeth Colwill for bringing this discussion to my attention.

36. Hunt, *Politics, Culture, and Class*, 75–81. See also Devocelle, 'D'un costume politique', 92–7.

37. *Réimpression de l'Ancien Moniteur*, 25: 763 (3eme jour complémentaire Year III/19 September 1795).

38. For brief remarks on this subject, see Madeleine Delpierre, 'Le Retour aux costumes de cour sous le Consulate et l'Empire', in Delpierre et al., *Modes*, 33–9, esp. 38.

39. The best brief account is in Ribeiro, *Fashion*, 115–17.

40. It should be noted, however, that in 1796 women also started wearing a version of the Titus cut. The mode lasted until 1809 with several variations. See Françoise Vittu, '1780–1804, ou vingt ans de "revolution des tetes francaises"', in Delpierre et al., *Modes*, 41–57, esp. 51. Most commentary on this fashion for women seems to have come after 1800.

41. *Journal de la mode*, 5 February 1792, 2.

42. *Journal de la mode*, 5 August 1790, 2.

43. *Le Mois*, [Prairial] Year VIII, 288 (at this time the journal no longer specified the month; Prairial was determined here by counting backward).

44. *Journal des modes et nouveautés*, 15 Ventôse Year VII, 516; 20 Frimaire Year VI, 10.

45. Charles Henrion, *Encore un tableau de Paris* (Paris, Year VIII), 125.

46. *Le Mois*, Germinal Year VII, 18; Prairial Year VIII, 283–5.

47. On the eighteenth-century journals, see Roche, *Culture des apparences*, 447–76.

48. *Le Mois*, Prairial Year VIII, 286–7, 302.

49. My account of the white dress comes from Mercier, quoted in Ribeiro, *Fashion* 128–9.

50. *Tableau général du goût, des modes et costumes de Paris* (Paris, Year VII/1799), 130. This is the only year of the journal I could locate at the Bibliothèque Nationale; the issue in question was supposed to be the first of a series, but many fashion journals disappeared after a year or so.

51. Akiko Fukai, 'Rococo and Neoclassical Clothing', in *Revolution in Fashion:*

European Clothing, 1715–1815, ed. Jean Starobinski et al. (New York: Abbeville Press, 1989), 109–17, esp. 116.

52. Anne Hollander, *Seeing Through Clothes* (New York: Avon, 1975), 385.
53. Ibid., 311.
54. 'What It All Means: Fashion? Well, Here's the Main Thing: It's Fun', *New Yorker*, 7 November 1994, 16.

7 Gender, Race, and Nation
The Comparative Anatomy of 'Hottentot' Women in Europe, 1815–1817

Anne Fausto-Sterling

A note about language use: Writing about nineteenth-century studies of race presents the modern writer with a problem: how to be faithful to the language usage of earlier periods without offending contemporary sensibilities. In this chapter I have chosen to capitalize words designating a race or a people. At the same time, I will use the appellations of the period about which I write. Hence I will render the French word *Negre* as Negro. Some nineteenth-century words, especially 'Hottentot', 'primitive', and 'savage', contain meanings that we know today as deeply racist. I will use these words without quotation marks when it seems obvious that they refer to nineteenth- rather than twentieth-century usage.

A note about illustrations: This chapter is unillustrated for a reason. The obvious illustrations might include drawings and political cartoons of Sarah Bartmann or illustrations of her genitalia. Including such visual material would continue to state the question as a matter of science and to focus us visually on Bartmann as a deviant. Who could avoid looking to see if she really was different? I would have had to counter such illustrations with an additional discussion of the social construction of visual imagery. But this essay is meant to focus on the scientists who used Bartmann. Thus an appropriate illustration might be the architectual layout of the French Museum, where Cuvier worked, or something of that order. Failing to have in hand a drawing that keeps us focused on the construction and constructors of scientific knowledge, I felt it would be better to have none at all. Readers who are dying to see an image of Bartmann may, of course, return to any of the original sources cited.

From Anne Fausto-Sterling, 'Gender, Race, and Nation' in *Deviant Bodies: Critical Perspectives on Difference in Science and Popular Culture*, eds. Jennifer Terry and Jacqueline Urla (Indiana University Press, 1995), pp. 19–48. Reprinted with permission.

....

INTRODUCTION

....

In 1816 Saartje Bartman, a South African woman whose original name is unknown and whose Dutch name had been anglicized to Sarah Bartmann, died in Paris. Depending upon the account, her death was caused by smallpox, pleurisy, or alchohol poisoning (Cuvier 1817; Lindfors 1983; Gray 1979). Georges Cuvier (1769–1832), one of the 'fathers' of modern biology, claimed her body in the interests of science, offering a detailed account of its examination to the members of the French Museum of Natural History. Although now removed, as recently as the early 1980s a cast of her body along with her actual skeleton could be found on display in case #33 in the Musée de l'Homme in Paris; her preserved brain and a wax mold of her genitalia are stored in one of the museum's back rooms (Lindfors 1983; Gould 1985; Kirby 1953).[1]

During the last several years Bartmann's story has been retold by a number of writers (Altick 1978; Edwards and Walvin 1983; Gilman 1985).[2] These new accounts are significant. Just as during the nineteenth century she became a vehicle for the redefinition of our concepts of race, gender, and sexuality her present recasting occurs in an era in which the bonds of empire have broken apart, and the fabric of the cultural systems of the nations of the North Atlantic has come under critical scrutiny. In this article I once again tell the tale, focusing not on Bartmann but on the scientists who so relentlessly probed her body. During the period 1814–70 there were at least seven scientific descriptions of the bodies of women of color done in the tradition of classical comparative anatomy. What was the importance of these dissections to the scientists who did them and the society that supported them? What social, cultural, and personal work did these scientific forays accomplish, and how did they accomplish it? Why did the anatomical descriptions of women of color seem to be of such importance to biologists of the nineteenth century?

The colonial expansions of the eighteenth and nineteenth centuries shaped European science; Cuvier's dissection of Bartmann was a natural extension of that shaping. (By 'natural' I mean that it seemed unexceptional to the scientists of that era; it appeared to be not merely *good* science; it was forward-looking.) But a close reading of the original scientific publications reveals the insecurity and angst about race and gender experienced by individual researchers and the European culture at large. These articles show how the French scientific elite of the early nineteenth century tried to lay their own fears to rest. That they did so at the expense of so many others is no small matter.

CONSTRUCTING THE HOTTENTOT BEFORE 1800

Several of the African women who ended up on the comparative anatomists dissecting tables were called Hottentots or, sometimes, Bushwomen. Yet the peoples whom the early Dutch explorers named Hottentot had been extinct as a coherent cultural group since the late 1600s (Elphick 1977). Initially I thought written and visual descriptions would help me figure out these women's 'true' race; I quickly discovered, however, that even the depictions of something so seemingly objective as skin color varied so widely that I now believe that questions of racial origin are like will-o-the-wisps. Human racial difference, while some sense obvious and therefore 'real', is in another sense pure fabrication, a story written about the social relations of a particular historical time and then mapped onto available bodies.

As early as the sixteenth century, European travelers circling the world reported on the peoples they encountered. The earliest European engravings of nonwhites presented idyllic scenes. A depiction by Theodor de Bry from 1590, for example, shows Adam and Eve in the garden, with Native Americans farming peacefully in the background. The de Bry family images of the New World, however, transformed with time into savage and monstrous ones containing scenes of cannibalism and other horrors (Bucher 1981). Similarly, a representation of the Hottentots from 1595 (Raven-Hart 1967) shows two classically Greek-looking men standing in the foreground, with animals and a pastoral scene behind. A representation from 1627, however, tells a different story. A man and woman with yellow brown skin stand in the foreground. The man's hair is tied in little topknots; his stature is stocky and less Adonis-like than before, and he looks angry. The woman, naked except for a loincloth holds the entrails of an animal in her hand. One of her breasts is slung backwards over her shoulder, and from it a child, clinging to her back, suckles. As we shall see, the drawings of explorers discussed here in turn became the working background (the cited literature) of the racial studies of the early nineteenth century, which are presented in a format designed to connote scientific certainty.

The Adamic visions of newly discovered lands brought with them a darker side. Amerigo Vespucci, whose feminized first name became that of the New World, wrote that the women went about 'naked and libidinous; yet they have bodies which are tolerably beautiful' (Tiffany and Adams 1985: 64). Vespucci's innocents lived to be 150 years old, and giving birth caused them no inconvenience. Despite being so at one with nature, Vespucci found Native American women immoral. They had special knowledge of how to enlarge their lovers' sex organs, induce miscarriages, and control their own fertility (Tiffany and Adams 1985). The early explorers linked the metaphor of the innocent virgin (both the women and the virgin land) with that of the wildly libidinous female. As one recent commentator puts it:

Colonial discourse oscillates between these two master tropes, alternately positing the colonized 'other' as blissfully ignorant, pure and welcoming as well as an uncontrollable, savage, wild native whose chaotic, hysterical presence requires the imposition of the law, i.e., suppression of resistance. (Shohat 1991: 55)

From the start of the scientific revolution, scientists viewed the earth or nature as female, a territory to be explored, exploited, and controlled (Merchant 1980). Newly discovered lands were personified as female, and it seems unsurprising that the women of these nations became the locus of scientific inquiry. Identifying foreign lands as female helped to naturalize their rape and exploitation, but the appearance on the scene of 'wild women' raised troubling questions about the status of European women. Hence, it also became important to differentiate the 'savage' land/woman from the civilized female of Europe. The Hottentot in particular fascinated and preoccupied the nineteenth-century scientist/explorer—the comparative anatomist who explored the body as well as the earth. But just who were the Hottentots?

In 1652 the Dutch established a refreshment station at the Cape of Good Hope, which not long after became a colonial settlement. The people whom they first and most frequently encountered there were pastoral nomads, short of stature, with light brown skin, and speaking a language with unusual clicks. The Dutch called these people Hottentots, although in the indigenous language they were called Khoikhoi, which means 'men of men'. Within sixty years after the Dutch settlement, the Khoikhoi, as an organized, independent culture, were extinct, ravaged by smallpox and the encroachment of the Dutch. Individual descendants of the Khoikhoi continued to exist, and European references to Hottentots may have referred to such people.

Nevertheless, nineteenth-century European scientists wrote about Hottentots, even though the racial/cultural group that late-20th-century anthropologists believe to merit that name had been extinct for at least three-quarters of a century. Furthermore, in the eighteenth and nineteenth centuries Europeans often used the word 'Hottentot' interchangeably with the word 'Bushman'.[3] The Bushmen, or Khoisan, or hunter-gatherer Khoi, were (and are) a physically similar but culturally distinct people who lived contiguously with the Khoikhoi (Elphick 1977; Guenther 1980). They speak a linguistically related language and have been the object/subject of a long tradition of cultural readings by Euro-Americans (Haraway 1989; Lewin 1988; Lee 1992). In this chapter I look at studies with both the word 'Bushman/Bushwoman' and the word 'Hottentot' in the titles. Cuvier, for example, argued vehemently that Sarah Bartmann was a Bushwoman and not a Hottentot. The importance of the distinction in his mind will become apparent as the story unfolds.

CONSTRUCTING THE HOTTENTOT IN THE FRENCH MUSEUM OF NATURAL HISTORY

The encounters between women from southern Africa and the great men of European science began in the second decade of the nineteenth century when Henri de Blainville (1777–1850) and Georges Cuvier met Bartmann and described her for scientific circles, both when she was alive and after her death (Cuvier 1817; de Blainville 1816). We know a lot about these men who were so needful of exploring non-European bodies. Cuvier, a French Protestant, weathered the French Revolution in the countryside. He came to Paris in 1795 and quickly became the chair of anatomy of animals at the Museum of Natural History (Appel 1987; Flourens 1845). Cuvier's meteoric rise gave him considerable control over the future of French zoology. In short order he became secretary of the Académie des Sciences, an organization whose weekly meetings attracted the best scientists of the city, professor at the museum and the College de France, and member of the Council of the University. Henri de Blainville started out under Cuvier's patronage. He completed medical school in 1808 and became an adjunct professor at the Faculté des Sciences, while also teaching some of Cuvier's courses at the museum. But by 1816, the year his publication on Sarah Bartmann appeared, he had broken with Cuvier.

After obtaining a new patron, he managed, in 1825, to enter the Academie and eventually succeeded Cuvier, in 1832, as chair of comparative anatomy.

Cuvier and de Blainville worked at the Musée d'Histoire Naturelle, founded in 1793 by the Revolutionary Convention. It contained ever-growing collections and with its 'magnificent facilities for research became the world center for the study of the life sciences' (Appel 1987: 11). Work done in France from 1793–1830 established the study of comparative anatomy, paleontology, morphology, and what many see as the structure of modern zoological taxonomy. Cuvier and de Blainville used the museum's extraordinary collections to write their key works. Here we see one of the direct links to the earlier periods of exploration. During prior centuries private collectors of great wealth amassed large cabinets filled with curiosities—cultural artifacts and strange animals and plants. It was these collections that enabled the eighteenth-century classifiers to begin their work.

Bruno Latour identifies this process of collection as a move that simultaneously established the power of Western science and domesticated the 'savage' by making 'the wilderness known in advance, predictable' (Latour 1987: 218). He connects scientific knowledge to a process of accumulation, a recurring cycle of voyages to distant places in which the ships returned laden with new maps, native plants, and sometimes even the natives themselves. Explorers deposited these mobile information bits at centers, such as museums or the private collections that preceded them. Scientists possessed unique knowledge merely by working at these locations, which enabled them literally to place the world before their eyes without ever leaving their place of employ. Latour writes: '[T]hus the history of science is in large part the history of the mobilization of anything that can be made to move and shipped back home for this universal census' (Latour 1987: 225). Cuvier literally lived, 'for nearly forty years, surrounded by the objects which engrossed so great a portion of his thoughts' (James 1830: 9). His house on the museum grounds connected directly to the anatomy museum and contained a suite of rooms, each of which held material on a particular subject. As he worked, he moved (along with his stove) from one room to the next, gathering his comparative information, transported from around the world to the comfort of his own home (Coleman 1964).

As centers of science acquired collections, however, they faced the prospect of becoming overwhelmed by the sheer volume of things collected. In order to manage the flood of information, scientists had to distill or summarize it. Cuvier, de Blainville, and others approached the inundation by developing coherent systems of animal classification. Thus the project of classification comprised one aspect of domesticating distant lands. The project extended from the most primitive and strange of animals and plants to the most complex and familiar. The history of classification must be read in this fashion; the attention paid by famous scientists to human anatomy cannot be painted on a separate canvas as if it were an odd or aberrant happening within the otherwise pure and noble history of biology.

During the French Revolution the cabinets of the wealthy who fled the conflict, as well as those from territories that France invaded, became part of the museum's collections. The cabinet of the Stadholder of Holland, for example, provided material for several of Cuvier's early papers. Appel describes the wealth of collected material:

> ... in 1822, the Cabinet contained 1500 mammals belonging to over 500 species, 1800 reptiles belonging to over 700 species, 5000 fishes from over 2000 species, 25,000 arthropods ... and an unspecified number of molluscs. ... (Appel 1987: 35–36)

Cuvier's own comparative anatomy cabinet contained still more. He championed the idea that, in order to classify the animals, one must move beyond their mere surface similarities. Instead, one must gather facts and measurements from all of the internal parts. Without such comparative information, he believed, accurate classification of the animals became impossible. By 1822, among the 11,486 preparations in Cuvier's possession were a large number of human skeletons and skulls of different ages and races.

The human material did not innocently fall his way. In fact he had complained unbelievingly 'that there is not yet, in any work, a detailed comparison of the skeletons of a Negro and a white' (Stocking 1982: 29). Wishing to bring the science of anatomy out of the realm of travelers' descriptions, Cuvier offered explicit instructions on how to procure human skeletons. He believed skulls to be the most important evidence, and he urged travelers to nab bodies whenever they observed a battle involving 'savages.' They must then 'boil the bones in a solution of soda or caustic potash and rid them of their flesh in a matter of several hours' (Stocking 1982: 30). He also suggested methods of preserving skulls with flesh still intact, so that one could examine their facial forms.

As we shall see, Egyptian mummies—both animal and human—supplied another significant source that Cuvier used to develop and defend his theories of animal classification. These he obtained from the travels of his mentor-turned-colleague, and eventual archenemy, Étienne Geoffroy Saint-Hilaire. Geoffroy Saint-Hilaire spent several years in Egypt as part of the young general Napoleon Bonaparte's expedition. Cuvier declined the opportunity, writing that the real science could be done most efficiently by staying at home in the museum, where he had a worldwide collection of research objects at his fingertips (Outram 1984).[4] In 1798 Bonaparte took with him the Commission of Science and the Arts, which included many famous French intellectuals. During his years in Egypt, Geoffroy Saint-Hilaire collected large numbers of animals and, of particular importance to this story, several human and animal mummies. By 1800, British armies had defeated the French in Egypt; the capitulation agreement stipulated that the British were to receive all of the notes and collections obtained by the French savants while in Egypt. But in a heroic moment, Geoffroy Saint-Hilaire refused. In the end he kept everything but the Rosetta stone, which now resides in the British Museum (Appel 1987). Once again we see how the fortunes of modern European science intertwined with the vicissitudes of colonial expansion.

Cuvier and de Blainville used the technologies of dissection and comparative anatomy to create classifications. These reflected both their scientific and their religious accounts of the world, and it is from and through these that their views on race, gender, and nation emerge. In the eighteenth century the idea of biologically differing races remained undeveloped. When Linnaeus listed varieties of men in his *Systema Naturae* (1758), he emphasized that the differences between them appeared because of environment. There were, of course, cross-currents. Proponents of the Great Chain of Being placed Hottentots and Negroes on a continuum linking orangutans and humans. Nevertheless, 'eighteenth-century writers did not conceptualize human diversity in rigidly hereditarian or strictly physical terms. . . .' (Stocking 1987: 18).

Cuvier divided the animal world into four branches: the vertebrates, the articulates, the molluscs, and the radiates. He used the structure of the nervous system to assign animals to one of these four categories. As one of his successors and hagiographers wrote, 'the nervous system is in effect the entire animal, and all the other systems are only there to serve and maintain it. It is the unity and the multiplicity of forms of the nervous system which defines the unity and multiplicity of the animal kingdom' (Flourens 1845: 98).[5] Cuvier expected to find similarities in structure within each branch of the animal world. He insisted, however, that the four branches themselves existed independently of one another. Despite similarities between animals within each of his branches, he believed that God had created each individual species (which he defined as animals that could have fertile matings). As tempting as the interrelatedness was to many of his contemporaries, Cuvier did not believe that one organism evolved into another. There were no missing links, only gaps put there purposely by the Creator. 'What law is there,' he asked, 'which would force the Creator to form unnecessarily useless organisms simply in order to fill gaps in a scale?' (Appel 1987: 137).[6]

Cuvier's emphasis on the nervous system makes it obvious why he would consider the skull, which houses the brain, to be of utmost importance in assigning animals to particular categories. It takes on additional significance if one remembers that, unlike present-day taxonomists, Cuvier did not believe in evolution. At least in theory, he did not build the complex from the primitive, although his treatment of the human races turns out to be more than a little ambiguous in this regard. Instead he took the most complex as the model from which he derived all other structures. Because humans have the most intricate nervous system, they became the model to which all other systems compared. In each of his *Leçons d'anatomie comparée*, he began with human structures and developed those of other animals by comparison (Coleman 1964). In this sense, his entire zoological system was homocentric.

Cuvier's beliefs about human difference mirror the transition from an eighteenth-century emphasis on differences in levels of 'civilization' to the nineteenth-century construction of race. His work on Sarah Bartmann embodies the contradictions such a transition inevitably brings. In 1790, for example, he scolded a friend for believing that Negroes and orangutans could have fertile matings and for thinking that Negroes' mental abilities could be explained by some alleged peculiarity in brain structure (Stocking 1982). By 1817, however, in his work on Sarah Bartmann, he brandished the skull of an Egyptian mummy, exclaiming that its structure proved that Egyptian civilization had been created by whites from whom present-day Europeans had descended (Cuvier 1817)[7]

Cuvier believed in theory that all humans came from a single creation, a view we today call monogeny. He delineated three races: Caucasians, Ethiopians (Negroes), and Mongolians. Despite uniting the three races under the banner of humanity (because they could interbreed), he found them to contain distinct physical differences, especially in the overall structure and shape of the head. One could not miss the invisible capabilities he read from the facial structures:

It is not for nothing that the Caucasian race had gained dominion over the world and made the most rapid progress in the sciences while the Negroes are still sunken in slavery and the pleasures of the senses and the Chinese [lost] in the [obscurities] of a monosyllabic and hieroglyphic language. The shape of their head relates them somewhat more than us to the animals. (Coleman 1964: 166)

Cuvier, it is worth noting, was opposed to slavery. His was 'a beneficent but haughty paternalism. . . .' (Coleman 1964: 167). In practice, however, his brother Frédéric, writing 'under the authority of the administration of the Museum' (i.e., brother Georges), would include Georges Cuvier's description of Sarah Bartmann as the only example of the human species listed in his *Natural History of the Mammals* (Geoffroy Saint-Hilaire and Cuvier 1824: title page). Accompanying the article were two dramatic illustrations similar in size, style, and presentation to those offered for each of the forty-one species of monkeys and numerous other animals described in detail. The Hottentots' inclusion as the only humans in a book otherwise devoted to mammalian diversity suggests quite clearly Cuvier's ambivalence about monogeny and the separate creation of each species. Clearly, his religious belief system conflicted with his role in supporting European domination of more distant lands. Perhaps this internal conflict generated some of the urgency he felt about performing human dissections.

Other scientists of this period also linked human females with apes. While they differentiated white males from higher primates, using characteristics such as language, reason, and high culture, scholars used various forms of sexual anatomy—breasts, the presence of a hymen, the structure of the vaginal canal, and the placement of the urethral opening—to distinguish females from animals. Naturalists wrote that the breasts of female apes were flabby and pendulous—like those in the travelers' accounts of Hottentots (Schiebinger 1993). Cuvier's description of Sarah Bartmann repeats such 'observations'. The Hottentot worked as a double trope. As a woman of color, she served as a primitive primitive: she was both a female and a racial link to nature—two for the price of one.

Although Cuvier believed that the human races had probably developed separately for several thousand years, there were others, who we today call polygenists, who argued that the races were actually separate species (Stepan 1982). Presentations such as those in the

Natural History of the Mammals provided fuel for the fire of polygeny. Cuvier's system of zoological classification, his focus on the nervous system, and his idea that species were created separately laid the foundations for the nineteenth-century concepts of race (Stocking 1982, 1987; Stepan 1982).

IN SEARCH OF SARAH BARTMANN

In contrast to what we know about her examiners, little about Bartmann is certain. What we do know comes from reading beneath the surface of newspaper reports, court proceedings, and scientific articles. We have nothing directly from her own hand. A historical record that has preserved a wealth of traces of the history of European men of science has left us only glimpses of the subjects they described. Hence, from the very outset, our knowledge of Sarah Bartmann is a construction, an effort to read between the lines of historical markings written from the viewpoint of a dominant culture. Even the most elementary information seems difficult to obtain. Cuvier wrote that she was twenty-six when they met and twenty-eight when she died, yet the inscription in the museum case that holds her body says that she was thirty-eight (Kirby 1949). She is said to have had two children by an African man, but de Blainville (1816) says that she had one child. One source says that the single child was dead by the time Bartmann arrived in Europe. According to some accounts, she was the daughter of a drover who had been killed by Bushmen. According to others, she was herself a Bushwoman (Altick 1978; Cuvier 1817). One London newspaper referred to her as 'a Hottentot of a mixed race', while a twentieth-century writer wrote that he was 'inclined to the view that she was a Bushwoman who possessed a certain proportion of alien blood' (Kirby 1949: 61).

Some sources state that Bartmann was taken in as a servant girl by a Boer family named Cezar. In 1810 Peter Cezar arranged to bring her to London, where he put her on exhibition in the Egyptian Hall of Picadilly Circus.[8] She appeared on a platform raised two feet off the ground. A 'keeper' ordered her to walk, sit, and stand, and when she sometimes refused to obey him, he threatened her. The whole 'performance' so horrified some that abolitionists brought Cezar to court, charging that he held her in involuntary servitude. During the court hearing on November 24, 1810, the following claims emerged: the abolitionists charged that she was 'clandestinely inveigled' from the Cape of Good Hope without the permission of the British governor, who was understood to be the guardian of the Hottentot nation 'by reason of their general imbecile state' (Kirby 1953: 61). In his defense, her exhibitor presented a contractual agreement written in Dutch, possibly after the start of the court proceedings. In it Bartmann 'agreed' (no mention is made of a signature, and I have not examined the original), in exchange for twelve guineas per year, to perform domestic duties for her master and to be viewed in public in England and Ireland 'just as she was'. The court did not issue a writ of habeas corpus because—according to secondhand accounts—Bartmann testified in Dutch that she was not sexually abused, that she came to London of her own free will in order to earn money, and that she liked London and even had two 'black boys' to serve her, but that she would like some warmer clothes. Her exhibition continued and a year later, on December 7, 1811, she was baptized in Manchester, 'Sarah Bartmann a female Hottentot of the Cape of Good Hope born on the Borders of Caffraria' (Kirby 1953: 61). At some point prior to 1814, she ended up in Paris, and in March of 1815 a panel of zoologists and physiologists examined her for three days in the Jardin du Roi. During this time an artist painted the nude that appears in Geoffroy Saint-Hilaire and Cuvier's tome (1824). In December of 1815, she died in Paris, apparently of smallpox, but helped along by a misdiagnosis of pleurisy and, according to Cuvier, by her own indulgence in strong drink.

Why was Bartmann's exhibition so popular? Prior to the nineteenth century there was a small population of people of color living in Great Britain. They included slaves, escaped slaves, and the children of freedmen sent to England for an education. Strikingly, the vast majority of the nonwhite population in England was male. Thus, even though people of color lived in England in 1800, a nonwhite female was an unusual sight (Walvin 1973). This, however, is an insufficient explanation. We must also place Bartmann's experiences in at least two other contexts: the London entertainment scene and the evolving belief systems about sex, gender, and sexuality.

The shows of London and those that traveled about the countryside were popular forms of amusement. They displayed talking pigs, animal monsters, and human oddities—the Fattest Man on Earth, the Living Skeleton, fire-eaters, midgets, and giants. Bartmann's exhibition exemplifies an early version of ethnographic displays that became more complex during the nineteenth century. After her show closed, 'the Venus of South America' appeared next. Tono Maria, a Botocudo Indian from Brazil, publically displayed the scars (104 to be exact) she bore as punishment for adulterous acts. In time, the shows became more and more elaborate. In 1822 an entire grouping of Laplanders shown in the Egyptian Hall drew 58,000 visitors over a period of a few months. Then followed Eskimos and, subsequently, a 'family grouping' of Zulus, all supposedly providing live demonstrations of their 'native' behaviors. Such displays[9] may be seen as a living, nineteenth-century version of the early-twentieth-century museum diorama, the sort that riveted my attention in the American Museum of Natural History when I was a child. The dioramas, while supposedly providing scientifically accurate presentations of peoples of the world, instead offer a Euro-American vision of gender arrangements and the primitive that serves to set the supposedly 'civilized' viewer apart, while at the same time offering the reassurance that women have always cooked and served, and men have always hunted (Haraway 1989).

Sometimes the shows of exotic people of color involved complete fabrication. A Zulu warrior might really be a black citizen of London, hired to play the part. One of the best documented examples of such 'creativity' was the performer 'Zip the What-is-it', hired and shown by P. T. Barnum. In one handbill, Zip was described as having been 'captured by a party of adventurers while they were in search of the Gorilla. While exploring the river Gambia . . . . they fell in with a race of beings never before discovered . . . in a PERFECTLY NUDE STATE, roving among the trees. . . . in a manner common to the Monkey and the Orang Outang' (Lindfors 1983: 96). As it turns out, Zip was really William Henry Johnson, an African American from Bridgeport, Connecticut. He made what he found to be good money, and in exchange kept mum about his identity. Interviewed in 1926, at the age of 84, while still employed at Coney Island, he is reported to have said, 'Well, we fooled 'em a long time, didn't we?' (Lindfors 1983: 98).

The London (and in fact European) show scene during the nineteenth century became a vehicle for creating visions of the nonwhite world.[10] As the century progressed, these visions 'grew less representative of the African peoples they . . . were meant to portray. . . . Black Africa was presented as an exotic realm beyond the looking glass, a fantasy world populated by grotesque monsters—fat-arsed females, bloodthirsty warriors, pre-verbal pinheads, midgets and geeks' (Lindfors 1983: 100). From this vision Britain's 'civilizing colonial mission' drew great strength. And it is also from this vision, this reflection of the other, that Europe's self-image derived; the presentation of the exotic requires a definition of the normal. It is this borderline between normal and abnormal that Bartmann's presentation helped to define for the Euro-American woman.

Bartmann's display linked the notion of the wild or savage female with one of dangerous or uncontrollable sexuality. At the 'perform-ance's' opening, she appeared caged, rocking back and forth to emphasize her supposedly wild and potentially dangerous nature. The *London Times* reported, 'She is dressed in a colour as nearly resem-bling her skin as possible. The dress is contrived to exhibit the entire frame of her body, and spectators are even invited to examine the peculiarities of her form' (Kirby 1949: 58). One eyewitness recounted with horror the poking and pushing Bartmann endured, as people tried to see for themselves whether her buttocks were the real thing. Prurient interest in Bartmann became explicit in the rude street ballads and equally prurient cartoons that focused on her steatopygous backside.[11]

According to the *Oxford English Dictionary*, the term *steatopygia* (from the roots for fat and buttocks) was used as early as 1822 in a traveler's account of South Africa, but the observer said the 'condition' was not characteristic of all Hottentots nor was it, for that matter, characteristic of any particular people. Later in the century, what had been essentially a curiosity found its way into medical textbooks as an abnormality. According to Gilman, by the middle of the nineteenth century the buttocks had become a clear symbol of female sexuality; and the intense interest in the backside, a displacement for fascination with the genitalia. Gilman concludes, 'Female sexuality is linked to the image of the buttocks, and the quintessential buttocks are those of the Hottentot' (Gilman 1985: 210).[12] Female sexuality may not have been the only thing at stake in all of the focus on Bartmann's backside. In this same historical period, a new sexual discourse on sodomy also developed. Male prostitutes, often dressed as women, walked the streets of London (Trumbach 1991), and certainly at a later date the enlarged buttocks became associated with female pros-titution (Gilman 1985). Until more historical work is done, possible relationships between cultural constructions of the sodomitical body and those of the steatopygous African woman will remain a matter of speculation.

Bartmann's story does not end in England. Her presentation in Paris evoked a great stir as well. There was a lively market in prints showing her in full profile; crowds went to see her perform. And she became the subject of satirical cartoons filled with not particularly subtle sexual innuendo. The French male's sexual interest in the exotic even became part of a one-act vaudeville play in which the male protagonist declares that he will love only an exotic woman. His good, white, middle-class cousin, in love with him, but unable to attract his attention, disguises herself as the Hottentot Venus, with whom he falls in love, making the appropriate mating, even after the fraud is revealed. (The full story has many more twists and turns, but this is the 'Cliff Notes plot' [Lindfors 1983: 100].)

Of all the retellings of Bartmann's story, only Gould's attempts to give some insight into Bartmann's own feelings. We can never see her except through the eyes of the white men who described her. From them we can glean the following: first, for all her 'savageness', she spoke English, Dutch, and a little French. Cuvier found her to have a lively, intelligent mind, an excellent memory, and a good ear for music. The question of her own complicity in and resistance to her exploitation is a very modern one. The evidence is scant. During her 'performances' 'she frequently heaved deep sighs; seemed anxious and uneasy; grew sullen, when she was ordered to play on some rude instrument of music' (Altick 1978: 270). Writing in the third person, de Blainville, who examined her in the Jardin du Roi, reported the following:

Sarah appears good, sweet and timid, very easy to manage when one pleases her, cantankerous and stubborn in the contrary case. She appears to have a sense of modesty or at least we had a very difficult time convincing her to allow herself to be seen nude, and she scarcely wished to remove for even a moment the handkerchief with which she hid her organs of generation. . . . [H]er moods were very changeable; when one believed her to be tranquil and well-occupied with something, suddenly a desire to do something else would be born in her. Without being angry, she would easily strike someone. . . . [S]he took a dislike to M. de Blainville, probably because he came too near to her, and pestered her in order to obtain material for his description; although she loved money, she refused what he offered her in an effort to make her more docile. . . . She appeared to love to sleep: she preferred meat, especially chicken and rabbit, loved (alcoholic) spirits even more and didn't smoke, but chewed tobacco. (de Blainville 1816: 189)

In this passage, de Blainville expressed the same conflicts evinced two centuries earlier by Vespucci. He found her to be modest, good, sweet, and timid (like any modern, 'civilized' Frenchwoman), but he could not reconcile this observation with what seemed to him to be the remnants of some irrational wildness (including habits such as chewing tobacco), which were out of line for any female he would wish to call civilized.

It is also worth comparing de Blainville's language to that used by Geoffroy Saint-Hilaire and F. Cuvier in the *Natural History of the Mammals*. In the section describing *Cynocephalus* monkeys (which follows immediately on the heels of Sarah Bartmann's description), they write that 'one can see them pass in an instant from affection to hostility, from anger to love, from indifference to rage, without any apparent cause for their sudden changes' (Geoffroy Saint-Hilaire and Cuvier 1824: 2). They write further that the monkeys are 'very lascivious, always disposed to couple, and very different from other animals, the females receive the males even after conception' (Geoffroy Saint-Hilaire and Cuvier 1824: 3). Clearly, de Blainville's language echoes through this passage framing the scientists' concerns about human animality and sexuality.

CONSTRUCTING THE (NONWHITE) FEMALE

Although a theater attraction and the object of a legal dispute about slavery in England, it was in Paris, before and after her death, that Bartmann entered into the scientific accounting of race and gender. This part of the story takes us from Sarah's meeting with scientists in the Jardin du Roi to her death, preservation, and dissection by Georges Cuvier—and to other scientific and medical dissections of nonwhites in the period from 1815 to, at least, the 1870s.[13]

The printed version of de Blainville's report to the Société Philomatique de Paris (given orally in December of 1815 and appearing in the Society's proceedings in 1816) offers two purposes for the publication. The first is 'a detailed comparison of this woman [Sarah Bartmann] with the lowest race of humans, the Negro race, and with the highest race of monkeys, the orangutan', and the second was to provide 'the most complete account possible of the anomaly of her reproductive organs' (de Blainville 1816: 183). De Blainville accomplished his first purpose more completely than his second. On more than four occasions in this short paper he differentiates Bartmann from 'Negroes', and throughout the article suggests the similarity of various body structures to those of the orangutan.

De Blainville began with an overall description of Sarah Bartmann's body shape and head. He then systematically described her cranium (one paragraph), her ears (two long, detailed paragraphs), her eyes (one paragraph), and other aspects of her face (five paragraphs, including one each devoted to her nose, teeth, and lips). In terms of printed space, her facial structure was the most important aspect. The final segment of his paper includes brief accounts (one paragraph each) of her neck, trunk, and breasts. In addition, he briefly described her legs, arms, and joints, devoting a full paragraph complete with measurements, to her steatopygous buttocks.

De Blainville's attempts to get a good look at her pudendum, especially at the 'Hottentot apron', which Cuvier finally succeeded in describing only after her death, were foiled by her modesty (see above). Despite this, de Blainville offers three full paragraphs of description. He verbally sketches the pubis, mentioning its sparse hair covering, and lamenting that, from a frontal view one cannot see the vaginal labia majora, but that, when she leaned over or when one watched from behind as she walked, one could see hanging append-ages that were probably the sought-after elongated labia minora.

De Blainville's ambivalences emerge clearly in the written text. He placed Bartmann among other females by reporting that she menstru-ated regularly, 'like other women', but noted that she wasn't really like white women because her periodic flow 'appear[ed] less abundant' (de Blainville 1816: 183). (Debates about menstruation from the turn of the eighteenth century considered menstruation a measure of full humanity; the heavier the flow, the higher one's place in nature [Schiebinger 1993].) Although the person who showed her in Paris claimed that she had a highly aggressive sexual appetite—one day even throwing herself on top of a man she desired—de Blainville doubted the truth of the specific incident. Not to have her too closely linked to European women, however, he also suggests that the modesty he observed might have resulted from her presence for some years among Europeans, conceding that, even after so many years, 'it is possible that there still remained something of the original' (de Blainville 1816: 183). Finally, de Blainville suggests 'that the extraordinary organiza-tion which this woman offers' (de Blainville 1816: 189) is probably natural to her race, rather than being pathological. In support of his contention, he cites travelers who found the same peculiarities—of jaws, buttocks, and labia—among 'natives' living in their home environments. Hence, he finishes with the assertion of natural racial difference.

In de Blainville's text different parts of the body carried specific meanings. To compare the Negro and the orangutan, he spent paragraphs on detailed descriptions of the head, face, jaws, and lips. He used these to link Hottentots to orangs, writing that the general form of the head and the details of its various parts, taken together, make clear that Hottentots more closely resemble orangs than they do Negroes. He repeatedly invoked Pieter Camper's facial angle (Gould 1981; Russett 1989), the shape and placement of the jaws, and—in somewhat excruciating detail—the arrangement and structure of the ears. These passages evoke the tradition of physiognomy elaborated by Lavater (1775–8), whose work, widely translated into French and other languages, offered a basis for Gall's phrenology and a method of using the face to read the internal workings of animals. Of humans Lavater wrote:

The intellectual life . . . would reside in the head and have the eye for its center . . . the forehead, to the eyebrows, [will] be a mirror . . . of the under-standings; the nose and cheeks the image of the moral and sensitive life; the mouth and chin the image of the animal life. . . . (Graham 1979: 48)

When de Blainville and then Cuvier offered detailed comparisons between Sarah Bartmann's cheeks and nose and those of Caucasians, they set forth more than a set of dry descriptions. Her 'moral and sensitive life' lay evident upon the surface of her face.[14]

It is to the description of the genitalia that de Blainville turns to place Bartmann among women. Here he balances his belief in the civilizing effects of Europe against a scarcely hidden savage libido. The gender norms of white women appear as a backdrop for the consideration of 'savage' sexuality. Although he gave detailed descriptions of most of her exterior, de Blainville did not succeed in fully examining Bartmann's genitalia. Where he failed on the living woman, Cuvier succeeded after her death. Clearly a full account of this 'primitive woman's' genitalia was essential to putting her finally in her appropriate place. By exposing them to what passed for scientific scrutiny, Cuvier provided the means to control the previously uncontrollable. Triumphantly, he opened his presentation to the French Academy with the following: 'There is nothing more celebrated in natural history than the Hottentot apron, and at the same time there is nothing which has been the object of such great argumentation' (Cuvier 1817: 259). Cuvier set the stage to settle the arguments once and for all.

Twentieth-century scientific reports open with an introduction that uses previously published journal articles to provide background and justification for the report to follow. In Cuvier's piece we see the transition to this modern format from an older, more anecdotal style.

Rather than relying on official scientific publications, however, Cuvier relied on travelers' accounts of the apron and the steatopygia. In later works, although these anecdotal, eyewitness testimonials fade from sight, they remain the source for knowledge incorporated into a more 'objective' scientific literature. (Sexologists William Masters and Virginia E. Johnson, for example, in their scientifically dispassionate work on the *Human Sexual Response*, include a claim that African women elongate their vaginal labia by physical manipulation; their cited source is a decidedly unscientific (by modern standards) compendium of female physical oddities that dates from the 1930s but draws on nineteenth-century literature of the sort discussed here. [Masters and Johnson 1966: 58].)

To set the stage for his revelations about the Hottentot apron, Cuvier first needed to provide a racial identity for his cadaver (which he referred to throughout the article as 'my Bushwoman'). Travelers' accounts indicated that Bushmen were a people who lived much deeper in 'the interior of lands' than did Hottentots. The apron and enlarged buttocks were peculiarly theirs, disappearing when they interbred with true Hottentots. Cuvier believed that the confusion between Bushmen and Hottentots explained the inconsistent nature of travelers' reports, since some voyagers to the Cape of Good Hope claimed sightings of the Hottentot apron, while others did not. Nevertheless, he had to admit that many people did not believe in the existence of a Bushman nation. Cuvier threw his weight behind what he believed to be the accumulation of evidence: that there existed 'beings almost entirely savage who infested certain parts of the Cape colony . . . who built a sort of nest in the tufts of the brush; they originated from a race from the interior of Africa and were equally distinct from the Kaffir and the Hottentot' (Cuvier 1817: 261). Cuvier believed that the Bushman social structure had degenerated, so that eventually 'they knew neither government nor proprieties; they scarcely organized themselves into families and then only when passion excited them . . . They subsisted only by robbery and hunting, lived only in caves and covered their bodies with the skins of animals they had killed' (Cuvier 1817: 261). By naming Bartmann as a Bushwoman, Cuvier created her as the most primitive of all humans—a female exemplar of a degenerate, barely human race. Despite his lack of belief in evolution, he constructed her as the missing link between humans and apes.

To the modern reader, several noteworthy aspects emerge from these introductory passages. First, Cuvier melds the vision of an interior or hidden Africa with the hidden or interior genitalia of the

Hottentot Venus. This becomes even clearer in subsequent passages in which, like de Blainville, he complains that when he examined her as a living nude in the Jardin du Roi in 1815 she 'carefully hid her apron either between her thighs or more deeply' (Cuvier 1817: 265). Second, he connected a hidden (and hypothetical) people from the deep African interior with an animal-like primitiveness. The passage about making nests from brush tufts evokes monkey and ape behaviors (chimps sleep each night in nests they weave from tree branches). Cuvier's goal in this paper was to render visible the hidden African nations and the hidden genitalia. By exposing them he hoped to disempower, to use observation to bring these unknown elements under scientific control. In the remainder of the account, Cuvier devoted himself simultaneously to the tasks of racial and sexual localization. Where among humans did these interior people belong, and what did their women conceal in their body cavities?

In his presentation to the members of the Museum of Natural History, Cuvier moved from a description of the exterior, living, and never quite controllable Bartmann (for he needed her permission to examine her hidden parts) to the compliant cadaver laid out before him, now unable at last to resist his deepest probings. In both life and death Sarah Bartmann was a vessel of contradictions. He found that her 'sudden and capricious' movements resembled those of a monkey, while her lips protruded like those of an orangutan. Yet he noted that she spoke several languages, had a good ear for music, and possessed a good memory. Nevertheless, Cuvier's vision of the savage emerged: belts and necklaces of glass beads 'and other savage attires' pleased her, but more than anything she had developed an insatiable taste for 'l'eau-de-vie' (Cuvier 1817: 263).

For fully one-fifth of the paper we read of her exterior. Cuvier paints what he clearly found to be a picture gruesome in its contradictory aspects. Only four and a half feet tall, she had enormous hips and buttocks, but otherwise normal body parts. Her shoulders and back were graceful, the protrusion of her chest not excessive, her arms slender and well made, her hands charming, and her feet pretty. But her physiognomy—her face—repelled him. In the jutting of the jaw, the oblique angle of her incisors, and the shortness of her chin, she looked like a Negro. In the enormity of her cheeks, the flatness of the base of her nose, and her narrow eye slits, she resembled a Mongol. Her ears, he felt, resembled those of several different kinds of monkeys. When finally, in the spring of 1815, she agreed to pose nude for a painting, Cuvier reported the truth of the stories about the enormity of her protruding buttocks and breasts—enormous hanging masses[15]—and her barely pilous pubis.

When she died, on December 29, 1815, the police prefect gave Cuvier permission to take the body to the museum, where his first task became to find and describe her hidden vaginal appendages. For a page and a half the reader learns of the appearance, folded and unfolded, of the vaginal lips, of their angle of joining, the measurements of their length (more than four inches—although Blumenbach reportedly had drawings of others whose apron extended for up to eight inches) and thickness, and the manner in which they cover the vulval opening. These he compared to analogous parts in European women, pointing out the considerable variation and stating that in general the inner vaginal lips are more developed in women from warmer climates. The variation in vaginal development had, indeed, been recognized by French anatomists, but a mere ten years earlier, medical writers failed to connect differences in vaginal structures to either southern races or nonwhite women. In a straightforward account of 'over-development' of vaginal lips, Dr M. Baillie, a British physician and member of the Royal Society of Medicine of London (whose book was translated into French 1807), wrote matter-of-factly of this variation, listing it among a number of genital anomalies, but not connected to non-European women (Baillie 1807). As Gilman (1985) points out, however, by the middle of the nineteenth century elongated labia had taken their place in medical textbooks alongside accounts of enlarged clitorises, both described as genital abnormalities, rather than as part of a wide range of 'normal' human variation.

Cuvier acknowledged the great variation in length of the inner vaginal lips found even among European women. But nothing, he felt, compared to those of 'negresses' and 'abyssynians', whose lips grew so large that they became uncomfortable, obliging their destruction by an operation carried out on young girls at about the same age that Abyssinian boys were circumcised. As an aside that served to establish a norm for vaginal structure and a warning to those whose bodies did not conform, we learn that the Portuguese Jesuits tried in the sixteenth century to outlaw this practice, believing that it was a holdover from ancient Judaism. But the now Catholic girls could no longer find husbands because the men wouldn't put up with such 'a disgusting deformity' (Cuvier 1817: 267), and finally, with the authorization of the Pope, a permission was made possible by a surgeon's verification that the elongated lips were natural rather than the result of manipulation, and the ancient custom resumed.

Cuvier contrasts the vaginal lips of Bushwomen with those of monkeys, the near invisibility of which provided no evidence to link them to these primitive humans. But the steatopygia was another matter. Bartmann's buttocks, Cuvier believed, bore a striking resemblance to the genital swellings of female mandrills and baboons, which grow to 'monstrous proportions' at certain times in their lives. Cuvier wanted to know whether the pelvic bone had developed any peculiar structures as a result of carting around such a heavy load. To answer the question, he made use of his well-established method of comparative anatomy, placing side by side the pelvises of 'his bushwoman', those of 'negresses', and those of different white women. In considering Bartmann's small overall size, Cuvier found her pelvis to be proportionally smaller and less flared, the anterior ridge of one of the bones thicker and more curved in back, and the ischial symphysis thicker. 'All these characters, in an almost unnoticeable fashion, resemble one another in Negro women, and female Bushwomen and monkeys' (Cuvier 1817: 269). Just as the differences themselves were practically imperceptible, amidst a welter of measurement and description, Cuvier impeceptibly separated the tamed and manageable European woman from the wild and previously unknown African.

But something worried Cuvier. In his collection he had also a skeleton of a woman from the Canary Islands. She came from a group called the Guanche (extinct since shortly after the Spanish settlement), a people who inhabited the islands before the Spanish and who, by all accounts, were Caucasians. An astonished Cuvier reported to his colleagues that he found the most marked of Bartmann's characters not in the skeleton of Negro women but in that of the Canary Islander. Since he had too few complete skeletons to assess the reliability of these similarities, he turned finally to more abundant material. In the last part of his account, he compares the head and skull (which 'one has always used to classify nations' [Cuvier 1817: 270]) of 'our Bushwoman' with those of others in his collection.

Bartmann's skull, he wrote, mixed together the features of the Negro and the Mongol, but, chiefly, Cuvier declared that he 'had never seen a human head more similar to those of monkeys' (Cuvier 1817: 271). After offering more detailed comparisons of various bones in the skull, Cuvier returned in the last few pages of his paper to the problem that concerned him at the outset—did the Bushmen really exist as a legitimate people, and just how far into the interior of Africa did they extend? Here he relied once more on travelers' reports. Although modern voyagers did not report such people in northern Africa,

Herodotus and others described a group that seemed in stature and skin color to resemble the Bushmen. According to some sources, these people invaded Abyssinia, although the evidence in Cuvier's view was too prescientific to rely on. But he could be sure of one thing: Neither the

Bushmen, nor any race of Negros, gave birth to the celebrated people who established civilization in ancient Egypt and from whom one could say that the entire world had inherited the principles of law, science and perhaps even religion. (Cuvier 1817: 273)

At least one modern author suggested that the ancient Egyptians were Negroes with wooly hair, but Cuvier could be sure that this, too, was in error. All he needed to do was compare the skulls of ancient Egyptians with those of the pretender races. One can picture him, as he spoke, dramatically producing from beneath his dissecting table the skulls of Egyptian mummies, those very same ones brought back by Geoffroy Saint-Hilaire from the Napoleonic incursion into Egypt.

Cuvier studied the skulls of more than fifty mummies. These, he pointed out, had the same skin color and large cranial capacity as modern Europeans. They provided further evidence for 'that cruel law that seems to have condemned to eternal inferiority those races with depressed and compressed crania' (Cuvier 1817: 273). And finally, he presented to his museum colleagues the skull of the Canary Islander whose skeleton had so troublingly resembled Bartmann's. This too 'announced a Caucasian origin' (Cuvier 1817: 274), which is the phrase that concludes his report. In this last section of his paper we watch him struggle with his data. First, he realized that he had a Caucasian skeleton that looked identical to Bartmann's. If he could not explain this away (what modern scientists call eliminating outliers—data points that don't neatly fit an expected graph line), his thesis that Bushmen represented a primitive form of humanity was in trouble. But that wasn't all that worried him: if his thesis was in trouble, so too was the claim of European superiority on which European and American colonization, enslavement, and disenfranchisement so depended. Thus, he went to considerable trouble to explain away the Guanche skeleton; ultimately he succeeded by using the scientific spoils of colonial expansion—the Egyptian mummies captured during Napoleon's Egyptian campaign.

ANNE FAUSTO-STERLING

CONCLUSION

This chapter places the scientific study of nonwhite women in several contexts. The investigations were, to be sure, part of the history of biology and, especially, a component of the movement to catalogue and classify all the living creatures of the earth. But this movement was in turn embedded in the process of European capitalist expansion. Not only did traders and conquerers, by collecting from around the world, create the need for a classification project, they also required the project to justify continued expansion, colonialism, and slavery. Further entangling the matter, the vast capital used to build the museums and house the collections came from the economic exploitation of non-European goods—both human and otherwise. This entire essay has been an argument against a narrowly constructed historiography of science; instead, I more broadly socialize the history of Euro-American biology in the first quarter of the nineteenth century by exposing its intersections with gender, race, and nation.

If one looks at the process less globally, one sees Cuvier and de Blainville as significant actors in a period of scientific change. From the perspective of the history of Euro-American biology, parochially extracted from its role in world expansion, one can say that the biologists of this period, and Cuvier in particular, made enormous scientific progress with the 'discovery of the great information content of the internal anatomy of the invertebrates' (Mayr 1982: 183). According to this view, Cuvier 'discovered' the importance of the nervous system as a way to organize animals. But 'Cuvier's vision of the animal world was deeply coloured by that of the human society in which he was forced to make his way' (Outram 1984: 65). Far from reflecting some underlying natural system, Cuvier's use of the nervous system in his classification schemes had a homocentric starting point. The ideas formed a meshwork. Cuvier gave the focus on the nervous system and brain (obtained from his conviction that classification should proceed from the most complex—in this case human—structure to the simplest) the status of scientific fact by developing a reasonably coherent story about how the structure of the nervous system enabled him to classify all animals. Once scientists agreed on the validity of Cuvier's animal classification scheme, it fed back on the question of human classification. It seemed only 'natural' to focus on the structure of the brain (as reflected in cranial and facial characteristics) to obtain evidence about the relative standing of the human races.

226

Sarah Bartmann's story is shocking to modern sensibilities. The racism of the period seems obvious—even laughable. But in the rush to create distance between nineteenth-century racist science and our modern, putatively less racist selves, even highly sophisticated scholars often lose sight of an important point. The loss becomes evident when I am asked (as I frequently am) what the *real* truth about Bartmann was. Just how big were those forbidden parts? The question reflects an ongoing belief in the possibility of an objective science. It suggests that, now that we have escaped all that silly racism of the nineteenth century, we ought to be able to get out our measuring tapes and find the real truth about other people's bodies. In this essay I argue that Bartmann's bodily differences were constructed using the social and scientific paradigms available at the time. The historical record tells us nothing about her agency; we can only know how Europeans framed and read her. Were she somehow magically alive today, contemporary biologists or anthropologists might frame and read her differently, but it would be a framing and reading, nevertheless. One contemporary difference might be that the varying worldwide liberation movements could offer her a context in which to contest the constructions of Euro-American science. In fact we see such contestations regularly in debates over such questions as brain size, race, and IQ (Maddock 1992; Schluter and Lynn 1992; Becker, Rushton, and Ankney 1992), brain shape and gender, and genetics and homosexuality (Fausto-Sterling 1992).

In *Playing in the Dark*, Toni Morrison (1992) makes her intellectual project 'an effort to avert the critical gaze from the racial object to the racial subject; from the described and imagined to the describers and imaginers. . . .' (Morrison 1992: 90). By analogy I look at the fears and anxieties of the scientists, rather than worrying about the (in)accuracies of their descriptions of Sarah Bartmann and other people of color. To quote further from Morrison:

The fabrication of an Africanist persona is reflexive; an extraordinary meditation on the self; a powerful exploration of the fears and desires that reside in the writerly conscious. It is an astonishing revelation of longing, of terror, of perplexity, of shame, of magnanimity. It requires hard work NOT to see this. (Morrison 1992: 17)

For our purposes we need only substitute the word 'scientific' for the word 'writerly'. What can we glean of the fears, desires, longings, and terrors that perfuse the works we've just considered? And how are race, gender, and nationality woven into the story? In the accompanying chart I have listed some of the paired contradictions that emerge from my reading of Cuvier and de Blainville.

The simultaneous anxiety about European women and the savage Other is especially clear in de Blainville's account. He identified Bartmann as a woman because she menstruated. But she also drank, smoked, and was alleged to be sexually aggressive—all masculine characteristics. And if Bartmann, a woman, could behave thus, why not French women? Furthermore, the soap opera dramas about Bartmann that played in contemporary Paris suggested that French men, despite their 'civilization', actually desired such women; civilization kept the European woman under control, decreasing the danger of rebellion, but thwarting male desire. Minute scientific observation converted the desire into a form of voyeurism, while at the same time confining it to a socially acceptable location.

Cuvier most clearly concerned himself with establishing the priority of European nationhood; he wished to control the hidden secrets of Africa and the woman by exposing them to scientific daylight. The French Revolution had frightened him, and certainly the prospect of resistance from other peoples must have seemed terrifying (Outram 1984; Appel 1987). Hence, he delved beneath the surface, bringing the interior to light; he extracted the hidden genitalia and defined the hidden Hottentot. Lying on his dissection table, the wild Bartmann became tame, the savage civilized. By exposing the clandestine power, the ruler prevailed. But one need only look at the list of anxieties

conquest	resistance
human	animal
surface	interior
tame	wild
sexually modest	libidinous
civilized	savage
compliant	angry
ruler	subject
powerlessness	hidden power
male	female
white	nonwhite
colonizer	colonized

glossed from the scientific literature to know how uneasy lay the head that wore a crown.[16]

Notes

Acknowledgments: This paper was written with the financial support of the National Science Foundation, Fellowship #DIR–9112556 from the Program in History and Philosophy of Science. I would like to thank Evelynn Hammonds, Joan Richards, Gregg Mitman, and Londa Schiebinger for reading and commenting on recent drafts of this paper. Londa Schiebinger also kindly shared with me drafts of chapters of her book *Nature's Body: Gender in the Making of Modern Science* (Beacon 1993).

1. In 1992 the Musée de l'Homme had removed the remnants of the Bartmann exhibit. In its place was a modern one entitled 'All relatives, all different', celebrating human genetic diversity. Discussion of Bartmann could still be found in a part of the exhibit devoted to the story of scientific racism.
2. There is also a book of poetry featuring the Venus Hottentot in the title poem. Elizabeth Alexander, *The Venus Hottentot* (Charlottesville. University Press of Virginia), 1990.
3. The Dutch word for Bushman is *bosjeman*, which translates as 'little man of the forest'. This is also the translated meaning of the Malay word *orangutan*.
4. This is in perfect accord with Latour's account of how scientific knowledge is constructed.
5. All translations from works cited in the original are mine.
6. In fact, de Blainville's break with Cuvier came over just this question. He devised a different classificatory system based on external, rather than internal characters, but he linked his divisions by creating intermediate groupings.
7. The question of the racial origins of European thought has been raised in our own era by the work of Martin Bernal (1987).
8. The detailed ins and outs of her sale and repurchase may be found in the references in note 11.
9. In contrast to the family groupings of Laps, Eskimos, and Zulus, the displays of Bartmann, Tono Maria, and Zip made no attempt to present a working culture.
10. Nonwhites were not the only 'others' constructed. I plan to address the use of 'freaks' in the construction of the Other in a book-length account of the construction of race and gender by biologists, anthropologists, and sociologists.
11. All the details cited here may be found in Altick (1978), Edwards and Walvin (1983), Gould (1985), Kirby (1949, 1953), and Lindfors (1983). Remarkably, prurient interest in the figure of the Hottentot continues to this day. Gould (1985) discusses a 1982 cover of the French magazine *Photo* that features a naked woman named 'Carolina, La Venus Hottentote de Saint-Domingue'. In the copy of the Geoffroy Saint-Hilaire and Cuvier held by the Brown Library, the frontal drawing of Bartmann (which exhibited her breasts in full form) has been razored out. The mutilation was first noticed by librarians in 1968. This is not the first time I have encountered such mutilation of material of this sort.

12. Although the bustle was not invented until 1869, various fashions in the eighteenth and nineteenth centuries accentuated the backside of middle- and upper-class white women (Batterberry and Batterberry 1977). The relationship between these fashions and scientific accounts of the body has yet to be detailed.

13. There were at least seven articles, falling into three chronological groupings, published in scientific journals in England, France, and Germany. The first two, by Henri de Blainville and Georges Cuvier, exclusively on Sarah Bartmann, were published in 1816 and 1817, respectively. The second group, containing two by German biologists, appeared in the 1830s. The first of these was written by Johannes Müller (1801–58) (Müller 1834), a physiologist and comparative embryologist, while the second, written by Frederick Tiedemann (1781–1861) (Tiedemann 1836), Professor of Anatomy and Physiology at the University of Heidelberg and Foreign Member of the Royal Society of London, appeared in 1836. Müller's article is about a Hottentot woman who died in Germany and is in the same scientific style as the French papers. Tiedemann's work, on the other hand, represents a scientific departure. Although Bartmann's is among a wide variety of brains obtained from museum collections, it is not the focus of the article. From a scientific point of view, Tiedemann's study represents a transition from a period in which scientists offered detailed examinations of the outside of the body, while focusing on a single individual and describing all body parts. Tiedemann awarded priority to one organ—the brain. A comparison of the brains of Europeans, Negroes, and orangutans convinced him that there was no difference among the humans. He used his results to condemn the practice of slavery. His method, though, is primitive compared to the approach of the scientists working in the 1860s (Marshall 1864; Flower and Murie 1867), whose work provides a useful contrast to the changing scientific and political times. In this paper I will consider the first two exemplars, reserving detailed examination of the other works for a future occasion.

14. Outram (1984) documents Cuvier's dispute with Franz Joseph Gall over the scientific nature of phrenology But Cuvier clearly believed in the principle that the face could be read for deeper meaning.

15. In the seventeenth century, breasts—as natural and social objects—had undergone a transformation, as male social commentators launched a successful campaign to do away with wetnursing and reestablish the breast as an object that connected women to nature through the act of nursing. For middle- and upper-class white women, doing the right thing with the right kind of breasts hooked them into a growing cult of domesticity, which exploded as the nineteenth-century ideal for gender relationships for the middle and upper classes in Europe and America. This naturalization of motherhood worked hand in glove with the desexualization of white women (Schiebinger 1993; Perry 1991). Perry cites Thomas Laqueur (1986) as explaining 'this cultural reconsideration of the nature of women's sexuality as part of a process ... committed to sweeping clean all *socially* determined differences among people' (Perry 1991: 212), instead relocalizing difference in the biological body. No part of the body escaped unscathed from this process.

16. In one of the lovely ironies of history, Cuvier himself was dissected when he died (in 1832), and his brain and head measurements were taken. In a ranking

of 115 men of note, Cuvier's brain weight came in third (Turgenev's was first). The French as a group ranked behind Americans and the British. The author of this 1908 paper concluded that 'the brains of men devoted to the higher intellectual occupations, such as the mathematical sciences . . . . [or] those of men who have devised original lines of research [Cuvier] and those of forceful characters, like Ben Butler and Daniel Webster, are generally heavier still. The results are fully in accord with biological truths' (Spitzka 1908: 215). In a second, larger sample, Spitzka included four women—mathematician Sonya Kovaleskaya, physician Caroline Winslow, actress Marie Bittner and educator and orator Madame Leblais—who ranked 134th–137th, in brain weight.

References

Altick, Richard D. (1978). *The Shows of London.* Cambridge, Mass: The Belknap Press of Harvard University.

Appel, Toby A. (1987). *The Cuvier–Geoffroy Debate: French Biology in the Decades before Darwin.* Oxford: Oxford University Press.

Baillie, Mathieu (1807). *Anatomie pathologique des organes les plus importans du corps humain.* Paris: Crochard.

Batterberry, Michael, and Batterberry, Ariane (1977). *Mirror Mirror: A Social History of Fashion.* New York: Holt, Rinehart and Winston.

Becker, Brent A., Rushton, J. Philippe, and Ankney, C. Davison (1992). 'Differences in Brain Size'. *Nature,* 358: 532.

Bernal, Martin (1987–91). *Black Athena: The Afroasiatic Roots of Classical Civilization,* vols. 1 and 2. New Brunswick, NJ: Rutgers University.

Bucher, Bernadette (1981). *Icon and Conquest: A Structural Analysis of the Illustrations of de Bry's* Great Voyages. Trans. Basia Miller Gulati. Chicago: University of Chicago Press.

Coleman, William (1964). *Georges Cuvier, Zoologist: A Study in the History of Evolution Theory.* Cambridge, Mass.: Harvard University Press.

Cuvier, Georges (1817) 'Faites sur le cadavre d'une femme connue à Paris et à Londres sous le nom de Vénus Hottentott'. Memoires *du Musée nationale d'histoire naturelle,* 3: 259–74.

de Blainville, Henri (1816). 'Sur une femme de la race Hottentote'. *Bulletin du Société philomatique de Paris,* pp. 183–90.

Edwards, Paul, and Walvin, James (1983). *Black Personalities in the Era of the Slave Trade.* Baton Rouge: Louisiana State University Press.

Elphick, Richard (1977). *Kraal and Castle: Khoikhoi and the Founding of White South Africa.* New Haven: Yale University Press.

Fausto-Sterling, Anne (1992). *Myths of Gender: Biological Theories about Women and Men.* 2nd edn. New York: Basic Books.

Figlio, Karl M. (1976). 'The Metaphor of Organization: An Historiographical Perspective on the Bio-Medical Sciences of the Early Nineteenth Century'. *History of Science,* 14: 17–53.

Flourens, P. (1845). *Cuvier. Histoire de ses travaux.* 2nd edn. rev. and corr. Paris: Paulin.

Flower, W. H., and Murie, James (1867). 'Account of the Dissection of a Bushwoman'. *Journal of Anatomy and Physiology*, 1: 189–208

Geoffroy Saint-Hilaire, Etienne, and Cuvier, Frédéric (1824). *Histoire Naturelle des Mammifères*, vols. 1 and 2. Paris: A. Belin.

Gilman, Sander L. (1985). 'Black Bodies, White Bodies: Toward an Iconography of Female Sexuality in Late 19th-Century Art, Medicine and Literature'. *Critical Inquiry*, 12: 204–42.

Gould, Stephen Jay (1981). *The Mismeasure of Man*. New York: Norton.

—— (1985) 'The Hottentot Venus', in Stephen Jay Gould, *The Flamingo's Smile: Reflections in Natural History*. New York: Norton, pp. 291–305.

Graham, John (1979). *Lavater's Essays on Physiognomy: A Study in the History of Ideas*. Berne: Peter Lang.

Gray, Stephen (1979). *Southern African Literature: An Introduction*. New York: Barnes and Noble.

Guenther, Mathias Georg (1980). 'From "Brutal Savage" to "Harmless People": Notes on the Changing Western Image of the Bushmen'. *Paideuma*, 26: 124–40.

Haraway, Donna (1989). *Primate Visions: Gender, Race, and Nature in the World of Modern Science*. New York: Routledge.

James, John Angell (1830). *Memoir of Clementine Cuvier, Daughter of Baron Cuvier*. New York: American Tract Society.

Kirby, Percival R. (1949). 'The Hottentot Venus'. *Africana News and Notes*, 6: 55–62.

—— (1953). 'More about the Hottentot Venus'. *Aftricana News and Notes*, 10: 124–34.

Laqueur, Thomas (1986). 'Orgasm, Generation, and the Politics of Reproductive Biology'. *Representations*, 14: 1–41.

Latour, Bruno (1987). *Science in Action: How to Follow Scientists and Engineers through Society*. Milton Keynes: Open University Press.

Lavater, J. C. (1775–8). *Physiognomische Fragmente zur Beförderung der Menschenkenntnis und Menschenliebe*. Leipzig: Weidmanns Erben und Reiche, H. Steiner und Companie.

Lee, Richard B. (1992). 'Art, Science, or Politics? The Crisis in Hunter-Gatherer Studies'. *American Anthropologist*, 94(1): 31–54.

Lewin, Roger (1988). 'New Views Emerge on Hunters and Gatherers'. *Science*, 240: 1146–8.

Lindfors, Bernth (1983). 'The Hottentot Venus and Other African Attractions in Nineteenth-Century England'. *Australasian Drama Studies* 1: 83–104.

Linnaeus (Carl von Linne) (1758). *Caroli Linnaei Systema Naturae. Regnum Animale*. 10th edn. Stockholm.

Maddock, John (1992). 'How to Publish the Unpalatable?' *Nature*, 358: 187.

Marshall, John (1864). 'On the Brain of a Bushwoman; and on the Brains of Two Idiots of European Descent'. *Philosophical Transactions of the Royal Society of London*, pp. 501–58.

Masters, William H., and Johnson, Viginia E. (1966). *Human Sexual Response*. Boston: Little, Brown.

Mayr, Ernst (1982). *The Growth of Biological Thought: Diversity, Evolution, and Inheritance*. Cambridge, Mass.: Belknap Press of Harvard University.

Merchant, Carolyn (1980) *The Death of Nature: Women, Ecology, and the Scientific Revolution*. San Francisco: Harper and Row.

Morrison, Toni (1992). *Playing in the Dark: Whiteness and the Literary Imagination*. Cambridge Mass.: Harvard University Press.

Müller, Johannes (1834). 'Ueber die äusseren Geslechtstheile der Buschmänninnen'. *Archiv fur Anatomie, Physiologie und Wissenschaftliche Medicin*, pp. 319–45.

Outram, Dorinda (1984). *Georges Cuvier: Vocation, Science, and Authority in Post-revolutionary France*. Manchester: Manchester University Press.

Perry, Ruth (1991). 'Colonizing the Breast: Sexuality and Maternity in Eighteenth-Century England'. *Journal of the History of Sexuality*, 2: 204–34.

Raven-Hart, Rowland (1967). *Before Van Riebeeck: Callers at South Africa from 1488 to 1652*. Cape Town: C. Struik.

Russett, Cynthia Eagle (1989). *Sexual Science: The Victorian Construction of Womanhood*. Cambridge Mass.: Harvard University Press.

Schiebinger, Londa (1993). *Nature's Body: Gender in the Making of Modern Science*. Boston: Beacon Press.

Schluter, Dolph, and Lynn, Richard (1992). 'Brain Size Difference'. *Nature*, 359: 181.

Shohat, Ella (1991). 'Imaging Terra Incognita: The Disciplinary Gaze of the Empire'. *Public Culture*, 3 (2): 41–70.

Spitzka, Edward Anthony (1908). 'A Study of the Brains of Six Eminent Scientists and Scholars Belonging to the American Anthropometric Society, together with Description of the Skull of Professor E. D. Cope'. *American Philosophical Society Transactions*, 21: 175–308.

Stepan, Nancy (1982). *The Idea of Race in Science: Great Britain, 1800–1960*. Hamden, Conn.: Archon.

Stocking, Jr, George W. (1982). *Race, Culture, and Evolution: Essays in the History of Anthropology*. Chicago: University of Chicago.

—— (1987). *Victorian Anthropology*. New York: Free Press.

Tiedemann, Frederick (1836) 'On the Brain of a Negro, Compared with That of the European and the Orang-outang.' *Philosophical Transactions of the Royal Society of London* pp. 497–558.

Tiffany, Sharon W., and Adams, Kathleen J. (1985). *The Wild Woman: An Inquiry into the Anthropology of an Idea*. Cambridge, Mass.: Schenkman.

Trumbach, Randolf (1991). 'Sex, Gender and Sexual Identity in Modern Culture: Male Sodomy and Female Prostitution in Enlightenment London'. *Journal of the History of Sexuality*, 2: 187–203.

Walvin, James (1973).*Black and White: The Negro and English Society, 1555–1945*. London: Allen Lane and Penguin.

8 Hard Labor
Women, Childbirth, and Resistance in British Caribbean Slave Societies

Barbara Bush

Over the past decade scholars have shown deepening interest in the lives of slave women, particularly of the antebellum South. Until recently, however, with the exception of Lucille Mathurin Mair's pioneering work on Jamaica, slave women of the Caribbean were given scant attention and analyses of their lives tended to reiterate the more popular misconceptions of contemporary observers.[1] The pioneering works on slave demography and medical treatment by Barry Higman and Richard Sheridan aroused new interest through their concern with a major enigma of Caribbean slave populations—their failure to reproduce naturally in comparison with the slave population of the antebellum South. The importance of fertility rates of black females in the demographic analyses led to research which increasingly focused on women's experience of childbirth and the adverse effects of sugar monoculture on the family and mating patterns of slaves.[2] These studies have yielded valuable insights into the lives of slave women, but they have only marginally addressed the issues of production and reproduction from the slaves' perspective. This chapter aims to integrate production and reproduction into studies of slave resistance.

Active struggle against slavery was an enduring and ever-present feature of slave life in the Caribbean. Resistance took many forms, from outright revolt to more subtle behavior. Women were no less prominent than men in resistance, and they may even have been in the vanguard, particularly in cultural resistance. As anthropologist Melville Herskovits noted in his study of a Trinidad village in the 1940s, a distinctive characteristic of black societies in the New World

From Barbara Bush, 'Hard Labor: Women, Childbirth, and Resistance in British Caribbean Slave Societies', in *More Than Chattel: Black Women and Slavery in the Americas*, eds. David Barry Gaspar and Darlene Clark Hine (Indiana University Press, 1996) pp. 193–217. Reprinted with permission.

was the part played by women as the 'principal exponents' and protectors of traditional African-derived culture.[3] This chapter argues that tensions inherent in slave women's 'dual burden' of production and reproduction, combined with attempts by slave masters to manipulate these women's cultural practices and fertility, strongly influenced the responses of slave women to childbirth and infant rearing at both conscious and unconscious levels. The discussion focuses on slave women who lived on large sugar plantations in the British territories during the later period of slavery.[4]

Women in the Plantation Economy

Women were valuable workers. A rough equality existed between slave men and women, particularly in field work on large plantations where they shared the arduous conditions of life and labor.[5] However, women played unique reproductive roles, and their lives were affected by the complex structures of African and European patriarchy which influenced the character of slave society. Patriarchal dominance was evident in the sexual division of labor in plantation and slave domestic production and in attempts by masters to control reproduction. In this context women's control over their bodies was arguably a major area of struggle involving power relations at a most basic level. According to Michel Foucault, where there is power, there is resistance, or to be more precise, a plurality of resistances. Power over women was exercised through control of their sexuality, a form of oppression rarely experienced to the same degree by slave men. As Arlette Gautier wrote, the appropriation of slave women's sexuality 'redoubled women's exploitation as workers', whereas male slaves could take refuge in 'the fantasies of their sexual power'.[6]

White men of British origin were the major owners of slaves and thus the wealthiest and most powerful persons in the British Caribbean. These men were also the most distant from the slave population, particularly during the late years of slavery, when many were absentee owners resident in Britain. The power of absentee owners was mediated through other white men or black male overseers, who frequently took sexual advantage of black women. Between 1807 and 1832 transient white men, particularly on the largest British Caribbean plantations of absentee owners, fathered numerous slave children. While some black women may have regarded sexual unions with whites as advantageous, providing privileges and possible manumission, such relations also represented a natural extension of the power of white over black.[7] If women resisted sexual advances, they risked physical cruelty and punishment. Power over the black woman's body in its productive capacity as an asexual labor machine was thus combined with sexual power to control both production and reproduction on slave plantations.

In this matrix of power, where patriarchal structures intermeshed with basic economic structures of labor exploitation, the position of white women was ambiguous. Many white women owned slaves, but rarely more than twenty, and these slaves were concentrated mostly in towns. Subordinate to white male authority and facing sexual competition from colored and black women for their husbands' favors, white women were arguably more brutal in their treatment of slaves than white men.[8] Yet, in the intimate area of childbirth and sexuality, white women were also subjected to the dictums on childbirth that resulted from the expropriation of the ancient art of female midwifery by male doctors. In addition, inadequate obstetric knowledge and unhygienic conditions rendered childbirth hazardous for all women, in Europe and the West Indies alike. At the same time, socially constructed distinctions based on race and class firmly separated black, colored, and white women. In contrast to white and free colored women, black female slaves were subjected to punishment and the rigors of the plantation work regime.[9] The formal plantation division of labor conflicted with the traditional African division of labor which defined the private sphere of slave women's lives. Within this sphere, women were primarily childbearers and mothers who bore sole responsibility for child care and food preparation. The conflicting demands of the plantation and household on slave women arguably placed them under psychic pressures and contradictions not experienced to the same degree by free women or, indeed, slave men.

Until near the end of slavery, planters of the British Caribbean paid scant regard to slave family bonds, and they undervalued the reproductive roles of slave women. Black women were valued mainly for their labor. From the earliest days of the slave trade, Europeans regarded women as eminently suited to fieldwork because of their perceived 'drudge' status in polygynous marriages. A large part of the labor on sugar estates consisted of digging holes for canes, hoeing, and weeding—tasks generally accepted in slaving circles as 'women's work' in Africa.[10] Planters professed a preference for males, and more males than females were brought to the Caribbean during the eighteenth century. However, in the hierarchical division of labor on large plantations, men were valued for crafts skills and for work in the semi-industrial process of the sugar mill, so that up to 50 percent of ordinary field gangs were made up of women, a pattern which was also evident in the French Caribbean.[11] As William Beckford, a prominent Jamaican planter, noted in the 1780s, 'A negro man is purchased for a trade, or the cultivation and different processes of the cane—the occupations of the women are only two, the house, with its several departments and supposed indulgences, or the field with its exaggerated labors. The first situation is the most honorable the last the most independent.' Apart from the chief housekeeper, who was almost always colored by the late eighteenth century, and the midwife and chief doctoress or hospital assistant, who were more highly valued after the introduction of pronatalist policies during the 1790s, men were the plantation elite. This had important implications for the working and living conditions, as well as for the health and reproductive potential of women.[12]

The most important slaves were the most robust males and females who worked the fields and sugar mills. The importance of women in the formal plantation economy (as opposed to the 'informal' economy of peasant cultivation and marketing, in which some women were also prominent) is reflected in the prices of prime female slaves. Between 1790 and the end of the slave trade in 1807, the approximate purchase price in Jamaica of a 'new' male slave was £50 to £70, while a healthy female brought from £50 to £60. Prices of creole slaves were roughly 20 percent higher. There is no indication that fertility increased the value of women—women of similar ages, with or without children, cost exactly the same. Men and women were often sold together in 'jobbing gangs'. A Jamaican advertisement in 1827, for instance, offered a 'small gang of effective and well-disposed slaves, 17 males and 17 females'.[13]

By the early nineteenth century at least 75 percent of the slaves in the Caribbean colonies were attached to sugar plantations. Detailed demographic analysis by Higman confirms the dominance of women in field gangs (on sugar plantations, 10 percent were domestics, compared with 20 percent on coffee and cotton plantations).[14] Although planters maintained that the general treatment of sugar slaves was 'mild and indulgent', Higman's calculations confirm that slave morbidity and mortality was highest, and the birth rate the lowest, on sugar plantations of the optimum size of 250 slaves. Next highest in mortality were coffee plantations, followed by cocoa, cotton, and pimento plantations, cattle ranches or pens, the towns, and, finally, marginal subsistence holdings. Sugar plantations were generally regarded as unhealthy locations where a number of factors contributed to high mortality. The labor regime itself ensured that women shared the same backbreaking work, miseries, and punishments as men. In crop time, between October and March, plantation slaves were turned out of their quarters at sunrise and worked till sunset, with little time to call their own. There was also extended night work during this period.[15]

Despite their economic value, fieldworkers were treated as the capital stock of the plantation, on par with the animals, and maintained at bare subsistence level. Though they performed the hardest labor and worked the longest hours, their living conditions were far inferior to those of domestics and skilled craftsmen and they suffered from greater ill health and higher mortality rates. To supplement the often inadequate diets provided by masters, they produced food in their free time on their provision grounds.[16] The rigors of fieldwork often led to low productivity, which was linked to various forms of resistance— from individual shirking and malingering to sabotage, arson, and more collective discontent in the gang. Control over productivity therefore became difficult. Women were prominent in such resistance.[17] As field slaves, women were subjected to the harshest conditions, but they retained greater cultural autonomy. According to Elsa Goveia, writing about the British Leeward Islands at the end of the eighteenth century, field slaves were allowed to retain Africanisms to underscore their inferiority.[18] In practice these characteristics formed a 'cultural shield' which helped field slaves to sustain their struggle against slavery. The barrier to assimilation or acculturation was more easily eroded among house slaves, who were in closer proximity to whites.

Accounts of plantation life confirm that women gave their labor unwillingly and were a constant source of frustration to managers and overseers. Plantation journals and punishment lists (required by law after slave registration in 1815) of absentee London merchants Thomas and William King, kept from the early 1820s to the beginning of apprenticeship in 1833, chart the deep level of everyday resistance that female slaves sustained. Women were far more often accused of insolence, 'excessive laziness', disobedience, quarreling, and 'disorderly conduct' than were male slaves. On the Kings' plantation, Good Success, there were 93 female slaves and 211 male, but the women were more consistently troublesome. As persistent offenders, women were punished on average more frequently than men.[19]

Faced with the intransigence of women, which ran to more serious crimes as flight, attempts to poison their masters, and 'exciting discontent' in the gang, beleaguered managers resorted to the whip. When legislation banning the whipping of female slaves was introduced in Trinidad in 1823, planters objected strongly. They complained that female slaves were 'notoriously insolent' and were kept in some 'tolerable order' only through fear of punishment, which they deserved more frequently than males.[20] Profit often could be extracted only through physical coercion, and, planters argued, without constant use of the whip it was impossible to work estates. Such punishment or correction had obvious implications for the well-being of women slaves, particularly during pregnancy.

Under the overseer's whip, 'neither age nor sex found any favor'.[21] For women, the degradation experienced by being whipped upon the bare buttocks while held down by other slaves was made worse because the black drivers who carried out the whippings, and who were eager to establish their privileged position, showed little lenience.[22] Allowance must be made for the propaganda of abolitionist writings, but records from the Kings' plantations suggest that even alternative punishments meted out to women were harsh. Women endured the public humiliation and discomfort of the 'hand and foot stocks' or solitary confinement, sometimes with the additional debasement of wearing a collar. Punishments lasted from a few hours to three days, occasionally longer in serious cases, and Sunday, the only full free day for slaves to cultivate their own plots, was a favorite day for confinement—presumably because it did not interfere with the plantation work regime.[23] Pregnancy did not guarantee immunity to such harsh punishments. A pregnant woman risked a flogging if she complained about work conditions. Until the last years of slavery, Jamaica laws limiting the number of lashes that could be inflicted on slaves made no special concessions for women, pregnant or not. Slaves of both sexes could receive up to ten lashes, except when an overseer was present, when thirty-nine lashes could be administered.[24]

Jamaica planters were renowned for their callous indifference to the special needs of pregnant women. Dr. John Williamson related how one woman, confined to the stocks for 'misconduct' and liberated only a few days before delivery, subsequently died of puerperal fever. The paternalist planter Matthew 'Monk' Lewis, on the basis of several adverse reports, concluded that 'white overseers and bookkeepers . . . [kicked] black women in the belly from one end of Jamaica to another', harming both the women and their unborn children.[25] For most planters, women remained, first and foremost, valuable workers and were rarely given preferential treatment. It was in slaveowners' interests, particularly during the late period of slavery when faced with abolitionist pressure, to conceal the degree of exploitation of and cruelty toward women. Written records can only provide limited insights into the punishments experienced by female slaves or the degree of resistance such treatment generated. However, by the 1790s, with the growing concern over the failure of the slave population to increase naturally, the maltreatment of women became, in principle at least, a seminal issue in the debate over the causes of low fertility. This debate brought more sharply into the focus the effect of field labor on the reproductive potential of slave women.

Female Slave Fertility: An Enduring Enigma

During the early period of slavery, planters tried to ensure an even balance between their male and female slaves and encouraged stable relationships, but conflicting accounts of women's fertility already existed. Sir Hans Sloane, recording impressions of his travels through the West Indies in the early eighteenth century, observed of female slaves that they 'are fruitful and go after the birth of their children to work in the fields with their little ones tied to their backs'. The seventeenth-century Barbados planter, Richard Ligon, complained, however, that 'though we breed both Negroes, Horses and Cattle . . . that increase will not supply the moderate decays which we find in all of those. . . .'[26] As sugar monoculture intensified, conditions deteriorated; however, the extremely high mortality rates, particularly of new, or 'saltwater', slaves, could be counteracted by cheap fresh imports. By the mid-eighteenth century, planters no longer encouraged fertility; indeed, the treatment of women possibly discouraged reproduction.

Pregnant women were often kept at fieldwork up to the last few weeks of pregnancy and were expected to return to work no later than three weeks after delivery. They suffered from many gynecological complaints, including early miscarriage and sterility, in addition to general ill health related to plantation life. Both Mair and Higman have found that women had a higher morbidity rate than men. This may explain alleged planter preference for male slaves.[27] In 1798 Barbados planters noted that slave women were 'very prone' to contract disorders of the reproductive system 'which will often last for their lives'. Edward Long believed that 'most black women' were 'subject to obstructions of the menstrua (monthly periods)'. This often resulted in 'incurable' sterility among Ebo women.[28] Such factors may explain why fertility rates of creole slaves were higher than those of African women in all the colonies from the registration period to abolition.

The enigma of the failure of slaves to increase naturally began to concern planters in the British and French colonies toward the end of the eighteenth century because of growing abolitionist pressure to end the slave trade. The causes of low fertility therefore became a major point of debate. Legislation designed to improve the well-being of slave women and promote a 'healthy increase' among slaves was introduced after 1790. Before such 'ameliorative' legislation was adopted, the slave laws of the British colonies did not discriminate between male and female slaves, offering at the same time no protection from sexual abuse, overwork, maltreatment during pregnancy, or the breakup of slave families, including the separation of mothers and children. The rare exception made for pregnant women in Jamaica was that they were 'respited . . . from execution until after their pregnancy'.[29] Under English law, in contrast to the laws governing French and Spanish slaves in the Caribbean, slaves were legally defined as chattel, in recognition of the sanctity of an Englishman's private property. Masters had complete control over their slaves and, as property owners, they were given wide discretion in enforcing subordination and control. Slaves could give evidence against each other, but not for or against free persons. This made it extremely difficult for slave women to protest maltreatment in pregnancy, sexual abuse, or rape.[30]

Reflecting the new pronatalist policies, incentives were introduced after 1790 to encourage women to have more children. Under the Leeward Islands Act of 1798 female slaves 'five months gone' were to be employed in light work. A 'roomy negro house' of two rooms was to be built for every slave woman pregnant with her first child, and rewards and bonuses were offered to slave women and midwives. Legislation passed in the Leewards and in Jamaica (1809) also included provisions that female slaves 'having six children living' should be exempt from hard labor and the owner exempted from taxation on such female slaves.[31] Laws were passed to encourage stable marriages, and gratuities were offered to slave parents to provide 'the several little necessaries wanted to keep infants clean and decent'. As it was generally held that sexual relations between white men and black women contributed in no small degree to the general 'immorality' of slave society, laws were passed which fined whites up to £100 for

241

'having criminal commerce' with any married female slave. A Jamaica law of 1826 introduced the death sentence for rape of female slaves or sexual abuse of slave girls under ten years old.[32]

Planters built additional plantation hospitals (hothouses) and lying-in houses for pregnant women. They also employed more European doctors to care for their slaves, and as a result there emerged a new interest in slave illnesses and the causes of low fertility and high infant mortality. One doctor, William Sells, advocated keeping detailed records of slave women's childbearing experiences. He also recommended careful medical attention, light employment after the fifth month, and use of better-educated black midwives, whom Europeans often regarded as incompetent or even dangerous. However, Sells warned against indulging slave women too much. He believed that they were fit to return to the fields a month to six weeks after delivery.[33]

New pronatalist policies nevertheless continued to reflect the racist view that African women, being nearer to the animal world than white women, gave birth painlessly and 'with little or no difficulty' and could be returned to hard labor soon after childbirth. In the seventeenth century Ligon noted that if the overseer was 'discreet', women slaves were allowed to rest a little more than ordinary, but if not, they were compelled 'to do as others' with 'times . . . of suckling their Children in the fields and refreshing themselves'. In contrast, Dr. Dancer, an influential physician practising in Jamaica at the beginning of the nineteenth century, exhorted white women to avoid 'all acts of exertion' after childbirth, even so light as bending down to open a drawer. According to Lady Nugent, wife of the governor of Jamaica, writing at the beginning of the nineteenth century, medical men in the colony believed that white and mulatto women had far more difficult pregnancies and more miscarriages than black women, although slave infant mortality was much higher.[34]

Anthropological studies of modern-day Africa challenge such wild and inaccurate generalizations rooted in racial justifications of slavery. Maria Cutrafelli argues that the apparent insensitivity of African women to pain in childbirth is a result of their socialization into stoically bearing pain, first experienced in clitoridectomy (female circumcision) at puberty. Such culturally determined behavior may have influenced African slave women, but it does not change the fact that childbirth was as hazardous for them as it was for white women.[35] It has already been shown that slave women suffered from many gynecological complaints and miscarriages associated with hard labor. Long ascribed their prevalence to the 'unskillfulness and absurd management of negro midwives'. William Beckford had a more honest answer. Reproaching his fellow planters for failing to provide better treatment for pregnant and nursing mothers, he accused them of not wishing their women to breed 'as thereby so much work is lost in attendance on their infant'.[36]

The abolition of the slave trade and the increasing intervention of the metropolitan government in internal island affairs led to further ameliorative legislation in the British Caribbean. The end of the slave trade gave new urgency to the pronatalist policies, but the majority of island laws retained their 'policing' character and were in many ways in conflict with the well-being of pregnant and nursing mothers. Abolitionists argued that concessions to the improved welfare of slaves were made 'grudgingly and of necessity' and were inoperative in practice, 'mere rags to cover the blotches and ulcers of the system'. Writing at the beginning of the nineteenth century, James Stephen observed that absentee proprietorship also contributed to the failure of new laws to improve female fertility.[37] Higman and Sheridan confirm that slave conditions in general deteriorated after 1800. The new interest in female slaves may even have rendered their lives more hazardous; in the French Caribbean, where similar measures were introduced to improve slave fertility, slave women and midwives were punished for infant deaths.[38]

Some paternalist planters were genuinely concerned about the treatment of female slaves, and this gave abolitionists a degree of optimism. John Jeremie argued that after punishment of slave women was reduced in St. Lucia, there was an increase in the slave birthrate. However, genuine attempts to improve 'breeding conditions' were more frequently unsuccessful, baffling well-intentioned liberal planters. Although Matthew Lewis stopped all punishment on his Jamaican plantation and provided all the 'comforts and requisites' deemed necessary to healthy childbirth by eighteenth-century European medical science, there was little improvement in the fertility rates of his plantation slave women.[39] Recently compiled demographic data confirm the failure of pronatalist policies. Higman has shown that, with the exception of Barbados, the British sugar colonies did not show an absolute increase in the slave population before 1832. Indeed, between 1807 and 1834 the total slave population declined from 775,000 to 665,000 at a time when adverse sex ratios of men to women—which some planters blamed for the low fertility of women—were evening out and there was a greater number of creole slaves, supposedly more adapted to plantation life.[40]

Harsh conditions of field labor undoubtedly contributed significantly to this demographic anomaly. Modern historians tend to favor the abolitionist view that slave women whose work was 'least and easiest' had more children. However, the correlation between hard labor and low fertility was not always so clear-cut. In its report to the Lords of Trade inquiry into the slave trade in 1789, the Barbados Committee admitted that on most plantations there were some hardworking females 'who breed very fast', while many others who labored less 'do not breed at all'. Higman's research suggests that fertility was relatively high on large-scale sugar plantations, almost the same as on those where cocoa was grown, and exceeded the rates for coffee and cotton plantations.[41] The causes of low fertility are obviously more complex than would appear on superficial examination of the evidence. The harsh conditions of sugar plantations may have had a more direct impact on mortality patterns than fertility (death rates were much higher on large-scale plantations). Planters themselves believed that they could more successfully manipulate mortality than fertility, and there is no hard evidence for the conscious breeding of slaves.[42]

Much of the scholarly discussion of the fertility rate has focused almost exclusively on external influences on slave women as passive subjects, not as active agents with a degree of control over their own bodies, despite the constraints of the system.[43] However, it is also necessary to examine the experience of childbirth and infant rearing from the perspective of the slave woman through consideration of the close link between culture, material conditions, and resistance. According to Higman, variations in levels of slave fertility were determined not only by work regimes but also by the cultural practices of slaves and the attempts of masters and missionaries to alter them.[44]

Slave Culture, Childbirth, and Resistance

If the abolitionists blamed harsh conditions, the proplanter faction preferred to blame 'immorality' and the inferior African cultural practices of slaves for the low fertility of women and the high infant mortality rate. Many planters believed that promiscuity of slave women and venereal disease were mainly responsible for the failure of Caribbean slaves to reproduce naturally. Edward Long declared that slaves would enjoy 'robust good health' if not 'prone to debauch'. In testimony before the 1789 inquiry into the slave trade, the Jamaican

Committee reported that 'negroes committed . . . foul acts of sensuality and Intemperance' and contracted venereal diseases; the women caught colds at 'nocturnal assemblies' and suffered from obstructed 'natural periods'. The 'barbaric' mating patterns of slaves, which included the practice of polygyny, were at the core of this general lack of morality. Slave women were thus 'rendered unprolific' through their own 'bad practices', while slaves of both sexes concealed venereal disease to avoid abstinence from pleasure during treatment. The committee further alleged that the medicines slave women took for venereal diseases produced sterility and killed their unborn children.[45]

It is difficult to determine the actual extent of venereal disease because contemporary medical opinion on the subject is conflicting. Sir Hans Sloane wrote that it was very common, especially among plantation slaves, and both gonorrhea and 'pox' (syphilis) were transmitted in the same way and had the same course as among Europeans. However, Dr. Thomas Dancer believed it took a different and milder form among blacks, who frequently transmitted it to their children.[46] Such pronouncements illustrate the confusion over definitions of diseases and diagnoses common in the field of tropical medicine at the time. In the context of slave reproduction, however, it is nevertheless important to challenge planters' moral diatribes against the general immorality of slaves. These opinions were based primarily on ethnocentric myths about the 'natural promiscuity' of Africans and the planters' need to confirm the 'social death' of the slave through negative stereotyping and denial of rights to family bonds. With respect to venereal disease, slaves probably suffered as much as the average inhabitant of eighteenth-century European cities, also allegedly rife with 'immorality' yet with booming birthrates. Moreover, planters like Long contradicted themselves in blaming polygamy for low fertility while citing the 'populousness' of West Africa, where the practice 'universally prevailed'.[47] Recent studies of slave family life suggest that wherever possible, slave men and women attempted to recreate African marriage patterns supported by strong African-derived moral codes that worked against 'promiscuity'.[48] They thus resisted plantocratic attempts to impose alien cultural practices upon them, and it was probably this resistance, not 'bad practices', which affected population growth.

Planters were particularly keen to change one aspect of slave childbearing—the late weaning of slave infants. This practice derived from Africa and was commonly blamed for low fertility. Research by Herbert S. Klein and Stanley L. Engerman and by Jerome S. Handler and Robert S. Corruccini suggests that breast-feeding practices may explain fertility differentials between female slaves in the Caribbean and those in the southern United States, where late weaning was rare. Late weaning was related to the two-year postnatal taboo on intercourse common to many African societies and may have provided limited contraception. In the Caribbean this custom was not confined to African-born women. It was also practiced by creoles. Thus, as Kenneth F. Kiple argues, it did not contribute to perceived differences in fertility between the two groups.[49] Late weaning, which provides strong evidence of the durability of African culture and resistance to imposed pronatalist policies, persisted in the Caribbean well into the twentieth century. Planters viewed it as another form of shamming and idling. (Women were frequently accused of citing a multitude of 'female complaints' to avoid work.) When weaning houses were introduced in Jamaica to cut down on extended suckling, women strongly resisted early separation from their infants. Lewis blamed this 'obstinacy' on women's desire to 'retain the leisure and other indulgencies . . . of nursing mothers'. Another planter, John Baillie, reached a similar conclusion after failing to get a single mother, from 1808 to 1832, to accept the premium of two dollars which he offered women to wean their children in twelve months.[50]

Women's reluctance to return to plantation work and their resistance to forcible separation from their infants were arguably rational responses to the problems associated with childbirth, including high infant mortality. Late weaning results in wide birth spacing and, in the absence of adverse influences, can improve infant and maternal well-being, but it cannot fully explain low fertility. In modern West African societies fertility rates remain very high despite long periods of breast-feeding, restrictions on intercourse, high fetal and infant mortality, and early sterility of women. Other explanations need to be explored. For Kiple, these are located in the high infant mortality and diet and disease patterns of Caribbean slavery. Richard Dunn argues, however, that 'eccentric' birth intervals are better explained by sexual abstinence, miscarriages, and abortions, suggesting a more active role for slaves in determining fertility.[51]

Certainly slave nutrition and work practices influenced childbirth patterns, particularly on large plantations. As Higman notes, few free populations of the New or Old World in the early nineteenth century were subjected to 'such a persistent combination of conditions unfavorable to population growth'. Dietary deficiencies could have affected slave women's ability to bear and raise healthy children. Kiple argues that women may have suffered calcium deficiencies leading to rickets (although Dr. Thomas Dancer said that this was 'rarely seen' in Jamaica) and were therefore frequently anemic. Kiple also notes that high frequency of stillbirths and toxemia in pregnancy may be linked to malnutrition. Delayed menarche and early menopause are also related to poor nutrition. Caribbean slaves experienced periodic famine, and there is a well-documented medical link between starvation and amenorrhea (cessation of menstruation leading to temporary sterility). In addition, endemic diseases, such as yaws, and epidemic diseases, such as measles and smallpox, may have increased the incidence of stillbirths and miscarriages. According to Michael Craton, such factors reduced the ability and willingness of women to bear children. The symbiotic relationship between nutrition, infection and fertility remains a controversial subject. Indeed, E. Van den Boogaart and P. Emmer's study of a model plantation in Surinam shows that the birthrate remained low even though the food provided was nutritionally excellent.[52] Thus other factors related more directly to the responses of slave women must be examined if high sterility and miscarriage rates are to be understood.

Planters cited sterility as a major reason for low fertility, particularly among African-born women. Creole slaves had a higher fertility rate. Kiple suggests that West African women who could not bear children may have fallen victim to the slave trade there. This is not an unreasonable proposition, given the high premium placed on female fertility in traditional African culture, but more supportive evidence is needed. Another factor which may have affected the fertility of African slave women was that they were more likely than creoles to be living alone. Women who lived in co-residential unions were significantly more fertile than those who lived alone. However, the presence of large numbers of African women in the slave population is, in itself, an insufficient explanation of low fertility.[53] Most African-born women were in the fertile age range when they arrived in the Caribbean, but the incidence of sterility among them was abnormally high. In African cultures sterility in mature women is regarded as a terrible stigma and social identity for women comes solely through motherhood. Newborn children are greeted with joy and celebration; prolific childbearing is honored. A childless couple will explore every possible means to overcome sterility.

A real desire for motherhood does not, however, mean that African women shunned birth control. They may have brought knowledge of abortion and contraception with them to the Caribbean. Apart from wide birth spacing through long lactation, ritual abstinence, abortion, and other elaborate forms of contraception are more widespread in traditional African societies than is generally recognized. Abortion is used when taboos are broken through adultery or in polygynous relationships where there is jealousy between co-wives. An almost universal reason for abortion in traditional African societies is unsanctioned pregnancy during the lactation period; it is also common to abort girls regarded as too young for pregnancy. Abortion allows women the only real choice in societies where female reproduction is subject to strict patriarchal control.[54] A stronger influence of African retentions among African-born women may thus explain fertility differentials. This operated perhaps on two levels: the psychological, where the impact of slavery weakened the desire to have children; and the practical, where the transmission of cultural knowledge about contraception and abortion came into play.

Plantation life provided little incentive for slave women to have children. Even after ameliorative legislation was introduced there remained insuperable difficulties attached to pregnancy and childbirth. Women who bore children continued to face the threat of separation from their infants or from their husbands and kin who provided emotional support. As early as 1789, planters in Barbados recognized that the specter of sale away from friends and relatives discouraged women from having children. In the Caribbean, it is possible that slave women would avoid unwanted pregnancies when, in the words of an eminent slave doctor, David Collins, the slave woman's life was 'upheld by no consolation, animated by no hope', her troubled pregnancy ending in the birth of a child 'doomed like herself to the rigors of eternal servitude'.[55]

Under extreme conditions the desire and ability of women to have children is reduced. The classic example is the concentration camp. Deportees and prisoners in World War II suffered terrible psychological conditions, anguish, and shock. Under such conditions, according to the French historian Le Roy Ladurie, amenorrhea could become a 'defense mechanism' reflecting the suppression of the 'luxury function' of reproduction in order to survive. The effects of physical starvation and hardship combined with psychological factors to reduce fertility. Such factors may help explain the high incidence of amenorrhea among slave women. Gynecologists now distinguish between 'emotional' amenorrhea, which can occur as a result of psychological disturbance, and 'secondary' amenorrhea, caused by illness or a change in environment. Slave women experienced conditions which increased their chances of developing both forms of this disorder.[56]

Caribbean planters frequently accused slave women of procuring abortions and frustrating their attempts to increase the slave population. Long linked abortion to promiscuity, arguing that slave women were no better than 'common prostitutes' who frequently took 'specifics' to abort so that they could resume their immoral activities 'without loss or hindrance to business'. Drs. John Quier and David Collins, who both practiced in Jamaica during the later period of slavery, gave their professional support to this view. They added that women induced miscarriage through 'violence' or the use of 'simples of the country . . . possessed of forcible power'. In 1826 Reverend Beame alleged that obeah practitioners in Jamaica administered herbs and powders known only to blacks to induce abortions—an indication of the durability of transmitted African knowledge at a time when creole slaves were in the majority.[57]

In determining the extent of such practices, however, it is necessary to distinguish between procured abortions and spontaneous miscarriages because no such distinction was made in contemporary accounts. Slave women almost certainly retained knowledge of such practices from Africa, and as primitive abortion techniques, mechanical or drug-based, can be dangerous, their use by slave women may have contributed to the high incidence of sterility and the slaves' gynecological complaints reported by whites.[58] In traditional African societies various techniques are commonly used to induce abortions in culturally prescribed circumstances. Major abortifacients used include infusions from herbs, leaves of special shrubs, plant roots, and the bark of some trees. Common plants used include manioc, yam, papaya, mango, lime, and frangipani. Mechanical means are less popular and rely mainly on the insertion of sharp sticks or stalks into the vaginal canal.[59]

Similarities between African and Caribbean practices included the use of some drugs and the important role played by 'medicine men and old women' (obeah practitioners). Some older women were skilled in techniques of midwifery and herbalism. They had carried their skills with them to the New World and were valued as doctoresses and midwives. They also provided postnatal care for mothers and infants. Some European doctors derided the folk medicine these women practised (possibly out of professional jealousy), but black healers and nurses were generally regarded as indispensable to the running of plantation hospitals, and more perceptive Europeans acknowledged the efficacy of many folk remedies derived from Africa.[60] In this context the practice of abortion by female slaves is a viable proposition.

On the slave plantation the formulae for herbal concoctions which induced abortions could have been passed on from mother to daughter, as in Africa. Some evidence exists that slave midwives administered abortifacients such as wild cassava and other substances. Dancer recorded the names of a number of plants indigenous to the West Indies used for 'promoting terms' in women. Besides cassava, they included cerasee (an emetic also mentioned by Dr. Barham), Barbados pride, wild passion flower, water germander, and wild tansey (a widely recognized abortifacient also used by slaves in the Old South). Sometimes strong emetics, such as the seed of the sandbox tree, were used to bring on menstruation. John Stedman referred to herbal remedies used in Surinam to induce abortion, including 'green pineapple', and he observed (as in traditional African cultures) young girls who reputedly aborted 'to preserve themselves as long as they were able'. Jealousy in polygamous marriages was another motive for abortion. Writing in the 1770s, Janet Schaw alleged that black women who mated with whites possessed knowledge of 'certain herbs and medicines', and in making use of them to abort, they damaged their health.[61]

This sparse contemporary evidence is strengthened by anthropological data. Among the Djukas of Surinam (culturally close to the societies which existed in eighteenth-century West Africa), abortion and contraception techniques similar to those reputedly used by slave women were still found in the 1930s. These included herbs and 'crude instruments' akin to the pointed sticks used in some African societies. Methods used by slave women have arguably been transmitted down the generations. Melville Herskovits noted that in Trinidad in the 1940s salt, green mangoes, and lime juice were used to successfully abort. Women in the modern Caribbean still buy herbal concoctions from old women to induce abortion.[62] In societies where contraceptive knowledge is poor, abortion is the only means available to women to control their reproduction.

High infant mortality was also cited as a major reason for the low rate of natural increase of Caribbean slave populations, and again, African cultural practices adopted by slave women were blamed for this. Most whites believed that the promiscuity of slave women led to 'a neglect and want of maternal affection towards children of former connexions', and even paternalistic planters like Lewis felt that slave women put pleasure before duty where care of children was concerned.[63] Such comments reflect Eurocentric, bourgeois attitudes toward slave motherhood. Slave women could not be good mothers in the modern sense of the word, but contemporary accounts also testify to the ways in which their strictness was tempered with affection and tenderness. Evidence of strong bonds between mothers and children exists in the ways women resisted separation from their children and attempts of planters to modify African-derived childrearing practices. Women were less frequently runaways because of the stronger ties they had to the plantation through children.[64]

Strong disincentives on Caribbean plantations toward raising children, including high infant mortality, may, however, have led to the supposed indifference of slave mothers. Jamaican planters reported that slave infants had a 'very precarious tenure' on life and that 'one-fourth perish within fourteen days of birth'. In the first nine days of life, slave infants were particularly vulnerable, and, according to Higman, contemporary estimates placed the mortality rate within this period at 25 and 50 percent of all live births. The biggest killers, according to contemporary sources, were peripneumonic fevers caused by damp air and infant tetanus, or *Trismus nascentium*, the 'jawfall', regarded by many Europeans as a major barrier to population growth. Predictably, slave doctors such as John Quier blamed such deaths on 'inadequate maternal attention' and 'want of cleanliness'.[65]

During the late period of slavery attempts were made to prevent infant deaths from tetanus. Dr. Dancer recommended applying laudanum and turpentine to the umbilicus at birth to prevent tetanus caused by the 'negro usage' of tying up the cut naval string with a burnt rag and leaving it for nine days without examination (a practice derived from Africa and still practiced in part of Haiti and Surinam). Dr. Sells advocated lying-in houses and 'properly instructed midwives', while the planter, Lewis, recommended plunging infants into cold baths at birth. Lewis was forced to abandon this practice, however, due to the 'obstinacy' of slave mothers who 'took a prejudice against it'. But 'care and kindness' and European medicine failed to check infant mortality from tetanus. Dancer admitted that 'no adequate solution' had been found and concluded that the cause of the disease depended on 'a certain state and condition peculiar to infants within [the nine-day] period'. John Quier argued that the 'lock-jaw' which affected infants was not tetanus.[66] One explanation of the apparent high death rate from infant tetanus was that it was mistaken for tetany, which causes convulsions and has been traced to a deficiency of calcium and other vital minerals during pregnancy.

However, there were perhaps other crucial variables which affected infant mortality, namely maternal attitudes. These may well have been influenced by the symbolic nature of the first nine days of a child's life in both West Africa (where neonatal tetanus was either rare or nonexistent) and Afro-Caribbean slave society.[67]

In West Africa a newborn infant is not regarded as part of this world until eight or nine days have passed, during which period it may be ritually neglected. The infant is regarded as no more than a 'wandering ghost', a capricious visitor from the underworld. Among the Akan of Ghana, a child remains within the spirit world until this period is over and it becomes a human being, recognized by its father. If a child dies before this time, it is considered never to have existed. Similar traditions are found among the Ga people.[68] The durability of West African practices relating to childbirth has already been observed. It could be argued that the nine-day period (when slave midwives reputedly held 'no hope' for infants) may have reflected African beliefs rather than, as Patterson suggests, deliberate neglect and fatalism because of the high risk of tetanus. Slave women were strongly attached to their 'old customs' where childbearing and rearing were concerned.[69]

Within the framework of cultural persistence, it cannot be discounted that slave babies may have been deliberately 'encouraged' to die. Dr. Robert Jackson, who practiced in Jamaica, argued that slave mothers were not naturally deficient in maternal affection, but 'hard usage' rendered them 'indifferent' or made them wish 'that their offspring may fail' rather than be subjected to the plantation regime.[70] It is very difficult to establish whether premeditated infanticide occurred, although this has existed in many diverse societies from time immemorial—particularly among non-Christian 'pagan' cultures where deformity, sickliness, or sheer strain on resources provided valid reasons. In traditional African culture, deformed infants and twins were commonly killed at birth. Infanticide is the natural corollary of abortion, but historians, sociologists, and demographers rarely consider this subject because of strong taboos in Western culture. For ethical reasons, infanticide is generally far less common than abortion, but where a 'strong desire' to limit infant numbers exists, it may be used in conjunction with abortion or contraception, or as a final resort if these measures fail.[71]

Eugene Genovese argued that slave women in the southern United States could successfully 'arrange' for infants to die soon after birth because infant deaths from natural causes were so common. Such unexplained deaths may have resulted from sudden infant death syndrome, which may be linked to mothers' labor in the fields rather than conscious attempts to deprive the system of slave infants. However, given the slaves' cultural beliefs and inherited knowledge of herbs, infanticide, like abortion, was arguably a valid response to enslavement.[72] If infanticide existed in the Old South where better material conditions prevailed, even stronger arguments apply for its practice in the Caribbean. In 'letting' their children die, women slaves would release them from a dismal future. West African religious beliefs provided the ethical rationale; an infant child, dead or alive, does not have any power for good or evil and its death is regarded as spiritually 'harmless'.[73] The infant mortality rate from natural causes was undoubtedly high in the Caribbean, but the unusually high death rate within the first week, not satisfactorily explained as caused by tetanus, may signify that women used preparations which effected apparently natural death. These could have been acquired from obeah men and women or herbalists, known for their dangerous knowledge of poisons of which whites were largely ignorant. In Europe in the early nineteenth century, 'artificial tetanus' resulting from strychnine poisoning was not unknown.[74]

Infanticide is a highly emotive word, and in the absence of evidence the arguments presented here can only remain speculative. However, as Eric Hobsbawn pointed out, there is a place for informed speculation and creative approaches in history. Indeed, in the study of oppressed social groups, this may be essential. If, as Sheridan claimed, there was a decline in infant deaths from tetanus by 1830, this certainly cannot be explained solely by better conditions, as planters and doctors tended to be disappointed by their efforts to cut infant mortality rates.[75] The decline in infant mortality may be explained more satisfactorily perhaps by the creolization of the population and greater impact of Christian beliefs which diluted the mystical justification of the nine-day period.

Women slaves had a number of powerful reasons for procuring abortion and releasing newly born infants from misery through 'letting' them die. Too many children can be an excessive burden when mothers have a hard and bitter existence. One contemporary observer argued that slaves 'refused to marry' in order 'to avoid generating a race of human beings to be enslaved to [brutal] masters'. Indeed, a paramount reason for the lack of will to have children, and hence the practice of abortion and infanticide, was the institution of slavery itself. Contemporary observers from both anti- and proslavery factions recognized this. James Stephen agreed with Bryan Edwards, who declared that slavery 'in its mildest form' was 'unfriendly to population', as the offspring of slaves were 'born but to perish'.[76]

After the abolition of slavery in 1838, Caribbean populations began to reproduce naturally, although conditions had barely improved and in some ways may have further deteriorated. This lends support for the proposition that the failure of Caribbean slave populations to reproduce cannot be explained solely by harsh conditions. The nature of slavery and the responses of slave women to their bondage must also be considered. There is no doubt that slave women loved and cared for their children, but they had no incentive to have large families. It has been argued here that there was a strong link between slave women's productive and reproductive roles and that the enigma of low fertility needs to be explored in the context of a wider slave resistance to the system. Among slave women, deliberate management of their fertility may have been a form of hidden, individual protest against slavery. Masters had less control over these forms of resistance than they had over the more overt forms of collective resistance. As Elizabeth Fox-Genovese has suggested, it is important to look not only at the specific experience of women as women but also at their struggle for an 'individual soul or consciousness' against objectification, alienation, and dehumanization.[77]

Slave women's labor on Caribbean plantations conflicted with their private domestic lives. This had important implications for the reproduction of the slave population. No 'normal' pattern of marriage or parenting was fully possible until slavery ended. Women had an enforced, alien division of labor imposed upon them which negated and undermined the traditional division of labor which was part of their African cultural inheritance. According to Jacqueline Jones, the 'schizoid' character of black women's lives under slavery was a product of white aspirations for blacks and blacks' aspirations for themselves. Nowhere was this more pertinent than in the intimate area of childbirth and infant rearing. After general emancipation, black women retreated, wherever possible, from plantation labor. They reclaimed their traditional role within the family and recreated a pattern of domesticity which had not been attainable under slavery. At this time the birthrate began to rise.[78]

In refusing to 'breed' when forced to perform hard labor on the plantations, slave women were protesting their slave status and the erosion of their African cultural heritage. Where sexuality and reproduction were concerned, slave women were quadruply burdened, by both black and white patriarchy and by both gender and racial oppression. Their material conditions of existence, hardly conducive to childbearing and rearing, arguably led them to seek to exercise a degree of conscious control over their own reproductive capacities which frustrated planters' attempts to naturally increase the slave population. Slave women's responses to childbirth may be viewed as part of a wider pattern of resistance informed by African cultural practices and the personal and institutional relations which developed in slave societies.

Notes

1. Lucille Mathurin Mair, 'A Historical Study of Women in Jamaica from 1655 to 1844', Ph.D. dissertation (University of the West Indies, 1974). For more recent works, see Deborah Gray White, *Ar'n't I a Woman: Female Slaves in the Plantation South* (New York: Norton, 1985); Jacqueline Jones, *Labor of Love, Labor of Sorrow: Black Women, Work and the Family from Slavery to the Present* (New York: Basic, 1985); Barbara Bush, *Slave Women in Caribbean Society, 1650–1838* (Bloomington: Indiana University Press, 1990).
2. Interest in slave demography developed in the 1970s. The main debates are summarized in Barry W. Higman, *Slave Populations of the British Caribbean, 1807–1834* (Baltimore: Johns Hopkins University Press, 1984), and Richard B. Sheridan, *Doctors and Slaves: A Medical and Demographic History of Slavery in the British West Indies, 1680–1834* (Cambridge: Cambridge University Press, 1985).
3. Melville J. Herskovits and Frances S. Herskovits, *Trinidad Village* (New York: Knopf, 1947), pp. 8–9. For a comprehensive study of slave resistance in the British Caribbean, see Michael Craton, *Testing the Chains: Resistance to Slavery in the British West Indies* (Ithaca, NY: Cornell University, Press, 1982).
4. Higman, *Slave Populations*, pp. 50–67. Such large-scale units, with the exception of Jamaica (53 percent), occupied 71–80 percent of all economically active slaves by 1830, and thus had a dominant role in determining the character of Caribbean slavery.
5. For detailed analyses of the living and working conditions of women field laborers, see Sheridan, *Doctors and Slaves*, pp. 178–90; Higman, *Slave Populations*, chap. 6.
6. Arlette Gautier, 'Les Esclaves femmes aux Antilles françaises, 1635–1884', *Reflexions Historiques*, 10 (Fall 1983), 409–35; Michael Foucault. *The History of Sexuality*, vol. 1, *An Introduction* (London: Allen Lane, 1978), trans. Robert Hurtley, pp. 96, 103–7. According to Foucault, sexuality should be described not merely as an instinctual drive but also as a 'dense transfer point' for relations of power between men and women.
7. Higman, *Slave Populations*, p. 150.
8. Gad J. Heuman, *Between Black and White: Race, Politics and the Free Coloreds in Jamaica, 1792–1865* (Westport, Conn.: Greenwood, 1981), p. 14. See also Patterson, *Slavery and Social Death*, p. 175. For contemporary comments, see

J. G. Stedman, *Narrative of a Five Years' Expedition against the Revolted Negroes of Surinam, 1772–1777* (London: J. Johnson and J. Edwards, 1796), vol. 1, p. 112; John Stewart, *A View of the Past and Present State of the Island of Jamaica* (Edinburgh: Oliver & Boyd, 1823), p. 170. The psychological basis of European women's racism and power in colonial societies is analyzed in a classic study by Octave Mannoni, *Prospero and Caliban: The Psychology of Colonization* (New York, 1964). Data relating to slave ownership are provided by Higman, *Slave Populations*, pp. 111–12.

9. Hilary Beckles, '"Black Men in White Skins": The Formation of a White Proletariat in West Indian Slave Society', *Journal of Imperial and Commonwealth History*, 15 (October 1986), 13. Even poor white women were not allowed to work as field hands.

10. Bryan Edwards, *The History, Civil and Commercial, of the British Colonies of the West Indies* (London, 1805), vol. 1, pp. 540–1; Edward Long, *The History of Jamaica* (1774; reprint, London: Frank Cass, 1970), vol. 1, pp. 304–404; John Adams, *Sketches Taken during Ten Voyages to Africa between the Years 1786 and 1800* (London, 1822), p. 8. Anthropological studies of the sexual division of labor in traditional African societies confirm that women grow most of the food crops for consumption (as opposed to export), but men also have clearly defined tasks, which include much of the heaviest labor. See, for instance, Maria Rosa Cutrafelli, *Women of Africa: Roots of Oppression*, trans. from Italian (London: Zed, 1983), chap. 2.

11. For data on sex ratios, see Sheridan, *Doctors and Slaves*, pp. 107–8; Higman, *Slave Populations*, pp. 117–18. David W. Galenson, in an econometric analysis, suggests that the higher prices of women in the internal African slave markets (in comparison to males) may have had an impact on the availability of female slaves for transport to the West Indies. See Galenson, *Traders, Planters and Slaves: Market Behavior in Early English America* (Cambridge: Cambridge University Press, 1986), pp. 105–7. The sexual division of labor on the sugar plantation is analyzed in some detail in Higman, *Slave Populations*, pp. 189–99. For comparative data on the French Caribbean, see Gautier, 'Les Esclaves femme', 410–11.

12. William Beckford, *Remarks upon the Situation of the Negroes in Jamaica* (London: Egerton, 1788), p. 13.

13. *The Royal Gazette*, Kingston, Jamaica, August 18, 1827. Contemporary estimates of slave prices are given in Edwards, *History*, vol. 2, p. 132, and 'Report of the Lords of Trade into the Slave Trade', *Parliamentary Papers*, Commons, 26 (1789), 646a, pt. 3, Jamaica, Ans. no. 29. Women were on average 81 percent as valuable as adult men in Barbados, 1673–1723 (Galenson, *Traders, Planters, and Slaves*, pp. 62–5).

14. Higman, *Slave Populations*, pp. 189–98, 224, 314–17. The dominance of women in fieldwork up to emancipation is confirmed by studies of individual plantations. See Michael Craton and James Walvin, *A Jamaican Plantation: The History of Worthy Park, 1670–1970* (London: W. H. Allen, 1970), p. 138, which gives the field gang composition for 1787–95; Richard S. Dunn, 'A Tale of Two Plantations: Slave Life at Mesopotamia in Jamaica and Mount Airy in Virginia, 1799–1828', *William and Mary Quarterly*, 3rd Series, 34 (January 1977), 32–65.

15. Higman, *Slave Populations*, pp. 183–4, 314–17. For a comparative description of the field slave's daily toil, see Beckford, *Negroes in Jamaica*, p. 44.

16. EIsa Goveia, *Slave Society in the British Leeward Islands at the End of the Eighteenth Century* (New Haven: Yale University Press, 1965), p. 234; Craton and Walvin, *A Jamaican Plantation*, pp. 103–4, 125; Higman, *Slave Populations*, p. 188.

17. The various forms of resistance employed by women, including verbal abuse, are discussed in Barbara Bush, 'Defiance or Submission? The Role of the Slave Woman in Slave Resistance in the British Caribbean', *Immigrants and Minorities*, 1 (March 1982), 16–39.

18. Goveia, *Slave Society*, pp. 244–45.

19. Baillies Bacolet Plantation Returns, 1820–1833; Punishment Record Books for Friendship Sarah and Good Success Plantations, 1823–1833 (British Guiana), Atkin's Slavery Collection, Wiberforce House, Hull. The intransigent nature of female field hands was also noted in the commentaries of individual planters. See Matthew Gregory Lewis, *Journal of a Residence among the Negroes of the West Indies* (London, 1845), pp. 93, 103.

20. Public Records Office, Kew, Colonial Office Series (C.O.) 295/60, 295/66, Commandant of Chaguanas to Governor Woodford, August 20, 1823; cited in Bridget Brereton, 'Brute Beast or Man Angel: Attitudes to the Blacks in Trinidad, 1802–1888', unpublished paper, Department of History, University of the West Indies, Trinidad, 1974. See also Long, *History of Jamaica*, p. 103.

21. Thomas Cooper, *Facts Illustrative of the Condition of the Negro Slaves in Jamaica* (London: Hatchard, 1824), pp. 17–18.

22. Higman, *Slave Populations*, p. 224. For comments on the power of black drivers, see, for instance, Stedman, *Narrative*, vol. 1, p. 117.

23. Baillies Bacolet Plantation Returns.

24. Cooper, *Facts Illustrative*, p. 20; 'Abstract from the Slave Laws of Jamaican Slave Law, 1826', no. 37 in Bernard Martin Senior, *Jamaica as It Was, as It Is, and as It May Be* (London: Hurst, 1835), p. 145; William Sells, *Remarks on the Condition of Slaves in the Island of Jamaica* (London: Richardson, Cornhill & Ridgways, 1823), p. 17.

25. John Williamson, *Medical and Miscellaneous Observations, Relative to the West Indian Islands* (Edinburgh, 1817), vol. 1, p. 191; Lewis, *Journal*, pp. 174–5.

26. Sir Hans Sloane, *A Voyage to the Islands Madera, Barbadoes, Nieves, S. Christophers and Jamaica* (London, 1707), vol. 1, p. lii; Richard Ligon, *A True and Exact History of the Island of Barbadoes* (London, 1657; reprint, London, 1970), p. 113.

27. Higman, *Slave Populations*, pp. 290–300; Mair, 'A Historical Study of Women in Jamaica', pp. 313–16.

28. 'Report . . . into the Slave Trade' (1789), pt. 3, Barbados, Ans. no. 16; Long, *History of Jamaica*, vol. 2, p. 341.

29. Long, *History of Jamaica*, vol. 2, p. 490. For an analysis of the slave laws from the abolitionist perspective, see John Jeremie, *Four Essays on Colonial Slavery* (London: J. Hatchard and Son, 1831).

30. EIsa Goveia, *The West Indian Slave Laws of the Eighteenth Century* (Barbados: Caribbean Universities Press, 1970), pp. 7, 25, 48; Michael Craton, *Sinews of Empire; A Short History of British Slavery* (London: Temple Smith, 1974), p. 175.

31. 'Abstract from the Leeward Islands Act 1798', Act no. 36, Clauses 37 & 38, Edwards, *History*, vol. 5, pp. 181–82, 185; 'Abstract of the Laws of Jamaica',

Long, *History of Jamaica*, vol. 2, p. 161; Goveia, *Slave Society*, p. 196; Stewart, *View of Jamaica*, p. 262.

32. 'Abstract from the Leeward Islands Act 1798', Act no. 36, Clauses 22–5, Edwards, *History*, vol. 5, pp. 183–5; 'Abstract from the Slave Laws of Jamaica, 1826', nos. 32 and 33; Senior, *Jamaica*, p. 144; Long, *History of Jamaica*, vol. 2, p. 440.

33. Sells, *Remarks*, pp. 15–18.

34. Ligon, *A True and Exact History*, p. 49; Thomas Dancer, *The Medical Assistant of Jamaica Practice of Physic, Designed Chiefly for the Use of Families and Plantations* (Kingston, 1809), pp. 263–4; Philip Wright, *Lady Nugent's Journal* (Kingston: Institute of Jamaica, 1966), p. 69. For racist attitudes, see Long, *History of Jamaica*, vol. 2, p. 385; Stedman, *Narrative*, vol. 2, p. 359. Long acknowledged that childbirth in the Caribbean was 'not so easy' as in Africa. In the early days slave women were reputedly 'left alone to God' in one room of their cabin and after two weeks they were back at work; see Ligon, *A True and Exact History*, p. 49; Sloane, *Voyage*, vol. 1, p. cxlvii.

35. Cutrafelli, *Women of Africa*, pp. 136–7.

36. Long, *History of Jamaica*, vol. 2, p. 436; Beckford, *Negroes in Jamaica*, pp. 24–5.

37. *Edinburgh Review*, 55 (1832), Article 7, p. 148, and 38 (1823), Article 8, p. 177; James Stephen, *The Crisis in the Sugar Colonies* (London, 1802), p. 32.

38. Sheridan, *Doctors and Slaves*, pp. 246–8; Higman, *Slave Populations*, pp. 67–8; Gautier, 'Les Esclaves femme', p. 417.

39. Lewis noted, 'The children do not come . . . despite encouragement'; *Journal*, pp. 45, 52. See also Jeremie, *Four Essays*, p. 97. Paternalistic planters may actually have been more far-sighted than other planters and were not necessarily 'uneconomic' in their policies to conserve slave life and encourage reproduction. See Daniel C. Littlefield, 'Plantations, Paternalism and Profitability: Factors Affecting African Demography in the Old British Empire', *Journal of Southern History*, 47 (May 1981), 171. However, such initiatives were arguably frustrated by the absentee system and increasingly uncertain economic conditions. The absence of white women to take a personal interest in pregnant and lying-in slave women may also have resulted in the neglect of these women.

40. Higman, *Slave Populations*, pp. 3, 73, 116–17. According to Higman's data, there were more women than men between 1817 and 1832 (p. 116), yet adverse sex ratios were often exaggerated by planters, particularly as abolitionist pressure built up at the end of the eighteenth century. Edwards maintained that the ratio on some Jamaican plantations was as high as five men to one woman; see Edwards, *History*, vol. 2, pp. 118, 132. See also Long, *History of Jamaica*, p. 385; Stewart, *View of Jamaica*, pp. 308–10. The sex ratio did vary from plantation to plantation, from island to island, and between rural and urban slaves but in itself is an adequate explanation of low fertility.

41. Sheridan, *Doctors and Slaves*, pp. 246–8, 'Report . . . on the Slave Trade', pt. 3, Barbados, Ans. no. 16; Higman, *Slave Populations*, pp. 361–2.

42. Higman, *Slave Populations*, pp. 259–61.

43. See, for example, Michael Craton, *Searching for the Invisible Man: Slaves and Plantation Life in Jamaica* (Cambridge, Mass.: Harvard University Press, 1978), pp. 102–12; Kenneth F. Kiple, *The Caribbean Slave: A Biological History* (Cambridge: Cambridge University Press, 1984), chap. 7.

44. Higman, *Slave Populations*, pp. 347–48, 362; Herbert S. Klein and Stanley L. Engerman also stress the need to account for the active agency of slaves in examining childrearing practices; see their 'Fertility Differentials between Slaves in the United States and the British West Indies: A Note on Lactation Practices and Their Possible Implications', *William and Mary Quarterly* 3rd series, 35 (April 1978), 374.

45. Long, *History of Jamaica*, vol. 2, pp. 412, 436–47; 'Report . . . into the Slave Trade', pt. 3, Jamaica, Ans. no. 11, no. 4. For a discussion of European attitudes to Afro-Caribbean medicine, see Sheridan, *Doctors and Slaves*, pp. 320, 330–7.

46. Sloane, *Voyage*, vol. 1, p. cxxvii; Dancer, *Medical Assistant*, pp. 212–13. Kiple notes that there was confusion over the diagnosis of syphilis and yaws in the Spanish islands, and modern research has confirmed a cross-immunity between the two diseases. It was widely believed that syphilis was a white man's disease and yaws a black disease. 'Pure' blacks were allegedly never stricken by syphilis, and Kiple argues that this disease became a problem for Caribbean blacks only when yaws disappeared in the twentieth century. He also maintains that tuberculosis of the urinary tract was frequently diagnosed as gonorrhea. See Kiple, *Caribbean Slave*, pp. 243–4, n. 29. Data on mortality indicate that the death rate from venereal disease was much lower than the rates for yaws and tuberculosis. See Sheridan, *Doctors and Slaves*, pp. 194–200; Higman, *Slave Populations*, pp. 339–47.

47. Long, *History of Jamaica*, vol. 2, p. 385. Comparative conditions in eighteenth-century Europe are discussed in M. C. Buer, *Health, Wealth and Population in the Early Days of the Industrial Revolution* (London: Routledge, 1926), chaps. 3, 6. The significance of negative stereotyping is discussed in Patterson, *Slavery and Social Death*, pp. 85, 100.

48. B. W. Higman, 'The Slave Family and Household in the British West Indies, 1800–1834', *Journal of Interdisciplinary History*, 6 (Autumn 1975), 261–87; Barbara Bush, 'The Family Tree Is Not Cut: Women and Cultural Resistance in Slave Family Life in the British Caribbean', *In Resistance: Studies in African, Caribbean and Afro-American History*, ed. Gary Y. Okihiro (Amherst: University of Massachusetts Press, 1986), pp. 117–33.

49. Klein and Engerman, 'Fertility Differential', p. 371; Jerome S. Handler and Robert S. Corruccini, 'Weaning among West Indian Slaves: Historical and Bioanthropological Evidence from Barbados', *William and Mary Quarterly*, 3rd ser., 43 (January l986), 111–17; Kiple, *Caribbean Slave*, p. 110. For contemporary comments, see Stedman, *Narrative*, vol. 2, p. 368; Dr. Collins, ('A Professional Planter'), *Practical Rules for the Management and Medical Treatment of Negro Slaves in the Sugar Colonies* (London: J. Barfield, 1803), p. 146. For observations on the practice in Africa, see Daryl Forde, 'Kinship and Marriage among the Matrilineal Ashanti', in *African Systems of Kinship and Marriage*, ed. A. R Radcliffe-Brown and D. Forde (London: Oxford University Press, 1950), pp. 262–3. The Ashanti prohibited intercourse for only eighty days after birth, which suggests that factors other than late weaning may have contributed to the low fertility rate where (as in the case of Jamaica) there was a high proportion of slaves of Ashanti (Akan) origin.

50. Lewis, *Journal*, pp. 145–6; Baillie, cited in Sheridan, *Doctors and Slaves*, p. 245.

51. Kiple, *Caribbean Slave*, pp. 113–15; Dunn, 'A Tale of Two Plantations', p. 61. Kiple suggests that slave infant mortality was the prime reason for the low rate

of natural increase in the Caribbean and argues, contrary to existing research, that slave fertility was actually high.

52. Higman, *Slave Populations*, pp. 373–6, 397; Kiple, *Caribbean Slave*, pp. 113–15; Dancer, *Medical Assistant*, pp. 232–3; Sheridan, *Doctors and Slaves*, pp. 158–9; Craton, *Searching for the Invisible Man*, p. 99; Dunn, 'A Tale of Two Plantations', pp. 62–3; E. Van Den Boogaart and P. Emmer, 'Plantation Slavery in Surinam in the Last Decade before Emancipation', *Annals of the New York Academy of Sciences*, 292 (June 1977), 205. For the link between starvation and amenorrhea, see E. Le Roy Ladurie, *The Territory of the Historian*, trans. from French (London, 1979), pp. 265–9.

53. Kiple, *Caribbean Slave*, pp. 107–98; Higman, *Slave Populations*, p. 372–3.

54. Cutrafelli, *Women of Africa*, pp. 133–5; Forde, 'Kinship and Marriage', pp. 262–6.

55. 'Report . . . into the Slave Trade', pt. 3, Barbados, Ans. no. 5; Collins, *Practical Rules*, p. 35.

56. Stanley M. Elkins, *Slavery: A Problem in American Institutional and Intellectual Life* (Chicago: University of Chicago Press, 1959). Ladurie traces the connection between low fertility and harsh conditions from the seventeenth to the twentieth century: *Territory of the Historian*, pp. 268–9. For definitions of amenorrhea, see *The British Medical Dictionary*, ed. Sir A. S. McNulty (London, 1961). Also relevant here is Terrence Des Pres's analysis of a World War II concentration camp; see Des Pres, *The Survivor: An Anatomy of Life in the Death Camps* (New York: Oxford University Press, 1976).

57. Long, *History of Jamaica*, vol. 2, p. 346; John Quier, 'A Slave Doctor's Views on Childbirth, Infant Mortality and the General Health of His Charges, 1788' in Report of the Assembly on Slave Issues, Jamaica House of Assembly, enclosed in Lt. Gov. Clarke's dispatch, November 20, 1788, no. 92, C.O. 137/88, Appendix C, p. 491; Collins, *Practical Rules*, p. 51; Reverend Henry Beame, cited in *Slavery, Abolition and Emancipation: Black Slaves and the British Empire*, ed. Michael Craton, James Walvin, and David Wright (London: Longman, 1976), p. 141.

58. For the dangerous nature of slave abortions, see Sloane, *Voyage*, p. cxlii. Cutrafelli notes that in Zaire a strong pepper (pili pili) is used, large doses of which can cause internal burns to the ovaries: *Women of Africa*, p. 141. Similarly the Guinea pepper used by the Efik of old Calabar (Nigeria) can cause 'organic lesions' and 'serious constitutional disturbances'; see George Devereux, *A Study of Abortion in Primitive Societies* (New York: Julian, 1955), p. 218.

59. Cutrafelli, *Women of Africa*, pp. 141–3; Devereux, *A Study of Abortion*, pp. 218, 289; R S. Rattray, *Religion and Art in Ashanti* (Oxford: Clarendon Press, 1927), p. 55; G. W. Harley, *Native African Medicine* (Cambridge, Mass.: Harvard University Press, 1941), p. 73; Melville T. Herskovits, *Dahomey: An Ancient West African Kingdom* (New York: J. J. Augustin, 1938), vol. 1, p. 268.

60. Sheridan, *Doctors and Slaves*, p. 74; Kiple, *Caribbean Slave*, pp. 152–4. For contemporary observations on slave women's knowledge of folk medicine, see Stewart, *View of Jamaica*, p. 312.

61. Dancer, *Medical Assistant*, pp. 263–4, 368, 381, 398; Stedman, *Narrative*, vol. 1, p. 334; *Journal of a Lady of Quality: Being the Narrative of a Journey from Scotland to the West Indies, North Carolina and Portugal in the Years 1774 to*

1776, ed. Evangeline Walker Andrews and Charles McLean Andrews (New Haven: Yale University Press, 1923), pp. 112–13. For transmission of folk knowledge, see Sheridan, *Doctors and Slaves*, pp. 77, 244.

62. Morton C. Kahn, *Djuka: The Bush Negroes of Dutch Guiana* (New York: Viking, 1931), pp. 127–28; Herskovits and Herskovits, *Trinidad Village*, p. 111; oral data collected by Barbara Bush.

63. Lewis, *Journal*, p. 65. For similar contemporary comments, see Edwards, *History*, vol. 2, p 148.

64. *Out of the House of Bondage: Runaways, Resistance and Marronage in Africa and the New World*, ed. Gad Heuman (London: Frank Cass, 1986), p. 6. For evidence of maternal tenderness, see Ligon, *A True and Exact History*, p. 51; Long, *History of Jamaica* vol. 2, p. 44; Sloane, *Voyage*, pp. lvi, lvii; Lewis, *Journal*, p. 90.

65. Quier, 'A Slave Doctor's Views', p. 490. See also the evidence of Jamaican planters in 'Report . . . into the Slave Trade', pt. 3, Jamaica, Ans. no. 16; Edwards, *History*, vol. 2, p. 140; Lewis, *Journal*, p. 97.

66. Dancer, *Medical Assistant*, pp. 257, 267, 278; Sells, *Remarks*, p. 18; Lewis, *Journal*, pp. 50, 141, 269; Quier, 'A Slave Doctor's Views', p. 490. For treatment of the umbilicus in modern Afro-Caribbean societies, see Kiple, *Caribbean Slave*, p. 121; Kahn, *Djuka*, p. 227. Statistics relating to infant mortality are given in B. W. Higman, *Slave Population and Economy in Jamaica, 1807–1834* (Cambridge: Cambridge University Press, 1976), pp. 112–13. Higman suggests that infant deaths were, if anything, under-recorded in the first few days.

67. Kiple, *Caribbean Slave*, pp. 120–5.

68. Forde, 'Kinship and Marriage', p. 67; Rattray, *Religion and Art*, pp. 57–8, 67; M. J. Field, *Religion and Medicine of the Ga People* (London: Oxford University Press, 1937), p. 214; G. Parrinder, *West African Religion* (London: Epworth, 1949), p. 214; Herskovits, *Dahomey*, vol. 1, pp. 266–7; Cutrafelli, *Women of Africa*, p. 133.

69. Orlando Patterson, *The Sociology of Slavery* (London: McGibbon and Kee, 1967), p. 155; Edward Brathwaite, *The Development of Creole Society in Jamaica, 1770–1820* (Oxford: Clarendon Press, 1971). For slave women's 'obstinacy', see Quier, 'A Slave Doctor's Views', p. 491.

70. Evidence of Dr. Jackson (Jamaica), House of Commons Report, 1791, cited in Sheridan, *Doctors and Slaves*, p. 228.

71. Herbert Apteker, *Anjea: Infanticide, Abortion and Contraception in Savage Society* (New York: W. Godwin, 1931), p. 151.

72. Eugene D. Genovese, *Roll, Jordan, Roll: The World the Slaves Made* (New York: Pantheon, 1974), p. 41; Michael P. Johnson, 'Smothered Slave Infants: Were Slave Mothers at Fault?' *Journal of Southern History*, 47 (November 1981), 510–15.

73. Rattray, *Religion and Art*, pp. 59–61.

74. Thomas B. Curling, *A Treatise on Tetanus* (London, 1836), p. 27. For slave knowledge of poisons, see Sloane, *Voyage*, vol. 2, pp. ix, xii; R. R. Madden, *A Twelve-month's Residence in the West Indies* (London: James Cochran, 1835), vol. 1, p. 187.

75. Sheridan, *Doctors and Slaves*, pp. 235–6; Eric Hobsbawm, quoted by Richard Gott in 'The History Man', *Guardian*, February 26, 1988.

76. Monsieur Bossue, cited in Long, *History of Jamaica*, vol. 2, p. 440; Edwards, *History*, vol. 2, p. 148; James Stephen, *The Slavery of the British West India Colonies Delineated* (London, 1824–30), vol. 1, p. 79. It may be argued that slavery in the West Indies existed in its most brutal and distorted form; see Orlando Patterson, *Slavery and Social Death: A Comparative Study* (Cambridge, Mass.: Harvard University Press, 1982), pp. 97–101.

77. Elizabeth Fox-Genovese, 'Strategies and Forms of Resistance: Focus on Slave Women in the United States,' in *In Resistance*, pp. 143–65. See also Sheridan, *Doctors and Slaves*, pp. 245, 339; Higman, *Slave Populations*, p. 366.

78. Jones, *Labor of Love*, p. 340. See also Sheridan, *Doctors and Slaves*, p. 340, for reversion to a traditional division of labor after emancipation. Gautier suggests that a similar reversion occurred in Saint Domingue (Haiti) after the revolution; see 'Les Esclaves femmes', p. 432.

Part III. Embodied Ideals

The Slipped Chiton

Marina Warner

Proposed by the jurist and historian Edouard de Laboulaye as a gesture of republican fellowship with the United States, the Statue of Liberty—standing today in New York's harbour—was formally accepted as a gift from the French-American Union in France by the US Congress in 1877. Funds had been raised by public subscription from all over France, on the patriotic basis that as the French *philosophes'* idea of Liberty had been exported to America and inspired the War of Independence, so it would be fittingly commemorated by a French statue. Only Liberty's arm with the beacon was finished in time to be exhibited at the world fair in Philadelphia in 1876, the centenary of independence. Her head followed, shown at the world fair in Paris two years later, but a problem developed about erecting her *in situ* at New York harbour, for the French gift did not include the cost of the pedestal. Joseph Pulitzer, the proprietor of the *World* newspaper, hearing of the hitch in Liberty's move to her island site, started a campaign in his newspaper and raised the necessary money over a startling 147 days. The pedestal was designed by the eclectic architect Richard Morris Hunt in the style of a classical mausoleum, and in 1885, the statue, which had towered over the roofs of Paris from the workshops of Gaget, Gauthier et Cie., where Auguste Bartholdi had assembled it, was shipped to New York in 214 crates. It was unveiled on 26 October 1886.

Although the political history of the statue, and its French origin, have not been erased from its story, Liberty is no longer La Liberté, but was identified from the start with an American ideal of democracy, now represented as an American gift to the world. Bartholdi created an allegory of the Republic in keeping with moderate rather than radical

From Marina Warner, *Monuments and Maidens: The Allegory of the Female Form* (Atheneum, New York, 1985), pp. 3–17, 227–293. Permission has been sought.

politics, to suit American taste and respectability in the late nineteenth century. His is a staid and matronly conception of the wild thing who surges across the barricade in Delacroix's *Liberty Guiding the People*, and could be guaranteed not to offend a people who had been dismayed by Horatio Greenough's marble statue of George Washington *all'antica*—naked under a toga.[1] He shifted the allusions of Liberty away from unbridled Nature in favour of an imagery of control and light, influenced by the symbols of the Freemasons, to whom Bartholdi belonged. However, as Ernst Gombrich has written, 'The balance of justice or the torch of Liberty . . . are not just fortuitous identification marks. . . . Their choice is rooted in the same psychological tendency to translate or transpose ideas into images which rules the metaphors of language.[2]

. . .

We can all take up occupation of Liberty, male, female, aged, children, she waits to enfold us in her meaning. But a male symbol like Uncle Sam relates to us in a different way, and the distinction between the two figures who have become emblematic of the United States indicates a common difference between male and female figures conveying ulterior meaning. The female form tends to be perceived as generic and universal, with symbolic overtones; the male as individual, even when it is being used to express a generalized idea.

'Uncle Sam', who appeared in New York State around 1812 as the owner of the initials U.S., stamped on government supplies to the army and other goods and equipment, stands to America as John Bull stands to England, an epitome of the nation's character, a collective caricature. John Bull too, has his humorous side; like Uncle Sam he is not thought to be based on anyone in particular, but was first born in print in Dr John Arbuthnot's pamphlet of 1712: 'Law is a Bottomless Pit, exemplified in the case of Lord Strutt, John Bull, Nicholas Frog, and Lewis Baboon, who spent all they had in a Law Suit.'[3]

John Bull typifies the Englishman; Uncle Sam and Brother Jonathan the US citizen. But Liberty can hardly be said to represent the typical American woman, or Britannia the Englishwoman of collective consciousness. Men are individual, they appear to be in command of their own characters and their own identity, to live inside their own skins, and they do not include women in their symbolic embrace: John Bull, however comic, can never be a cow. But the female form does not refer to particular women, does not describe women as a group, and often does not even presume to evoke their natures. We can all live inside Britannia or Liberty's skin, they stand for us regardless of sex, yet we cannot identify with them as characters. Uncle Sam and John Bull are popular figures; they can be grim, sly, feisty, pathetic, absurd, for they have personality. Liberty, like many abstract concepts expressed in the feminine, is in deadly earnest and one-dimensional. Above all, if John Bull appears angry, it is his anger he expresses; Liberty is not representing her own freedom. She herself is caught by the differences, between the ideal and the general, the fantasy figure and the collective prototype, which seem to hold through the semantics of feminine and masculine gender in rhetoric and imagery, with very few exceptions.

· · ·

Classical allegories, using the female form, abound in the United States' public places. In New York, Justice, holding her scales, stands over the law courts. The chapel of St Paul in the Civic Centre district, founded in 1764, is the oldest church in the city, an example of British classical architecture, stranded like an architect's model between the dizzy columns of Wall Street. Inside it displays inside the arms of the State of New York, painted by a naïve eighteenth-century artist in imitation of the highest Western models. The bearers of the arms are Justice and Freedom, the latter in rosy décolletage and carrying on her spear an uncensored liberty cap. And at night, on the downtown skyline, a single feminine form lights up and dominates the view of TriBeCa and the business district by her singularity and her height, as she stands in a gentle amber aureole among the glinting shafts of Wall Street and the Trade Centre. The windows around wink and flash in cryptograms, a galactic console of messages coded in light, in diamonds and topazes and amber. High up on the pinnacles of the grey pyramidal mass of the Civic Centre's Municipal Building there appears this single shape that recalls the human form and not the electronic impulse or the illuminated and abstract perforation. It is gilded, sinuous (especially in the context of Lower Manhattan's perpendiculars and horizontals) and represents Civic Fame, under the aspect of a classical figure of Victory, bearing a palm to the winner. The statue was sculpted by Adolph Alexander Weinman, and erected a generation after Liberty, two generations after Freedom Triumphant on the Capitol.[4]

· · ·

The most famous prototype for Bartholdi's Statue of Liberty is Delacroix's *Liberty Guiding the People*, of fifty years before, now become a universal image of revolution, in its glory and its terror (Fig. 14).[5] Delacroix painted the canvas after a short-lived enthusiasm for the 'Trois Glorieuses'—27, 28 and 29 July—of the 1830 Revolution that brought down the Bourbons.

Caught up in the capital's excitement, Delacroix volunteered for the Garde Nationale. But in February of the following year, when the bourgeoisie had succeeded in placing Louis-Philippe on the throne, he was ambivalent about his own *engagement*. In an eloquent letter, he wrote:

We are living, my dear friend, in a discouraging time. You have to have virtue to make beauty your only God. Ah well, the more people hate it, the more I adore it, I'll finish up believing that there's nothing true in the world except our illusions. . . . Come back quickly . . . come and taste the sweet pleasures of the national guard which one fine day like an idiot I got myself stuffed into. Come above all to kiss us and speak badly of the human species, of the times, of this and that, of ourselves, of women above all.[6]

That autumn, Delacroix exhibited his famous picture of the July Days in the Salon. It possesses the full charge of disquiet he shows in his correspondence, the clash of cynicism and optimism in his feelings. The painting is smaller than the impression it gives in reproduction (2.60 metres high and 2.25 metres wide) and the paint is applied more thinly than expected, rather as if Delacroix dashed it off; the different characters—the gamin brandishing pistols, the imploring worker at Liberty's feet, the guard in the top hat to his right—have not fused completely with the dramatic scene at the barricades Delacroix has imagined. They bear the traces of the studio sitting in their fixed stance and studied expressions. Yet the image as a whole survives such criticisms; its essential tragic character makes them fribbling. Delacroix may have painted with brio, but the composition is very carefully considered; he has cast each type of citizen worker who rose in the July Days, and distinguished them by different headgear: a student's velvet cap on the pistol waving boy, the Garde Nationale's shako on the boy to the far right, the labourer's beret on the man with the sabre, the agricultural worker's kerchief on the figure kneeling, the top hat of the town-dwelling artisan (not a bourgeois as is usually thought) for the man who holds a double-barrelled shotgun. On her head Liberty is wearing a Phrygian cap, and it is chief among the signs of her character, marking Delacroix's brief partisanship with the Left, and setting Liberty herself apart, in a place of ideal difference, as we shall see.

FIG. 14.   *Liberty Leading the People* by Eugène Delacroix recalls the July uprising of 1830 by introducing the goddess into an acutely observed crowd of participants (Louvre, Paris). By permission of Giraudon/Art Resource, New York.

In one hand Liberty brandishes the tricolour and in the other she holds a flintlock, but otherwise she is in the classical costume of a goddess of victory, and her lemony chiton has slipped off both shoulders. Her breasts, struck by the light from the left, are small, firm and conical, very much the admired shape of a Greek Aphrodite, and her profile with its shallow indentation at the bridge of her nose is clearly inspired by a classical goddess' head. Her feet, like an ethereal being of the sky, are bare. But there her resemblance to antiquity ends, and Delacroix's brilliant invention begins. That bare foot is large, the ankle is thick, the toes grip the rubble of the barricade over which Liberty leaps; her arms are muscly, her big fists grip the flag and the gun, she has a smudge of hair in her armpit. She surges forward over a heap of bodies, and in her heedless pace Delacroix has precipitated the tragic essence of this image: her robust, hearty undress underscores the pathos of the fallen man who, literally *sans culottes*, lies on the barricade to Liberty's right; his thin, stringy, limp legs lying in her way, his head flung back without vigour. Dying or dead, he has been stripped from the waist down; one sock only remains, there is a pubic shadow. Like a figure of Hector dragged behind Achilles' chariot, he lies stricken, prone, debased in despoliation and in this he acts as an almost exact reversal of Liberty's shameless, noble, invulnerable exposure: he is a victim of his mortality, whereas she cannot die.

Liberty's characterization was fraught with tension for the painting's first audience. A mere suggestion of body hair was intolerable to the public of 1831, reared in the glabrous convention of David and Canova, and Delacroix's image was roundly denounced. The word *sale* (dirty) recurs in the abuse. He had made Liberty look like a filthy creature, a *poissarde* (a fishwife), a whore, ugly, ignoble, *populacière* (of the rabble).[7] Delacroix was concerned to paint the ideal realistically; he had succeeded only too well. Moreover, in 1831, the *bonnet rouge* was a subversive emblem of the radicals and its presence on Delacroix's Liberty's head may account in part for the disappearance of the painting after the Salon. For although it was bought by the Minister of the Interior, it was not put on show, and was returned to Delacroix in 1839. It re-emerged twenty-four years later, after Delacroix had petitioned the Emperor Louis Napoleon to allow him to exhibit it again. During the next republican regime, in 1874, it was moved to the Louvre, where it now hangs. It is even possible that Delacroix painted over the original scarlet colour of the liberty cap. It is now a subdued brown, in no way picking up the brilliant red of her sash or of the tricolour Liberty waves to rally her followers.

Delacroix's dynamic vision dips its roots in a direct memory of the July Days. The barricades had not been seen in Paris for two hundred years, for the French Revolution took place without them: but the bourgeois revolution of the July Days gloried in the romance of the blocked city streets. Delacroix was influenced by the eager renditions of street fights from print-makers, engravers and others who worked quickly to celebrate the events.[8] The exploits of Marie Deschamps, a working woman who fought at the barricades, may have inspired his Liberty's courage in part; or the more distant bravery of Jeanne Laisné, nicknamed 'Hachette', on account of the weapon she wielded in the siege of Beauvais in 1472, when she rallied the townspeople to drive Charles le Téméraire from the walls. She had been celebrated in a painting of the battle by Jules Le Barbier of 1778 which since 1826 had hung in the Town Hall of Beauvais, where Delacroix might have seen it. The composition bears striking similarities to Delacroix's *Liberty*, with a heaped mass of mingled bodies and fighters rushing upon us and rising to the apex of a pyramid at the tip of the tricolour waved by the allegorical figure (in a slipped chiton) who leads Jeanne with her 'little axe' over the tumbled stones of the town's fortifications.[9]

But Delacroix's more immediate historical source was an earlier famous incident which had taken place in Greece, during the prolonged War of National Liberation against the Turks. A number of sketches, dating from before 1830, executed with Delacroix's magnificent assurance, provide models for the Liberty and other figures in *Liberty Guiding the People*. In 1792, the partisans of the small republic of Souli were rallied to withstand the troops of Ali Pasha by a Souliot woman, Moscho, wife of their chief Tzavellas, who seized pistol and sabre and led them to victory; the incident might have gone unnoticed if the Souliots, nine years later, had not been finally overwhelmed by the Turks, whereupon the womenfolk threw their babies into the ravine below the town and hurled themselves after. The mass suicide caught the imagination of the Romantics and turned Delacroix's eyes to their heroine Moscho.[10]

But even in the early sketches of the principal figure, the allegorical lineaments of Liberty's body are already marked.[11] Plucked from the history of women's courage, Moscho has been transmuted according to the precepts of the classical tradition, retrieved and reshaped during the years before and during the French Revolution.

The Phrygian cap, which Delacroix's Liberty wears on her head, significantly combines two different types of antique headgear, each emblematic of an essential aspect of the freedom represented by the *bonnet rouge*. One type of hat indicated a foreigner, who had come from afar—from Phrygia—from an exotic and different place, charged with the exotic energy of that difference; the other was worn by the Roman slave released from bondage; both, in their way, represented another order of things, brought in from the outside by the strangers, inaugurated from the inside in the case of the freedman; both also participated in the symbolism of the unfettered. The foreigner is not bound by the same laws, the slave is set free from old ones; both could therefore stand for change.

. . .

Containing a sequence of inversions, the iconography of Liberty operates on the central premise that signs of Otherness can be recuperated to express an ideal. The galley slaves' headgear becomes a cap of freedom; the capitulating barbarians' cap becomes a sign of Liberty; a woman, empirically less enfranchised than the man in nineteenth-century France, and even ridiculed when she wore the cap of liberty in real life, occupies the space of autonomy and pre-eminence. Above all, Liberty's exposed breast stands for freedom because thereby a primary erotic zone is liberated from eroticism. As eroticism is a condition of the depicted female body, a semi-naked figure, who is no longer constrained by it, becomes free.[12] The slipped chiton is a most frequent sign that we are being pressed to accept an ulterior significance, not being introduced to the body as person.

The allegorical female body either wears armour, emblematic of its wholeness and impregnability, or it proclaims its virtues by abandoning protective coverings, to announce it has no need of them. By exposing vulnerable flesh as if it were not so, and especially by uncovering the breast, softest and most womanly part of woman, as if it were invulnerable, the semi-clad female figure expresses strength and freedom. The breast that it reveals to our eyes carries multiple meanings, clustered around two major themes. It presents itself as a zone of power, through a primary connotation of vitality as the original sustenance of infant life, and secondly, though by no means secondarily, through the erotic invitation it extends, only to deny. The first theme relates to the Christian iconography of Charity, and the eighteenth-century cult of Nature, as we shall see; the second to classical mythology about virgin goddesses and warrior Amazons. The slipped chiton rarely reveals the compliant body of Aphrodite or Venus or her train; goddesses are naked to the thighs, or completely nude; it is the asymmetry of the undress worn by Delacroix's figure that is essential to the imagery, for it suggests both the breast's function as nourishment (one at a time is revealed to feed) and the ardent hoyden, so caught up in the action of the moment that she has no thought for her person.

The semi-nakedness of Delacroix's Liberty is emblematic. Those polished breasts which still recall the stoniness of their possible original have nothing in common with the genital pathos of the foreground corpse. As Anne Hollander has written in her book *Seeing Through Clothes*: 'Her exposed bosom could never have been denuded by the exertions of the moment; rather, the exposure itself, built into the costume, is an original part of her essence—at once holy, desirable, and fierce.'[13] The eager partisans Liberty guides reflect her heedlessness; the boy with the pistols surges towards us, his mouth open, regardless of the bodies that lie collapsed and dishevelled in his path, under the barricade.

That heedlessness is itself a reflection of Liberty's sincerity; this spirit of France has become like the French people themselves, the Franks, who are *franc*, full of *franchise*, that necessary quality of chivalry that goes with being free, en*franc*hised, and not in chains.[14] The outward sign of this inward and crucial pun in Liberty's bared breast. The female breast, which we so quickly and reductively think of as only sexual, is as much the seat of honesty, of courage and feeling, as is the male. For both sexes it is the place of the heart, held to be the fountainhead of sincere emotion in both classical culture and our own (unlike the Chinese, who grant the head and liver the powers of affect we associate with the heart). Odysseus for instance communes with his *thymos*, his inner feelings, sited in his chest or midriff just as we 'consult our hearts' and enjoy 'bosom friendships';[15] in Latin, *pectus* carries the same metaphorical value, as the seat of feelings whose source lies in the heart (*cor*) it holds inside it. In French—as Liberté's bosom is in question—*le sein* can be used of the bosom of a masculine confidant, for instance. In English, the words 'breast' and 'bosom' were not sex-specific either until possibly very recently.[16]

In Greek art, the goddess who most frequently appears in a slipped chiton is Artemis, the goddess of the hunt. On a beautiful amphora in the British Museum, she appears with a panther skin slung over her bare shoulders and another panther tame at her side, facing her brother Apollo.[17] But as Artemis is the fierce virgin who orders Actaeon's death after he has seen her bathing, her undress issues no erotic invitation, however wondrously it works on Actaeon's senses. Of all the Greek pantheon, she is most associated with dedicated virginity; in Euripides' *Hippolytus* the eponymous hero turns to her, not to Athena, when he spurns the love of women and the pleasures of the flesh.[18] She orders Callisto her favourite's death after Callisto has lost her maidenhead to a disguised Zeus.[19] Artemis belongs to the wild, to the forest outside Athena's polis, where she has dominion over animals who are her companions and her quarry. In one of the statues of her that have come down from antiquity, the Artemis Bendis, she is even wearing a Phrygian cap, sign of the outsider.[20] She is the patroness of unmarried girls, who on marriage pass out of her domain into the tutelage of other, less farouche, goddesses. The Amazons worship her, and model their lives on hers, hunting, forswearing friendship with men. At Ephesus, in Asia Minor, Artemis/ Diana had her principal shrine; it was also the mythical capital of the Amazons' terrain, and an exquisite frieze of the fourth century BC from her temple there celebrates their strength and their courage in battle.[21] Just as Artemis lives outside the Greek definition of normal womanhood in her virginity, so Amazons live beyond the border of civilization which made them the subject of so much fantasy.

The Amazons' appearance is often close to the delirious undress of the Maenads, followers of Bacchus, and thus suggests their abandonment, their whole-hearted ardour.[22] On the mid-fourth-century BC Mausoleum of Halicarnassus, one of the seven Wonders of the World, the Amazons were carved in relief, dressed in Greek tunics over their naked bodies, heroically resisting the onslaught of the hero Heracles and his warriors. The Amazon frieze on the Mausoleum forms a pendant to another, on the other side, depicting the battle of the Lapiths and Centaurs, and both myths, which were immensely popular in architectural Greek art, celebrate the triumph of civilization, championed by the Greeks, against barbarians, represented by the outsiders, the code-breakers who come in from the wild: Centaurs who were so overwhelmed by lust they attempted to abduct the Lapith brides from their own wedding feast,[23] and the equally unbridled Amazons who also repudiated natural law, by refusing sexual union and living beyond the pale.[24] While conquering Centaurs and Amazons, the exemplars of Greek mores show disapproval and at the same time hankering; the frequency with which both representatives of the outlandish turn up in their art, their very role in putting heroes like Theseus and Heracles to the test, establishes these imagined pariahs as central to Greek identity and self-definition.[25]

In Roman political iconography, the mirror image provided by the figure of the Greek Amazon became a true image: the inversion slides into the position of the ideal when Rome herself is personified on coins. It is often impossible to distinguish the principle Virtus from the goddess Roma, in the figure of an armed maiden, in short tunic like an Amazon.[26] Virtus, Virtue, assumes this form when she attends a great man at his great deeds; her iconographic counterparts are active, strong, independent heroines of myth, Atalanta the runner and huntress, Hippolyta the Amazon, and the goddess Diana of the chase. The rhumb linking the freedom of the chase and the run of the forests, to the unattached virgin, arrives at its resolution in the figure who personifies the claims of the State to be free, Rome itself. The goddess of such a place had to appear heroic, resisting, chaste and strong.

At the Temple of Roma and Augustus at Ostia, the cult statue of Roma showed the goddess of the city wearing a long tunic, slipped off one shoulder in the asymmetrical style of the Amazon; on the Severan Arch in Leptis Magna, Roma attends an imperial commemoration in the form of a young Amazon, in short tunic this time, a drape knotted on her shoulder and held by a baldric slung sideways across her chest with a helmet on her head. But Hadrian set up a new cult statue, representing Roma enthroned like a Tyche, in the Temple of Venus Felix and Roma Aeterna, in Rome around AD 135–6, and afterwards the Amazon-style goddess who resembled Atalanta or Hippolyta began to be effaced by the much more sedate and respectable matronly Tyche of the city; yet even she, as she appears on the coins of Nero and Vespasian, bears traces of her Amazonian past, in her short tunic and high buskins and crested helmet. Her chiton slips sideways to reveal a breast on the coins issued by Nero and Galba, and sometimes she carries a bow and quiver.[27]

The image of Liberty, developed in France from the French Revolution onwards, descends from this classical figure of the free state, with her overtones of Amazonian virginity and independence. Although Delacroix's cluster of identifying signs—from the Phrygian cap to that naturalistic armpit which so upset his contemporaries—comes to life through the vigour of his execution, his protagonist is appropriately female not only because *eleutheria* and *libertas* are feminine in gender in Greek and Latin but because lack of constraint, liberation, *wildness*, are unconsciously aligned with the wild itself with the world outside society as manifest in cities; 'cities' stand for civilization and culture, and its exclusion zones stand for 'nature'. Those who dwell in that natural domain, like huntresses and Amazons, develop a closer association with the wild and its energies.[28]

As the anthropologist Edwin Ardener has suggested, in a classic paper in *Perceiving Women*, human beings bond themselves together as a whole to distinguish between humanity and non-humanity, according to one set of criteria; but within human society itself, in the same way as men and women together look upon animals as different, we distinguish between men and women, and bond with our own sex to see the other as different.[29] But because men have generally controlled the processes of communication and the male view dominated social perceptions, it is women who have chiefly suffered from the imaginary overlap between sexual difference and non-humanity, 'because their [the men's] model for *mankind* is based on that for *man*, their opposites, *women* and *non-mankind* (the wild) tend to be ambiguously placed'.[30] From this ambiguity arise women's 'sacred and polluting aspects', so visible in the polyvalent significance of the bared breast. As Ardener goes on to say, 'Women accept the implied symbolic content by equating *womankind* with the men's wild',[31] although they know themselves to be threatened, not only by that characterization itself, but by the wilderness which men occupy *vis-à-vis* women: it is only when women find their voices that they conjure openly the defilement of their sphere by men's dangerous pursuits (as the sustained anti-nuclear demonstrations at Greenham Common). Otherwise they mutely preserve their own outlook and the anthropologist has an uncommonly hard task to break through woman's 'muted'-ness.

In her nineteenth-century form, the figure of Liberty reveals that the female who enjoyed and suffered the special stigma which associated her with the wild, could exercise in a time of ferment a positively perceived and desirable potency. The breast of a woman as distinct from the generic human bosom acts as a sign of nature and its wild connotations in visual imagery that complements and reinforces the magic outsider status of the Amazon. Of all the Virtues, only Charity, as noted before, perhaps possesses a non-virginal body which is capable of emitting and flowing without pollution. The infants who drink from her breasts and who are thereby kept in life are Charity's attributes, the equivalent of Egalité's carpenter's level during the French Revolution or Temperance's pitchers in the early mediaeval period. Her breast becomes the mark of her natural, motherly relation to those who come into contact with her. Adapted to political imagery, the exposed breast denoting the wild thing can also mutate her into a Tyche, a matron and nurturer, and a type of the protective state. The oscillation between these two dominant meanings of the breast is constant after the first Revolution in France, and it reflects swings between accepting woman as an active agent of change or desiring her to remain a passive source of strength.

In Greek, the root *mamm-e* gives both the word for the breast, the word for a child's cry for the breast, and the name of mother, as it still does in English, and the Romance languages. In Greek literature, the revealed breast often designates the claim of a mother's love upon a hero, the bond that still joins the private and the public worlds.

At the end of the *Iliad*, for instance, Hecuba implores Hector not to go out to the field and fight against Achilles. Priam his father has failed to move him with his chiliastic visions of his own terrible end, so Hecuba takes up the thread:

And now his mother in her turn began to wail and weep. Thrusting her dress aside, she exposed one of her breasts in her other hand and implored him, with the tears running down her cheeks. 'Hector, my child,' she cried, 'have some regard for this, and pity me. How often have I given you this breast and soothed you with its milk! Bear in mind those days, dear child. Deal with your enemy from within the walls, and do not go out to meet that man in single combat. He is a savage; and you need not think that, if he kills you, I shall lay you on a bier and weep for you, my own, my darling boy; nor will your richly dowered wife; but far away from both of us, beside the Argive ships, you will be eaten by the nimble dogs.'

Thus they appealed in tears to their dear son. But all their entreaties were wasted on Hector.[32]

In Aeschylus' *Oresteia*, a different mother makes the same gesture to entreat her son. But in this case, Clytemnestra sues for her life, not his, when she reminds Orestes that he is about to murder the woman who gave him life and nourished him at her breast:

> Wait, son—no feeling for this, my child?
> The breast you held, drowsing away the hours,
> Soft gums tugging the milk that made you grow?

Orestes wavers, but only for a moment; he remembers the fate of his father and his mother's love of Aegisthus, and forces her into the palace to kill her.[33] As Froma Zeitlin has commented: 'Clytemnestra's breast is the emblem of the basic dilemma posed by the female—[it symbolizes] the indispensable role of women in fertility for the continuity of the group by reason of its mysterious sexuality, and the potential disruption of that group by its free exercise.'[34]

In ancient funeral rites it is the women who mourn, by unfastening their dress and baring their breasts and scoring their cheeks and chest with their nails. Although a very early Roman law, of around 450 BC, forbade the practice,[35] women mourners appear in sculpture and painting, as in the magnificent sarcophagus of the second century AD now in the Louvre showing the funeral of Hector. Hector's naked body is raised up over the shoulders of two Trojans, his head slumped back in a death's mask. His state arouses a wild lament among the women who follow behind his corpse; with their hair and their clothes loosed, and their arms flung wide, holding on to one another and imprecating against the heavens, they express vividly to the beholder the unrestrained nature of their grief.

The maternal body, in the disarray of sorrow or the undress of nurture, charged semi-nakedness positively even within the Christian tradition which had so obsessively focused on sexuality as an occasion of sin, and symbolized the Fall by the discovery of nakedness and physical shame. As the justifying function of women in the postlapsarian world was motherhood, childbirth and nursing gave legitimacy to a body that was otherwise a source of peril. The Virgin Mary gives suck to the infant Jesus both as his historical mother and as the metaphysical image of nourishing Mother Church.[36] The image provided artists with a chance to meditate on a mystic source of spiritual life or to evoke the solemn unity of the holy family during Jesus' infancy, as in the beautifully calm and yet mysterious *Rest on the Flight into Egypt* in the Birmingham City Museum by Orazio Gentileschi. The painter suggests Mary's primacy in the midst of her humbleness by depicting her nursing the naked child at her breast and seated barefooted on the earth, while Joseph lies absent in heavy slumber from the spiritual communion of mother and child. The apex of the triangular composition, binding Joseph and Mary together, is formed by the massive and faithfully painted head of the Holy Family's donkey, framed against an ultramarine evening sky, beyond a dilapidated wall under the lee of which they lie.

Orazio Gentileschi's painting condenses with elegant restraint many of the themes of lowliness and renunciation that the nursing Virgin often expresses; in her acceptance of her creatureliness, Mary demonstrates her Christian virtue of humility as well as her charity, ever since the Franciscans first developed the icon of the Madonna of Humility for contemplation.[37] Gentileschi's donkey, the broken wall, the homely grey bundle of belongings on which Joseph sleeps, Jesus' bare body, and Mary's bare feet all suggest to us the rugged and simple conditions the incarnate godling had willingly accepted. And these poor conditions are themselves outside urban civilization, just as the stable where he was born is normally the resting place of animals; according to Christian piety it is the humility of Jesus and Mary's life which inspires awe, not their splendour, and that humility resides in their physical hardships, which associate them with animals, and in their acceptance of basic—animal—biological human nature.

In the iconography of intercession, Mary the mother of Jesus makes the same gesture as Hecuba when she pleads for sinners' salvation before the judgement seat.[38] 'Happy the womb that bore you, and the breasts you sucked!' cried out a woman in the crowd around Jesus, in a scriptural salutation which has passed into the liturgy of Mary (Luke 11: 27–8). But the theme was subject to fluctuations in ideas about decorum. In twentieth-century Spain, for instance, Mary sets an up-to-date—and very much more acceptable—example to poor carnal women, when, in the carving by Eusebio Arnau over the doorway of the Casa de Lactancia in Barcelona, she feeds the baby with a bottle.[39]

The allegorical figure of Charity, who often replaces Mary and the child as the expression of artists' pleasure in the subject, sometimes echoes the classical Abundance, with her horn of plenty, brimming cup, bunches of grapes as well as nurslings, as in Botticelli's most beautiful drawing in the British Museum.[40] But the materiality of Mary forms a thin source of her cult; it would be stretching the evidence to claim otherwise. A link does however exist between her nourishing the child Jesus at her breast like any ordinary mortal mother and the exalted nakedness of allegorical figures like Liberty; both achieve an unusual effect on the beholder's senses, and one which is genuinely helpful on a humanist level: they are stating that nature is good. Mary may only give suck to Jesus because she is so humble, and so apt to consent to an activity which the upper strata of society have frequently eschewed as fit only for hirelings,[41] but by doing so, she affirms fruitfulness and our animal condition at the same time, and she transforms the erotic dangerousness of the breast in Christian imagery to a symbol of comfort, of candour, of good.

The acceptance wrought by the maternal theme spreads through secular imagery: King Charles VII's beloved mistress Agnès Sorel may be commemorated as the Madonna in the altarpiece commissioned by the royal treasurer Etienne Chevalier from Jean Fouquet, where scarlet and cobalt cherubim flock around the luscious young woman's uncovered form, promising all kinds of bounty other than spiritual.[42] There is no such intended *double entendre* in Rubens' 1615–22 celebration of his patron Marie de Médicis in the colossal sequence of paintings in the Louvre, where her status as allegory—as the earthly exemplar of Fecundity, of Truth and the repository of the Regency's prosperity—is marked by the sign of a single bare breast. Rubens borrowed established visual language to proclaim the Regent's virtues, using the available repertory of ripeness and nakedness, as he did in his wise and human meditations on the griefs of war and the blessings of peace.[43] After the seventeenth century it becomes almost routine for noblewomen to be depicted as aspects of Charity by baring their breast, frequently asymmetrically, the pose that often indicates nourishment as well as selfless ardour.[44]

In France, visual representation explored the wider meaning of the revealed breast in playful fantasies which cautioned against the dangers of sexual love. In a marble sculpture like Jean-Pierre-Antoine Tassaert's *Sacrifice of the Arrows of Love on the Altar of Friendship*, a grieving Cupid looks on while a tender young woman (Friendship,

Amitié) chidingly snaps his arrows in two on her votive fire. She is wearing a slipped chiton, to reveal her sincere and generous heart.[45]

Madame de Pompadour, when she was represented in the role of Friendship by the great Pigalle, appeared in a similar state of undress, indicating that softest spot with one hand and beckoning with the other.[46] Marie Leczinska, the wife of Louis XV of France, commissioned a statue of herself as Charity from Augustin Pajou, for her mausoleum. She had borne ten children herself, but she stipulated that orphans should appear at her knees. It is unthinkable today that any queen be portrayed in a symbolic state of undress, let alone uncover herself in life, but Marie Leczinska found it appropriate vesture for the statement she wished to make about herself for posterity.[47]

In the same century, Grand Tourists and classical cultists sometimes even daringly undraped themselves *all'antica*, as if they were ideal works of art. At the Grand Jubilee Ball of 1749, Miss Chudleigh, a maid of honour to the Princess of Wales, appeared in the costume of Iphigenia on her way to the altar of sacrifice. An anonymous caricaturist drew her in rather unclassical petticoats, and Mary Wortley Montagu remarked that the high priest would have had no difficulty inspecting the entrails of his victim.[48] Her costume, credited to a collaboration between herself and Mrs (Susanna Maria) Cibber, the actress, created a stir, but with its learned claims, did not upset her social standing.

In the fascinating and often beautiful engraving in the *Almanach Iconologique*, published 1766–74, where the artists Gravelot and Cochin devised influential figures for the Virtues, the Sciences and other abstractions, the pastoral, charming interpretation of La Nature shows a young woman standing with her arms held out almost in the act of proffering herself. She is set in an Edenic wilderness of flowers and animals, with a many-breasted effigy of Diana of the Ephesians behind her; her own breasts are leaking milk.[49]

This identity of motherhood and nature is present in the propaganda of the French Revolution; the Catholic iconography of Charity and the classical vocabulary of motherly gesture pervades its rituals, even though through those same rituals it aimed to inaugurate the world afresh and break with that past, especially the Catholic past. Its propaganda, inspired by Rousseau's *Emile*, also strived to break with the widespread custom of wet-nursing, and encourage mothers to feed their own babies.[50]

Rousseau had shrewdly intimated that the secular State should generate its own symbolism and ritual to replace the Church's role as moral arbiter over the masses, and during the 1790s the new Government's elaborate and didactic pageants and processions dramatized its shifting aims and ideology.[51] In these remarkable experiments in the invention of mythology and the control of thought, participants sometimes played themselves, as in sacred liturgy, and were cast as communicants in the mass ritual of the Revolution: in the Fête de l'Unité of 1793, eighty-six deputies to the national Convention drank from the spouting breasts of the goddess of nature, an Egyptian style statue surmounting a Fountain of Regeneration, while, like devotees in a procession on a patron saint's feast day, the crowd solemnly marched behind a triumphal chariot where Liberté, rather than an effigy of a saint or the Virgin, was enthroned. For the Feast of the Supreme Being, of 1794, a procession made its way to the top of an artificial mountain which had been created for the occasion on the Champ de Mars, and, as cannons volleyed, the participants exchanged kisses beside the Tree of Liberty planted there, rather like concelebrating priests embracing one another at the conclusion of the Mass.

But the most notorious Fête staged by the Revolution, which has come down in popular memory as a mobsters' orgy, was the Festival of Reason of the year before (1793), when the Goddess Reason was enthroned in Notre-Dame. The imitation of ecclesiastical rites was here overt; like the ceremony of the crowning of Mary's statue on the feast of her queenship in May, La Raison, presiding genius of the atheist radicals, was enthroned and garlanded. The blasphemousness of this rite was underscored, not just by the nature of the usurping goddess, but by her presence in flesh and blood; as Maurice Agulhon had pointed out, live allegories characterize radicalism.[52] To place a real woman in the place of the ideal challenges the ever-elusive character of the ideal itself, and says something unequivocal about women. The ceremony was not however licentious at all, but solemn, administered by young women in white robes with tricolour belts and cockades, and torches emblematic of truth. The hymn to Liberty was written by André Chénier, later to die himself under the guillotine during the Terror. Reason was 'the very picture of Beauty herself'[53] and after her investiture in the cathedral of Mary, she was carried by 'four stalwarts from the market' to the Convention, where the deputies paid homage to her. In front of the President, the procession's leader declared the principles Liberty and Reason combined in the goddess and joyfully orated: 'Here we have abandoned inanimate idols for this animate image, a masterpiece of nature.'[54] The living woman was not acting Liberté: she embodied her. She was, the Jacobin

Hébert said, 'a charming woman, as beautiful as the goddess she represented'.[55]

The interpenetration of actual and symbolic planes has rarely been as full as during the Revolution's early years, and such a convergence can present tremendous possibilities for emancipation. But the emphasis in the female allegories of 1794 and afterwards returned to Rousseau's ideals of virtue, purity and decorum, even bashfulness, to motherhood and nubility; and so women were not enfranchised or enabled by them. Although real women incarnated the presiding principle of the Revolution, they did not recall the historical participation of women in its early tumult; though personifications lived and moved, they projected a carefully considered ritual fiction. A veil was drawn over the unruly elements against whom Burke for one had inveighed in his *Reflections on the Revolution in France* (1790), the Parisiennes who, with 'horrid yells, and shrilling screams, and frantic dances, and infamous contumelies, and all the unutterable abominations of the furies of hell, in the abused shape of the vilest of women', had surged to Versailles to bring the King and Queen back to the city in 1789.[56] As Lynn Hunt has acutely observed, 'the collective violence of seizing Liberty and overthrowing the monarchy was effaced behind the tranquil visage and statuesque pose of an aloof goddess'.[57] The brave call that a partisan like Olympe de Gouges made for equality in property, education, authority and employment 'according to their abilities, and without distinctions other than their virtues and their talents',[58] was not answered, and the ineffectiveness of such proposals can be felt when we read about these solemn pageants, where women still occupied the site of remote fantasy, not agency or power, unless they fulfilled their 'natural' role, as mothers to the new citizens.

The Republic declared itself as belonging to the order of Nature; it had rebelled against the old Christian God and had so declared itself heathen, 'pagan'. Just as 'paganus' originally meant a country-dweller, so the new regime saw itself as living outside the polis of the former state, in a newly discovered 'natural' world. It was 'human nature' to be born free, as Rousseau had so proudly said,[59] and when the revolutionaries swept away the old order, they abolished the Christian calendar too, with its personal stories about individuals, and replaced it with months called after the natural procession of the seasons and the characteristics of the months: Floréal, Thermidor, Brumaire, Ventôse, and so forth, installing the weather of the countryside at the centre of daily life.[60] The Feast of the Supreme Being was also dedicated to La Nature, and pregnant women were specifically summoned to take part in the processions.[61] Though the high status accorded to motherhood was a precious right, the equation between the divine Nature and women's maternity reflected the restrictions on women in other areas of life.

In terms of Gravelot and Cochin's iconography, the French Revolution's female allegory tended to prefer La Nature to La Liberté who, in their version, numbers a cat among her attributes, as cats will not be tamed.[62] By contrast, the breasts of La Nature were a domesticated, unthreatening and wholesome sign of the wild side of human nature brought under social control.

The cult of Nature also influenced a profound revolution in dress, and the changing fashions reflect faithfully, with a small time lag, the changing definitions of women's roles, according to the official ideology, expressed in the festivals. In 1794, the Société Populaire et Républicaine des Arts demanded that if the heroes of the Revolution were to be immortalized in monuments, they should be liberated from the 'paralysing' clothes of the *ancien régime*. The society called for a costume which would reveal them in all 'their brilliant beauty, in all their natural grace'.[63] When David was commanded by the Committee of Public Safety that year to design an appropriate costume for the French people, he reverted to the imagined simplicity of classical republican dress for the new *citoyenne*, and thus started a fashion of unprecedented revealingness.[64]

The new look liberated women of all classes from the constraints of earlier dress. The style was patriotically termed 'en gaule', and was created by muslin-like shifts, usually white to connote purity, which clung to the legs and hips and were caught at the waist by a sash.

But later, to emphasize the relation of woman's naturalness to her role as nurturer, the waistline crept up to under the breasts; it was quite in order for the rather niminy-piminy young woman who sat to a painter of David's circle around 1800 to show her bosom very plainly through the transparent upper bodice of her classical tunic.[65] In these fashions, the freedom which had been gained on behalf of the French woman was relayed to her dress by a kind of pun: she was relieved of restrictive clothing, and taboos on bodily display were lifted, just as she had been in theory relieved of restrictive government. But the predominant message was still conservative, stressing that women's potency was founded in her capacity to bear and to nurse, and forgetful of the activist and independent contribution of women to the early years in roles other than that of mother.

When the actions of women in the 1789 uprising were recalled, it was with disapproval; and the memory, compounded by the evasions of republican propaganda itself, effectively stigmatized the female revolutionaries as harpies and whores. In 1848, a Breton doctor characteristically decried the pageants of the First Republic: 'On the very altars where in the past you used to come to worship the true God, she [the Republic] exposes herself, half-naked, to the eyes of libertines who smile, and to those of the people as a whole who groan in the thrall of the Terror, and to those of a few admiring fools.'[66] In the literature of the nineteenth century, the women who call to mind revolutionary Liberté are often seen as sexual beings, driven by passion, not political consideration. In his novel of 1845, *Les Paysans*, Balzac's Catherine is a tigress, who 'resembles in every way those girls whom sculptors and painters took as a model of Liberty, as they used to of the Republic'; with her eyes 'flecked with fire' and her 'nearly ferocious smile', she is 'the image of the People',[67] of 'this Robespierre with one head and twenty million pairs of arms'—the peasantry whom the Revolution had rendered ungovernable, he claimed.[68] In a later novel, *L'Education sentimentale*, Flaubert describes the sack of the Tuileries in 1848, and there sees a 'fille publique' (a woman of the streets) 'en statue de la Liberté—immobile, les yeux grands ouverts, effrayante' (like a statue of Liberty—motionless, her eyes wide open, terrifying).[69]

The propaganda of the 1790s, by failing to grasp and accept the implications of women's actions, had maintained the split between the two faces of Liberty, the matron and the virago, the nice and the not-so-nice, and ratified the former in order to obscure the latter. The effect of this was to intensify the field of energy around the wild, dishevelled, and asymmetrically bared Amazon; through her heterodoxy, she could represent an original freedom still beyond the freedom promised by the first Revolution. The slipped chiton of Delacroix's Liberty of the 1830 Revolution gathered up this energy by replacing the emphasis on the unaffiliated and active character of the goddess, portraying her as feline, untamed, a Fury rather than a Tyche. But the constant swing between Liberté and Nature continued, and the next Republic rebelled again against the Amazon, in favour of the Tyche. When Daumier submitted an oil sketch to the 1848 competition for an image of the Republic, he showed her as a monumental Charity, sitting in majesty, while two sturdy children drank from large breasts and another read at her knee. Though the work was not commissioned, the study exactly catches the tenor of the Second Republic. Gentler, motherly, caring, the self-image of the 1848 Republic declared its disagreement with the frenzy of eighteen years before, expressed then just as clearly through the differently applied sign of the female breast, by the slipped chiton of Delacroix's Liberty.[70]

In 1863, Baudelaire published a homage to the painter whom he admired all his life with unswerving passion. To Baudelaire, Delacroix was 'the most suggestive of all painters, the one whose works make one think more than any other's',[71] and his death the year before had left the poet desolate. He was a man he loved as a friend, as well as an artist who, through the powers of a colossal imagination, could faithfully capture the 'greatness and the native passion of the universal man'.[72] The same year, Baudelaire wrote a prose poem, 'La Belle Dorothée', which evokes a woman as a living Liberty, and reveals to us how the classical antecedents of the Republic's solemn allegories and Delacroix's vital reinterpretation had migrated, in the bourgeois capital of the Second Empire, into the private domain of affect and eros, and thrived there on the ancient symbolic associations of nature, fertility and paganism.

A paean to the vigour and splendour of a 'natural' woman, the poem evokes one of Baudelaire's 'Vénus noires', symbolizing the energy of the exotic and pagan. The imagery of sculpture from the city he wrote about so obsessively recurs, bringing now the art to life, now turning Dorothée herself into a work of art, against the background of a southern sea:

Meanwhile Dorothée, strong and proud as the sun, goes up the deserted street, the only living thing at this hour beneath the immense blue heaven, and casting a stain against the light like a burst of blackness.

On she goes, softly swinging her torso, so thin, on her hips, so wide. Her clinging silk dress, of a clear pink colour cuts vividly across the shadows of her skin and moulds the exact shape of her long waist, her hollow back and her pointed breasts. . . .

Now and then the sea breeze lifts the corner of her floating skirt and reveals her leg, gleaming and splendid; and her foot, like the feet of those marble goddesses which Europe keeps closed up in its museums, prints its form faithfully on the fine sand. For Dorothée is so prodigiously flirtatious that her pleasure at being admired surpasses the pride of the freed slave, and even though she is free, she walks without shoes. . . . At the hour when even dogs groan in pain under the sun which bites them, what powerful motive moves her to make her way like this, indolent Dorothée, beautiful and cold as bronze?[73]

In Delacroix's painting, the real and the ideal overlapped; and when Baudelaire's Dorothée takes off her shoes, she can adopt with impunity this badge of dispossession and servitude, he says, partly because his word picture has established her so vividly as a type of sculpted and deified Liberty, a goddess of marble or bronze, *affranchie,* freed from the museums where works of art are confined.

The cipher of the female breast still tells us of revolutionary valour in Soviet propaganda images, like the colossal stone sculpture of *The Motherland* in the former Stalingrad, clearly influenced by Delacroix[74] or in Scandinavian images of nationalism, like the haggard champion who stamps on her enemy's grimacing head in Karlstad in Sweden, to commemorate the fiftieth anniversary of the 1905 treaty dissolving the union between Sweden and Norway.[75] A bared female torso—with nipples sculpted in an emphatic state of arousal—could still be considered eloquent of freedom in a socialist country—in 1955. When the women's movement urged loose clothing in the sixties, and when the newspapers spread the slogan 'Burn your bra', both feminists and their detractors were deeply inspired by the ancient significations of the breast and its release.

The State erects statues of Liberty, or appropriates painted images of them, even when the impetus to create them has been personal in origin, as in the case of Delacroix, but the State can only offer freedom under the law. It promises to guarantee individuals' rights by binding them in differing ways to varying degrees. In France, in the first half of the nineteenth century, an image of Liberty emerged to express this promise, and within it developed a symbolism, focused on the breast. It sometimes continued to represent the natural function of breast-feeding, in order to promise the State's protection as provider, comforter, and nourisher. But Nature was perceived as possessing another aspect, epitomized by the autonomous and virginal Amazon, who neither consents to live within the purlieus of conventional, law-abiding cities, nor to perform the socialized female function of child-bearing within marriage. By harnessing the figure of this outlaw, through the connotations of the slipped chiton, the bare foot and the Phrygian cap, the Liberty image brings her under control; the containment of so many multilayered images of natural, wild, womanly processes paradoxically empties them of resonance.

If even an anarchic Amazon heroine can be incorporated into the State's self-image, then nothing remains beyond its reach, it encompasses all within its bounds. But every society will define its own 'wild' differently and then try to annex it,[76] until there can be no freedom outside its definition, and Liberty herself becomes captive in the image's frame of reference. However, by creating such an image of the Otherness that can be encompassed, the signs of Otherness become self-perpetuating, through the dissemination of the figure and its symbolic elements: if a wild natural woman stands for this, she can be condemned to remain so.

It is because women continue to occupy the space of the Other that they lend themselves to allegorical use so well; in spite of the convergence which occurred in the late eighteenth and nineteenth centuries, when real women represented the ideals of the Revolution, we would fail to respond to the symbolic content of such images as Delacroix's *Liberty* if we perceived her as an individual person. If women had had a vote or a voice, Marianne would have been harder to accept as a universal figure of the ideal.

Cast as a wild thing, who breaks the bonds of normal conventions, Liberty prolongs the ancient associations of women with Otherness, outsiderdom, with carnality, instinct and passion, as against men, endowed with reason, control and spirituality, who govern and order society. But she also subverts the value normally ascribed to these categories and, in so doing, she places women in a different relation to civilization, to its content and happiness as well as its discontents. La Belle Dorothée, who like a figure of Greek Victory and Delacroix's Liberty steps out in bare feet in swirling thin silks and pointed breasts, belongs to that unruly land beyond society's borders, where the wearers of Phrygian caps also dwell; this is also the conventional ascribed site of women, close to the natural processes, those mysteries of death and birth, which refuse to yield up to reason and social control.

Yet the figure of Liberty lives up to her name by affirming nature within culture itself, as a necessary and intrinsic part of it. From the Amazon to Marianne, the female body's bounty and its ardour, often denoted by the bare breast, has been seen to possess the energy a society requires for that utopian condition, lawful liberation. But it has done so only by recapitulating the ancient and damaging equivalences between male and culture, female and nature. Otherness is a source of potential and power; but it cannot occupy the centre.[77]

Notes

1. See Taft (1930), pp. 37–53, for Greenough's *Washington.*
2. Gombrich (1965), p. 35.
3. John Arbuthnot, 'Law is a Bottomless Pit . . .', London, 1712.
4. *WPA Guide to New York*, p. 101.

5. For Delacroix's *Liberty Guiding the People*, see Hobsbawm (1978); Hadjin-icolaou (1979); Toussaint (1982); Hollander (1975), pp. 199–202; Honour (1979), pp. 235–6.
6. Letter to Félix Guillemardet, 15 February 1831, quoted Toussaint (1982), pp. 6–7.
7. Hadjinicolaou (1979), pp. 23–5.
8. T. J. Clark (1973), p. 17.
9. Toussaint (1982), pp. 36–8.
10. Ibid., pp. 18–19.
11. Ibid., p. 19: *Les Femmes Souliotes*, pp. 20–1; *Etudes de guerrière* and *Projet pour une composition célébrant la Grèce*, pp. 21–6.
12. I am grateful to all those who took part in the discussion of this chapter at a meeting of the History Workshop, 1984, for analysis of these inversions.
13. Hollander (1975), p. 202.
14. Keen (1984), p. 149, on the association of the Franks, the knight crusaders, with the French and their *franchise*, or frankness.
15. *Odyssey*, 5. 298, p. 96; Garnier (1982), p. 184, says that the hand indicating the breast is a sign of sincerity, generosity and acceptance, and present in images of Christ as intercessor.
16. I am grateful to Susannah Clapp for reminding me of Keats's 'Ode to Autumn'.
17. Red-figured amphora, BM E256, found at Vulci, 510–500 BC, Birchell and Corbett (1974), pl. 6. See also BM 21 55, mid-fourth century BC. Compare two celebrated Dianas of antiquity, much reproduced later: the *Diane de Versailles*, which shows her as a huntress in short tunic (but not slipped) in the Louvre, and the *Diane de Gabies* (also Louvre) in which she is either fastening or unfastening her long robe (Haskell and Penny (1981), pp. 196–9).
18. Euripides, *Hippolytus*, trans. R. Warner, pp. 78–9, for Aphrodite's rage against the hero for worshipping Artemis 'the virgin goddess'. For Actaeon's sad tale, see Ovid, 3, pp. 77–8.
19. Ibid., 2, pp. 61–2.
20. Robertson (1975), 1, p. 378.
21. In Kunsthistorisches Museum, Vienna.
22. Euripides, *Bacchae*, pp. 215–16.
23. The battle of the Lapiths and Centaurs is the probable subject of the Parthe-non metopes in the British Museum, reproduced Brommer (1979), pp. 23–30; see Graves (1966), i, pp. 360–2 for the myth.
24. For Amazons, see Warner (1981), pp. 203–5, n. 29, p. 314. It is a late accretion to their myth, found in Diodorus Siculus, III. 52ff., trans. C.H. Oldfather (London, 1935), that they severed a breast in order to draw the bow. The mutilation is not shown in Greek art.
25. See Merck (1978), pp. 99–106.
26. Cf. silver multiple of Priscus Attalus (*c.*409–410) in *Wealth of the Roman World*, pl. 569.
27. Vermeule (1959), *passim*.
28. King (1983), pp. 111–13.
29. E. Ardener (1975), pp. 14–15.
30. Ibid., p. 14.

31. Ibid.

32. *Iliad*, 22. 79–89, p. 399. Hecuba uses the word *mastos*, a sex-specific word for the breast.

33. Aeschylus, *Oresteia, Libation Bearers*, 883–9, pp. 216–17.

34. Zeitlin (1978), p. 158.

35. One of the Twelve Tables of the Republic, in Lefkowitz and Fant (1982), p. 175.

36. See Warner (1976), pp. 192–205.

37. Ibid., pp. 181–3.

38. Arnaud de Chartres (d. 1156) used Hecuba's gesture to illuminate Mary's powers of intercession as the Mother of Jesus in a sermon. See Warner (1976), pp. 199–200, and fig. 26 for an anonymous painting of *c*.1402 showing Mary interceding with God the Father and showing her breast like Christ showing his wounds (in the Cloisters Museum, New York).

39. In A. Cirici Pellier, *El Arte Modernista Catalan* (Barcelona, 1951), p. 155.

40. Reproduced in Marle (1932), 2, p. 177.

41. See Warner (1976), p. 197 for St Bernard's indignation at the custom; see Sussman (1982) for an account of wet-nursing practices from 1700 to 1900.

42. The Melun Diptych, *c*.1450. The Virgin panel is in the Antwerp Museum.

43. Heilbrunn (1977), *passim*; cf. *Minerva Protects Pax from Mars*, in NGL, and the *Consequences of War*, in the Pitti Palace, Florence.

44. Princess Charlotte, the beloved daughter of George IV, whose death in childbirth in 1817 eventually led to Princess Victoria's accession, is commemorated in St George's Chapel, Windsor, by a remarkable monument by Matthew Wyatt, showing her soul ascending to heaven attended by angels, while draped mourners weep around her shrouded corpse. The Princess' ascending figure is wearing a slipped chiton.

45. In the Philadelphia Museum of Art.

46. Beaulieu (1981), p. 4. In the Louvre.

47. Also in the Louvre, see Beaulieu (1981), p. 3.

48. Rudofsky (1971), pp. 42–3; two drawings in the British Museum, *Iphigenia* (1749) and *Miss Ch-dly* (1749), the latter exhibited in *Masquerade*, Museum of London, July–October 1983.

49. Gravelot and Cochin (1768).

50. Sussman (1982), pp. 109–10 gives figures for babies put out to nurse from Paris; see also pp. 27–9.

51. 'Lettre à d'Alembert', 1758, in Rousseau (1820), pp. 86–7, 187–96; Mosse (1971), p. 170. For David's *Fêtes*, see Dowd (1948), *passim*; Ozouf (1981), *passim*; Henderson (1912), pp. 356ff; Brookner (1980), pp. 100–7, 119–20.

52. Agulhon (1979), p. 88.

53. Gombrich (1979), pp. 188ff; Starobinski (1979), p. 102; Ozouf (1981), pp. 119ff., 135ff.

54. Gombrich, 1979), pp. 188–9, quoting M.A. Thiers, *Histoire de la Révolution française* (Brussels, 1845), I, p. 449.

55. Gombrich (1979), p. 189, quoting Aulard (1892), p. 83.

56. E. Burke (1790), p. 69.

57. Hunt (1980), p. 14.

58. Article VI, Declaration of the Rights of Women and Citizens, 1790, in Riemer and Fout (1983), pp. 63–7; for the role of women in the Revolution, see R.B.

Rose, *The Making of the Sans-Culottes. Democratic Ideas and Institutions in Paris 1789–92* (Manchester, 1984).
59. Rousseau (1962), 'De l'Inégalité parmi les hommes', in *Du Contrat social*, p. 55.
60. Paulson (1983), p. 14.
61. Ozouf (1981), p. 135.
62. Gravelot and Cochin (1767).
63. Harris (1981), pp. 283–313.
64. *La Révolution française* (1982), pp. 36–7, reproduces his male designs.
65. In NGW, reproduced Hollander (1975), p. 65.
66. Agulhon (1979), p. 29.
67. Balzac, *Les Paysans*, p. 392.
68. Ibid., p. 233.
69. Flaubert (1964), p. 291.
70. Daumier, *La République*, now in the Louvre. See Boime (1971), pp. 76–7; T.J. Clark (1973), pp. 107–8, reproduced on the cover of the 1982 edition.
71. 'L'Œuvre et la vie d'Eugène Delacroix', in Baudelaire (1968), p. 531.
72. Ibid.
73. 'La belle Dorothée', in *Petits Poèmes en prose*, Baudelaire (1968), p. 165.
74. Bruce Chatwin, 'The Banks of the Volga', *Observer Magazine*, 10 June 1984.
75. I'm most grateful to Philip French for bringing this grisly example to my attention.
76. See Sanday for an illuminating general study of different societies' views of women and nature, and the corresponding higher or more equal position of women among people who do not depreciate what they consider to be 'nature'.
77. See Daly (1984), for a radical acceptance of women's closeness to the wild and nature and Otherness.

References

Aeschylus (1979). *The Oresteia.* Trans. Robert Fagles. PC.

Agulhon, Maurice (1979). *Marianne into Battle.* Trans. Janet Lloyd. Cambridge, 1981.

Ardener, Edwin (1972, 1975). 'Belief and the Problem of Women', in Shirley Ardener (1975).

Ardener, Shirley (ed.) (1975). *Perceiving Women.* London.

Aulard, Alphonse (1892). *Le Culte de la raison et le culte de l'Etre suprême.* Aalen, 1975.

Balzac, Honoré de (1870). *Œuvres complètes.* Paris.

Baudelaire, Charles (1968). *Œuvres complètes.* Ed. Marcel A. Ruff. Paris.

Beaulieu, Michèle (1981). *Le Portrait sculpté au XVIIIe siècle.* PGGM, 85. Paris.

Birchall, Ann, and Corbett, P.E. (1974). *Greek Gods and Heroes.* London.

Boime, Albert (1971). 'The Second Republic's Contest for the Figure of the Republic'. *AB* 53 (March).

Brommer, Frank (1979). *The Sculptures of the Parthenon.* London.

Brookner, Anita (1980). *Jacques-Louis David.* London.

Burke, Edmund (1790). *Reflections on the Revolution in France*. London, 1912.

Cameron, Averil, and Kuhrt, Amélie (eds.) (1983). *Images of Women in Antiquity*. Beckenham, Kent.

Clark, T.J. (1973). *The Absolute Bourgeois: Artists and Politics in France 1848–1851*. London, 1982.

Daly, Mary (1984). *Pure Lust: Elemental Feminist Philosophy*. London.

Dowd, David Lloyd (1948). *Pageant-Master of the Republic: J.L. David and the French Revolution*. Lincoln, Nebr.

Euripides. *The Bacchae and Other Plays* (*Ion, The Women of Troy, Helen*). Trans. Philip Vellacott (1954). PC, 1984.

—— *Medea. Hippolytus. Helen*. Trans. Rex Warner (1944, 1950, 1951). New York, 1958.

Flaubert, Gustave (1964). *L'Education sentimentale*. Paris.

Garnier, François (1982). *Le Langage de l'image au moyen-âge: Signification et symbolique*. Paris.

Gombrich, E.H. (1965). 'The Use of Art for the Study of Symbols'. *American Psychologist*, 20 (January): 34–50.

—— (1979). 'The Dream of Reason: Symbolism in the French Revolution'. *BJES* 2: 3 (Autumn): 187–204.

Gravelot, H., and Cochin, N. *Almanach iconologique*. Paris, 1766 (Des Sciences); 1767 (Des Vertus); 1768 (Etres métaphysiques); 1772 (L'Homme); 1773 (Etres moraux); 1774 (Des Sciences); 1775 (Les Vertus et les Vices).

Graves, Robert (1955). *The Greek Myths*. 2 vols. Harmondsworth, rev. 1966.

Hadjinicolaou, Nicos (1979). '*La Liberté guidant le peuple de Delacroix devant son premier public*'. *ARSC* 28 (June).

Harris, Jennifer (1981). 'The Red Cap of Liberty'. *BJES* 14: 283–313.

Haskell, Francis, and Penny, Nicholas (1981). *Taste and the Antique*. New Haven.

Heilbrunn, Françoise (1977). *Galerie Médicis de Rubens*. PGGM, 34. Paris.

Henderson, Ernest F. (1912). *Symbol and Satire in the French Revolution*. New York.

Hobsbawm, Eric (1978). 'Man and Woman in Socialist Iconography'. *HWJ* 6: 121–38.

Hollander, Anne (1975). *Seeing Through Clothes*. New York.

Honour, Hugh (1979). *Romanticism*. London.

Hunt, Lynn (1980). 'Engraving the Republic: Prints and Propaganda in the French Revolution'. *History Today*. October: 11–17.

Keen, Maurice (1984). *Chivalry*. New Haven.

King, Helen (1983). 'Bound to Bleed, Artemis and Greek Women', in Cameron and Kuhrt, pp. 109–27.

Lefkowitz, Mary R., and Fant, Maureen B. (1982). *Women's Life in Greece and Rome*. London.

Marle, Raymond Van (1932). *Iconographie de l'art profane*. 2 vols. The Hague.

Merck, Mandy (1978). 'The Patriotic Amazonomachy and Ancient Athens', in *Tearing the Veil*, ed. Susan Lipshitz. London.

Mosse, David (1971). 'Caesarism, Circuses and Monuments'. *JCH* 6 (2): 167–82.

Ovid. *Metamorphoses*. Trans. Mary M. Innes. PC, 1955.

Ozouf, Mona (1981). *La Fête révolutionnaire*. Paris.

Paulson, Ronald (1983). *Representations of Revolution 1789–1820*. New Haven.

La Révolution française/Le Premier Empire. Dessins du Musée Carnavalet (1982). Musée Carnavalet, Paris, 22 February–22 May.

Riemer, Eleanor S., and Fout, John C. (eds.) (1983). *European Women: A Documentary History 1789–1945*. Brighton.

Robertson, M. (1975). *A History of Greek Art*. 2 vols. Cambridge.

Rousseau, Jean-Jacques (1820). 'Lettre à d'Alembert', in *Œuvres de J.-J. Rousseau*, vol. 2. Paris.

—— (1962). *Du Contrat social ou Principes du droit politique*. Paris.

Rudofsky, Bernard (1971). *The Unfashionable Human Body*. New York.

Sanday, Peggy Reeves (1981). *Female Power and Male Dominance: On the Origins of Sexual Inequality*. Cambridge.

Starobinski, Jean (1979). *1789: Les Emblèmes de la Raison*. Paris.

Sussman, George D. (1982). *Selling Mother's Milk*. London.

Taft, Lorado (1930). *The History of American Sculpture*. New York.

Toussaint, Hélène (ed.) (1982). *'La Liberté guidant le peuple' de Delacroix*. (Catalogue). Paris.

Vermeule, Cornelius C. (1959). *The Goddess Roma in the Art of the Roman Empire*. Cambridge, Mass..

The WPA Guide to New York City (1982). Intro. William H. Wyte. New York.

Warner, Marina (1976). *Alone of all her Sex: The Myth and Cult of the Virgin Mary*. London.

—— (1981). *Joan of Arc: The Image of Female Heroism*. London.

Wealth of the Roman World: Gold and Silver AD 300–700 (1977). Catalogue by J.P.C. Kent and K.S. Painter. BM.

Zeitlin, Froma I. (1978). 'The Dynamics of Misogyny: Myth and Mythmaking' in *The Oresteia'. Arethusa*, 11: 1–2 (Spring and Fall).

10 The Development of Horticulture in the Eastern Woodlands of North America

Women's Role

Patty Jo Watson and Mary C. Kennedy

Introduction

We begin with the words of some famous anthropologists:

> The sound anthropological position is that certain sex-linked behaviors are biologically based, although subject to cultural modifications within limits. (Hoebel 1958: 391)

> A limited number of sex-associated characteristics also appear to be transmitted at the genetic level, such as an apparent tendency shared with many other animals for dominance and passivity in the male and female, respectively . . . (Keesing 1966: 75)

> The community recognizes that women must be accompanied by their babies wherever they go; hence they cannot hunt or fish as efficiently as the unencumbered males. Males are therefore free to be mobile and active while females have been accorded, by nature, a prior responsibility or obligation to rear additional members of the community in the only way this can be done. Hence the community assigns work involving more mobility to men and work involving less mobility to women. (Jacobs and Stern 1952: 145–6)

> Man, with his superior physical strength, can better undertake the more strenuous physical tasks, such as lumbering, mining, quarrying, land clearance, and housebuilding. Not handicapped, as is woman, by the physiological burdens of pregnancy and nursing, he can range farther afield to hunt, to fish, to herd, and to trade. Woman is at no disadvantage, however, in lighter tasks which can be performed in or near the home, e.g., the gathering of vegetable

From Patty Jo Watson and Mary C. Kennedy, 'The Development of Horticulture in the Eastern Woodlands of North America: Women's Role' in *Engendering Archaeology: Women and Prehistory*, eds. Joan Gero and Margaret Conkey (Blackwell, 1991), pp. 225–75. Reprinted with permission.

products, the fetching of water, the preparation of food, and the manufacture of clothing and utensils. All known human societies have developed special-ization and cooperation between the sexes roughly along this biologically determined line of cleavage. (Murdock 1949: 7)

Up to about nine thousand years ago all human populations lived by hunting and most of them also by fishing, supplemented by the picking of berries, fruits and nuts, and the digging of roots and tubers. Perhaps the first division of labor between the sexes was that the male became the hunter and the female the food-gatherer. (Montagu 1969: 134)

Even in those societies where there are no professional or semi-professional artisans, all ordinary manufactures are delegated either to men or to women. Moreover, this sex division of labor is much the same wherever it occurs. Such universal patterns derive from universally present facts, such as the greater size and strength of the male, and his greater activity based on the differing roles of the two sexes in connection with the production and care of children. These factors unquestionably led to the earliest differentiation in food gather-ing activities. This must have begun at an extremely remote period. The males became the main providers of animal foods, since they were able to run down their prey and engage it in combat. The females being hampered throughout most of their adult lives by the presence either of infants *in utero* or in arms, were unable to engage in such active pursuits, but were able to collect vege-table foods and shellfish. . . . Thus to this day in the American family dinner the meat is placed in front of the father to be served and the vegetables in front of the mother. This is a folk memory of the days when the father collected the meat with his spear and the mother the vegetables with her digging stick. (Linton 1955: 70–1)

Thus, the sexual division of labor is neatly laid out, and simply and cogently explained. Men are strong, dominant protectors who hunt animals; women are weaker, passive, hampered by their reproductive responsibilities, and hence, consigned to plant gathering. Not only is that the case for every ethnographically observed society, but it is carried back to 'extremely remote' periods. This received view could be schematized as:

men>hunt>animals>active
women>gather>plants>passive

All the introductory texts from which these excerpts were taken were written before the women's movement and the reorganization of contemporary American life that made women working outside the domestic sphere an inescapable reality.

Introductory texts are much more cautious these days. Conspicuously absent is the explicit male/active, female/passive dichotomy (e.g. Harris 1987: 127; Oswalt 1986: 104–5; Peoples and Bailey 1988: 254–60). Current texts acknowledge that in every known society there is a sexual division of labor; that men hunting and women gathering seems almost always to be the case, but that beyond this there is tremendous variation in which labors a particular society assigns to a particular sex. The received view today is that in foraging societies:

men>hunt>animals
women>gather>plants

For purposes of argument we accept this premise and attempt to formalize this very division of labor for a particular time, region, and cultural historical process: the origin and early development of plant cultivation and domestication in the Eastern Woodlands of North America. We draw upon three lines of evidence: archaeological data, ethnohistoric data, and general schemes of human social organization derived from ethnography.

Data Sources and Arguments of Relevance

Archaeological and archaeobotanical data

The time period relevant to the origin and early development of horticulture in the Eastern Woodlands is approximately 7000–2000 BP. Primary evidence for plant use during these five millennia comes from a variety of archaeological contexts, but only two basic categories: charred and uncharred plant remains. Charred plant remains are usually recovered by flotation-water separation systems; uncharred plants are recovered from dry caves and rockshelters; both categories are then analysed by paleoethnobotanists (Hastorf and Popper 1988; Pearsall 1989).

The evidence to date suggests three different episodes of domestication in the Eastern US (Smith 1987*a*; Watson 1989; Yarnell 1986). The first began about 7000 BP when a gourd-like cucurbit (*Cucurbita* sp.; perhaps *C. pepo*, perhaps *C. texana*, or even *C. foetidissima*) and bottle gourd (*Lagenaria siceraria*) begin to appear in archaeological deposits in the Eastern US. The second is from 3500 BP onward when domesticated forms of the weedy plants sumpweed, chenopod, and sunflower begin to appear. The third is the development of varieties of maize specific to the requirements of the Eastern US, a process that took place between 2000 and 1000 BP.

We are in the pioneer phase of knowledge expansion about prehistoric plant use. One characteristic of this pioneer phase is a scramble by interested scholars to synthesize the new data as they become accessible, not only in annotated inventories of the primary evidence (Yarnell 1977, 1983, 1986, forthcoming), but also in comparative discussions of regional developments (Asch and Asch 1985; Chapman and Shea 1981; Watson 1985, 1988, 1989), and in more general theoretical formulations (e.g. Chomko and Crawford 1978; Crites 1987; Lathrap 1987; Smith 1987a, 1987b). It seems desirable to launch an inquiry at yet another level: that of the women and men involved in the events and processes. While a great deal is being written about the evidence, it is, for the most part, gender-neutral writing; when actors are mentioned they are 'people', or 'humans', or 'individuals'. These accounts tend to be discussions of the archaeological evidence, the plant remains, rather than the people who manipulated the plants. We depart from that pattern here.

The ethnohistoric record

Although the archaeological record of plant use has only recently been sought, information about plant use by living peoples in the Eastern Woodlands has been available since the time of European entry in the sixteenth century. Thus, one source of data is historical or ethnohistorical such as that provided by Dye (1980), Hudson (1976), Le Page du Pratz (n.d.), Parker (1968), Swanton (1948), Will and Hyde (1917), Willoughby (1906), Wilson (1917), and Yarnell (1964). For example, in an often-quoted passage, Le Page du Pratz (n.d.: 156–7) describes the planting of *choupichoul* (Smith 1987c), an Eastern Woodlands cultigen being grown in Louisiana by the Natchez at the time of European entry into the Southeast: 'I have seen the Natchez, and other indians, sow a sort of grain, which they called Choupichoul, on these dry sand-banks. This sand received no manner of culture; and the *women and children* covered the grain any how with their feet, without taking any great pains about it' (emphasis added). Throughout the ethnohistoric and ethnographic literature of the Eastern US are similar examples of women planting, reaping, collecting, and processing plants.

License to use the ethnographic and ethnohistorical information for archaeological inference would presumably be granted because it fulfills the criteria for the use of ethnographic analogy (the most comprehensive recent discussion is Wylie 1985, but see also Ascher 1961; Salmon 1982: 57–83; Watson, LeBlanc, and Redman 1971: 49–51, 1984: 259–63). That is, the information in question comes from the same or a very similar physical environment, and from people who are closely related physically and culturally. Therefore, the ethnographic/ethnohistoric information carries a certain degree of prior probability, of plausibility or likelihood, with respect to the archaeological materials. That does not mean, however, that it is to be accepted uncritically.

Social organization

There is another necessary source, and that is the more abstract and theoretical literature in anthropology about the organization and functioning of human groups at various levels of technological complexity. For present purposes we are content to evoke Fried (1967), Friedl (1975), Murdock and Provost (1973), Sahlins (1963, 1968), Service (1962, 1971), Steward (1948), and White (1959) and refer to them in general as the authority for some basic assumptions:

1. In small-scale, non-food-producing, egalitarian societies, subsistence activities are divided on the basis of age and sex.

2. For biological reasons relating to gestation and lactation, adult women are primarily responsible for nourishing and socializing infants and small children, although various others can assist in these tasks.

3. For biological reasons relating to greater physical strength and hormone levels, adult men are charged with the primary responsibility for safeguarding the social units in which children are born and reared, and—in general—with tasks that require sudden bursts of energy, such as running after game.

4. Because of these biological constraints on men and women, groups tend to divide labor between the sexes so that women are responsible for activities that do not interfere with childcare and that can be performed near the habitation—cooking and 'domestic activities'—as well as the collecting of stationary resources such as plants and firewood. Men are responsible for exploiting mobile resources, primarily the hunting of game, as well as for defense, and a variety of other such tasks.

Using these assumptions about the sexual division of labor, as well as evidence from the archaeological record and the ethnohistoric/ ethnographic record, we depart from the usual gender-neutral perspective to discuss one of the most recent and most comprehensive theoretical treatments we know for the development of horticulture in the Eastern Woodlands: Bruce Smith's (1987a) paper, 'The Independent Domestication of Indigenous Seed-Bearing Plants in Eastern North America'.

The Domestication of Plants

Women and coevolution in the Eastern Woodlands: Weedy plant domestication

Smith's interpretive formulation may be rendered schematically as follows: at about 8000 BP, the beginning of the Middle Archaic period, human populations in the Eastern Woodlands are thought to have been small, few, and dispersed rather widely across the landscape. Their subsistence systems—hunting–gathering–foraging with some emphasis on deer and several kinds of nuts, especially hickory and acorn—were further characterized by residential mobility, probably cycling through similar or the same series of locales, season after season, year after year. Occupation sites were small, although summer camps on river terraces or levees were probably considerably larger than winter residential units in the uplands.

Geological studies indicate significant changes in Mid-Holocene (8000–5000 BP) fluvial systems throughout the East, partly as a result of a long drying trend, the Hypsithermal. Rivers stabilized and aggraded so that previously ephemeral or rare features such as meanders and oxbow lakes, bars and shoals, sloughs and backwater lagoons became much more common and long-lasting than in the previous post-Pleistocene millennia. As a result of these changes, human subsistence-settlement patterns also changed. When slack-water habitats, shoals, etc. took shape as relatively permanent features, then the abundant flora (edible bulbs, rhizomes, shoots, seeds) and fauna (many different kinds of fish, amphibians, mollusks, and waterfowl) characterizing them became readily accessible on a predictable, seasonal, long-term basis. Human settlements—in the form of base camps occupied from late spring to summer and on through fall, and oriented to these resource points—increased in size and were permanently occupied for at least four to five months each year for hundreds of years. This process resulted in the first recognizable anthropogenic locales in the archaeological record of the Eastern US; i.e. these sites represent the first long-lasting impact on this physical environment produced by human activity. Archaic shell mounds and midden mounds are then the scene where the rest of the story unfolds: these 'domestilocalities', as Smith calls them, are the crux of his formulation.

The mounds and middens are significant and long-lived disturbed areas, highly congenial to the weedy species ancestral to the earliest cultivated and domesticated food plants. Smith discusses a series of four important factors in the coevolutionary interplay between human and plant populations at the domestilocalities: sunlight, fertile soil, continually disturbed soil, and continual introduction of seeds. He also stresses that some of the selective pressures operating on the weedy plants colonizing such locales are congruent with the best interests of humans who harvest them for food, most significantly big seeds and thin seed coats.

Intense competition among pioneer species in these rich openings favors seeds that sprout quickly and grow quickly. These traits translate botanically into seeds with reduced dormancy (a thin seed coat is one good means of effecting reduced dormancy) and large endosperm (food reserves to sustain rapid, early spring growth), the two morphological characteristics enabling identification of the earliest domesticates: sumpweed and sunflower (bigger seeds than in wild populations) and *Chenopodium berlandieri* ssp. *jonesianum* (seeds with thinner seed coats than in wild *C. berlandieri).*

Finally, Smith outlines the main stages or steps in the general coevolutionary trajectory between *c.*6500 and 3500 BP in numerous places in the Eastern Woodlands.

At about 6500 BP. In the first stage, domestilocalities are inhabited by humans and by a series of weedy, colonizing, or pioneering plant species for several months during each growing season. Natural selective pressures operate on the plants in the directions just noted to produce big seeds and thin seed coats.

In the second stage, humans tolerate the useful edible species, but ignore, or even occasionally remove the useless or harmful species.

In stage three, humans actively encourage the useful species (which have gradually become even more useful), and while systematically harvesting them also systematically remove competing non-useful plants. Thus, the incidental gardens of stage 2 become true managed gardens, the proceeds of which are stored to augment the winter and early spring diet.

In stage four, humans deliberately plant seeds of the useful species each year, carefully tending and caring for the resulting crops.

At about 3500 BP. In stage five, plants emerge that are clearly recognizable morphologically as domesticates.

The entire process is quite low-key, and there is no drastic alteration in the diet as a result of it, but there is an increase in dependable plant resources.

If we populate Smith's evolutionary stages with gendered human beings chosen to accord with the four operating assumptions for the division of labor already noted, and with the ethnographic record for the Eastern Woodlands, then we must conclude that the adult women are the chief protagonists in the horticultural drama of the domestilocalities. Although the entire human group contributes to the sunlight and soil fertility factors, it is the women who are primarily responsible for soil disturbance and continual introduction of seeds. Smith lists as examples of disturbance: the construction of houses, windbreaks, storage and refuse pits, drying racks, earth ovens, hearths. Most of these probably represent women's work as do the majority of other examples he mentions: primary and secondary disposal of plant and animal debris, and a 'wide range of everyday processing and manu-facturing activities'.

As to the continual introduction of seeds, Smith notes harvesting plants for processing and consumption at the domestilocality. Seed loss during processing, storage, and consumption (plus defecation subsequent to consumption) continually introduce seeds to the fertile soil of the domestilocality. Once again—although everyone joins in consumption and defecation—it is the women who are responsible for processing, and for food preparation and storage.

Have we not then definitively identified woman the gatherer, har-vester, and primary disturber of domestilocalities in the prehistoric Eastern Woodlands as woman the cultivator and domesticator? Yes, we have. But anyone persuaded by Smith's or similar coevolutionary constructions would doubtless respond 'So what?' Our conclusion, with which they would probably readily agree, is at best anti-climactic because the coevolutionary formulation downplays stress, drive, intention, or innovation of any sort on the part of the people involved, in this case the women. The coevolutionary formulation highlights gradualness; the built-in mechanisms adduced carry plants and people smoothly and imperceptibly from hunting–gathering–foraging to systematic harvesting to at least part-time food production with little or no effort on anyone's part. The plants virtually domesticate themselves.

While a number of the initial and ongoing selective pressures acting on these plants within such disturbed habitats were clearly related to human activities, these activities were unintentional and 'automatic' rather than the result of predetermined and deliberate human action toward the plant species in question.
... It is this simple step of planting harvested seeds, even on a very small scale, that if sustained over the long-term marks both the beginning of cultivation, and the onset of automatic selection within affected domestilocality plant populations for interrelated adaptation syndromes associated with domestication.
... This continuing evolutionary process did not require any deliberate selection efforts on the part of Middle Holocene inhabitants of domestilocalities in the Eastern Woodlands. All that was needed was a sustained opportunistic exploitation and minimal encouragement of what were still rather unimportant plant food sources. (Smith 1987a: 32, 33, 34)

This is in keeping not only with the current scheme for division of labor

women>gather>plants,

but also with the earlier (Keesing, Linton, et al.) scheme

women>gather>plants>passive.

Are we to be left with such a muted and down-beat ending to the Neolithic Revolution in the Eastern Woodlands?

Shaman the cultivator: Gourd domestication

The domestication of the native cultigens described by Smith was apparently preceded by the introduction of another type of domesticate, *Cucurbita* gourd and bottle gourd, in various parts of the Eastern US beginning about 7000 years ago. In an article entitled, 'The Origins of Plant Domestication in the Eastern United States: Promoting the Individual in Archaeological Theory,' Guy Prentice (1986) constructs a scenario for that earlier transformation.

Prentice first details the evidence for the tropical squash, *Cucurbita pepo*, in archaeological deposits dating from 7000–3500 BP, some of the earliest evidence for domesticated plants in Eastern North America. Investigators agree that *Cucurbita pepo* fruits would have been used primarily as containers, or perhaps as rattles, rather than food (Prentice 1986: 104). He then argues (ibid. 106) that the species was probably introduced through some form of trade with the tropical areas in which it grows naturally. The sites at which the earliest evidence for *Cucurbita pepo* is found are those of Archaic period, hunting and gathering, band-level societies. He notes that authoritarian controls in such societies would have been weak, and that shamans and headmen were probably exercising the strongest control within these groups (ibid. 107). Prentice presents information from studies indicating that change is not an automatic process in human societies, that certain conditions must be met before an innovation is accepted (ibid. 108–11). An innovation will take hold if it is introduced by an individual of high status, a specialist, an ambitious person who is in contact with outsiders and who is oriented toward commerce rather than subsistence:

> By postulating a ritual use for cucurbits during the Archaic period, one is led to conclude that it would be the shaman who would be most likely to adopt cucurbit agriculture. He would be the one most interested in new religious paraphernalia. He would have the greatest knowledge of plants. He would have been in communication with other shamans and probably exchanging plants and plant lore. If gourds were introduced as magical rattles and ritualistic containers for serving stimulants and medicines, he would have gained a very impressive 'medicine' in the eyes of his patients. In fact, the gourd itself may have provided the medicine. (Prentice 1986:113)

Here is an instance in which at least one archaeologist is not arguing for

women>gather>plants.

Why not? Perhaps because this is a discussion of innovation, and

women>gather>plants>passive

— although it might lead to dinner—would not lead to *innovation*. Rather, this is a scenario for

man>trade>ceremony>active>innovation>cultivation>domestication.

Perhaps it really was like that (see Decker 1988 and Smith 1987*a* for alternative views on the development of *Cucurbita* cultivation in eastern North America), but we are leary of explanations that remove women from the one realm that is traditionally granted them, as soon as innovation or invention enters the picture.

Women and maize agriculture in the Eastern Woodlands: The creation of northern flint

Maize was the most important domesticate among the horticultural societies of North America at the time of European contact. Although there was enormous variety among the indigenous subsistence economies in the Eastern and Western United States, wherever crops were grown maize was central, often being literally deified (e.g. Cushing 1974; Hudson 1976; Munson 1973; Stevenson 1904; Swanton 1948). Yet, as the latest evidence makes clear (Chapman and Crites 1987; Conard et al. 1984; Doebly, Goodman, and Stuber 1986; Ford 1987; Yarnell 1986), maize is a rather late entrant into the Eastern Woodlands, probably introduced from the Southwest, and appearing about 1800 BP The development of horticulture in the Southwest seems to have been very different from that in the East (Minnis 1985, in prep.; Wills 1988), and maize is much earlier there, apparently present by ca.4000 BP.

There is great unclarity in the literature about the exact definitions and time–space distributions of contemporary maize varieties, but there is some consensus that the earliest kinds known in the East (and grown together with sunflower, sumpweed, chenopod, maygrass, knotweed, etc.) are of a type with 10 to 12 rows of kernels, and called Chapalote, Tropical Flint, North American Pop, and/or Midwestern 12-row. These varieties were developed from the earliest Maize, which was originally created from wild populations occurring in Mesoamerica, Central America, or South America (or all three), but exactly where and how has been hotly debated for some 20 years (e.g. Galinat 1985*b*; Lathrap 1987).

Our concern here, however, is with the transition in the more northerly Eastern Woodlands from the early, higher row-number maize to a lower row-number variety variously known as Maiz de Ocho, Northern Flint, or Eastern 8-row maize, which appeared in the Northeast around AD 800–900, became quite standardized, and was the dominant agricultural crop of this region from approximately AD 1000 to historic times (Wagner 1983, 1987). There are at least three alternatives as to how this happened: (1) Eastern 8-row was created in the Eastern Woodlands from the Earlier Midwestern 12-row varieties; or (2) it, like the older form itself, was developed somewhere south of the Border and later diffused to the Eastern US; or (3) both (1) and (2) are too simple, and the origin of Northern Flint/Eastern 8-row involved more complicated combinations of both southerly (the Tropical Flint or Chapalote type) and northerly (the Northern Flint/ Eastern 8-row type) maize varieties originating in several parts of northern, central, and southern America. Lathrap (1987) provides a comprehensive presentation of the second alternative (see also Upham et al. 1987), and Bruce Smith favors the third (Smith in prep.). Although there is now a very solid corpus of information on Fort Ancient maize (Wagner 1983, 1987), primary evidence about Middle Mississippian maize is only beginning to be available (Blake 1986; Fritz 1986; Scarry 1986), so it is not yet possible to assess definitively the relative merits of these three suggestions.

At the moment one can suggest, without contravening the scanty available evidence and in fact remaining congruent with it, that the original form of Northern Flint (Eastern 8-row) was developed from a Chapalote (Midwestern 12-row, Tropical Flint) type of maize at one or several places in the northerly portions of Eastern North America. The earliest date now known for Northern Flint is c. AD 800–900 from one site in western Pennsylvania and two near the north shore of Lake Erie (Blake 1986; Blake and Cutler 1983; Stothers 1976). Northern Flint is present by AD 1000 in the cultural context called Fort Ancient located in the Ohio River drainage of southern Ohio and Indiana, northern Kentucky, and northern West Virginia (Wagner 1983, 1987). On the present evidence, robust forms of Northern Flint appear later or not at all as one moves west and south from Fort Ancient territory (Blake 1986; Fritz 1986; Johannessen 1984; Scarry 1986). Thus, we believe we are justified in accepting, at least for purposes of our argument here, that the Northern Flint variety of maize was developed indigenously in northeastern North America from an older, Chapalote form that came into the east 1800 or 1900 years ago.

We assume that the first Chapalote or Chapolote-like forms of this tropical cultigen to enter northern latitudes in the Eastern U.S. were not well suited to that physical environment, even if the plant diffused northward gradually from Mesoamerica. Day-length, annual tempera-ture and moisture cycles, growing-season length, and substrate charac-teristics probably all differed significantly from those of the locales where Chapalote was initially grown. Hence, the Middle Woodland groups who planted and hoped to harvest this novel crop in the most northerly North American regions may have had more failures and near failures than successes. Lower row numbers on maize cobs are thought to be a botanical reflection of poor growing conditions such as short growing seasons, drought, or even unchecked competition from weeds (Blake and Cutler 1983: 83–4). Thus, it is possible that adverse climate was compounded by neglect in the development of Northern Flint.

We think it more likely, however, that cultivators in the northeasterly portions of North America actively encouraged, against environmental odds, the new starchy food source. Accepting the AD 800–900 dates from the Lake Erie and western Pennsylvania sites as the first establishing of Northern Flint, and noting the rapidity with which Northern Flint agriculture spread throughout the central Ohio River drainage (Fort Ancient) area immediately thereafter, we conclude that deliberate nurturing of maize in an inhospitable environment is a more plausible interpretation than is neglect in the development of this hardy variety.

Two points are implied from the above discussion: (1) maize acceptance and cultivation north of the Border was purposeful and deliberate; and (2) it was surely the women sunflower-sumpweed-chenopod gardeners in Middle and Late Woodland communities who worked (with varying success and interest) to acclimatize this imported species, by planting it deeper or shallower, earlier or later, in hills or furrows, and who crossed varieties to obtain or suppress specific traits. From c. AD 1100–1200 to the time of European contact in the sixteenth century, Northern Flint was the main cultivated food of the hamlets, villages, towns, and chiefdoms that arose in the Ohio River Valley and the vast region north to the Great Lakes. To the west and south, in the Mississippi River and its tributary drainages, Northern Flint was also sometimes grown but in combination with other varieties having higher row-numbers (Blake 1986; Fritz 1986; Johannessen 1984; Scarry 1986; Smith in prep.) Thus Northern Flint, together with pumpkins, squashes, sunflowers, and a long list of other cultigens, planted, tended, harvested, and processed by the women agriculturalists (Hudson 1976; Parker 1968; Swanton 1948; Will and Hyde 1917; Willoughby 1906; Wilson 1917), supported many thousands of people each year for hundreds of years. The accomplishments of these women cultivators is even more impressive when one realizes that their creation, Northern Flint, is the basis (together with Southern Dent, a maize variety that entered the southeastern United States somewhat later) for all the modern varieties of hybrid 'Corn Belt Dent' maize grown around the world today (Doebley et al. 1986; Galinat 1985*a*).

....

CONCLUSIONS
....

We close with a few further thoughts about the first women gardeners in the Late Archaic domestilocalities. Their contribution of domesticated sumpweed, sunflower, and chenopod (and possibly maygrass as well) to the diet and the archaeological record of initial Late Holocene human populations in the Eastern Woodlands may not have been so automatic a process with so insignificant a result as the coevolutionary formulation makes it out to be.

In the first place, the natural history, natural habitat and distribution, ecology, and genetic structure of most of the Late Archaic/Early Woodland cultigens and domesticates are not well understood. On closer inspection, it may turn out to be the case that some if not all the species initially domesticated would have required special, self-conscious, and deliberate treatment to convert them to garden crops, and to cause the very significant and progressive changes in seed size that at least two of them (sumpweed and sunflower) exhibit. Sunflower and maygrass were apparently being grown outside their natural ranges by 3000–2500 BP, and this must have been done purposefully.

Secondly, the best and most comprehensive dietary evidence for the early horticultural period comes from the long series of human paleofecal and flotation derived remains in Salts Cave and Mammoth Cave, west central Kentucky. The fecal evidence dates to 2800–2500 BP and is quite clear and consistent. Over 60 per cent of the plant foods consumed were seeds of indigenous domesticates and cultigens: sunflower, sumpweed, and chenopod (Gardner 1987; Marquardt 1974; Stewart 1974; Watson and Yarnell 1986; Yarnell 1969, 1974, 1977, 1983, 1986, forthcoming). If maygrass, whose status is uncertain but which is here beyond its natural range, is added, then the total proportion of indigenous cultigens rises to well over two-thirds. This single and well-established datum for a period relatively early in the history of the indigenous domesticates might be taken to cast some doubt on the generalization that the addition of the domesticate species had only a slight dietary impact. The doubt is strengthened by corroborating evidence from Newt Kash (Jones 1936; Smith and Cowan 1987), Cloudsplitter (Cowan 1985), and Cold Oak (Gremillion 1988; Ison 1986; Ison and Gremillion 1989) shelters in eastern Kentucky (see also Smith 1987*b*).

A third matter to think about is the fact that—quite apart from all other considerations—the women plant collectors and gardeners of 3500–2500 BP were the first to devise and use techniques of tilling, harvesting, and processing the new domesticates. These same techniques must have been in use throughout the later periods, and were then applied to the production and processing of maize as well as the older cultigens. As Bruce Smith (1987*a*) points out, more than 60 years ago Ralph Linton described significant differences in the tools and techniques used for maize production and processing in Eastern North America vs. those of the Southwest and Mexico, and suggested that in the East maize 'was adopted into a preexisting cultural pattern which had grown up around some other food or foods' (Linton 1924: 349).

The fourth and last point is somewhat more tenuous, but we think it is important to consider the implications of one issue unanimously demonstrated by all the relevant ethnographic and ethnohistorical literature: the extensive and intensive botanical and zoological knowledge possessed by people in hunting–gathering–foraging societies. Botanical knowledge is (and would have been in prehistory) greatest among the women who gather, collect, harvest, and process plant resources. Such knowledge goes far beyond foodstuffs to include plants and plant-parts useful for dyes and for cordage and textile manufacture, as well as a vast array of medicinal leaves, bark, roots, stems, and berries. The ethnographically-documented women who exploited these various plant resources knew exactly where and when to find the right plant for a specific purpose. Surely their prehistoric predecessors controlled a similar body of empirical information about their botanical environments, and were equally skilled at using it for their own purposes. Viewed against such a background, the image of unintentional and automatic plant domestication by Late Archaic women pales considerably.

We think that archaeologists operate under at least two different schemes for explaining gender and labor in prehistoric foraging groups. The first is based upon the assumption that women are seriously encumbered and disadvantaged by their reproductive responsibilities and that men are unencumbered by theirs. In this scheme, these physical limitations are combined or conflated with certain personality traits that are thought by some to apply universally to the sexes. This is the scheme of Linton, Montagu, Hoebel, Keesing, *et al.*; in it women cannot be responsible for culture change because they are not men and therefore they are not active:

$$men>hunt>animals>active$$
$$women>gather>plants>passive$$

The second scheme is based upon a universal sexual division of labor for hunter-gatherers derived from available ethnographic evidence, but does not suppose that any innate psychological characteristics or activity levels separate males and females:

$$men>hunt>animals$$
$$women>gather>plants$$

We do not know who domesticated plants in the prehistoric Eastern Woodlands, but faced with a choice between an explanation that relies on scheme number one and one that relies upon scheme number two, we prefer the alternative we have presented: based on available ethnographic evidence for the Eastern United States in particular and the sexual division of labor in general, women domesticated plants. We would like to think that they domesticated them on purpose because they were bored, or curious, or saw some economic advantage in it, that they acted consciously with the full powers of human intellect and that their actions were a significant contribution to culture change, to innovation, and to cultural elaboration. We prefer this explanation because it makes explicit a formulation that anyone who has ever studied anthropology has to some degree absorbed, i.e. that food plants in foraging societies are women's business. Neither Prentice nor Smith argues that women did *not* domesticate plants in prehistoric North America, yet Prentice does argue that a particular group of men was responsible for this major innovation and Smith argues that the innovation was not consciously achieved. It may be the case that shamans were responsible for the introduction of horticulture. It may be that the invention of horticulture was largely unintentional, or passive. But until there is convincing evidence for either of these hypotheses, we prefer to pursue a third alternative: prehistoric women were fully capable not only of conscious action, but also of innovation.

References

Asch, David L., and Asch Nancy B. (1985). 'Prehistoric Plant Cultivation in West-Central Illinois', in *Prehistoric Food Production in North America*, ed. Richard Ford. Ann Arbor: University of Michigan Museum of Anthropology, Anthropological Papers No. 75, 149–204.

Ascher, Robert (1961). 'Analogy in Archaeological Interpretation', *Southwestern Journal of Anthropology*, 17: 317–25.

Blake, Leonard (1986). 'Corn and Other Plants from Prehistory into History in the Eastern United State', in *The Protohistoric Period in the Mid-South: 1500–1700. Proceedings of the 1983 Mid-South Archaeological Conference*, ed. D. Dye and R. Bristler. Mississippi Department of Archives and History, Archaeological Report 18, 3–13.

Blake, Leonard, and Cutler, Hugh C. (1983). 'Plant Remains from the Gnagey Site (36SO55)'. Appendix II, in R. George, 'The Gnagey Site and the Monongahela Occupation of the Somerset Plateau'. *Pennsylvania Archaeologist*, 53: 83–8.

Chapman, Jefferson, and Crites, Gary (1987). 'Evidence for Early Maize (*Zea mays*) from the Icehouse Bottom Site, Tennessee'. *American Antiquity*, 52: 352–4.

Chapman, Jefferson, and Shea, Andrea Brewer (1981). 'The Archaeological Record: Early Archaic to Contact in the Lower Little Tennessee River Valley'. *Tennessee Anthropologist*, 6: 64–84.

Chomko, Stephen A., and Crawford, Gary W. (1978). 'Plant Husbandry in Prehistoric Eastern North America: New Evidence for its Development'. *American Antiquity*, 43: 405–8.

Conard, N., Asch, D., Asch, N., Elmore, D., Gove, H., Rubin, M., Brown, J., Wiant, M., Farnsworth, K., and Cook, T. (1984) 'Accelerator Radiocarbon Dating of Evidence of Prehistoric Horticulture in Illinois', *Nature*, 308: 443–6.

Cowan, C. Wesley (1985). 'From Foraging to Incipient Food-Production: Subsistence Change and Continuity on the Cumberland Plateau of Eastern Kentucky', Ph.D. dissertation (University of Michigan).

Crites, Gary (1987). 'Human–Plant Mutualism and Niche Expression in the Paleoethnobotanical Record: A Middle Woodland Example', *American Antiquity* 52: 725–40.

Cushing, Frank H. (1974). *Zuni Breadstuff*, New York: Museum of the American Indian Hey Foundation, Indian Notes and Monographs 8 (reprint edn.).

Decker, Deena (1988). 'Origin(s), Evolution, and Systematics of *Cucurbita pepo* (Cucurbitaceae)'. *Economic Botany*, 42: 4–15.

Doebley, J., Goodman, M., and Stuber, C. (1986). 'Exceptional Genetic Divergence of Northern Flint Corn'. *American Journal of Botany*, 73: 64–9.

Dye, David (1980). 'Primary Forest Efficiency in the Western Middle Tennessee Valley'. Ph.D. dissertation (Department of Anthropology, Washington University, St Louis).

Ford, Richard I. (1987). 'Dating Early Maize in the Eastern United States'. Paper read at the 10th Annual Conference of the Society of Ethnobiology, Gainesville, Fla.

Fried, Morton H. (1967). *The Evolution of Political Society*. New York: Random House.

Friedl, Ernestine (1975). *Women and Men: An Anthropologist's View*. New York: Holt, Rinehart, and Winston. Reprinted 1984 by Waveland Press, Prospect Heights, Ill.

Friedman, J. and Rowlands, M. J. (eds.) (1977) *The Evolution of Social Systems*. *London: Duckworth*.

Fritz, Gayle (1986). 'Prehistoric Ozark Agriculture: The University of Arkansas Rockshelter Collections'. Ph.D. dissertation (Department of Anthropology, University of North Carolina, Chapel Hill).

Galinat, Walton C. (1985*a*). 'Domestication and Diffusion of Maize', in *Prehistoric Food Production in North America*, ed. Richard Ford. Ann Arbor: University of Michigan Museum of Anthropology, Anthropological Papers No. 75, 245–78.

—— (1985*b*) 'The Missing Links between Teosinte and Maize: A Review'. *Maydica*, 30: 137–60.

Gardner, Paul S. (1987). 'New Evidence Concerning the Chronology and Paleoethnobotany of Salts Cave, Kentucky'. *American Antiquity* 52: 358–67.

Gremillion, Kristin J. (1988). 'Preliminary Report on Terminal Archaic and Early Woodland Plant Remains from the Cold Oak Shelter, Lee County, Kentucky'. Report submitted to Cecil R. Ison, USDA Forest Service Station, Stanton Ranger District, Stanton, Kentucky.

Harris, Marvin (1987). *Cultural Anthropology*, 2nd edn. New York: Harper & Row.

Hastorf, Christine, and Popper, Virginia (eds.) (1988). *Current Paleoethnobotany: Analytical Methods and Cultural Interpretations of Archaeological Plant Remains*. Chicago: University of Chicago Press.

Hoebel, E. Adamson (1958). *Man in the Primitive World: An Introduction to Anthropology*. New York: McGraw-Hill.

Hudson, Charles (1976). *The Southeastern Indians*. Knoxville: University of Tennessee Press.

Ison, Cecil R. (1986). 'Recent Excavations at the Cold Oak Shelter, Daniel Boone National Forest, Kentucky'. Paper presented at the Kentucky Heritage Council Annual Conference, Louisville.

Ison, Cecil R., and Gremillion, Kristin J. (1989). 'Terminal Archaic and Early Woodland Plant Utilization along the Cumberland Plateau'. Paper presented at the Society for American Archaeology Annual Meeting.

Jacobs, Melville, and Stern, Bernhard (1952). *General Anthropology*. New York: College Outline Series, Barnes and Noble. Reprinted 1964.

Johannessen, Sissel (1984). 'Paleoethnobotany', in *American Bottom Archaeology*, ed. C. Bareis and J. Porter. Urbana and Chicago: University of Chicago Press, 197–214.

Jones, Volney (1936). 'The Vegetal Remains of Newt Kash Hollow Shelter', in *Rock Shelters in Menifee County, Kentucky*, eds. William Webb and W. Funkhouse. Lexington: University of Kentucky, Reports in Archaeology and Anthropology 3.

Keesing, Felix M. (1966). *Cultural Anthropology: The Science of Custom*. New York: Holt, Rinehart, and Winston.

Lathrap, Donald W. (1987). 'The Introduction of Maize in Prehistoric Eastern North America: The View from Amazonia and the Santa Elena Peninsula', in *Emergent Horticultural Economies of the Eastern Woodlands*, ed. William Keegan. Carbondale: Center for Archaeological Investigations, Southern Illinois University, Occasional Paper No. 7, 345–71.

Le Page du Pratz, Antoine Simon (n.d.). *The History of Louisiana*. Pelican Press, Inc.

Linton, Ralph (1924). 'The Significance of Certain Traits in North American Maize Culture'. *American Anthropologist*, 26: 345–9.

—— (1955) *The Tree of Culture*. New York: Alfred A. Knopf.

Marquardt, William H. (1974). 'A Statistical Analysis of Constituents in Paleofecal Specimens from Mammoth Cave', in *Archaeology of the Mammoth Cave Area*, ed. Patty Jo Watson. New York: Academic Press, 193–202.

Minnis, Paul (1985). 'Domesticating Plants and People in the Greater American Southwest', in *Prehistoric Food Production in North America*, ed. Richard Ford. Ann Arbor: Museum of Anthropology, University of Michigan, Anthropological Papers No. 75, 309–40.

—— (in prep.). 'Earliest Plant Cultivation in Desert North America', in *Agricultural Origins in World Perspective*, eds. Patty Jo Watson and C. W. Cowan. MS chapter. For submission to Smithsonian Institution Press.

Montagu, Ashley (1969). *Man, His First Two Million Years: A Brief Introduction to Anthropology*. New York: Columbia University Press.

Munson, Patrick J. (1973). 'The Origins and Antiquity of Maize Beans-Squash Agriculture in Eastern North America: Some Linguistic Implications', in *Variation in Anthropology: Essays in Honor of John C. McGregor*, eds. D. Lathrap and J. Douglas. Urbana: Illinois Archaeological Survey, 107–35.

Murdock, George P. (1949). *Social Structure*. New York: Free Press.

Murdock, George P. and Provost, Caterina (1973). 'Factors in the Division of Labor by Sex: A Cross Cultural Analysis'. *Ethnology*, 12: 203–25.

Oswalt, Wendell H. (1986). *Life Cycles and Lifeways: An Introduction to Cultural Anthropology*. Palo Alto, Calif.: Mayfield Publishing.

Parker, Arthur C. (1968). 'Iroquois Uses of Maize and Other Plant Foods', in *Parker on the Iroquois*, ed. W. Fenton. Syracuse: Syracuse University Press, 1–119.

Pearsall, Deborah (1989). *Paleoethnobotany: Reconstructing Interrelationships between Humans and Plants from the Archaeological Record*. San Diego, Academic Press.

Peoples, James, and Bailey, Garrick (1988). *Humanity: An Introduction to Cultural Anthropology*. St Paul, Minn.: West Publishing.

Prentice, Guy (1986). 'Origins of Plant Domestication in the Eastern United States: Promoting the Individual in Archaeological Theory'. *Southeastern Archaeology*, 5: 103–19.

Sahlins, Marshall (1963). Poor Man, Rich Man, Big-Man, Chief: Political Types in Melanesia and Polynesia'. *Comparative Studies in Society and History*, 5 285–303.

—— (1968). *Tribesmen*. Englewood Cliffs, NJ: Prentice-Hall.

Salmon, Merrilee (1982). *Philosophy and Archaeology*. New York: Academic Press.

Scarry, C. Margaret (1986). 'Change in Plant Procurement and Production During the Emergence of the Moundville Chiefdom'. Ph.D. dissertation (Department of Anthropology, University of Michigan, Ann Arbor).

Service, Elman (1962). *Primitive Social Organization*. New York: Random House.

—— (1971). *Primitive Social Organization*, 2nd edn. New York: Random House.

Smith, Bruce D. (1987*a*). 'The Independent Domestication of the Indigenous Seed-Bearing Plants in Eastern North America', in *Emergent Horticultural Economies of the Eastern Woodlands*, ed. William Keegan. Carbondale: Center for Archaeological Investigations, Southern Illinois University, Occasional Paper No. 7, 3–47.

—— (1987*b*). 'Hopewellian Farmers of Eastern North America'. Paper presented at the 11th International Congress of Prehistoric and Protohistoric Science, Mainz, West Germany.

—— (1987*c*). 'In Search of Choupichoul, the Mystical Grain of the Natchez'. Keynote Address, 10th Annual Conference of the Society of Ethnobiology, Gainesville, Florida.

—— (in prep.). 'Prehistoric Plant Husbandry in North America', in *Origins of Agriculture in World Perspective*, eds. Patty Jo Watson and C. W. Cowan. MS chapter. For submission to Smithsonian Institution Press.

Smith, Bruce D., and Cowan, C. Wesley (1987). 'The Age of Domesticated *Chenopodium* in Prehistoric North America: New Accelerator Dates from Eastern Kentucky'. *American Antiquity*, 52: 355–7.

Stevenson, Matilda G. (1904). *The Zuni Indians*. Washington: Annual Report of the Bureau of American Ethnology 1901–1902, vol. 23.

Steward, Julian (1948). *Patterns of Cultural Change*. Urbana: University of Illinois Press.

Stewart, Robert B. (1974). 'Identification and Quantification of Components in Salts Cave Paleofeces, 1970–1972', in *Archaeology of the Mammoth Cave Area*, ed. Patty Jo Watson. New York: Academic Press, 41–8.

Stothers, David M. (1976). 'The Princess Point Complex: A Regional Representative of the Early Late Woodland Horizon in the Great Lake Area', in *The Late Prehistory of the Lake Erie Drainage Basin*, ed. David Brose. Cleveland: Cleveland Museum of Natural History, 137–61.

Swanton, John R. (1948). *The Indians of the Southeastern United States*. Washington: Bureau of American Ethnology Bulletin 137. Reprinted 1979 by Smithsonian Institution Press.

Upham, S., MacNeish, R. S., Galinat, W. C., and Stevenson, C. M. (1987). 'Evidence Concerning the Origin of Maiz de Ocho'. *American Anthropologist*, 89: 410–19.

Wagner, Gail E. (1983). 'Fort Ancient Subsistence: The Botanical Record'. *West Virginia Archaeologist*, 35: 27–39.

—— (1987). 'Uses of Plants by Fort Ancient Indians'. Ph.D. dissertation (Department of Anthropology, Washington University, St Louis).

Watson, Patty Jo (1985). 'The Impact of Early Horticulture in the Upland Drainages of the Midwest and Midsouth', in *Prehistoric Food Production in North America*, ed. Richard Ford. Ann Arbor: University of Michigan Museum of Anthropology, Anthropological Papers No. 75, 73–98.

—— (1988). 'Prehistoric Gardening and Agriculture in the Midwest and Midsouth', in *Interpretation of Culture Change in the Eastern Woodlands During the Late Woodland Period*, ed. R. Yerkes. Columbus: Ohio State University, Department of Anthropology, Occasional Papers in Anthropology No. 3, 38–66.

—— (1989). 'Early Plant Cultivation in the Eastern Woodlands of North America', in *Foraging and Farming: The Evolution of Plant Exploitation*, eds. D. Harris and G. Hillman. London: Allen and Hyman, 555–70.

Watson, Patty Jo, and Yarnell, Richard A. (1986) 'Lost John's Last Meal'. *Missouri Archaeologist*, 47: 241–55.

Watson, Patty Jo, LeBlanc, Steven A., and Redman, Charles L. (1971). *Explanation in Archaeology: An Explicitly Scientific Approach*. New York: Columbia University Press.

—— (1984). *Archaeological Explanation: The Scientific Method in Archaeology*. New York: Columbia University Press.

White, Leslie (1959). *The Evolution Culture*. New York: McGraw-Hill.

Will, George F., and Hyde, George E. (1917). *Corn among the Indians of the Upper Missouri*. Lincoln: University of Nebraska Press.

Willoughby, Charles C. (1906). 'Houses and Gardens of the New England Indians'. *American Anthropologist*, 8: 115–32.

Wills, W. H. (1988). *Early Prehistoric Agriculture in the American Southwest*. Sante Fe, N. Mex.: School of American Research Press.

Wilson, Gilbert L. (1917). *Agriculture of the Hidatsa Indians, an Indian Interpretation*. University of Minnesota Studies in the Social Sciences 9. Reprints in Anthropology 5 (May 1977), J&L Reprint Co. Lincoln, Nebr.

Wylie, Alison (1985). 'The Reaction Against Analogy', in *Advances in Archaeological Method and Theory*, vol. 8, ed. Michael Schiffer. Orlando, Fla.: Academic Press, 63–111.

Yarnell, Richard A. (1964). *Aboriginal Relationships between Culture and Plant Life in the Upper Great Lakes Region*. Ann Arbor: University of Michigan, Museum of Anthropology, Anthropological Papers No. 23.

—— (1969). 'Contents of Human Paleofeces', in *The Prehistory of Salts Cave, Kentucky*, ed. Patty Jo Watson. Springfield: Illinois State Museum. Reports of Investigations No. 16, 41–54.

—— (1974). 'Plant Food and Cultivation of the Salt Cavers', in *Archaeology of the Mammoth Cave Area*, ed. Patty Jo Watson. New York: Academic Press, 113–22.

—— (1977). 'Native Plant Husbandry North of Mexico', in *Origins of Agriculture*. ed. C. Reed. The Hague: Mouton, 861–75.

—— (1983). 'Prehistory of Plant Foods and Husbandry in North America'. Paper presented at the Annual Meeting of the Society for American Archaeology, Pittsburgh.

—— (1986). 'A Survey of Prehistoric Crop Plants in Eastern North America'. *The Missouri Archaeologist*, 47: 47–59.

—— (forthcoming) 'Sunflower, Sumpweed, Small Grains, and Crops of Lesser Status', in *Handbook of North American Indians*, ed. W. Sturtevant. Washington, DC: Smithsonian Institution Press.

Part IV. Masculinities

11 I Could Have Retched All Night
Charles Darwin and His Body

Janet Browne

Celebrated Victorian thinkers usually knew how to arrange their ill health. A day in the life of one prominent man went something like this:

His custom was to work in his official room from 9 to about 2.30, though in summer he was frequently at work before breakfast. He then took a brisk walk, and dined at about 3.30. This early hour had been prescribed and insisted upon by his physician, Dr Haviland of Cambridge, in whom he had great confidence. He ate heartily, though simply and moderately, and slept for about an hour after dinner. He then had tea, and from about 7 to 10 he worked in the same room with his family. . . . He would then play a game or two at cards, read a few pages of a classical or historical book, and retire at 11. . . . He was very hospitable, and delighted to receive his friends in a simple and natural way at his house. . . . But he avoided dinner parties as much as possible—they interfered too much with his work—and with the exception of scientific and official dinners he seldom dined away from home. His tastes were entirely domestic, and he was very happy in his family. With his natural love of work, and with the incessant calls upon him, he would soon have broken down had it not been for his system of regular relaxation.[1]

We could be forgiven for thinking that this was a pretty accurate description of Charles Darwin grinding through his days at Down House in Kent. On the contrary, however, it was George Biddell Airy, the astronomer royal: the man who ran professional astronomy in the British empire, president of the Royal Society in 1871, Plumian Professor at Cambridge, director of the university observatory, and author of eleven books; a man who was often extremely ill but whose illnesses barely figure in our collective historical memories.

From Janet Browne, 'I Could Have Retched All Night: Charles Darwin and His Body', in *Science Incarnate: Historical Embodiments of Natural Knowledge*, eds. Christopher Lawrence and Steven Shapin (University of Chicago Press, 1998), pp. 240–87. Reprinted with permission.

Anyone can find similar passages in the period's voluminous sets of lives and letters. A great number of Victorian scientists were unwell, some of them worse than Darwin, some less so. All of them, like Darwin, had to negotiate ways to work while suffering from ill health. He was not alone. Yet even during his own lifetime Darwin's illnesses became something special, something unusual. How many contemporaries worried about Airy's physical troubles, for example? Or Thomas Henry Huxley's? Or Adam Sedgwick's? Darwin was not so much an invalid as a *famous* invalid. More than this, his fame became closely intertwined with what people thought about his invalidism and his intellect.

These intertwinings are still manifest. Few modern readers need reminding of the extensive literature published during recent decades on Darwin's illnesses, ranging from the flurry of interest in the 1950s in possible biological conditions such as Chagas' disease to the psychological and nervous conundrums of the late 1970s and 1980s. All of these come together as a decided genre in medical and historical writing.[2]

Nevertheless, much of the work on Darwin's medical state tends to limit itself, for one reason or another, to identifying or discussing the conditions from which he may have been suffering. Even the well-known notable exceptions, such as Ralph Colp, George Pickering, and Adrian Desmond and James Moore, focus mainly on the complex interrelations between ill health and Darwin's extraordinarily fertile inner life. Yet the whole question surely cries out for some broader attention to his ailing body as a cultural phenomenon along the lines set out in the other essays in this volume. How did ill health, celebrity status, and brains interlock in the nineteenth century, for example, and how did Darwin's very public life of the shawl mesh with Victorian cultural commitments of wider relevance? Some of the simplest inquiries along these wider lines can be revealing. When did the ordinary man or woman in the street, for instance, realize that Darwin was an ill man? Because their purpose mainly lies elsewhere, neither Pickering's *Creative Malady* nor Colp's *To Be an Invalid* can tell us. However, the first public announcement of his unhealthy condition, appropriately enough, seems to be in Darwin's *On the Origin of Species* (1859). In the introduction, after speaking of his return from the *Beagle* voyage and of the years spent puzzling over species, the mystery of mysteries, Darwin said his work was nearly finished. 'But as it will take me two or three more years to complete it, and as my health is far from strong, I have been urged to publish this abstract.'[3] Deeply symbolic in his choice of moment, Darwin made his bad health a primary reason for publishing the *Origin*, putting his precarious physical state well before any remarks about Alfred Russel Wallace's having arrived at 'almost exactly the same general conclusions'. Illness therefore became an integral part of the radical text that followed, a persuasive device of the first order. Such an announcement in the *Origin*—on the first page of the *Origin* no less—can take on real meaning in the light of the cultural uses of medicine.

This larger question about Darwin's health and his public renown surely hinges on the way his illnesses were simultaneously experienced, presented, and interpreted—on the way notable men and women apparently integrated their faltering states into more comprehensive campaigns for engaging the attention of the Victorian community. Darwin was certainly, on many occasions, horribly ill. But he was also adept at deploying nearly everything that came to hand for promoting evolutionary theory. This chapter therefore suggests a few avenues that might be explored by considering Darwin's sick body as one further professional resource in a rich repertoire of resources: not so much in the individual sense, where continued unwellness undoubtedly etched his character and notion of self-identity and contributed significantly to his dogged determination to publish, although these were profoundly significant in Darwin's case; and not really in the context of his immediate family environment, which positively reveled in ill health; but much more in relation to the demanding and multifaceted public eye.[4] In other words, how did Darwin's afflictions enter into the construction of social relations with his scientific colleagues, with his doctors, and with his readers? How did he feel about it all? Oddly enough, as Dorinda Outram points out in a recent essay, the one thing that is frequently ignored in accounts of bodies is the owner's subjective experience of just such an item.[5] Darwin seems to have put his body under the Victorian spotlight just as concretely as he presented his mind through the *Origin of Species*. In the process, it seems that this 'public' body gradually came to evoke the disembodied quality of thought. Darwin—and Darwin's body—consequently offer a good point of entry into some of the complex forces at play when we try to talk about the presentation of intellectual authority, authority that invariably reaches far beyond the usual framework of printed books.

The Illnesses

In common with most sick people, Darwin was noticeably conscious of his outer frame and never shy of describing its miseries to intimates. Though his symptoms came and went over the years and varied in intensity, they remained more or less within the same parameters. Putting it succinctly, these were predominantly gastrointestinal. 'For 25 years,' he wrote, 'extreme spasmodic daily & nightly flatulence; occasional vomiting, on two occasions prolonged during months. . . . All fatigue, especially rocking, brings on the head symptoms . . . cannot walk above ½ mile—always tired—conversation or excitement tires me most.'[6]

The truth was that, as he said to Joseph Hooker and Thomas Henry Huxley, he suffered from incessant retching or vomiting, usually brought on by fatigue; and from painful bouts of wind that churned around after meals and obliged him to sit quietly in a private room until his body behaved more politely. Reading between the lines, his guts were noisy and smelly. 'I feel nearly sure that the air is generated somewhere lower down than stomach,' he told one doctor plaintively in 1865, 'and as soon as it regurgitates into the stomach the discomfort comes on.'[7] He was equally forthright with his cousin William Darwin Fox: 'all excitement & fatigue brings on such dreadful flatulence that in fact I can go nowhere.' When he did go somewhere, he needed privacy after meals, 'for, as you know, my odious stomach requires that.'[8]

He also had trouble with his bowels, frequently suffering from constipation and vulnerable to the obsession with regularity that stalked most Victorians. He developed crops of boils in what he called 'perfectly devilish attacks' on his backside, making it impossible to sit upright, and occasional eczema. There were headaches and giddiness. He probably had piles as well.[9]

Not surprisingly, when these debilitating signs of weakness arrived in a batch, Darwin felt terribly dejected, almost as if his physical shell was taking over. Despite all the care and attention he lavished on it, the body rebelled: illness was an alien presence that robbed him of his power over himself. 'I shd suppose few human beings had vomited so often during the last 5 months,' he gasped early in 1864.[10] 'It is astonishing the degree to which I keep up some strength. . . . I have had a bad spell vomiting every day for eleven days and some days many times after every meal.'[11] His body was not particularly pleasant for him to be with. Considerately, he tried to keep it out of other people's way as well. But he was often preoccupied with it to the exclusion of almost everything else in ordinary life. He was sure that such prolonged misery indicated a physiological disorder.

It fell to Joseph Hooker, his botanical friend at Kew Gardens who was previously trained as a physician, to ask the obvious question. 'Do you actually throw up, or is it retching?' It was both, Darwin replied, but food hardly ever came up. 'You ask after my sickness—it rarely comes on till 2 or 3 hours after eating, so that I seldom throw up food, only acid & morbid secretion.' 'What I vomit [is] intensely acid, slimy (sometimes bitter), corrodes teeth.' 'Doctors puzzled,' he added defiantly.[12]

Hooker did not venture a diagnosis. Nor did Huxley, or any other of Darwin's closest medically trained friends, although they offered constant sympathy and practical advice. They kept themselves out of what might be a difficult situation. Instead, they suggested the names of leading doctors he might wish to consult. Acting on this advice over the years, Darwin sought out a number of London physicians, most of whom he knew by reputation or through his scientific work; the two exceptions were his father, Robert Darwin of Shrewsbury, when he was alive, and Henry Holland, a second cousin on the Wedgwood side. With the usual prerogative of the wealthy classes, he tended to choose doctors with a reputation for having studied some topics in greater detail than usual. In fact, the variety of Darwin's physicians through the 1860s and 1870s is an interesting theme in itself, one that the continuing publication of his correspondence will reveal in detail.[13] Darwin also had more doctors than might be expected, another perquisite of the rich. Like many patients searching for an acceptable diagnosis, he moved constantly among different medical men, trying their remedies and their diets, their purges, mineral acids, and magnesium salts for a couple of months before giving up and turning in despair to another expert and another treatment. In a medical world barely beginning to fragment into specialties, this movement was perhaps inevitable. It was certainly a regular feature of Darwin's life and also of others'.

Such constant medical attention was addictive. In his time, Darwin sought out physicians with an interest in stomachs, skin, urine, blood, nerves, and gout, and on one occasion the entrepreneurial publisher of the *Westminster Review*, John Chapman, who qualified at Saint Andrews medical school when his publishing business tottered and claimed to have found a cure for seasickness and nausea in icebags applied to the spine.[14] Darwin was very taken with Chapman and his therapy and was sorry when a month's application of icebags along his spine (three times a day for an hour and a half at a time) made no difference to his retching. 'We liked Dr. Chapman so very much we were quite sorry the ice failed for his sake as well as ours,' wrote Emma Darwin.[15] Apart from all the other things, Darwin's body was starting to represent an expensive medical investment.[16]

These doctors mostly agreed that Darwin suffered from an intestinal or stomach disorder of a chronic recurrent nature, probably involving the nerves supplying the gut. Darwin's rapid trajectory through them reveals something of his own belief that his nervous system, his brain, and his stomach were uniformly implicated—if one doctor neglected to include all the elements he felt were failing he soon moved elsewhere for another opinion. From very early on, he believed that too much work brought on bouts of vomiting: that 'the noddle and the stomach are antagonistic powers.' As a Darwin aunt said, 'his health is always affected by his nerves.'[17] Some physicians, like William Brinton, the eminent physiologist, called his condition dyspepsia and prescribed magnesium to counteract excess acid secretions.[18] Others, like Henry Bence Jones, thought the problem more to do with the physiology of digestion indicating an imbalance of acids and alkalies in the blood—what Bence Jones and Henry Holland called 'suppressed gout', a state defined by them as too much uric acid remaining in the blood.[19] So Darwin had his urine tested, followed Bence Jones's special diet, and dosed himself with colchicum, a dangerously corrosive specific for gout.[20] Dr. Chapman treated the base of his spinal cord with ice. Dr. Gully at the Malvern water cure treated the top of it with cold water.[21] Dr. Lane at Moor Park and then Dr. Smith at Ilkley sat him in freezing hip baths.[22] Dr. William Jenner prescribed podophyllum and other drastic purgatives. Dr. George Busk thought the problem was primarily mechanical—the stomach did not push on its contents as rapidly as it ought to.[23] On the other hand, Dr. Engleheart, the physician in Down village, told Darwin to look to his drains. Almost despairing of a cure, he eventually sent some vomit on a slide to John Goodsir for him to search for pathogenic vegetable spores and was disappointed to hear there were none.[24] It seems probable that toward the end he was suffering just as much from overmedication and a surfeit of conflicting advice as from his own special combination of physical disorders and medical neuroses.

Darwin's doctors also acknowledged the importance of his nervous system and tactfully dealt with his delicate mental constitution as well as they could. In the post-*Origin* years, they knew they were dealing with a famous thinking man, however modest and unassuming in personal demeanor. It was important for both physician and patient to reach a diagnosis they felt mutually comfortable with; and that therapy should err on the side of professional caution. No one wanted to go down in history as the man who killed the Newton of nineteenth-century biology. It was not so many years earlier that Sir Richard Croft, accoucheur to Princess Charlotte, had committed suicide after the princess's unfortunate death in childbirth in 1818.[25] Bence Jones, for example, recommended the traditional upper-class remedy of a change of scene. Failing that, he said, Darwin should 'get a pony and be shaken once daily to make the chemistry go on better.'[26] This Darwin did, and enjoyed the exercise until a riding accident required that he should call it a day. Perhaps some yachting, Bence Jones blithely proposed a year or two later. Mental distraction, he thought, was a crucial part of the answer. Andrew Clark reiterated the same instructions in a different form. 'Do not notice your own sensations . . . struggle to avoid self-scrutiny & self-consciousness.'[27] Above all, he said, try to relax and take time off from the punishing self-imposed schedule of scientific work. But Darwin dismissed these suggestions. His work, he claimed, was the only thing that took his mind off his sickness; and in later life, a game of billiards.

However prominent they were, these physicians evidently found it difficult to be blunt with him; in Darwin's case the customary negotiation between doctor and patient had to accommodate his intellectual status as much as anything else.[28] And naturally enough, the doctors wanted to keep Darwin's custom. Andrew Clark, for instance, in his eagerness to become a great man's physician, nearly overstepped the mark early in their medical relationship. In 1873 he felt it necessary to write an abject letter to Darwin apologizing for 'pushing too close'. Nevertheless, he said, he hoped one day to be of service.[29] During the late 1870s he achieved that aim in becoming Darwin's primary physician, warmly praised by Francis Darwin in his edition of Darwin's *Life and Letters*, and eventually attending Darwin on his deathbed. It must have been daunting, furthermore, for these men to find themselves simultaneously sucked into Darwin's scientific projects. 'Can you persuade the resident doctor in some hospital,' Darwin asked James Paget when consulting him on his own behalf, 'to observe a person retching violently, but throwing nothing from the stomach.' Darwin wanted to know whether tears came to their eyes, a trait he was investigating for the *Expression* book. 'From my own personal experience I do think that this is the case.'[30] In his usual methodical way, he made his illnesses, and his intellectual project, inseparable from the more general problem of being treated by medical experts.

Still, as Pickering, Colp, and others have noted, Darwin's illnesses usually took the external form that would be most useful.[31] A weak stomach was a very good reason for avoiding dinner parties, much better than an attack of rheumatism or unmentionable boils. Repeated vomiting after a train journey was well suited to avoiding trips to London. A night of retching after ten minutes' talk was a valid obstacle to prolonged social activity: half an hour with Ernst Haeckel, or even close friends like Hooker, Huxley, or Lyell, or any selection of agreeable Down House neighbors, could literally make Darwin sick. Many of Darwin's disabilities were, in this sense, socially relevant ones. Much of their circumstantial value lay in their diversity, applicability, and lack of diagnosis. Such illnesses, as novelists like Jane Austen and Elizabeth Gaskell readily recognized, act as a mode of social circulation as well as instruments of domestic tyranny. Perpetually ailing and complaining, *Emma*'s Mr. Woodhouse was the focus of some of Austen's most pointed observations.

Darwin took advantage of his versatile failings. In the most general manner, of course, his avowed need to stay quiet allowed him to keep apart from the controversies surging around the *Origin*. Ill health permitted him to discourage all but the most wanted visitors to Down House and to choose whom he saw when he went into London. It allowed him to fall asleep during piano recitals and novel readings. It excused him from boring evenings at scientific societies. It sanctioned his retreat after dinner (with the ladies) instead of sitting up with cigars and wine in masculine company. In these subtle ways, he let ill health carry the brunt of displaying a preoccupation with other, more intellectual concerns: a nineteenth-century counterpart to ascetic philosophers of the seventeenth century who gave themselves up wholly to the search for truth. Like them, Darwin's belly was at the opposite pole to mentality. The poet William Allingham got it just right when, after meeting Darwin in 1868, he said that: 'he has his meals at his own times, sees people or not as he chooses, has invalid's privileges in full, a great help to a studious man.'[32] Other passing acquaintances were less impressed. 'Why drat the man,' said old Mrs. Grote at Chevening Court, 'he's not as bad as I am.'[33]

Ill health also helped turn Darwin's absences into a statement. It provided a reason for not going into London to receive the Royal Society's Copley Medal in 1864, the award that generated intense debate in the council on whether the *Origin* should be acknowledged or not.[34] If it was acknowledged, as Darwin's friends in the council demanded, Huxley believed there would be a public reprimand from the president about the book's dangerous opinions and promised Darwin that he would provide a spirited defense. Too ill to go, Darwin sent Hugh Falconer (his proposer) and his brother Erasmus Alvey Darwin along instead. They all knew why. 'What a pity you can't be there,' Erasmus wrote sardonically. 'And yet if you were it could not be done so well.'[35]

Darwin employed the same panic-stricken tactic when his old friend and professor, John Stevens Henslow, was at death's door.

I write now only to say that if Henslow . . . would really like to see me I would of course start at once. The thought had [at] once occurred to me to offer, & the sole reason why I did not was that the journey, with the agitation, would cause me probably to arrive utterly prostrated. I shd be certain to have severe vomiting afterwards, but that would not much signify, but I doubt whether I could stand the agitation at the time. I never felt my weakness a greater evil. I have just had a specimen, for I spoke a few minutes at Linnean Society on Thursday & though extra well, it brought on 23 hours vomiting. I suppose there is some Inn at which I could stay, for I shd not like to be in house (even if you could hold me) as my retching is apt to be extremely loud.[36]

Not many people could persist in asking Darwin to visit after receiving a letter like that. By staying away, the implication goes, he was helping Henslow far more than if he arrived. In both instances, if only for a critical moment, his body's significance lay in its absence. There are many other examples in his correspondence.

It appears too easy then to write off all these complaints as mere hypochondria, although a strong dash of it definitely ran in the family. Darwin was as much aware of the family foible as anyone and perfectly capable of joking about it with his wife and relatives. His sister Caroline, he would say, was intensely irritating when she became heroic about her illnesses: he and Emma much preferred people speaking up.[37] Equally, it was Erasmus Alvey Darwin, Darwin's older brother, who was considered the irretrievable family hypochondriac, not Darwin. This Erasmus Darwin was a witty, cynical man, enjoying what he called his misanthropy in Marlborough Street. 'I have been lying on the sofa in a state of utter torpor,' he wrote once. 'I mean to go out today to see if I am well or not . . . If the present beautiful weather continues I shall be compelled to go and be happy in the country but at present I prefer being miserable in London.'[38]

Darwin's bulletins about his own symptoms need to be fitted into this gently self-mocking and intelligent family pattern. 'Charles came up yesterday,' said Erasmus, 'and went out like a dissipated man to a tea party.' We need to remember that people can have a sense of humor about illness, which they sometimes direct at themselves. When Darwin signs himself at the end of a letter as 'your insane and perverse friend,' he does not mean it literally.

It appears almost too easy, as well, to assume that all these illnesses were somehow a consequence of Darwin's deep-seated anxieties about evolutionary theory and its religious consequences. We seem to expect individuals with radical new ideas to be tormented by them to the exclusion of other anxieties—other anxieties we retrospectively perceive as having lesser importance. We are probably right to expect a great deal of that mental conflict to emerge as illness. But for the sake of argument, Darwin's nights of retching might just as well be related to financial preoccupations as to any known personal crisis about the metaphysical implications of his theories. Although he was rich, he invariably worried about where his next penny was coming from. The ups and downs of his investments in railway stock, the tortuous ramifications of family trust funds, and profit-sharing arrangements with the publisher John Murray, for example, often coincided with his best-documented bouts of sickness. One very bad attack came a year or two after establishing his eldest son, William, as a partner in a privately owned bank in Southampton, which had required Darwin, as his father, to promise £10,000 as security against a run on deposits. In 1863, when Darwin became ill, an act of Parliament was passed allowing the establishment of joint-stock banking concerns that spread the risk and promised only limited liability for its members, a new state of affairs which jeopardized William's old-style partner-led syndicate. At the same time, railway amalgamations were creating vast, unregulated monopolies by swallowing up many of the smaller companies in which Darwin had a stake; and interest rates on cash capital that year were particularly unstable.[39] Darwin was a cautious investor, keen to avoid risks. The prospect of gambling his substantial fortune, and the future inheritances of his children, on the vagaries of the City of London and nebulous entities like public confidence was, to him, a very serious question. Only a few years before, his own London bank, the Union Bank, temporarily collapsed after an immense fraud carried out by a clever clerk.[40]

Alternatively, or simultaneously, his illnesses may well have acted as a mediator in married life—something we are closely attuned to in accounts of literary couples like Robert and Elizabeth Browning,

Thomas and Jane Carlyle, and Charles and Catherine Dickens but seemingly ignore when dealing with an evolutionist.[41] He might have been dismayed about getting old, or going bald, or about the lack of future occupations for his unhealthy sons and daughters. There was the Civil War in North America and the implications of continued slavery to worry about. If Darwin had been a famous *female* invalid, moreover, like Harriet Martineau, Florence Nightingale, or Ada Lovelace, our interpretation of the disorders would also be very different.[42] If he had been plain old Professor Airy, his illnesses would not be interesting in the wider sense at all. And have we really eliminated the possibility of a long-lasting subclinical problem or constellation of problems? It is not doing full justice to the richness and complexity of Darwin's situation to opt for one or another 'cause' without careful analysis of what we really wish to claim about medical embodiment, for richly diverse sets of discourses can envelop one and the same individual.

The Body

Nonetheless, Darwin was acutely aware of his body and all its failings. It was the primary focus of his attention. It was the focus of his friends' and doctors' attention. It dominated the domestic arrangements of his wife and family. It was something he told the readers of the *Origin* about. And he found it a useful device for avoiding tiresome social obligations and unpleasant scientific controversy. Darwin's ill health was doing a lot of 'work' in the modern sense. Yet at the same time it remains difficult to say precisely how this ill health materially contributed to his special genius. Personally, he recognized that too much study made him ill. But equally, he claimed that he was capable of dissociating himself from his physical disorders only through abstract thought.[43] In terms of the simple dualism that attracted and helped many Victorian invalids through their daily lives, Darwin, so to speak, invariably rose above the malfunctions of the flesh to engage in what his contemporaries celebrated as an extraordinarily active life of the mind. Where some intellectuals displayed their intellectuality by neglecting their bodies, Darwin's 'disembodiment' emerged out of a determined, and in time heroic, conquering of his inadequate frame. As all the essays in this volume variously show, this too is what we would call 'work'.[44]

Even so, few historians ask how far that idea of personal dissoci-
ation, the triumph of mind over matter, filtered into the public realm
and became part of what people thought about Darwin. What were
Victorian men and women offered in the way of visual information
about his mind and body? Did they see intellect or illness?

Judging from the mass-reproduced photographs available in archive
collections, they principally saw a well-to-do gentleman in dark, sober
suits, a gentleman with little regard for fashionable taste (Figs. 15 and
16). He always wears warm clothes, sometimes a cape or an overcoat
on top, a waistcoat underneath, perhaps a scarf draped over the shoulders, a felt hat. He is mostly sitting down. There are no symbolic props
to supply clues about the figure's special calling: no books open on the
knee, no microscope, no dogs or plants beside his chair, no spectacles
on the nose. His accoutrements, or his lack of accoutrements, tenta-
tively suggest a philosopher—a careworn and modest philosopher at
that.[45] Yet he could as easily be a member of any one of a number of
solidly prosperous Victorian professions: a university don or school-
teacher, a member of Parliament, a banker, a lord, or a country
gentleman.[46]

Such photographs are not often reproduced in twentieth-century
histories and biographies—they are not sufficiently striking to appeal
to modern tastes or perhaps do not fully resonate with what is now
expected of a Darwin illustration. But they were how Darwin presented
himself to his public. Figures 15 and 16, for example, were taken
expressly to cater to the surge of contemporary interest in portrait
photography and served a specific public purpose. Darwin, as much as
anyone, was gripped by the craze for exchanging cartes de visite with
his correspondents. By 1865 or so photographic technology in Britain,
France, and America had diversified sufficiently to allow the mass
production of studio portraits on small cards, often incorporating a
facsimile of the sitter's signature. Darwin posed several times for cards
like these and made use of them as a kind of autograph to send
through the post.

What is perhaps less well known is just how much commercial
activity surrounded the carte de visite business.[47] Professional photographers naturally supplied the sitter with a few packets of cards for
private use. But they also sold them for profit in their shops. The
Photographic Journal for 1862 recorded that one London studio was
selling £50 worth of portrait cards daily and that more than fifty
thousand items passed through the hands of another dealer in a single
month.[48] 'The public are little aware,' said a surprised author in *Once a*

ELLIOTT & FRY   Copyright   55, BAKER ST
PORTMAN SQ⁵

FIG. 15.   Charles Robert Darwin, Carte de Visite by Elliott and Fry, *c*.1878. These cartes de visite became hugely popular in the 1860s and brought likenesses of scientific authors (and many other individuals) before the Victorian public. The carte was sold in bulk to Darwin for his personal use, and was also available for public sale in the photographer's shop or studio. Note the copyright label. Courtesy of the Wellcome Trust Library, London.

FIG. 16.   Commercial photograph of Darwin taken by the London Stereoscopic and Photographic Company, *c.*1866. Darwin was ill when this photograph was taken. He started growing the beard in 1862. Again, his dress and pose indicate his respectability and gentlemanly status. The light catching the dome of his skull, his vigorous beard, and his contemplative look nevertheless suggest that the sitter is also a great thinker. Courtesy of the Wellcome Trust Library, London.

Week, 'of the enormous sale of the *cartes de visite* of celebrated persons. An order will be given by a wholesale house for 10,000 of one individual—thus £400 will be put into the lucky photographer's pocket who happens to possess the negative.'[49] Not surprisingly, a photograph of Darwin after the *Origin of Species* was published, was a sound commercial proposition. In October 1862, the photographer Polyblank (of Maull and Polyblank) wrote to Darwin, via Erasmus, asking for general permission to reproduce and sell the one taken by the firm some years earlier. Erasmus reported to Darwin that 'Polyblank says that for some he has a general order to sell & for others he requires special permission so I shd. think you might as well give a general order as it is a good photograph.'[50]

Furthermore, the shop windows where these pictures were displayed, said one literary magazine, were better than the National Portrait Gallery (itself only opened in 1859)—more egalitarian, for one thing, where an engineer could be seen beside the queen of Naples, or Mrs. Fry cheek by jowl with Lord Brougham. Or Huxley and Samuel Wilberforce, if we did but know it. Coming to the same conclusion from a different angle, the *London Review* ran an article entitled 'The New Picture Gallery' criticizing the lack of artistic merit in such cartes de visite.

Darwin resisted the craze for several years before capitulating to having his own cartes de visite made. Before then he sent out copies of a photograph taken by his son William, a keen amateur photographer whose hobby was financed by Darwin. But as the swapping and requests for pictures accumulated, he had his cartes made and updated them every few years thereafter. He also started his own album in 1864 for mounting the photographs sent in return by scientific friends.[51] In the process, Darwin became aware of his own value as a public image, though unassuming and diffident about it on most occasions.

Even the earliest nonphotographic portrait of him was relatively widely distributed. The well-known study by Thomas Maguire, a lithographic print taken from life in 1849, was from the outset a commercial project (Fig. 17).[52] This was one of a series of fifty or so portraits of scientists taken in honor of a British Association meeting at Ipswich in 1851. The series was conceived by George Ransome, the head of the agricultural machinery company and local secretary of the association meeting, as a paying venture to mark the Ipswich occasion and the opening of the Ipswich Museum.[53] Some sitters were drawn twice; and multiple prints and sets were offered for sale both at that time and later. Prints were still circulating in the commercial sense as

FIG. 17.   Lithograph of Darwin by Thomas Maguire for a series made for the British Association meeting at Ipswich 1851. Prints were produced in sufficient numbers to make the project a commercial one. Darwin's clothes and his pose emulate the masculine sobriety of his friend Charles Lyell, also pictured in the same series. Courtesy of the Wellcome Trust Library, London.

FIG. 18. Charles Lyell. Lithograph by Thomas Maguire for the British Association series. In these Maguire portraits Darwin and Lyell reveal little of their scientific calling except for the eyeglass each wears on a ribbon around his neck. Courtesy of the Wellcome Trust Library, London.

late as 1864 when Darwin was offered two 'most beautiful' proof impressions of his own portrait for 7s. 6d.[54]

In the Maguire portrait Darwin appears prosperous and confident, not at all ill in his outward appearance. However, more modish philosophical gentlemen of the period looked quite different, usually sporting a fashionably 'lank' hairstyle, a shortened form of frock coat, and a stock fastened with a tiepin.[55] In the same Maguire series, Edward Forbes, a romantic philosophical naturalist, adopted the latter style as a badge of his poetic predisposition, possibly also as a sign of his French and German scientific affiliations. In choosing the clothes and masculine pose that he did, Darwin patently aligned himself with the sturdy, no-nonsense faction of nineteenth-century scientific life represented by men such as Henslow, Sedgwick, and Lyell (Fig. 18).

The first commercial photograph of Darwin was taken in the studio of Henry Maull in 1855 or so—the first readily reproduced photographic image of him, so to speak. This was semipublished in the sense that Maull released it as part of a set of photographs under the title of *Literary and Scientific Portrait Club* (published in parts from around 1854). Darwin thought it made him look 'atrociously wicked',[56] and it is not known if he ever ordered any duplicates for private use: there are none in the Darwin archive at Cambridge, for example, and very few sets of the original publication have survived. The second photograph, taken a short while later by the same firm, was generally the one he preferred and this, once taken, became something to send out to friends. The general effect of the second photograph is strikingly polychromatic. Darwin wears a necktie, waistcoat, and trousers in the noisy checks which were greatly favored toward 1860, usually dubbed 'Great Exhibition' checks. Once again he did not bother with scientific props. Other men, perhaps less certain of their status, or with more avowedly polemic things to express, were less retiring. Richard Owen, the comparative anatomist, posed with a bone, Michael Faraday with his scientific apparatus and bench, Alexander von Humboldt in his study with traveling boxes and a large map of the world, Alfred Russel Wallace with a globe, and Ernst Haeckel amid a plentiful supply of natural history apparatus.

After the *Origin*, and with the major technical advances of the 1860s, commercial reproductions of photographs of Darwin proliferated. 'Such heaps of people want to know what you are like,' said Hooker. Yet he thought in general 'the photographs are not pleasing.'[57]

In 1866, for example, Darwin was asked if he would sit for Ernest

Edwards for a series of photographs and biographical memoirs edited by Lovell Reeve (a series soon taken over by Edward Walford, the genealogist from Balliol), called *Portraits of Men of Eminence* (1863–7), and he said he would be proud to do so. This book of portrait photographs was one of a number of lavishly produced albums issued around this time, mostly reissues and compilations of texts and previous studio studies by Walford and others.[58] He sat again for another photograph (or possibly it was taken at the previous sitting) to be included in a selection reprinted by Walford in 1868, called *Representative Men in Literature, Science and Art*, each print available separately priced at 1s. 6d. This one is interestingly different from the pre-*Origin* photographs. Darwin looked less confident, less well dressed, more anxious, more like an invalid, especially when the handle of the walking stick is glimpsed on the left. Eventually Darwin became quietly aware of the fact that photographers hoped to make money out of his face. In 1869, when George Charles Wallich proposed that he should be included in a new edition of his *Eminent Men of the Day*, he politely refused.[59]

Julia Margaret Cameron nevertheless managed to make capital out of him. Her famous photograph of Darwin, and its less famous mates (one forms the frontispiece to Richard Freeman's *Companion*), were taken by Cameron in July 1868, when Darwin was on holiday with his family and brother Erasmus on the Isle of Wight.[60] The Darwins occupied a villa rented from the Camerons in Freshwater Bay, the fashionable artistic center that Julia Cameron and Alfred Tennyson had between them created. The holiday was hardly a secluded rest. During those six weeks Darwin was visited by Tennyson, Longfellow, and Thomas Appleton, as well as socializing with the Camerons themselves. Mrs. Cameron often took the opportunity of photographing any celebrated visitors. 'She thinks it is a great honour to be done by her,' said one guest. 'Sitting to her was a serious affair, not to entered lightly upon . . . she expected much from her sitters.'[61] Mrs. Cameron also photographed Erasmus Darwin and Horace Darwin, Darwin's youngest son.

One of these photographs was the one Darwin liked better than any other portrait of him and he wrote a sentence to that effect. Eventually, either he or Cameron included the remark as a mechanically reproduced inscription at the bottom.[62] But Cameron was notorious for forcing her sitters into some kind of public endorsement along these lines, which she then used to promote sales of authenticated prints through Colnaghi's London gallery. The Cameron photograph was consequently just as much in the public arena as the cartes de visite Darwin employed—although more expensive and more artistic. It should further be noted that Darwin paid £4 7s. for this photograph, and other sums later on for various reproductions.[63] During her time in England, Cameron's photographic sales were virtually her only means of supporting herself and her husband.

They were all excellent portrait studies. Julia Cameron subscribed to the 'men of genius' school of thought, and her studies of other Victorian thinkers like Tennyson and Herschel showed a deep appreciation of masculine intellect. Her portraits of women and children made the reverse point rather more vividly, in that these rarely carried any proper names and the sitters were usually dressed or posed for some allegorical purpose—'Alethea', perhaps, or 'The gardener's daughter'. These sentimental *tableaux vivants* were often criticized and ridiculed in her own day. Furthermore, she tended to use a light color wash and give full rein to her special trick of fuzziness: 'very daring in style', said the *Photographic News* reviewing her first exhibition in 1864; 'out of focus', complained the *British Journal of Photography*.[64] These images are a marked contrast to her other, far more rugged and individually named projections of notable men.

Yet despite the air of biographical transparency, Cameron's male sitters were just as carefully posed as her female subjects—she made John Herschel wash and fluff up his hair for his sitting, and draped Robert Browning in a velvet cloak.[65] Darwin's costume was evidently his own and mostly conveys his careful precautions against the cold, especially on holiday on the British south coast in July. His dress was that of a respectable middle-aged gentleman: a man whose clothes signaled that, beyond going to a good tailor, he was relatively uninterested in clothes. On the whole, however, they are but a minor part of the composition. Because of her careful use of ceiling light, one could almost say that there is nothing in the photographs except Darwin's massive forehead, top-lit to emphasize the great dome of his skull, the brow creased in thought, and his luxuriant beard. As in classical paintings, the effect was of a softened, extremely wise, subject. More than anyone else Cameron created the visual image of Darwin as a great abstract mind.

Darwin's beard is one of the most interesting aspects of this kind of public representation. He grew it in the late summer of 1862 with the expressed intention of soothing his eczema: 'Mamma [Emma] says I am to wear a beard,' he told William in July. Constant shaving irritated his face. 'Charles, Emma and Lenny slept here on Monday,' Erasmus

Darwin wrote to Fanny Wedgwood some months later: 'Emma in a splendid wig, Lenny bald & Charles in a fine grey beard.'[66]

Its impact, however, was much more subtle and pervasive than a mere family event. When Darwin distributed a photograph of himself with this new asset (taken by his son William), Hooker replied immediately. 'Glorified friend! Your photograph tells me where Herbert got his Moses for the fresco in the House of Lords—horns & halo & all. . . . Do pray send me one for Thwaites, who will be enchanted with it. Oliver is calling out too for one.'[67] Funnily enough, said Darwin, his sons declared it made him look like Moses too. The botanist Asa Gray agreed. 'Your photograph with the venerable beard gives the look of your having suffered, and perhaps, from the beard, of having grown older. I hope there is still much work in you—but take it quietly and gently!'[68]

It was a philosopher's beard, as Cameron, Hooker, and Gray plainly saw, with strongly religious overtones. Darwin was delighted with the idea: 'Do I not look reverent?' he teased relatives. For himself, he hardly gave the possible motives for growing it or the symbolism of such a patristic outgrowth a second thought.[69] Yet at some fairly obvious level, it must have served as an external disguise on a par with his notorious personal shyness—a form of evading difficult confrontations by hiding behind a smokescreen of hair, not just a literal disguise, but a metaphysical one as well. The beard helped keep many of his thoughts private in the same way as his autobiographical writings avoided any penetrating self-analysis.[70] Such a beard allowed him, if he wished, to become a sage or a prophet. Or a sphinx.

This beard came to be featured more and more in the photographs and their subsequent reproductions in magazines through the 1870s and early 1880s (Figs. 15 and 16). As it got larger, and more patriarchal generally, it began to codify many of the things Victorians were told or thought about Darwin: it represented precisely the paradoxical fact that he was simultaneously a gentleman and a revolutionary, with distinguished philosophical antecedents in Plato and Socrates.[71] Such a beard made it clear that he was no fresh-faced radical, no dangerously groomed Frenchman.[72] It was reassuring in its religious demeanor, its benevolence, its suggestion of a wise father and patient friend. Such a beard hinted at hermits and holy men, even the apostles. Popular depictions of ancient Greek philosophers invariably emphasized the same features of beard and expansive forehead. Nor is it too far fetched to allude to Father Christmas, at the start of his mythical existence in Victoria and Albert's England. It conveyed sagacity. It was, moreover, a dramatically masculine beard, a very visual symbol of the real seat of Victorian power, and one of the most obvious outward results of what Darwin went on to describe as sexual selection among humans.[73] It was a gift to the cartoonists when they got to work on his theories of monkey ancestry. Above all, it signified intensely deep qualities of mind—those qualities of the Victorian masculine intellect that a mere set of whiskers could never hope to represent in a similar context.

By the 1880s, and the last two years of Darwin's life, virtually all that the public saw in published photographs and photogravures were his beard, his hat, and his eyes. Darwin, as a physical presence, had almost disappeared. All that was left was the intense impression of mind. The final photograph of his life, probably taken by Clarence E. Fry, the senior photographic partner of the firm of Elliot and Fry,[74] who must have visited Down House in 1880 or so and taken at least three different portrait shots of Darwin on the veranda, speaks powerfully of wisdom and frailty combined, a last evocative statement in the gradual, progressive sequence of Darwin's disengagement from his malfunctioning body.

The process of focusing in, as it were, on mind, or what Christopher Shilling calls an absent presence,[75] reached its apogee with representations of Darwin's study. The place where his knowledge was created seems to have become as interesting to Victorians as the mental attributes and personality of the man himself; and magazine articles about his life and times often featured pictures of his house, garden, greenhouse, and study, usually devoid of human figures. While making allowance for stylistic conventions in conveying domestic interiors, these empty places or spaces for generating knowledge suggest that Darwin's intellect was, by then, seen by the public as almost entirely disembodied. The room did not have Darwin in it. Instead, it is filled with signs of his mind at work—the plants, the papers, the books, the prints of scientific friends and family on the walls; and with signs of his unhealthy body—the fire, the shawl on the chair, the chaise longue. It would be quite a different kind of picture if Darwin were present, as in the engravings published in the *Illustrated London News* of Lubbock or Hooker bristling with activity at their desks. This picture depends on his absence.[76]

Through a long and arduously medicalized life, Darwin had created an idea of himself in which, at the end, he could be recognized—and venerated—by an empty room.

Notes

I gratefully acknowledge permission to reproduce material from the Syndics of Cambridge University Press, the publishers of the *Correspondence of Charles Darwin*, and from the Syndics of Cambridge University Library, where the vast majority of Darwin's papers and letters are held. The Darwin Correspondence Project has also kindly allowed me to use material that will be published in future volumes of the *Correspondence*. I am extremely grateful to the Wellcome Trust Library, London, for permission to cite manuscripts and to reproduce illustrative material from its collection; and to the Royal Society of London and to John Johnson Ltd.

1. Airy, *Autobiography*, 8–9.
2. Any survey would include the psychological and psychosomatic interpretations as presented by Hubble, 'Darwin and Psychotherapy'; idem, 'Life of the Shawl'; Keith, *Darwin Revalued*; Pickering, *Creative Malady;* Kempf, 'Charles Darwin'; and Colp's fine study, *To Be an Invalid*, which additionally surveys the preexisting literature. In *Charles Darwin*, John Bowlby opts for hyperventilation. Johnston, 'Ill-Health', discusses neurasthenia. Darwin's relations with his father are examined by Good, 'Life of the Shawl', and Greenacre, *Quest for the Father*. Chagas' disease is discussed by Adler, 'Darwin's Illness', and Bernstein, 'Darwin's Illness', and disputed by Woodruff, 'Darwin's Health', and Keynes, *Beagle Diary*, 263, 315. Multiple allergy is proposed by Smith, 'Darwin's Ill Health' and 'Darwin's Health Problems'; hypoglycemia by Roberts, 'Reflections'. Darwin would make a good subject for a historical analysis along the lines of Bynum and Neve, 'Hamlet on the Couch'.
3. Darwin, *Origin of Species*, 1. This sentence remained unchanged through all subsequent editions; see Peckham, *Origin: A Variorum Text*, 71. Ill health is not mentioned again in any of his publications until a footnote to the introduction to his *Variation under Domestication*, 1:2: 'the great delay in publishing this work has been caused by continued ill health.' Otherwise, Darwin rarely discussed his condition except in private correspondence, e.g., Burkhardt and Smith, *Correspondence*, vols. 1–11, *passim*. It is clear that he did not deliberately employ illness as a means of engaging public sympathy for his ideas, although the friends to whom he wrote undoubtedly made up a large proportion of the influential, elite scientific community. The image of him as an invalid seems to have been constructed more by the interweaving of private information, his own statements, his acquaintances' statements, and public interest. Reviewers certainly referred to his health in writing about the *Origin:* for example, W. B. Carpenter, 'Darwin on the Origin of Species', in Hull, *Darwin and His Critics*, 92 ('as soon as his imperfect health should permit'); Heinrich Bronn, review of the *Origin of Species*, in ibid., 124 ('The author's poor health'); and an anonymous reviewer in the *Athenaeum*, 19 November 1859, 659 ('sicklied o'er with the pale cast of thought'). Such ill health was codified, as it were, for the public in Darwin's autobiography, in which he talked frankly about continued sickness, first published in F. Darwin, *Life and Letters*, 1:26–107. This was accompanied by Francis Darwin's reminiscences, which also include references to his father's ill health, ibid., 108–60, esp. 159–60.
4. Described, for example, in Morris, *Culture of Pain*, and Taylor, *Sources of the*

Self. For the Darwin family's ill health, see F. Darwin, *Life and Letters*, 1:159–60, and Raverat, *Period Piece.*

5. Outram, 'Body and Paradox'. See also her *Body and the French Revolution.*

6. Medical notes supplied by Darwin to Dr. John Chapman, 16 May 1865, University of Virginia Library, transcribed in Colp, *To Be an Invalid,* 83–4. By 'rocking' Darwin means the motion of horse-drawn carriages and railway trains: a nineteenth-century version of travel sickness.

7. Darwin to Chapman, 7 June 1865, University of Virginia Library.

8. Darwin to W. D. Fox, 24 October 1852, transcribed in Burkhardt and Smith, *Correspondence,* 5:100; Darwin to J. D. Hooker, 17 June 1847, ibid., 4:51.

9. Darwin's symptoms, and the various remedies and treatments he tried over the years, are fully described by Colp, *To Be an Invalid,* esp. 109–44.

10. Darwin to Hooker, 27 January 1864, Cambridge University Library, Darwin Archive (hereafter DA) 115:217. For illness as an alien presence, see particularly Leder, *Absent Body,* and Shilling, *Body and Social Theory.*

11. Darwin to Hooker, 5 December 1863, DA 115:213.

12. Hooker to Darwin, 5 February 1864, DA 101:180; Darwin to Hooker, DA 115:219.

13. Burkhardt and Smith, *Correspondence,* vols. 1–9, which cover the years 1821 to 1861. The entire correspondence is listed in summary form in Burkhardt and Smith, *Calendar.* See also F. Darwin, *Life and Letters,* and Darwin and Seward, *More Letters.*

14. Nausea (including seasickness), according to Chapman's theories, was caused by a rush of blood to the spinal cord. See Chapman, *Sea Sickness* and *Neuralgia,* for what he calls neuro-dynamic medicine, and Haight, *George Eliot and John Chapman,* 113–16.

15. Emma Darwin to Hooker, 18 July 1865, DA 115:272v.

16. From September 1862 to September 1863, for example, his expenditure on medical treatment was £129 11s. 6d. Compare this with around £50 for 'Science', £10 for 'Books', and £158 for 'Manservants'. The following year was less expensive at £56 4s. 10d. Darwin's Classified Account Books, Down House Archives, Kent.

17. Litchfield, *Emma Darwin,* 2:142.

18. Brinton was prominent in the field of stomach disorders. Darwin recorded paying ten guineas for a consultation on 22 November 1862 (Account Book, Down House Archives). See also Darwin to Hooker, 10 November 1863, DA 115:208.

19. Jones, *On Gravel,* discusses the role of diet in diminishing nonnitrogenous principles in the blood. See also his *Animal Chemistry.* Holland, *Medical Notes,* 239–69, discusses gout more generally. 'Latent' gout was a common diagnosis in midcentury and was understood as a metabolic disorder characterized by raised uric acid in the blood with few obvious clinical features; see Colp, *To Be an Invalid,* 109–12, 236, and Porter and Rousseau, *Gout.*

20. Holland, *Medical Notes,* 258–69, and Scudamore, *Colchicum Autumnale.* On colchicum's injurious effects, see Rennie, *Observations on Gout.* William Jenner prescribed 'enormous quantities of chalk, magnesia & carb. of ammonia'; Darwin to Hooker, 13 April 1864, DA 115:229.

21. Gully, *The Water Cure,* and Browne, 'Spas and Sensibilities'.

22. Colp, *To Be an Invalid,* and Burkhardt and Smith, *Correspondence,* vols. 6

and 7, *passim*. For general accounts of hydrotherapy, see Metcalfe, *Hydropathy*, Turner, *Taking the Cure*, and Rees, 'Water as a Commodity'.

23. Busk to Darwin, October 1863, DA 170.

24. Goodsir to Darwin, 21 August 1863, 26 August 1863, and 28 August 1863, DA 165.

25. Price, *Critical Inquiry*.

26. Henry Bence Jones to Darwin, 10 February 1866, DA 168. See also Jones to Emma Darwin, 1 October 1867, and Jones to Darwin, 2 August 1870, DA 168.

27. Clark to Darwin, 8 July 1876, DA 210.21.

28. Detailed analyses of the interrelations between doctors and patients are given by Peterson, *The Medical Profession*; Porter, *Patients and Practitioners*; Digby, *Making a Medical Living*; and Oppenheim, 'Shattered Nerves'.

29. Clark to Darwin, 3 September 1873, DA 161. In the same letter he reports that Darwin's urine was loaded with uric acid but showed no albumin—a favorable diagnosis.

30. Darwin to Paget, 4 June 1870, Wellcome Trust Library, London, Western MS 5703, item 38. For Darwin's theories of expression, see his *Expression of the Emotions*.

31. Pickering, *Creative Malady*, 71, 77–80, Colp, *To Be an Invalid*, 122–6, 141–4. Useful accounts in this area are given by Berrios, 'Obsessional Disorder'; Bynum, 'Rationales for Therapy'; Inglis, *Diseases of Civilization*; Lopez Pinero, *Neurosis*; and Sicherman, 'Uses of Diagnosis'. See also Wiltshire, *Jane Austen and the Body*, and Bailin, *The Sickroom in Victorian Fiction*.

32. Allingham, *Diary*, 185.

33. George Darwin's reminiscences, DA 112 (ser. 2): 23.

34. Bartholomew, 'Copley Medal', and MacLeod, 'Of Medals and Men'. See also F. Darwin, *Life and Letters*, 3:27–28; and Erasmus Alvey Darwin to Darwin, 9 November 1863, DA 105 (ser. 2): 13.

35. E. A. Darwin to Darwin, undated 1864, DA 105 (ser. 2): 33. The proposal and citation were prepared by Falconer, who wished to mention Darwin's illness: 'Dr Falconer . . . wants dates of your voyage. Also, what I should think would not be judicious to bring forward, when your sickness came on & how long it lasted & whether in consequence of it FitzRoy persuaded you to give up the voyage.' E. A. Darwin to Darwin, 27 June 1864, DA 105 (ser. 2): 28. Darwin's health was not mentioned in the published citation for the award.

36. Darwin to Hooker, 23 April 1861, DA 115:98.

37. Burkhardt and Smith, *Correspondence*, 2:314.

38. E. A. Darwin to Frances Wedgwood. Wedgwood/Mosely Collection, Keele University Library.

39. English law confined the number of partners in a bank to six until the Joint Stock Banks and Companies Act of 1863 allowed some redistribution of the risk in the wake of the crashes of 1857–8. See Anderson and Cottrell, *Money and Banking*. For railways, see Barker and Savage, *An Economic History of Transport*, 87.

40. These financial details are drawn from the *Annual Register*, 1863.

41. See, for example, Markus, *Dared and Done*, and Checkland, *The Gladstones*. General analyses are in Wohl, *The Victorian Family*; Graham, *Women, Health and the Family*; Peterson, *Family Love*; Davidoff and Hall, *Family Fortunes*. In some ways Darwin and his wife reversed the customary roles of the 'weak'

woman and 'strong' man while still expressing much of their relationship through the preoccupations of invalid and nurse. For comments on the ways in which ideas of male health were based on self-control and of female ill health on an inability to control the body, see Shuttleworth, 'Female Circulation'.

42. For example, Cooter's study of Martineau, 'Dichotomy and Denial', Pickering on Nightingale, *Creative Malady*, 122–77, and Micale, 'Hysteria Male/Hysteria Female'. See also Edel, *Diary of Alice James*, and Trombley, *All That Summer She Was Mad*. Interpretations of women's diseases are discussed in Bailin, *The Sickroom in Victorian Fiction*, 17–19; Digby, 'Women's Biological Strait-jacket'; Wood, 'Fashionable Diseases'; Ehrenreich and English, *Complaints and Disorders*; and Figlio, 'Chlorosis and Chronic Disease'. See also Bynum, 'The Nervous Patient'; Micale, *Approaching Hysteria*; and Oppenheim, '*Shattered Nerves*'. Authoritative studies of bodies and gender can be found in Ehrenreich and English, *For Her Own Good*; Schiebinger, *Nature's Body*; Vicinus, *Suffer and Be Still*; Gallagher and Laqueur, *The Making of the Modern Body*; Laqueur, *Making Sex*; and Jordanova, *Sexual Visions*.

43. Leder, *Absent Body*.

44. See particularly Haley, *Healthy Body*; Goffman, *Presentation of Self*; Gilman, *Disease and Representation*; and idem, *Health and Illness*. General studies of the field are Turner, *Body and Society*; and Porter, 'History of the Body'. See also Outram, 'Body and Paradox'.

45. Clarke, *The Portrait in Photography*; Linkman, *The Victorians*; Piper, *Personality and the Portrait*; Fyfe and Law, *Picturing Power*; and Tagg, *The Burden of Representation*. Other forms of visual presentation are discussed in Adler and Pointon, *The Body Imaged*; Stafford, *Body Criticism*; Porter, 'Bodily Functions'; Lynch and Woolgar, *Representation in Scientific Practice*; Goffman, *Presentation of Self*; Gilman, *Seeing the Insane*; and Cowling, *The Artist as Anthropologist*. Symbolism in early portraiture is analyzed in Gent and Llewellyn, *Renaissance Bodies*. Photographic imagery is thoroughly discussed in Fox and Lawrence, *Photographing Medicine*; Edwards, *Anthropology and Photography*; and Weaver, *British Photography*.

46. Cunnington and Cunnington, *Handbook of English Costume*. The clothes of less prosperous groups are illustrated in Cunnington and Lucas, *Occupational Costume*. See also Hollander, *Seeing through Clothes*; and Harvey, *Men in Black*.

47. The rise of commercial portrait photography is discussed by Bolas, *The Photographic Studio*; Prescott, 'Fame and Photography'; and Darrah, *Cartes de Visite*. See also Lee, 'Victorian Studio'; Pritchard, 'Commercial Photographers'; and Welford, 'Cost of Photography'. In 1856 Maull and Polyblank charged 5s. for an albumin print, 8×6 inches. Three cartes de visite cost 2s. 6d. from Ernest Edwards in the mid-1860s.

48. 'Miscellanea', *The Photographic Journal*, 15 March 1862, 21.

49. Wynter, 'Cartes de Visite', 376. See also 'Commercial Photography', *British Journal of Photography*, 2 January 1867, 47. According to anecdote, five thousand portraits of John Wilkes Booth were sold after the assassination of Abraham Lincoln.

50. E. A. Darwin to Darwin, undated, October 1862, DA 105 (ser. 2): 9. It is not clear to which photograph he refers although judging from the wide

circulation of a number of subsequent reissues, it was probably the one with checked trousers. See note 56.

51. The whereabouts of this album is unknown.

52. T. H. Maguire, the Irish painter and lithographer, was appointed lithographer to the queen in 1851 and subsequently took portraits of Prince Albert and the royal children. Throughout his career he exhibited genre scenes at the Royal Academy, eventually issuing *The Art of Figure Drawing* in 1869.

53. There is no record of a direct payment to Maguire in Darwin's Account Book of 1849–51, although he recorded payment, on subscription, to the portraits of the bishop of Norwich and George Ransome in the same series. A sum of £1 6d. was paid to Lovell Reeve, the publishers of the series, on 17 November 1849.

54. E. A. Darwin to Darwin, 9 April 1864, DA 105 (set. 2): 25.

55. Cunnington and Cunnington, *Handbook of English Costume*; and Lurie, *The Language of Clothes*. Shortland, 'Bonneted Mechanic', addresses the theme of dress in greater detail.

56. This set of photographs was issued on subscription, loose in a folder. A copy is preserved at the National Portrait Gallery, London. The exact date of the first photograph is difficult to ascertain. Freeman, *Companion*, 97, gives a tentative date of c. 1854. However, a sitting to Maull in 1855 is apparently confirmed in Burkhardt and Smith, *Correspondence*, 5:339, with a payment to Maull and Polyblank of £3 1s. 6d. in Darwin's Account Book, 31 December 1855, Down House Archives. The set is discussed in Prescott, 'Fame and Photography'. Some later engravings of this photograph incorporate the date 1854: this date cannot be independently substantiated. Francis Darwin thought the second photograph was also taken in 1854 and reproduced it as such as the frontispiece to *Life and Letters*, vol. 1. For the dates of Maull's various partnerships, see Pritchard, *Directory of London Photographers*.

57. Hooker to Darwin, 24 January 1864, DA 101:176.

58. A copy of this Edwards photograph is in the National Portrait Gallery, London, neg. 28523. Prescott describes the publication of several such series, e.g., Mason and Co., *The Bench and the Bar,* and *The Church of England Portrait Gallery*, from 1858 to 1861; and J. E. Mayall, *Royal Album*, 1860. Darwin's picture is in Reeve and Walford, *Portraits of Men of Eminence*, vol. 5, opposite p. 49. A fee of £1 for Ernest Edwards, the photographer for this volume, is recorded in Darwin's Account Book, 2 March 1866, and another sum of £3 8s. 6d. on 5 September.

59. Wallich, *Eminent Men*, issued in 1870. Wallich knew Darwin through his natural-history researches; see Darwin to Wallich, 18 April 1869, American Philosophical Society Library. Later, Darwin asked Wallich for photographs that he could use in his *Expression of the Emotions*; Darwin to Wallich, 24 February 1872, Cleveland Health Sciences Library, Ohio; and Darwin to Wallich, 20 March 1872, Northumberland Record Office. These arrangements suggest a certain amount of collaboration between sitters and photographers/publishers. Celebrity photography clearly generated its own rules.

60. F. Darwin, *Life and Letters*, 3:92, 102; and Litchfield, *Emma Darwin*, 2:220–2.

61. Hopkinson, *Julia Margaret Cameron*; and Hinton, *Immortal Faces*, 33. See also Cameron, *Alfred, Lord Tennyson and His Friends*; Weaver, *Julia Margaret Cameron*; and Woolf and Fry, *Victorian Photographs*. For Cameron's

photographing of Erasmus Alvey Darwin and Horace Darwin, see Litchfield, *Emma Darwin*, 2:220. Cameron was not the only art photographer to photograph Darwin. In 1871 Darwin approached Oscar Rejlander for help with his studies on the expression of the emotions and established a close working relationship with him. See Browne, 'Darwin and Expression of the Emotions', and Jones, *Father of Art Photography*. Rejlander made a fine portrait study of Darwin in or around 1871, ultimately engraved and reproduced in *Nature*, 4 June 1874. Rejlander taught Cameron and C. L. Dodgson something of their technique.

62. The Royal Society copy, on the original Colnaghi mount, with blind stamp, is the only copy I have seen with this mechanically reproduced text at the bottom.

63. F. Darwin, *Life and Letters*, 3:102; and Classified Account Books, 19 August 1868, Down House Archives.

64. *Photographic News*, 8 (3 June 1864): 266; *British Journal of Photography*, 11(1864): 261.

65. Hopkinson, *Cameron*, 68.

66. Darwin to William Darwin, 4 July 1862, DA, 210.6; and E. A. Darwin to F.M. Wedgwood, 1 October 1862, Wedgwood/Mosely Collection, Keele University Library. Although Darwin may well have shaved intermittently, he had a full beard when his friends saw him again in 1864. He earlier sported a large black beard on the *Beagle* while surveying in Tierra del Fuego; see Browne, *Charles Darwin*, 217, 246. With regard to the wig, Emma and Leonard Darwin suffered from scarlet fever that summer and had been shaved during the skin-peeling period.

67. Hooker to Darwin, 11 June 1864, DA 101:225. The attribution to William is in a letter from Darwin to Asa Gray, 28 May 1864, Harvard University, Gray Herbarium, 79.

68. As a Gray to Darwin, 11 July 1864, DA 165.

69. As they became more common after the Crimean War, beards were often discussed in Victorian literature, e.g., Hannay, 'The Beard'. Something of the history of beards is given by Asser, *Historic Hairdressing*; Cooper, *Hair*; Corson, *Fashions in Hair*, esp. 398–461; and Reynolds, *Beards*.

70. See Neve, *Charles Darwin's Autobiography*, introduction; and Colp, '"I Was Born a Naturalist"'; and idem, 'Notes on Darwin's *Autobiography*'. Berg, *Unconscious Significance of Hair*, summarizes the psychoanalytic view.

71. Constable, 'Beards in History', surveys beards and hair from Antiquity to the Middle Ages in the West. See also Reynolds, *Beards*, 48–9. Pliny speaks of the respect and fear inspired by the beard of Euphrates, a Syrian philosopher. The Roman satirists more usually ridiculed the relationship between beards and wisdom, e.g., Herod Atticus, 'I see the beard and the cloak, but I do not see the philosopher.' Williams, *The Hairy Anchorite*, discusses the religious symbolism.

72. Reynolds, *Beards*, 267–8, on the seditious mustache; and Pointon, 'Dirty Beau'. In 1854, Hannay, 'The Beard', 49, stated that the beard was at that time the symbol of 'revolution, democracy and dissatisfaction with existing institutions . . . only a few travellers, artists, men of letters and philosophers wear it.' Francis II of Naples forbade beards because of their association with Garibaldi.

73. Darwin, *Descent of Man*, 2:317–23, 372, 379–80. See also Mangen and Walvin, *Manliness and Morality*; Roberts, 'Paterfamilias'; and Shortland, 'Bonneted Mechanic'. Masculinity in art is discussed by Kestner, *Mythology and Misogyny*; and idem, *Masculinities in Victorian Painting*; and in science by Richards, 'Darwin and the Descent of Woman'; and idem, 'Huxley and Women's Place in Science'.

74. Hillier, *Victorian Studio Photographs*.

75. Shilling, *Body and Social Theory*. The idea of 'absent presence' relates in some degree to Foucault, *The History of Sexuality*, introduction; and Ostrander, 'Foucault's Disappearing Body'. The theme of death in visual culture is discussed in Llewellyn, *The Art of Death*. See also Elias, *The Loneliness of the Dying*.

76. One comparable scene would be Freud's study, of which photographs were issued in 1938, reproduced in Engleman, *Berggasse 19*. Discussions of the creation of intellectual spaces can be found in Smith, *Making Space*; Livingstone, 'Spaces of Knowledge'; Shapin, 'The Mind Is Its Own Place'; and Ophir and Shapin, 'Place of Knowledge'. The history of studies as places for generating knowledge is much neglected in the literature, although it is addressed briefly in Ophir and Shapin; Girouard, *Life in the English Country House*; and Thornton, *Authentic Decor*. Thomson, 'Some Reminiscences', reproduces interesting photographs. Such private masculine spaces (for smoking, business papers, writing, and reading) were seemingly a Victorian development that went hand in hand with the diversification of room use and the division of labor in larger country houses, and are not as clearly related to the traditional use of academic spaces like libraries, monastic cells, and college rooms as we might perhaps expect. For Darwin's use of his study as a place of experiment, see Chadarevian, 'Laboratory Science'. See also Marsh, *Writers and Their Houses*.

References

Adler, Kathleen, and Pointon, Marcia, eds. *The Body Imaged: The Human Form and Visual Culture since the Renaissance*. Cambridge: Cambridge University Press, 1993.

Adler, Saul. 'Darwin's Illness'. *Nature*, 184 (1959): 1102–3.

Airy, George B. *Autobiography of Sir George Biddell Airy*. Ed. Wilfred Airy. Cambridge: Cambridge University Press, 1896.

Allingham, William. *A Diary*. Ed. Helen Allingham and D. Radford. London: Macmillan, 1907.

Anderson, Bruce L., and Cottrell, Philip L. *Money and Banking in England: The Development of the Banking System, 1694–1914*. Newton Abbot: David and Charles, 1974.

Asser, Joyce. *Historic Hairdressing*. London: Sir Isaac Pitman, 1966.

Bailin, Miriam. *The Sickroom in Victorian Fiction: The Art of Being Ill*. Cambridge: Cambridge University Press, 1994.

Barker, Theodore C., and Savage, Christopher I. *An Economic History of Transport in Britain*. London: Hutchinson, 1974.

Bartholomew, Michael. 'The Award of the Copley Medal to Charles Darwin'. *Notes and Records of the Royal Society of London*, 30 (1976): 209–18.

Benjamin, Marina, ed. *Science and Sensibility: Gender and Scientific Enquiry, 1780–1945*. Oxford: Blackwell, 1991.

Berg, Charles. *The Unconscious Significance of Hair*. London: George Allen and Unwin, 1951.

Bernstein, Ralph B. 'Darwin's Illness: Chagas Disease Resurgens'. *Journal of the Royal Society of Medicine*, 77 (1984): 608–9.

Berrios, Germen E. 'Obsessional Disorders during the Nineteenth Century: Terminological and Classificatory Issues', in Bynum, Porter, and Shepherd, 1:166–87.

Bolas, Thomas. *The Photographic Studio*. London: Marion and Co., 1895.

Bowlby, John. *Charles Darwin: A Biography*. London: Hutchinson, 1990.

Browne, Janet. *Charles Darwin: A Biography*. Vol. 1, *Voyaging*. London: Jonathan Cape; New York: Alfred Knopf, 1995.

—— 'Darwin and the Expression of the Emotions', in *The Darwinian Heritage*, ed. David Kohn, 307–26. Princeton: Princeton University Press in association with Nova Pacifica, 1985.

—— 'Spas and Sensibilities: Darwin at Malvern', in *The Medical History of Waters and Spas*, ed. William F. Bynum and Roy Porter. *Medical History*, supplement, 10 (1990): 102–13.

Burkhardt, Frederick H., and Smith, Sydney, eds. *Calendar of the Correspondence of Charles Darwin*. Rev. edn. Cambridge: Cambridge University Press, 1994.

—— eds. *The Correspondence of Charles Darwin*. 11 vols. to date. Cambridge: Cambridge University Press, 1983– .

Bynum, William F. 'The Nervous Patient in Eighteenth and Nineteenth Century Britain: The Psychiatric Origins of British Neurology', in Bynum, Porter, and Shepherd, 1:88–102.

—— 'Rationales for Therapy in British Psychiatry, 1785–1830'. *Medical History*, 18 (1974): 317–34.

Bynum, William F., and Neve, Michael. 'Hamlet on the Couch', in Bynum, Porter, and Shepherd, 1:289–304.

Bynum, William F., Porter, Roy and Shepherd, Michael, eds. *The Anatomy of Madness: Essays in the History of Psychiatry*. 3 vols. London: Tavistock, 1985.

Cameron, Henry H. H. *Alfred, Lord Tennyson and His Friends*. With 25 portraits by Julia Cameron. London: T. Fisher Unwin, 1893.

Chadarevian, Soraya de. 'Laboratory Science versus Country House Experiments: The Controversy between Julius Sachs and Charles Darwin'. *British Journal for the History of Science*, 29 (1996): 17–41.

Chapman, John. *Neuralgia and Kindred Diseases of the Nervous System*. London: Churchill, 1873.

—— *Sea Sickness: Its Nature and Treatment*. London: Trubner, 1864.

Checkland, Sydney G. *The Gladstones: A Family Biography, 1764–1851.* Cambridge: Cambridge University Press, 1971.

Clarke, Graham, ed. *The Portrait in Photography.* London: Reaktion Books, 1994.

Colp, Ralph. '"I Was Born a Naturalist": Charles Darwin's 1838 Notes about Himself'. *Journal of the History of Medicine,* 35 (1980): 8–39.

—— 'Notes on Charles Darwin's *Autobiography*'. *Journal of the History of Biology,* 18 (1985): 357–401.

—— *To Be an Invalid: The Illness of Charles Darwin.* Chicago: University of Chicago Press, 1977.

Constable, Giles. 'Beards in History', in *Apologia duae,* ed. Robert B. C. Huygens, 47–56. Corpus Christianorum. Continuatio Medievalis 62. Turnhout, Belgium: Brepols, 1985.

Cooper, Wendy. *Hair: Sex, Society, Symbolism.* London: Aldus Books; New York: Stein and Day, 1971.

Cooter, Roger. 'Dichotomy and Denial: Mesmerism, Medicine and Harriet Martineau', in Benjamin, 144–73.

Corson, Richard. *Fashions in Hair: The First Five Thousand Years.* London: Peter Owen, 1965.

Cowling, Margaret. *The Artist as Anthropologist: The Representation of Type and Character in Victorian Art.* Cambridge: Cambridge University Press, 1989.

Cunnington, Cecil W., and Cunnington, Phillis. *Handbook of English Costume in the Nineteenth Century.* 3d edn. London: Faber and Faber, 1970.

Cunnington, Phillis, and Lucas, Catherine. *Occupational Costume in England.* London: Black, 1967.

Dale, Philip M. *Medical Biographies: The Ailments of Thirty-Three Famous Persons.* Norman: University of Oklahoma Press, 1952.

Darrah, William C. *Cartes de Visite in Nineteenth Century Photography.* Gettysburg, Pa.: W. C. Darrah, 1981.

Darwin, Charles. *The Descent of Man and Selection in Relation to Sex.* 2 vols. 1871. Facsimile, with an introduction by John T. Bonner and R. M. May, Princeton: Princeton University Press, 1981.

—— *The Expression of the Emotions in Man and Animals.* London: John Murray, 1872. Reprinted with an introduction by Konrad Lorenz, Chicago: University of Chicago Press, 1965.

—— *On the Origin of Species by Means of Natural Selection: A Facsimile of the First Edition.* 1859. Reprinted with an introduction by Ernst Mayr, Cambridge, Mass.: Harvard University Press, 1964.

—— *Variation of Animals and Plants under Domestication.* 2 vols. London: John Murray, 1868.

Darwin, Francis, ed. *The Life and Letters of Charles Darwin, Including an Autobiographical Chapter.* 3 vols. London: John Murray, 1887.

Darwin, Francis, and Seward, Albert C. eds. *More Letters of Charles Darwin.* 2 vols. London: John Murray, 1903.

Davidoff, Leonore, and Hall, Catherine. *Family Fortunes: Men and Women of the English Middle Class 1780–1850*. Chicago: University of Chicago Press, 1987.

Desmond, Adrian, and Moore, James. *Darwin*. London: Michael Joseph, 1992.

Digby, Anne. *Making a Medical Living: Doctors and Patients in the English Market for Medicine, 1720–1911*. Cambridge: Cambridge University Press, 1994.

—— 'Women's Biological Straitjacket', in *Sexuality and Subordination: Interdisciplinary Studies of Gender in the Nineteenth Century*, ed. Susan Mendus and Jane Rendell, 192–220. London: Routledge, 1989.

Edel, Leon, ed. *The Diary of Alice James*. Harmondsworth: Penguin, 1982.

Edwards, Elizabeth, ed. *Anthropology and Photography, 1860–1920*. New Haven: Yale University Press, 1992.

Ehrenreich, Barbara, and English, Deidre. *Complaints and Disorders: The Sexual Politics of Sickness*. London: Writers and Readers Publishing Cooperative, 1973.

—— *For Her Own Good: 150 Years of the Experts' Advice to Women*. London: Pluto, 1979.

Elias, Norbert. *The Loneliness of the Dying*. Oxford: Blackwell, 1985.

Engleman, Edmund. *Berggasse 19: Sigmund Freud's Home and Offices, Vienna 1938*. New York: Basic Books, 1976.

Figlio, Karl. 'Chlorosis and Chronic Disease in Nineteenth-Century Britain: The Social Constitution of Somatic Illness in a Capitalist Society'. *Social History*, 3 (1978): 167–97.

Foucault, Michel. *The History of Sexuality*. Vol. 1, *Introduction*. Harmondsworth: Penguin, 1981.

Fox, Daniel, and Lawrence, Christopher. *Photographing Medicine: Images and Power in Britain and America since 1840*. New York: Greenwood Press, 1988.

Freeman, Richard. *Charles Darwin: A Companion*. Folkestone: Dawson, 1978.

Fyfe, Gordon, and Law, John eds. *Picturing Power: Visual Depiction and Social Relations*. Sociological Review Monographs, no. 35. London: Routledge, 1988.

Gallagher, Catherine, and Laqueur, Thomas eds. *The Making of the Modern Body: Sexuality and Society in the Nineteenth Century*. Berkeley and Los Angeles: University of California Press, 1987.

Gent, Lucy, and Llewellyn, Nigel eds. *Renaissance Bodies: The Human Figure in English Culture c. 1540–1660*. London: Reaktion Books, 1990.

Gilman, Sander. *Disease and Representation: Images of Illness from Madness to AIDS*. Ithaca, NY: Cornell University Press, 1988.

—— *Health and Illness: Images of Difference*. London: Reaktion Books, 1995.

—— *Seeing the Insane*. New York: Brunner, Mazel, 1982.

Girouard, Mark. *Life in the English Country House*. Harmondsworth: Penguin, 1980.

Goffman, Erving. *The Presentation of Self in Everyday Life*. 2nd edn. London: Allen Lane, 1969.

Good, Rankine. 'The Life of the Shawl'. *Lancet*, i (1954): 106–7.

Graham, Hilary. *Women, Health and the Family*. Brighton: Harvester Press, 1984.

Greenacre, Phyllis. *The Quest for the Father: A Study of the Darwin–Butler Controversy as a Contribution to the Understanding of the Creative Individual*. New York: International Universities Press, 1963.

Gully, James M. *The Water Cure in Chronic Disease*. London: J. Churchill, 1842.

Haight, Gordon. *George Elliot and John Chapman: With Chapman's Diaries*. New Haven: Yale University Press, 1940.

Haley, Bruce. *The Healthy Body and Victorian Culture*. Cambridge, Mass.: Harvard University Press, 1978.

Hannay, James. 'The Beard'. *Westminster Review*, 62 (1854): 48–67.

Harvey, John. *Men in Black*. London: Reaktion Books, 1995.

Hillier, Bevis. *Victorian Studio Photographs from the Collections of Studio Bassano and Elliott and Fry*. Boston: D. R. Godine, 1976.

Hinton, Brian. *Immortal Faces: Julia Margaret Cameron on the Isle of Wight*. Newport, Hants: Isle of Wight County Press, 1992.

Holland, Henry. *Medical Notes and Reflections*. 3rd edn. London: Longman, 1855.

Hollander, Anne. *Seeing through Clothes*. Berkeley and Los Angeles: University of California Press, 1975.

Hopkinson, Amanda. *Julia Margaret Cameron*. London: Virago, 1986.

Hubble, Douglas. 'Charles Darwin and Psychotherapy'. *Lancet*, i (1943): 129–33.

—— 'The Life of the Shawl'. *Lancet*, ii (1953): 1351–4.

Hull, David L., ed. *Darwin and His Critics: The Reception of Darwin's Theory of Evolution by the Scientific Community*. Cambridge, Mass.: Harvard University Press, 1973.

Inglis, Brian. *The Diseases of Civilization*. London: Hodder and Stoughton, 1981.

Johnston, W. W. 'The Ill-Health of Charles Darwin: Its Nature and Its Relation to His Work', *American Anthropologist*, n.s., 3 (1901): 139–58.

Jones, Edgar Yoxall. *Father of Art Photography: O. G. Rejlander, 1813–1875*. Newton Abbot: David and Charles, 1973.

Jones, Henry Bence. *On Animal Chemistry in its Application to Stomach and Renal Diseases*. London: John Churchill, 1850.

—— *On Gravel, Calculus and Gout*. London: Taylor and Walton, 1842.

Jordanova, Ludmilla. *Sexual Visions: Images of Gender in Science and Medicine between the Eighteenth and Twentieth Centuries*. Madison: University of Wisconsin Press, 1989.

Keith, Arthur. *Darwin Revalued*. London: Watts, 1955.

Kempf, Edward. 'Charles Darwin: The Affective Sources of His Inspiration and Anxiety Neurosis'. *Psychoanalytic Review*, 5 (1918): 151–92.

Kestner, Joseph. *Masculinities in Victorian Painting*. London: Scolar Press, 1995.

—— *Mythology and Misogyny: The Social Discourse of Nineteenth-Century British Classical-Subject Painting*. Madison: University of Wisconsin Press, 1989.

Keynes, Richard D., ed. *Charles Darwin's Beagle Diary*. Cambridge: Cambridge University Press, 1988.

Laqueur, Thomas. *Making Sex: Body and Gender from the Greeks to Freud*. Cambridge, Mass.: Harvard University Press, 1990.

Leder, Drew. *The Absent Body*. Chicago: University of Chicago Press, 1990.

—— ed. *The Body in Medical Thought and Practice*. Dordrecht: Kluwer Academic Press, 1992.

Lee, David. 'Victorian Studio'. *British Journal of Photography*, 133 (1986): 152–65, 188–99.

Linkman, Audrey E. *The Victorians: Photographic Portraits*. London: Tauris Parke Books, 1990.

Litchfield, Henrietta E., ed. *Emma Darwin, Wife of Charles Darwin: A Century of Family Letters*. 2 vols. Cambridge: privately printed at Cambridge University Press, 1904.

Livingstone, David. 'The Spaces of Knowledge: Contributions towards a Historical Geography of Science'. *Environment and Planning D: Society and Space*, 13 (1995): 5–34.

Llewellyn, Nigel. *The Art of Death: Visual Culture in the English Death Ritual, c.1500–c.1800*. London: Reaktion Books, 1994.

Lopez Pinero, Jose M. *Historical Origins of the Concept of Neurosis*. Trans. D. Berrios. Cambridge: Cambridge University Press, 1983.

Lurie, Alison. *The Language of Clothes*. New York: Random House, 1981.

Lynch, Michael, and Woolgar, Steve eds. *Representation in Scientific Practice*. Cambridge, Mass.: MIT Press, 1990.

MacLeod, Roy. 'Of Medals and Men: A Reward System in Victorian Science, 1826–1914'. *Notes and Records of the Royal Society of London*, 26 (1971): 81–105.

Mangan, James, and Walvin, James eds. *Manliness and Morality: Middle-Class Masculinity in Britain and America 1800–1940*. Manchester: Manchester University Press, 1987.

Markus, Julia. *Dared and Done: The Marriage of Elizabeth Barrett and Robert Browning*. London: Bloomsbury, 1995.

Marsh, Kate, ed. *Writers and Their Houses*. London: Hamish Hamilton, 1993.

Maull, Henry, and Polyblank. *Literary and Scientific Portrait Club*. London: Maul and Polyblank c.1855.

Metcalfe, Richard. *The Rise and Progress of Hydropathy in England and Scotland*. London: Simpkin Marshall, Hamilton, Kent and Co., 1906.

Micale, Mark. *Approaching Hysteria: Disease and Its Interpretations*. Princeton: Princeton University Press, 1995.

—— 'Hysteria Male/Hysteria Female: Reflections on Comparative Gender Construction in Nineteenth-Century France and Britain', in Benjamin, 200–39.

Morris, David B. *The Culture of Pain*. Berkeley and Los Angeles: University of California Press, 1991.

Neve, Michael, ed. *Charles Darwin's Autobiography: With an Introduction*. Harmondsworth: Penguin, forthcoming.

Ophir, Adi, and Shapin, Steven. 'The Place of Knowledge: A Methodological Survey'. *Science in Context*, 4 (1991): 3–21.

Oppenheim, Janet. '*Shattered Nerves': Doctors, Patients, and Depression in Victorian England*. New York: Oxford University Press, 1991.

Ostrander, G. 'Foucault's Disappearing Body', in *Body Invaders: Sexuality and the Postmodern Condition*, ed. Arthur Kroker and Marilouise Kroker, 169–82. Basingstoke: Macmillan Education, 1988.

Outram, Dorinda. 'Body and Paradox', *Isis*, 84 (1993): 347–52.

—— *The Body and the French Revolution: Sex, Class and Political Culture*. New Haven: Yale University Press, 1989.

Peckham, Morse, ed. *The Origin of Species by Charles Darwin: A Variorum Text*. Philadelphia: University of Pennsylvania Press, 1959.

Peterson, M. Jeanne. *Family Love and Work in the Lives of Victorian Gentlewomen*. Bloomington: Indiana University Press, 1989.

—— *The Medical Profession in Mid-Victorian London*. Berkeley and Los Angeles: University of California Press, 1978.

Pickering, George W. *Creative Malady: Illness in the Lives and Minds of Charles Darwin, Florence Nightingale, Mary Baker Eddy, Sigmund Freud, Marcel Proust, Elizabeth Barrett Browning*. London: George Allen and Unwin, 1974.

Piper, David. *Personality and the Portrait*. London: BBC Publications, 1973.

Pointon, Marcia. 'The Case of the Dirty Beau: Symmetry, Disorder and the Politics of Masculinity', in Adler and Pointon, 175–89.

Porter, Roy. 'Bodily Functions'. *Tate: The Art Magazine*, 3 (1994): 42–7.

—— 'History of the Body', in *New Perspectives on Historical Writing*, ed. Peter Burke, 206–32. Cambridge: Polity Press, 1991.

—— ed. *Patients and Practitioners*. Cambridge: Cambridge University Press, 1985.

Porter, Roy, and Rousseau, George. *Gout: The Patrician Malady*. Princeton: Princeton University Press, 1997.

Prescott, Gertrude. 'Fame and Photography: Portrait Publications in Great Britain, 1856–1900'. Ph.D. thesis, University of Texas, Austin, 1985.

Price, Rees. *A Critical Inquiry into the Nature and Treatment of the Case of H.R.H. the Princess Charlotte*. London: Chapple, 1817.

Pritchard, Michael. 'Commercial Photographers in Nineteenth Century Britain'. *History of Photography*, 11 (1988): 213–15.

Pritchard, Michael. *A Directory of London Photographers, 1841–1908.* Rev. edn. Watford: PhotoResearch, 1994.

Raverat, Gwen. *Period Piece: A Cambridge Childhood.* London: Faber and Faber, 1952.

Rees, Kelvin. 'Water as a Commodity: Hydropathy at Matlock', *Studies in the History of Alternative Medicine,* ed. Roger Cooter, 28–45. London: Macmillan, 1988.

Reeve, Lovell Augustus, and Walford, Edward. *Portraits of Men of Eminence in Literature, Science and Art.* With Biographical Memoirs by Edward Walford. 6 vols. London: Lovell Reeve, 1863–7.

Rennie, Alexander. *Observations on Gout . . . with Practical Remarks on the Injurious Effects of Colchicum.* London: Underwood, 1825.

Reynolds, Reginald. *Beards: An Omnium Gatherum.* London: George Allen and Unwin, 1950.

Richards, Evelleen. 'Darwin and the Descent of Women', in *The Wider Domain of Evolutionary Thought,* ed. David Oldroyd and Ian Langham, 57–111. Dordrecht: D. Reidel, 1983.

—— 'Huxley and Women's Place in Science: The "Woman Question" and the Control of Victorian Anthropology', in *History, Humanity and Evolution: Essays for John C. Greene,* ed. James Moore, 253–84. Cambridge: Cambridge University Press, 1989.

Roberts, David. 'The Paterfamilias of the Victorian Governing Classes', in Wohl, 59–81.

Roberts, Hyman J. 'Reflections on Darwin's Illness'. *Journal of Chronic Diseases,* 19 (1966): 723–5.

Schiebinger, Londa. *Nature's Body: Gender in the Making of Modern Science.* Boston: Beacon Press, 1993.

Scudamore, Charles. *Observations on the Use of the Colchicum Autumnale in the Treatment of Gout.* London: Longman, 1825.

Shapin, Steven. '"The Mind Is Its Own Place": Science and Solitude in Seventeenth-Century England'. *Science in Context,* 4 (1991): 191–218.

Shilling, Christopher. *The Body and Social Theory.* London: Sage, 1993.

Shortland, Michael. 'Bonneted Mechanic and Narrative Hero: The Self-Modelling of Hugh Miller', in *Hugh Miller and the Controversies of Victorian Science,* ed. Michael Shortland, 14–75. Oxford: Clarendon Press, 1996.

Shuttleworth, Sally. 'Female Circulation: Medical Discourse and Popular Advertising in the Mid-Victorian Era', in *Body/Politics: Women and the Discourses of Scientific Knowledge,* ed. Sally Shuttleworth, Mary Jacobus, and Evelyn Fox Keller, 47–68. London: Routledge, 1990.

Sicherman, Barbara. 'The Uses of Diagnosis: Doctors, Patients, and Neurasthenia'. *Journal of the History of Medicine,* 32 (1977): 33–54.

Smith, Crosbie, and Agar, Jon, eds. *Making Space for Science.* Basingstoke: Macmillan, 1998.

Smith, Fabienne. 'Charles Darwin's Health Problems: The Allergy Hypothesis'. *Journal of the History of Biology*, 25 (1992): 285–306.

—— 'Charles Darwin's Ill Health'. *Journal of the History of Biology*, 23 (1990): 443–59.

Stafford, Barbara. *Body Criticism: Imaging the Unseen in Enlightenment Art and Medicine*. Cambridge, Mass.: MIT Press, 1991.

Tagg, John. *The Burden of Representation: Essays on Photographies and Histories*. Amherst: University of Massachusetts Press; London: Macmillan, 1988.

Taylor, Charles. *Sources of the Self: The Making of Modern Identity*. Cambridge, Mass.: Harvard University Press, 1989.

Thomson, Joseph John. 'Some Reminiscences of Scientific Workers of the Past Generation and Their Surroundings'. *Proceedings of the Physical Society*, 48 (1936): 217–46.

Thornton, Peter. *Authentic Decor: The Domestic Interior, 1620–1920*. London: Weidenfeld and Nicolson, 1984.

Trombley, Stephen. *'All That Summer She Was Mad': Virginia Woolf and Her Doctors*. London: Junction Books, 1981.

Turner, Bryan S. *The Body and Society: Explorations in Social Theory*. Oxford: Blackwell, 1984.

Turner, Ernest S. *Taking the Cure*. London: Michael Joseph, 1967.

Vicinus, Martha, ed. *Suffer and Be Still: Women in the Victorian Age*. Bloomington: Indiana University Press, 1972.

Walford, Edward. *Photographic Portraits of Living Celebrities; Executed by Maull and Polyblank, With Biographical Notices*. London: Maull and Polyblank, 1856–59.

—— *Representative Men in Literature, Science, and Art. The Photographic Portraits from Life by Ernest Edwards*. London: A. W. Bennett, 1868.

Wallich, George C. *Eminent Men of the Day. Photographs by G. C. Wallich*. London: Van Voorst, 1870.

Weaver, Michael. *British Photography in the Nineteenth Century: The Fine Art Tradition*. Cambridge: Cambridge University Press, 1989.

—— *Julia Margaret Cameron, 1815–1879*. London: Herbert Press, 1984.

Welford, Samuel. 'The Cost of Photography in the Period 1850–1897'. *PhotoHistorian*, 92 (1991): 15–17.

Williams, Charles A. *Oriental Affinities of the Legend of the Hairy Anchorite*. Urbana: University of Illinois Press, 1925.

Wiltshire, John. *Jane Austen and the Body: 'The Picture of Health'*. Cambridge: Cambridge University Press, 1992.

Winter, Alison. 'Harriet Martineau and the Reform of the Invalid in Victorian England'. *Historical Journal*, 38 (1995): 597–616.

Wohl, Anthony, ed. *The Victorian Family: Structure and Stresses*. London: Croom Helm, 1978.

Wood, Ann Douglas. 'The Fashionable Diseases: Women's Complaints and Their Treatment in Nineteenth Century America', in *Clio's Consciousness*

Raised: New Perspectives on the History of Women, ed. Mary S. Hartmann and Lois Banner, 1–22. New York: Harper Colophon Books, 1974.

Woodruff, Alan W. 'Darwin's Health in Relation to His Voyage to South America'. *British Medical Journal*, i (1965): 745–50.

Woolf, Virginia, and Fry, Roger. *Victorian Photographs of Famous Men and Fair Women*. London: Harcourt Brace, 1926. Reprint, London: Hogarth Press, 1973.

Wynter, Andrew. 'Cartes de Visite'. *Journal of the Photographic Society of London*, 7 (1862): 375–77; also in *Once a Week*, 6 (1862): 1134–7.

Zanker, Paul. *The Mask of Socrates: The Image of the Intellectual in Antiquity*. Berkeley and Los Angeles: University of California Press, 1995.

more at risk for certain types of disease than others. In the world of nineteenth-century medicine, this difference becomes labeled as the 'pathological' or 'pathogenic' qualities of the Jewish body. What this essay will examine is a 'footnote' to the general representation of the pathophysiology of the Jew: the meaning attributed to the Jewish foot in the general and medical culture of the late nineteenth century.

The idea that the Jew's foot is unique has analogies with the hidden sign of difference attributed to the cloven-footed devil of the Middle Ages. That the shape of the foot, hidden within the shoe (a sign of the primitive and corrupt masked by the cloak of civilization and higher culture) could reveal the difference of the devil, was assumed in early modern European culture.[2] But the association between the sign of the devil and the sign of disease was well established in the early modern era. The view which associates the faulty gait of the Jew with disease caused by demonic influence is at least as old as Robert Burton's *Anatomy of Melancholy*, where Burton writes of the 'pace' of the Jews, as well as 'their voice, . . . gesture, [and] looks', as a sign of 'their conditions and infirmities'.[3] Johann Jakob Schudt, the seventeenth-century Orientalist, commented on the 'crooked feet' of the Jews among other signs of their physical inferiority.[4] By the nineteenth century the relationship between the image of the Jew and that of the hidden devil is to be found not in a religious but in a secularized scientific context. It still revolves in part around the particular nature of the Jew's foot—no longer the foot of the devil but now the pathog-nomonic foot of the 'bad' citizen of the new national state. The polit-ical significance of the Jew's foot within the world of nineteenth-century European medicine is thus closely related to the idea of the 'foot'-soldier, of the popular militia, which was the hallmark of all of the liberal movements of the mid-century. The Jew's foot marked him (and the Jew in this discussion is almost always the male) as con-genitally unable and, therefore, unworthy of being completely inte-grated into the social fabric of the modern state. This was echoed, for example, in Johann David Michaelis's answer in 1783 to Christian Wilhelm Dohm's call for the civil emancipation of the Jews. Michaelis argued that Jews could not become true citizens because they were worthless as soldiers due to their physical stature.[5] (That the Jewish woman had a special place in the debate about the nature of the Jewish body is without a doubt true, but it was not in regard to her role as a member of the body politic. In the context of nineteenth-century science it was assumed that she could not function in this manner).[6]

As early as 1804, in Joseph Rohrer's study of the Jews in the Austrian monarchy, the weak constitution of the Jew and its public sign, 'weak feet', were cited as 'the reason that the majority of Jews called into military service were released, because the majority of Jewish soldiers spent more time in the military hospitals than in military service.'[7] This link of the weak feet of the Jews and their inability to be full citizens (at a time when citizenship was being extended piecemeal to the Jews) was for Rohrer merely one further sign of the inherent, intrinsic difference of the Jews. What is of interest is how this theme of the weakness of the Jews' feet (in the form of flat feet or impaired gait) becomes part of the necessary discourse about Jewish difference in the latter half of the nineteenth century (Fig. 19).[8]

There is an ongoing debate throughout the late nineteenth century and well into the twentieth century which continues the basic theme which Rohrer raised in 1804. The liberal novelist and journalist, Theodor Fontane felt constrained to comment in 1870 on the false accusation that Jews were 'unfitted' for war in his observations about the role which Jewish soldiers played in the Seven Weeks War of 1866 (which led to the creation of the dual monarchy and the liberal Austrian constitution). His example is telling to measure the power of the legend of the Jewish foot: 'Three Jews had been drafted as part of the reserves into the first battalion of the Prince's Own Regiment. One, no longer young and corpulent, suffered horribly. His feet were open sores. And yet he fought in the burning sun from the beginning to the end of the battle of Gitschin. He could not be persuaded to go into hospital before the battle'.[9] For Fontane, the Jewish foot serves as a sign of the suffering that the Jew must overcome to become a good citizen. This becomes part of the liberal image of the Jew. In America, Mark Twain finds it necessary to place special stress on the role of the Jew as soldier in his 1898 defense of the Jews, noting that the Jew had to overcome much greater difficulties in order to become a soldier than did the Christian.[10]

In 1867 Austria institutionalized the ability of the Jews to serve in the armed services as one of the basic rights of the new liberal constitution. This became not only one of the general goals of the Jews but also one of their most essential signs of acculturation. But, as in many other arenas of public service, being 'Jewish' (here, espousing the Jewish religion) served as a barrier to status. Steven Beller points out that at least in Austria there was 'in the army evidence [that there was] a definite link between conversion and promotion.'[11] István Deák observed that 'of the twenty-three pre-1911 Jewish generals and colonels in the army's combat branches, fourteen converted at some

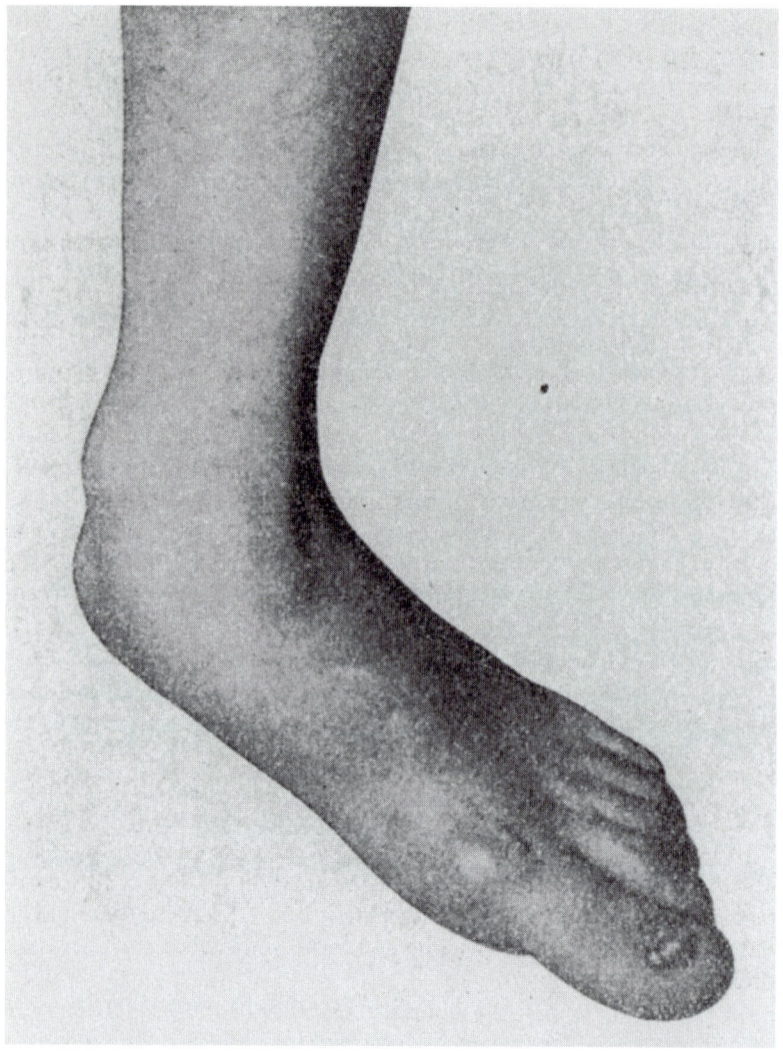

FIG. 19.   The image of a flat foot illustrating Gustav Muskat's essay on the diseases of the feet in Max Joseph, *Handbuch der Kosmetik* (Leipzig: Veit & Co., 1912). Private Collection, Chicago.

point during their career, although some only at the very end.'[12] Indeed, one of the legends which grew up about the young, 'liberal' monarch Franz Joseph of the 1850s, was that he promoted a highly decorated corporal from the ranks when he discovered that the soldier had not been previously promoted because of the fact that he was a Jew. Franz Joseph is said to have remarked: 'In the Austrian army there are no Jews, only soldiers, and a soldier who deserves it becomes an officer'.[13] Evidently it was difficult but not impossible to achieve the rank of officer. And Jews wanted to be officers. Theodor Fontane observed that 'it seemed as if the Jews had promised themselves to make an end of their old notions about their dislike for war and inability to engage in it.'[14] The status associated with the role of the Jew as soldier was paralleled by the increasingly intense anti-Semitic critique of the Jewish body as inherently unfit for military service. This critique became more and more important as the barriers of Jewish entry into the armed services in Germany and Austria were lessened in the closing decades of the nineteenth century. What had been an objection based on the Jew's religion came to be pathologized as an objection to the Jewish body. Images of Jewish difference inherent within the sphere of religion become metamorphosed into images of the Jew within the sphere of public service.

In 1893 H. Nordmann published a pamphlet on 'Israel in the Army' in which the Jew's inherent unfittedness for military service is the central theme.[15] The title image stressed the Jew's badly formed, unmilitary body. At the same time a postcard showing an ill-formed 'little Mr. Kohn', the eponymous Jew in German caricatures of the period, showing up for his induction into the military, was in circulation in Germany.[16] And in the Viennese fin-de-siècle humor magazine, *Kikeriki*, the flat and malformed feet of the Jew served as an indicator of the Jewish body almost as surely as the shape of the Jewish nose.[17] By the 1930s the image of the Jew's feet had become ingrained in the representation of the Jewish body. The Nazi caricaturist Walter Hofmann, who drew under the name of 'Waldl', illustrated how the Jew's body had been malformed because of the Jew's own refusal to obey his own God's directives.[18] Central to his representation are the Jew's flat feet. The foot became the hallmark of difference, of the Jewish body being separate from the real 'body politic'. These images aimed at a depiction of the Jew as unable to function within the social institutions, such as the armed forces, which determined the quality of social acceptance. 'Real' acceptance would be true integration into the world of the armed forces.

The Jewish reaction to the charge that the Jew cannot become a member of society because he cannot serve in the armed services is one of the foci of the Jewish response to turn-of-the-century anti-Semitism. The 'Defense Committee against Anti-Semitic Attacks in Berlin' published its history of 'Jews as soldiers' in 1897 in order to document the presence of Jews in the German army throughout the nineteenth century.[19] After the end of World War I, this view of Jewish non-participation became the central topos of political anti-Semitism. It took the form of the 'Legend of the Stab in the Back' (*Dolchstosslegende*) which associated Jewish slackers (war profiteers who refused to serve at the front) with the loss of the war.[20] In 1919 a brochure with the title 'The Jews in the War: A Statistical Study using Official Sources' was published by Alfred Roth which accused the Jews of having systematically avoided service during the war in order to undermine the war effort of the home front.[21] On the part of the official Jewish community in Germany, Jacob Segall in 1922 provided a similar statistical survey to the 1897 study in which he defended the role of the Jewish soldier during the World War I against the charge of feigning inabilities in order to remain on the home front.[22] In the same year, Franz Oppenheimer drew on Segall's findings in order to provide an equally detailed critique of this charge, a charge which by 1922 had become a commonplace of anti-Semitic rhetoric.[23]

There were also attempts on the part of Jewish physicians to counter the argument about the weakness of the Jewish body within the body politic. Their arguments are, however, even more convoluted and complicated, because of the constraints imposed by the rhetoric of science. As Jewish scientists, they needed to accept the basic 'truth' of the statistical arguments of medical science during this period. They could not dismiss published statistical 'facts' out of hand and thus operated within these categories. Like Segall and Oppenheimer, who answer one set of statistics with another set of statistics, the possibility of drawing the method of argument into disrepute does not exist for these scientists, as their status as scientists rested upon the validity of positivistic methods. And their status as scientists provided a compensation for their status as Jews. As Jews, they were the object of the scientific gaze; as scientists, they were themselves the observing, neutral, universal eye.

German-Jewish scientists attempted to resolve this dilemma. In 1908 the German-Jewish eugenist Dr. Elias Auerbach of Berlin undertook a medical rebuttal, in an essay entitled 'The Military Qualifications of the Jew', of the 'fact' of the predisposition of the Jew for certain disabilities which precluded him from military service.[24] Auerbach begins by attempting to 'correct' the statistics which claimed that for every 1,000 Christians in the population there were 11.61 soldiers, but for 1,000 Jews in the population there were only 4.92 soldiers. His correction (based on the greater proportion of Jews entering the military who were volunteers and, therefore, did not appear in the statistics) still finds that a significant portion of Jewish soldiers were unfit for service (according to his revised statistics, of every 1,000 Christians there were 10.66 soldiers; of 1,000 Jews, 7.76). He accepts the physical differences of the Jew as a given, but questions whether there is a substantive reason that these anomalies should prevent the Jew from serving in the military. He advocates the only true solution that will make the Jews of equal value as citizens: the introduction of 'sport' and the resultant reshaping of the Jewish body.[25] Likewise, Heinrich Singer attributes the flat feet of the Jew to the 'generally looser structure of the Jews' musculature'.[26]

More directly related to the emblematic nature of the Jewish foot is the essay by Gustav Muskat, a Berlin orthopedist, which asks the question of 'whether flat feet are a racial marker of the Jew?'[27] He refutes the false charge that 'the clumsy, heavy-footed gait of the Semitic race made it difficult for Jews to undertake physical activity, so that their promotion within the military was impossible.' While seeing flat feet as the 'horror of all generals', he also refutes the charge that Jews are particularly at risk as a group from this malady. Like Auerbach, Muskat sees the problem of the weaknesses of the Jewish body as a 'real' one. For him, it is incontrovertible that Jews have flat feet. Thus the question he addresses is whether this pathology is an inherent quality of the Jewish body which would preclude the Jew from becoming a full-fledged member of secular (i.e., military) society. For Muskat the real problem is the faulty development of the feet because of the misuse of the foot. The opinion that it is civilization and its impact on the otherwise 'natural' body which marks the Jew becomes one of the major arguments against the idea of the sign value of the Jewish foot as a sign of racial difference.[28] The Jew is, for the medical literature of the nineteenth century, the ultimate example of the effect of civilization (i.e., the city and 'modern life') on the individual.[29] And civilization in the form of the Jewish-dominated city, 'is the real center for the degeneration of the race and the reduction of military readiness.'[30] The Jewish foot, like the foot of the criminal and the epileptic, becomes a sign of Jewish difference. The Jew is both the city dweller par excellence as well as the most evident victim of the city.[31] And the occupation of the Jew in the city, the Jew's role as merchant, is the precipitating factor for the shape of the Jew's feet.[32]

For nineteenth-century medicine cities are places of disease and the Jews are the quintessential city dwellers. It is the 'citification' of the Jew, to use Karl Kautsky's term, which marks the Jewish foot.[33] The diseases of civilization are the diseases of the Jew. In 1940 Leopold Boehmer can speak of the 'foot as the helpless victim of civilization'.[34] The shape of the Jew's foot is read in this context as the structure of the Jewish mind, the pathognomonic status of the Jew's body as a sign of the Jew's inherent difference. The Jew's body can be seen and measured in a manner which fulfills all of the positivistic fantasies about the centrality of physical signs and symptoms for the definition of pathology. It can be measured as the mind cannot.

Muskat's argument is a vital one. He must shift the argument away from the inherited qualities of the Jewish body to the social anomalies inflicted on the body (and feet) of all 'modern men' by their lifestyle. He begins his essay with a refutation of the analogy present within the older literature which speaks about the flat foot of the black (and by analogy of the Jew) as an atavistic sign, a sign of the earlier stage of development (in analogy to the infant who lacks a well-formed arch). Muskat notes that flat feet have linked the black and the Jew as 'throwbacks' to more primitive forms of life. He quotes a nineteenth-century ethnologist, Karl Hermann Burmeister, who in 1855 commented that 'Blacks and all of those with flat feet are closest to the animals.' Muskat begins his essay by denying that flat feet are a racial sign for any group, but rather a pathological sign of the misuse of the feet. (But he carefully avoids the implications that this misuse is the result of the urban location of the Jew or the Jew's inability to deal with the benefits of civilization.) He cites as his authorities a number of 'modern' liberal commentators such as the famed Berlin physician-politician Rudolf Virchow, who had examined and commented on the feet of the black in order to refute the implication that flat feet are a sign of racial difference. Like other Jewish commentators, Muskat is constrained to acknowledge the 'reality' of the 'flat feet' of both the Jew and the black, but cites the renowned orthopedist Albert Hoffa[35] to the effect that only 4% of all flat feet are congenital. Flat feet are not a racial sign, they are 'merely' a sign of the abuse of the foot. That 25% of all recruits in Austria and 30% in Switzerland were rejected for flat feet is a sign for Muskat that the Jewish recruit is no better nor worse than his Christian counterpart. (He never cites the rate of the rejection of Jewish recruits.) Muskat's rebuttal of the standard bias which sees the foot as a sign of the racial difference of the Jew still leaves the Jew's foot deformed. What Muskat has done is to adapt the view of the corruption of the Jew by civilization (and of civilization by the Jew) to create a space where the Jew's foot has neither more nor less signification as a pathological sign than does the flat foot in the general population. This attempt at the universalization of the quality ascribed to the Jewish foot does not however counter the prevailing sense of the specific meaning ascribed to the Jew's foot as a sign of difference.

Moses Julius Gutmann repeats the association between the classical image of the flat foot and the new neurological syndrome in his 1920 dissertation which attempted to survey the entire spectrum of charges concerning the nature of the Jewish body. Gutmann dismisses the common wisdom that ' "all Jews have flat feet" as excessive'.[36] He notes, as does Muskat, that Jews seem to have a more frequent occurrence of this malady (8 to 12% higher than the norm for the general population). And his sources, like that of Muskat, are the military statistics. But Gutmann accepts the notion that Jews have a peculiar pathological construction of the musculature of the lower extremities.

In a standard handbook of eugenics published as late as 1940, the difference in the construction of the musculature of the foot is cited as the cause of the different gait of the Jew.[37] The German physician-writer Oskar Panizza, in his depiction of the Jewish body, observed that the Jew's body language was clearly marked: 'When he walked, Itzig always raised both thighs almost to his mid-rift so that he bore some resemblance to a stork. At the same time he lowered his head deeply into his breast-plated tie and stared at the ground. Similar disturbances can be noted in people with spinal diseases. However, Itzig did not have a spinal disease, for he was young and in good condition.'[38] The Jew looks as if he is diseased, but it is not the stigmata of degeneracy which the observer is seeing, but the Jew's natural stance.

This image of the pathological nature of the gait of the Jews is linked to their inherently different anatomical structure. Flat feet remain a significant sign of Jewish difference in German science through the Nazi period. And it is always connected with the discourse about military service. According to the standard textbook of German racial eugenics by Baur, Fischer and Lenz in 1936: 'Flat feet are especially frequent among the Jews. Salaman reports during the World War that about a sixth of the 5000 Jewish soldiers examined had flat feet while in a similar sample of other English soldiers it occurred in about a fortieth.'[39] In 1945 Otmar Freiherr von Verschuer can still comment without the need for any further substantiation that great numbers of cases of flat feet are to be found among the Jews.[40]

The debate about the special nature of the Jew's foot and gait enters into another sphere, that of neurology, which provides a series of links between the inherent nature of the Jew's body and his psyche. This concept, too, has a specific political and social dimension. The assumption, even among those physicians who saw this as a positive quality was that the Jews were innately unable to undertake physical labor:

In no period in the history of this wonderful people since their dispersion, do we discover the faintest approach to any system amongst them tending to the studied development of physical capacity. Since they were conquered they have never from choice borne arms nor sought distinction in military prowess; they have been little inducted, during their pilgrimages, into the public games of the countries in which they have been located; their own ordinances and hygienic laws, perfect in other particulars, are indefinite in respect to special means for the development of great corporeal strength and stature; and the fact remains, that as a people they have never exhibited what is considered a high physical standard. To be plain, during their most severe persecutions nothing told so strongly against them as their apparent feebleness of body.[41]

It is within such a medical discourse about the relationship between the Jews' inability to serve as a citizen and the form of the Jews' body that the debate about Jews and sport can be located. Elias Auerbach's evocation of sport as the social force to reshape the Jewish body had its origins in the turn-of-the-century call of the physician and Zionist leader Max Nordau for a 'new Muscle Jew'.[42] This view became a commonplace of the early Zionist literature, which called upon sport, as an activity, as one of the central means of shaping the new Jewish body.[43] Nordau's desire was not merely for an improvement in the physical wellbeing of the Jew, but rather an acknowledgment of the older German tradition which saw an inherent relationship between the healthy political mind and the healthy body. It was not merely *mens sana in corpore sano*, but the sign that the true citizen had a healthy body which provided his ability to be a full-scale citizen, itself a sign of mental health. Nordau's cry that we have killed our bodies in the stinking streets of the ghettoes and we must now rebuild them on the playing fields of Berlin and Vienna, is picked up by the mainstream of German-Jewish gymnastics.

In Vienna, at the third 'Jewish Gymnastics Competition' a series of lectures on the need for increased exercise among the Jews departed from the assumption of a statistically provable physical degeneration of contemporary Jewry.[44] The medicalization of this theme is continued by M. Jastrowitz of Berlin in the *Jewish Gymnastics News*, the major Jewish newspaper devoted to gymnastics in 1908.[45] Jastrowitz accepts the basic premise of Nordau's conviction, that the Jewish body is at risk for specific diseases, and attempts to limit and focus this risk. For Jastrowitz the real disease of the Jews, that which marks their bodies, is a neurological deficit which has been caused by the impact of civilization. Jastrowitz, like most of the Jewish physicians[46] of the fin de siècle, accepts the general view that Jews are indeed at special risk for specific forms of mental and neurological disease. He warns that too great a reliance on sport as a remedy may exacerbate these illnesses. For Jastrowitz, the attempt to create the 'new muscle Jew' works against the inherent neurological weaknesses of the Jew. This is the link to the general attitude of organic psychiatry of the latter half of the nineteenth century which saw the mind as a product of the nervous system and assumed that 'mind illness is brain illness.' Thus the improvement of the nervous system through training the body would positively impact on the mind (so argued Nordau). Jastrowitz's view also assumes a relationship between body and mind and he fears that, given the inherent weakness of the Jewish nervous system, any alteration of the precarious balance would negatively impact on the one reservoir of Jewish strength, the Jewish mind. The Jew could forfeit the qualities of mind which have made him successful in the world by robbing his brain of oxygen through overexercise. For Nordau and Jastrowitz the relationship between the healthy body, including the healthy foot and the healthy gait, and the healthy mind is an absolute one. The only question left is whether the degeneration of the Jewish foot is alterable. For neurologists this problem could not simply be limited to orthopedic theory and treatment. More than that was at stake, as is clear from the debate (in which Sigmund Freud was involved) over the new diagnostic category of intermittent claudication.

'Claudication intermittente' was created by Jean Martin Charcot at the beginning of his medical career in 1858.[47] (Charcot taught not only Freud but also was Max Nordau's doctoral supervisor.) This diagnostic category was described by Charcot as the chronic reoccurrence of pain and tension in the lower leg, a growing sense of stiffness and finally a total inability to move the leg, which causes a marked and noticeable inhibition of gait. This occurs between a few minutes and a half hour after beginning a period of activity, such as walking. It spontaneously vanishes only to be repeated at regular intervals.

Charcot determined that this syndrome seems to be the result of the reduction of blood flow through the arteries of the leg leading to the virtual disappearance of any pulse from the four arteries which provide the lower extremity with blood. The interruption of circulation to the feet then leads to the initial symptoms and can eventually lead to even more severe symptoms such as spontaneous gangrene. Charcot's diagnostic category was rooted in work done by veterinarians, such as Bouley and Rademacher, who observed similar alterations in the gait of dray horses.[48] Charcot did not himself speculate on any racial predisposition for this syndrome, as he did on the origin of hysteria and diabetes.[49] Nevertheless, the image of the Jew's foot as an atavistic structure similar to the flat feet of the horse soon appeared at the very heart of neurological research on this syndrome. Like flat feet, intermittent claudication is a 'reality,' i.e., it exists in the real world, but, like flat feet, it was placed in a specific ideological context at the turn of the century.

What is vital is that this diagnostic category soon after Charcot's identification of it became one of the markers in neurology for the difference between the Jewish foot and that of the 'normal' European. Intermittent claudication became part of the description of the pathological difference of the Jew. And it was, itself, differentiated from other 'racially' marked categories which evoked the impairment of gait as part of their clinical presentation. Charcot clearly differentiated intermittent claudication from the chronic pain associated with the diabetic's foot (a diagnostic category so closely associated with Jews in nineteenth-century medicine that it was commonly called the 'Jewish disease'[50]). In 1911 Dejerine also differentiated this syndrome from 'spinal intermittent claudication.'[51] This was one of the syndromes associated with syphilis, a disease which also had a special relationship to the representation of the Jew in nineteenth-century and early twentieth-century medicine. What is clear is that the sign of the 'limping Jew' was read into a number of diagnostic categories of nineteenth-century neurology.

Very quickly intermittent claudication became one of the specific diseases associated with Eastern European Jews. H. Higier in Warsaw published a long paper in 1901 in which he summarized the state of the knowledge about intermittent claudication as a sign of the racial makeup of the Jew.[52] The majority of the 23 patients he examined were Jews and he found that the etiology of the disease was 'the primary role of the neuropathic disposition [of the patients] and the inborn weakness of their peripheral circulatory system'. By the time Higier published his paper at the turn of the century this was a given in the neurological literature. The debate about the flat feet of the Jews as a marker of social stigma gave way to the creation of a scientific discourse about the difference of the Jew's feet, a discourse which does not merely rely on the argument of atavism (which had been generally refuted in the neurological literature of the fin de siècle), but on the question of the relationship between the Jew's body and the Jew's mind through the image of the deficits of the neurological system. Intermittent claudication became a sign of inherent constitutional weakness, so that it was also to be found as a sign for the male hysteric.[53] Hysteria was, of course, also understood as a neurological deficit, one which, it was believed in the fin de siècle, was primarily to be found among Eastern European Jewish males. (The association of hysteria with impairment of gait has remained a truism of medical science for many decades. It can, in fact, be traced back to the eighteenth century.[54]) The hysterical and the limping Jew are related in the outward manifestation of their illness: both are represented by the inability of the limbs to function 'normally', by the disruption of their gait, as in Sigmund Freud's case of 'Dora' (1905 [1901]).[55]

The link between the older discussion of 'flat feet' and the new category of intermittent claudication was examined by a number of sources. H. Idelsohn in Riga made the association between the Jews, Charcot's category of intermittent claudication, and flat feet overt when he examined his Jewish patients to see whether there was any inherent relationship between the fabled Jewish flat feet and inherent muscular weakness.[56] Idelsohn placed the discussion of the special nature of the Jewish foot into the context of a neurological deficit. While he did not wish to overdetermine this relationship (according to his own statement), he did find that there was reason to grant flat feet a 'specific importance as an etiological moment'. He described the flat foot, citing Hoffa, as 'tending to sweat, often blue colored and cold, with extended veins . . . People with flat feet are often easily tired, and are incapable of greater exertion and marches'.[57] He saw in this description a visual and structural analogy to Charcot's category of intermittent claudication. Heinrich Singer picked up Idelsohn's and Higier's views concerning the relationship between intermittent claudication and flat feet and repeated them as proof of the 'general nervous encumbrance born by the Jewish race. . . .'[58] This conviction is echoed by Gustav Muskat in a paper of 1910, in which he made the link between the appearance of intermittent claudication and the preexisting pathology of flat feet.[59]

One of Idelsohn's major sources for his attitude was a paper by

Samuel Goldflam in Warsaw.[60] Goldflam was one of the most notable neurologists of the first half of the century and the co-discoverer of the 'Goldflam-Oehler sign' (the paleness of the foot after active movement) in the diagnosis of intermittent claudication.[61] Goldflam stressed the evident predisposition of Jews for this syndrome. What was also noteworthy in Goldflam's discussion of his patients was not only that they were all Eastern Jews, but that they almost all were very heavy smokers. He does attribute a role in the etiology of the disease to tobacco intoxication.[62] Idelsohn argued that since *all* of Goldflam's patients were Jews, it is clear that intermittent claudication was primarily a Jewish disease and that this is proven by the relationship between the evident sign of the difference of the Jewish foot, the flat foot, and its presence in a number of his own cases. Its absence in the clinical description given by Goldflam (and others) was attributed by him to its relatively benign and usual occurrence, which is often overlooked because of the radical problems, such as gangrene, which occur with intermittent claudication.

In a major review essay on the 'nervous diseases' of the Jews, Toby Cohn, the noted Berlin neurologist (long the assistant of the noted Emanuel Mendel at the University of Berlin), included intermittent claudication as one of his categories of neurological deficits.[63] While commenting on the anecdotal nature of the evidence, and calling on a review essay by Kurt Mendel (who does not discuss the question of 'race' at all),[64] he accepted the specific nature of the Jewish risk for this syndrome while leaving the etiology open. Two radically different etiologies had been proposed: the first, as we have noted in Higier, Idelsohn and Singer, reflected on the neuropathic qualities of the Jewish body, especially in regard to diseases of the circulatory system. (Hemorrhoids, another vascular syndrome, were also identified as a 'Jewish' disease during this period). The second, noted by Goldflam and Cohn, did not reflect on the inherent qualities of the Jewish foot, but on the Jewish misuse of tobacco and the resulting occlusion of the circulatory system of the extremities. It is tobacco which, according to Wilhelm Erb, played a major role in the etiology of intermittent claudication.[65] In a somewhat later study of forty-five cases of the syndrome, Erb found, to his own surprise, that at least thirty-five of his patients showed an excessive use of tobacco.[66] (This meant the consumption of 40–60 cigarettes or 10–15 cigars a day). Indeed, the social dimension which the latter provide in their discussion of the evils of tobacco misuse supplies both an alternative and an explanation for the neurological predisposition of the Jew's body to avoid

368

military service.[67] But according to all of these Western (or Western-trained) Jewish physicians, the misuse of tobacco is typical of the Eastern Jew, not of the Western Jew. Some years earlier, the noted Berlin neurologist Hermann Oppenheim also made the East–West distinction when he observed that of the cases of intermittent claudication in his practice (48 cases over five years) the overwhelming majority, between 35 and 38, were Russian Jews.[68] For Oppenheim and many others, the Eastern Jew's mind is that of a social misfit and his body reifies this role, but this is not a problem of Western Jewry—except by extension.

Here the parameters of the meaning ascribed to the Jewish foot are set: Jews walk oddly because of the form of their feet and legs. This unique gait represents the inability of the Jew to function as a citizen within a state which defines full participation as including military service. Jewish savants who rely on the status of science as central to their own self-definition, who cannot dismiss the statistical evidence of Jewish risk as nonsense, seek to 'make sense' of it in a way which would enable them (as representatives of the authoritative voice of that very society) to see a way out. The 'way out' is, in fact, the acceptance of difference through the attribution of this difference to social rather than genetic causes and its projection on to a group labeled as inherently 'different', the Eastern Jews. This is an important moment in the work of these scientists—for the risk which they see lies in the East, lies in the 'misuse' of tobacco by Eastern Jews. It is not accidental that the major reports on the nature of the Jew's gait come from Eastern Europe and are cited as signs of the difference in social attitudes and practices among Eastern European Jews. The rhetorical movement which these scientists undertook, implied an inherently different role for the Western Jew, serving in the armed forces of the Empire, whether German or Austrian. This movement was important, because the clean line between 'nature' and 'nurture' was blurred in nineteenth-century medicine. The physician of the nineteenth and early twentieth century understood that the appearance of signs of degeneration, such as the flat foot, may have been triggered by aspects in the social environment but was, at its core, an indicator of the inherent weakness of the individual. The corollary to this is the inheritance of acquired characteristics—that the physiological changes which the impact of environment triggers become part of the inheritance of an individual. This view can be found in the work of C. E. Brown-Séquard, who, as early as 1860, had argued that there were hereditary transmissions of acquired injuries, as in the case of 'animals born of parents having been rendered epileptic by an injury to the

12. István Deák, *Jewish Soldiers in Austro-Hungarian Society*, Leo Baeck Lecture 34 (New York: Leo Baeck Institute, 1990), p. 14.
13. Cited by Max Grunwald, *Vienna* (Philadelphia: Jewish Publication Society of America, 1936), p. 408.
14. Cited by Grunwald, *Vienna*, p. 177.
15. H. Naudh [i.e., H. Nordmann], *Israel im Heere* (Leipzig: Hermann Beyer, 1893). On the background, see Horst Fischer, *Judentum, Staat und Heer in Preussen im frühen 19. Jahrhundert: Zur Geschichte der staatlichen Judenpolitik* (Tübingen: J. C. B. Mohr, 1968).
16. See Dietz Bering, *Der Name als Stigma: Antisemitismus im deutschen Alltag 1812–1933* (Stuttgart: Klett/Cotta, 1987), p. 211.
17. Eduard Fuchs, *Die Juden in der Karikatur* (Munich: Langen, 1921), p. 200.
18. Walter Hofmann, *Lacht ihn tot! Fin tendenziöses Bilderbuch von Waldl* (Dresden: Nationalsozialistischer Verlag für den Gau Sachsen, n.d.), p. 23.
19. *Die Juden als Soldaten* (Berlin: Sigfried Cronbach, 1897).
20. See Joachim Petzold, *Die Dolchstosslegende: Eine Geschichtsfälschung im Dienst des deutschen Imperialismus und Militarismus* (Berlin [East]: Akadamie Verlag, 1963) which discusses the statistical arguments.
21. Otto Armin [i.e., Alfred Roth], *Die Juden im Heere, eniee statistische Untersuchung nach amtlichen Quellen* (Munich: Deutsche Volks-Verlag, 1919).
22. Jacob Segall, *Die deutschen Juden als Soldaten im Kriege 1914–1918* (Berlin: Philo Verlag, 1922).
23. Franz Oppenheimer, *Die Judenstatistik des preußischen Kriegsministeriums* (München: Verlag für Kulturpolitik, 1922).
24. Elias Auerbach, 'Die Militärtauglichkeit der Juden', *Jüdische Rundschau*, 50 (11 December 1908), pp. 491–2.
25. See John Hoberman, *Sport and Political Ideology* (Austin: University of Texas Press, 1984).
26. Heinrich Singer, *Allgemeine und spezielle Krankheitslehre der Juden* (Leipzig: Benno Konegen, 1904), p. 14.
27. Gustav Muskat, 'Ist der Plattfuss eine Rasseneigentümlichkeit?', *Im deutschen Reich* (1909), 354–8. Compare J. C. Dagnall, 'Feet and the Military System', *British Journal of Chiropody*, 45 (1980), p. 137.
28. In other contexts the atavistic foot is taken to be a sign of insanity. See Charles L. Dana, 'On the New Use of Some Older Sciences: A Discourse on Degeneration and its Stigmata', *Transactions of the New York Academy of Medicine*, 11 (1894), pp. 471–89, here pp. 484–5. See also my discussion in *Difference and Pathology: Stereotypes of Sexuality, Race, and Madness* (Ithaca, NY: Cornell University Press, 1986), p. 155.
29. George L. Mosse, *Germans and Jews: The Right, the Left, and the Search for a 'Third Force' in Pre-Nazi Germany* (Detroit: Wayne State University Press, 1987), pp. 14–15.
30. Moritz Alsberg, *Militäruntauglichkeit und Grossstadt-Einfluss: Hygienisch-volkswirtschaftliche Betrachtungen und Vorschläge* (Leipzig: B. G. Teubner, 1909), p. 10. Compare the discussion of the rate of military readiness in the various sections of Vienna during this period: Victor Noach, 'Militärdiensttauglichkeit und Berufstätigkeit, soziale Stellung und Wohnweise in Österreich-Ungarn, insbesondere in Wien', *Archiv für soziale Hygiene und Demographie*, 10 (1915), pp. 77–128.

31. See my discussion in *Difference and Pathology* as well as *Disease and Representation: Images of Illness from Madness to AIDS* (Ithaca, NY: Cornell University Press, 1988). Compare to D. B. Larson et al., 'Religious Affiliations in Mental Health Research Samples as Compared with National Samples', *Journal of Nervous and Mental Disease*, 177 (1989), pp. 109–11.

32. Gustav Muskat, 'Die Kosmetik des Fusses', in Max Joseph (ed.), *Handbuch der Kosmetik* (Leipzig: Veit & Co., 1912), pp. 646–64, here, p. 662.

33. Karl Kautsky, *Rasse und Judentum*, 2nd edn. (Stuttgart: Dietz, 1921), p. 62. This was first published in 1914. On the general context of anti-urbanism and its relationship to proto-Fascist thought, see George L. Mosse, *The Crisis of German Ideology: Intellectual Origins of the Third Reich* (New York: Grosset and Dunlap, 1964).

34. Leopold Boehmer, 'Fussschäden und schwingedes Schuhwerk', in *Zivilizationsschäden am Menschen*, ed. Heinz Zeiss and Karl Pintschovius (Munich/Berlin: J. F. Lehmann, 1940), p. 180. Compare J. Swann, 'Nineteenth-Century Footwear and Foot Health', *Cliopedic Items*, 3 (1988), pp. 1–2.

35. Hoffa is the author of the authoritative study *Die Orthopädie im Dienst der Nervenheilkunde* (Jena: Gustav Fischer, 1900) and the compiler (with August Blencke) of the standard overview of the orthopedic literature of the fin de siècle, *Die orthopädische Literatur* (Stuttgart: Enke, 1905).

36. M. J. Gutmann, *Über den heutigen Stand der Rasse- und Krankheitsfrage der Juden* (Munich: Rudolph Müller & Steinicke, 1920), p. 38.

37. Günther Just (ed.), *Handbuch der Erbbiologie des Menschen: I. Erbbiologie und Erbpathologie Körperlicher Zustände und Funktionen: Stützgewebe, Haut, Auge*, 3 vols. (Berlin: Julius Springer, 1939–40), p. 39.

38. The translation is by Jack Zipes, Oskar Panizza, 'The Operated Jew', *New German Critique*, 21 (1980), pp. 63–79, p. 64. See also Jack Zipes, 'Oscar Panizza: The Operated German as Operated Jew', *New German Critique*, 21 (1980), pp. 47–61 and Michael Bauer, *Oskar Panizza: Ein literarisches Porträt* (Munich: Hanser, 1984).

39. Erwin Baur, Eugen Fischer, and Fritz Lenz (eds.), *Menschliche Erblehre und Rassenhygiene: I: Menschliche Erblehre* (Munich: J. F. Lehmann, 1936), p. 396.

40. Otmar Freiherr von Verschuer, *Erbpathologie: Ein Lehrbuch für Ärtze und Medizinstudierende* (Dresden/Leipzig: Theodor Steinkopff, 1945), p. 87.

41. Benjamin Ward Richardson, *Diseases of Modern Life* (New York: Bermingham and Co., 1882), p. 98.

42. Max Nordau, *Zionistische Schriften* (Cologne: Jüdischer Verlag, 1909), pp. 379–81. This call, articulated at the second Zionist Congress, followed his address on the state of the Jews which key-noted the first Congress. There he spoke on the 'physical, spiritual and economic status of the Jews'. On Nordau, see P. M. Baldwin, 'Liberalism, Nationalism, and Degeneration: The Case of Max Nordau', *Central European History*, 13 (1980), pp. 99–120 and Hans-Peter Söder, 'A Tale of Dr. Jekyll and Mr. Hyde? Max Nordan and the Problem of Degeneracy', in Rudolf Käser and Vera Pohland (eds.), *Disease and Medicine in Modern German Cultures* (Ithaca, NY: Western Societies Program, 1990), pp. 56–70.

43. Hermann Jalowicz, 'Die körperliche Entartung der Juden, ihre Ursachen und ihre Bekämpfung', *Jüdische Turnzeitung*, 2 (1901), pp. 57–65.

44. Isidor Wolff (ed.), *Die Verbreitung des Turnens unter den Juden* (Berlin: Verlag der Jüdischen Turnzeitung, 1907).

45. M. Jastrowitz, 'Muskeljuden und Nervenjuden', *Jüdische Turnzeitung*, 9 (1908), pp. 33–6.

46. I am using the term 'Jewish physician' to refer to those physicians who either self-label themselves as Jews or are so labeled in the standard reference works of the time.

47. His first paper on this topic is Jean Martin Charcot, 'Sur la claudication intermittente', *Comptes rendus des séances et mémoires de la société de biologie* (Paris) 1858, Mémoire 1859, 2 series, 5: 25–38. While this is not the first description of the syndrome, it is the one which labels this as a separate disease entity. It is first described by Benjamin Collins Brodie, *Lectures Illustrative of Various Subjects in Pathology and Surgery* (London: Longman, 1846), p. 361. Neither Brodie not Charcot attempt to provide an etiology for this syndrome. Compare M. S. Rosenbloom et al., 'Risk Factors Affecting the Natural History of Intermittent Claudication', *Archive of Surgery*, 123 (1989), pp. 867–70.

48. The work, dated as early as 1831, is cited in detail by Charcot, 'Sur la claudication intermittente', pp. 25–6.

49. See my *Difference and Pathology*, p. 155 and Toby Gelfand, 'Charcot's Response to Freud's Rebellion', *Journal of the History of Ideas*, 50 (1989), p. 304.

50. Gutmann, *Über den heutigen Stand*, p. 39. Compare Dr. M. Kretzmer, 'Über anthropologische, physiologische und pathologische Eigenheiten der Judern', *Die Welt*, 5 (1901), pp. 3–5 and Dr. Hugo Hoppe, 'Sterblichkeit und Krankheit bei Juden und Nichtjuden', *Ost und West*, 3 (1903), pp. 565–8, 631–8, 775–80, 849–52.

51. G. Steiner, 'Klinik der Neurosyphilis', in *Handbuch der Haut-und Geschlechtskrankheiten*, ed. Josef Judassohn et al.: vol. 17 1, ed. Gustav Alexander (Berlin: Springer, 1929), p. 230.

52. H. Higier, 'Zur Klinik der angiosklerotischen paroxysmalen Myasthenie ("claudication intermittente" Charcot's) und der sog. spontanen Gangrän', *Deutsche Zeitschrift für Nervenheilkunde*, 19 (1901), pp. 438–67.

53. P. Olivier and A. Halipré, 'Claudication intermittente chez un homme hystérique atteint de pouls lent permanent', *La Normandie Médcale*, 11 (1896), pp. 21–8.

54. Louis Basile Carré de Montgeron, *La Verité des miracles operés par l'intercession de M. de Pâris et autres appellans demontrée contre M. L'archevêque de Sens . . .* 3 vols. (Cologne: Chez les libraires de la Campagnie, 1745–7).

55. On the representation of the alteration of gait in the case of Dora, see the complex distinction made between the meaning of impaired gait in the syphilitic and the hysteric which Freud documents in Sigmund Freud, *Standard Edition of the Complete Psychological Works of Sigmund Freud*, ed. and trans., J. Strachey, A. Freud, A. Strachey, and A. Tyson, 24 vols. (London: Hogarth, 1955–74), 7: 16–17, n. 2. On the background to Freud's training in neurology, see Henri Ellenberger, *The Discovery of the Unconscious: The History and Evolution of Dynamic Psychiatry* (New York: Basic Books, 1970). Compare J. H. Baker and J. R. Silver, 'Hysterical Paraplegia', *Journal of Neurology, Neurosurgery and Psychiatry*, 50 (1987), pp. 375–82 and J. R. Keane, 'Hysterical Gait Disorders: 60 Cases', *Neurology*, 39 (1989), pp. 586–9.

56. H. Idelsohn, 'Zur Casuistik und Aetiologie des intermittierenden Hinkens', *Deutsche Zeitschrift für Nervenheilkunde*, 24 (1903), pp. 285–304.

57. Idelsohn, 'Zur Casuistik', p. 300.
58. Singer, *Krankheitslehre der Juden*, pp. 124–5.
59. Gustav Muskat, 'Über Gangstockung', *Verhandlungen des deutschen Kongresses für innere Medizin*, 27 (1910), pp. 45–56.
60. Samuel Goldflam, 'Weiteres über das intermittierende Hinken', *Neurologisches Centralblatt*, 20 (1901), pp. 197–213. See also his 'Über intermittierende Hinken ("claudication intermittente" Charcot's) und Arteritis der Beine', *Deutsche medizinische Wochenschrift*, 21 (1901), pp. 587–98. On tobacco misuse as a primary cause of illness, see the literature overview by Johannes Bresler, *Tabakologia medizinalis: Literarische Studie über den Tabak in medizinischer Beziehung*, 2 vols. (Halle: Carl Marhold, 1911–13).
61. See Enfemiuse Herman, 'Samuel Goldflam (1852–1932)', in Kurt Kolle (ed.), *Grosse Nervenärtze*, 3 vols. (Stuttgart: Thieme, 1963), 3: 143–9.
62. Samuel Goldflam, 'Zur Ätiologie und Symptomatologie des intermittierenden Hinkens', *Neurologisches Zentralblatt*, 22 (1903), pp. 994–6.
63. Toby Cohn, 'Nervenkrankheiten bei Juden', *Zeitschrift für Demographie und Statistik der Juden*, New Series 3 (1926), pp. 76–85.
64. Kurt Mendel, 'Intermitterendes Hinken', *Zentralblatt für die gesamt Neurologie und Psychiatrie*, 27 (1922), pp. 65–95.
65. Wilhelm Erb, 'Über das "intermittirende Hinken" und andere nervöse Störungen in Folge von Gefässerkrankungen', *Deutsche Zeitschrift für Nervenheilkunde*, 13 (1898), pp. 1–77.
66. Wilhelm Erb, 'Über Disbasia angiosklerotika (intermittierendes Hinken)', *Münchener medizinische Wochenschrift*, 51 (1904), pp. 905–8.
67. Compare P. C. Waller, S. A. Solomon, and L. E. Ramsay, 'The Acute Effects of Cigarette Smoking on Treadmill Exercise Distances in Patients with Stable Intermittent Claudication', *Angiology*, 40 (1989), pp. 164–9.
68. Hermann Oppenheim, 'Zur Psychopathologie und Nosologie der russisch-jüdischen Bevölkerung', *Journal für Psychologie und Neurologie*, 13 (1908), p. 7.
69. C. E. Brown-Séquard, 'On the Hereditary Transmission of Effects of Certain Injuries to the Nervous System', *The Lancet* (2 January 1875), pp. 7–8.

The Ideal Couple
A Question of Size?

Sabine Gieske

Translated from the German by Barbara Harshav

We know how it is to feel short or to be tall. We know the feeling of dependence on tall people so well because every one of us has gone through a stage of being short. Ever since our childhood we have known the feeling of helplessness and weakness, which we have carried in our 'biographical bones' ever since.[1] The experience of being at the mercy of someone is registered early and hence very intensely in our memory. At the same time, when a person is taller, stronger, and thus in charge, a small person may feel protected, for physical hierarchy can also offer a sense of security.

There are many examples in cultural history of ritualized forms of making oneself short or tall. Even though acts such as bowing, doffing one's hat, or curtseying are rare today, they are rituals that diminish body size. They also suggest deference and respect, but are principally associated with expressions of humiliation and submission. Sitting on a raised throne, speaking from a lectern, living in rooms with high ceilings and grandiloquent doors, all indicate an attempt to enhance corporeal size. These acts elevate the social rank of the person in question and express power and control.

In seventeenth- and eighteenth-century Europe, royalty collected the extremely tall and extremely short in the form of court giants or dwarfs. At court, 'giants' were produced as spectacles, but were also employed as bodyguards and personal servants. Standing behind the throne of the ruler, they accented his power and greatness. Friedrich Wilhelm I, King of Prussia, had his palace in Potsdam guarded by 2,400 'tall chaps', whom he had collected from all over Europe for his guard battalion.[2] And politicians today are still aware of the power of physical

From Sabine Gieske, ed. *Jenseits vom Durchschnitt* (translated from the German by Barbara Harshav), (Marburg, 1998), pp. 61–94. Reprinted with permission.

height. In the American presidential election debates of 1976, Jimmy Carter stood on a stool to appear equal in height to his opponent.

Theorist Erving Goffman has noted that in advertising, power, authority, office, and fame, hence, social standing, are expressed by elevating the central figure in the picture.[3] The message of the ad depends on the rarely expressed cultural conviction that there is a correlation between physical height and social position.[4] Economists Gail Tom and Jeanne Shevell have also demonstrated a connection between height and status. Several studies have demonstrated that tall men are trusted more on average than their shorter colleagues, and enjoy preference in hiring.[5] The role of physical height as a symbol of cultural prestige in interhuman communication cannot be underestimated. In a few seconds, speakers assess the body of their interlocutors, measure their stature, and position themselves accordingly, forming a judgement on the basis of looks, gestures, or skin colour, and also on size and body build. The hierarchical classification involved with physical height seems clearly to hold for many cultures. In his article, 'Die Körper des Fremden', sociologist Rudolf Stichweh emphasized: 'Physical height seems to be one of the few universal classifications, and as such, in many known societies, it is directly connected with attributing social rank to the person in question.'[6]

Given the importance of height, it is curious that there have been few studies of the relative height of men and women in couples. Why, in contemporary Western culture, is the male partner so often taller than the female? What is the tacit rule prescribing that the man be taller and the woman shorter? Psychologists Ellen Berscheid and Elain H. Walster noted in 1974: 'It is exceedingly curious that, despite the voluminous literature on the determinants of date and mate selection, no investigation has focussed upon the fact that a cardinal principle of date selection is that the man must be as tall or taller than the woman.'[7] In the famous film, 'Casablanca', Humphrey Bogart and Ingrid Bergmann were made to embody the ideal couple: while the film was shot, the short Bogart was required to stand on a stool.

The Ideal Couple

My thesis is that what we regard as an ideal couple—i.e. a taller man with a shorter woman—is simply a matter of taste. As I will argue, Western ideas about the normal appearance of a couple express middle-class ideals and are culturally transmitted. These sex-specific height relations are not well understood because they function for the most part unconsciously.[8] The selection of partners demonstrates clearly that these ideals ensure 'that very nearly every couple will exhibit a height difference in the expected direction, transforming what would otherwise be a statistical tendency into a near certitude.'[9]

The unconscious production of this specific bodily regime in everyday life means that this social compulsion has gradually turned into a self-compulsion. The requirement to couple properly is present, is sensed, and is enforced by emotional reactions, grins, gazes, the inability to take our eyes off couples in which the man is shorter than the woman at his side. The unspoken subject, the threatening fantasy, which determines our aesthetic sensibilities, is produced by the very specific picture of the short man and the tall woman, who represent the reverse of the ideal middle-class couple. This 'unnatural couple' evokes shame because it poses a symbolic challenge to assumed middle-class gender relations. Both the man and the woman in this picture disobey the roles assigned to them. While the woman seems to demand something that is not hers by right—i.e., to be tall, secretly to be a man and compete in a man's world—the man seems to let himself be dominated, to relinquish power to a woman. The man appears to be a weakling and a loser. But, of course, he is also someone who is so sexually attractive that he has been chosen. This couple does not fulfil the designated ideal and must therefore be punished with looks of contempt; and the couple may, in fact, react with shame because they feel the pressure of social expectations. In this couple, the woman appears as a man and the man as a woman: a reversal of relations, a symbolic sex change, an attack on middle-class values and rules. That is the unspoken subject, the fantasy and ideology, reliably operating deep inside us, in our bodies.

Rules of Seeing

We do not notice the commonplace, for it is inconspicuous, not startling. The commonplace confirms what we already know, or at least what we think we know. We order and classify according to rules and methods we have learned to accept as yardsticks. Although we move unwittingly, as if naturally, our bodies also express social positioning and cultural knowledge, and are used primarily for differentiating and making hierarchical classifications—also in relation to gender.

We have learned to see through the lens of gender and to classify alleged physical differences: men are tall, broad-shouldered, and strong; while women are shorter, thinner, and weaker. Polarities, dualism, and contrasts are obviously important for making gender distinctions, attributing meaning to the presentation of femininity and masculinity, and for providing security in everyday life. 'One of the stages of individual development is marked by a penchant for hackneyed symbolic meanings in one's native culture. That is, individuals assume available stereotypes, using them to decode their environment and to lend "meaning" to their individual experiences.'[10] The more directed, judgemental and simplistic our view of the world, the less we can see. 'The fact that we are culturally shaped to "see" in gender-polarizing ways makes us "blind" in a way to the fact that there are always women who are stronger, more powerful, and taller than many representatives of the "stronger sex," or that many men and women exhibit a combination of "typical" female and "typical" male physical characteristics.'[11]

In other words, a middle-class discourse of bipolarity, dichotomy, and hierarchy continues to shape our perception, is reproduced invisibly in our body, and even influences our emotions. 'The height ratio between the sexes is a central element of heterosexual attraction. Couples are usually formed of men and women whose individual physical dimensions symbolically portray male superiority and a corresponding female inferiority.'[12] The ideal of the large male and the small female is also projected onto the past, as shown by an archaeological reconstruction recently described by archaeologist Margret Conkey: in the volcanic ashes of the excavation of Laetoli in Tanzania, two footprints from 3.6 million years ago have survived: two big feet and two small feet walked side by side. The position of the big toe and the instep indicate that this is the imprint of erect, human-like creatures—nothing more can be read from the excavation. Nevertheless, in the graphic reconstruction a familiar couple emerges: a taller male protectively embraces a shorter female.[13]

This reconstruction indicates the rigidity and precision of our ideas about couples. Of course, statistically, men are taller than women, but note that the formation of couples benefits from that fact and dictates that the only acceptable form is that of the taller man with the shorter woman. The opposite, which is certainly plausible, is culturally inadmissible and hence seldom encountered. Clearly a couple is poured into a symbolic mould that confirms the stereotype of the man who is physically superior (with intellectual superiority close at hand) to the shorter and more delicate woman assigned to him, who classically looks up to him. But scenes of aristocratic culture show clearly that couples can be presented in other symbolic terms. Women with colossal skirts and metre high wigs attract attention by their actual physical height,[14] and also through the cultural production of their physical size. Yet since the eighteenth century, this financially independent, educated, and politically influential woman, not yet trapped in what later came to be debilitating domesticity, has been deemed socially obsolete and unfeminine. She was swept away by a new, middle-class image of women, and the relationship of couples was newly defined in terms of an intimacy that applied to all social circles—even if this intimacy varied according to time and place.[15]

This change is artistically expressed in portraits of aristocratic couples who often reverse the height ratio between man and woman now perceived as normal. In upper-class circles at that time, women, whose physical bulk was produced and reinforced by fashionable accoutrements, towered over the men at their sides. This was seen as neither incongruous nor inappropriate.

But that pattern collapsed, and the middle-class arrangement that replaced it produced different cultural ideals and forms, which are also expressed graphically. First, the old-fashioned model of the shorter man with a taller woman is seldom portrayed and if it is, it is usually the sort of popular social caricature of the late eighteenth and nine-teenth centuries that allowed smirking and sneering. These drawings convey an unambiguous message, showing who is progressive and who is not. They reflect an urgent desire for orientation, illustrating an up-to-date, i.e. middle-class, presentation of the body. The norm was set by reference to the deviant, the unsanctioned. The point of view may vary, but these caricatures and illustrations must have been accepted, and are to be understood as signs that not only reproduce reality, but reflect it.[16] What produced this collapse?

The Middle-Class Gender Project

The eighteenth century in Western Europe was marked by political and social tensions, a new economic order, luxury and poverty; in short, by changes and revolutions that shaped the everyday life of growing numbers of people, and forced a renegotiation of social rules.

Around mid-century, values gradually shifted from the old to the new order, from collectivity to individuality.[17] New lifestyles were designed and gender relations were no longer determined by tradition and family interests in the old sense. The ideal union was two people who were emotionally compatible and who, most importantly, were in love. According to cultural historian Ulrike Prokop, 'The aristocratic model of separating passion and marriage is rejected by the middle-class as cold and artificial. For the middle-classes, self-determination and individual freedom required strict self-control and a strong sense of morality. Passion, courtly marriage, and adventure belonged to an aristocratic or feudal lifestyle; they were useless for the complicated mediation between extreme individualism and law developed by the middle-class.'[18] Novels and guidebooks discussed the dimensions of this new ideal relationship between the sexes within the revolutionary ideas of liberty, equality, and fraternity, distinguishing it from the obsolete courtly relationship of couples. The utopian design posited by these books soon became a prototype for everyday experiment.

Ulrike Prokop noted that women participated in this ideological modification, but that they did not actively structure the regime of rights and political power that emerged towards the end of the eighteenth century.[19] She goes on to state: 'In the eighteenth century, the connections between female experience and self-understanding disintegrated, while male forms of cooperation were organized as centres of economic and political power. In a culture trained in literacy and a scientific worldview, female traditions were rejected and eventually eradicated.'[20] For uneducated women, wrenched out of the old traditions of home, family,[21] and neighbourhood, this development entailed a fixation on the breadwinner, the man who acted in the public realm. For women, unlike men, marriage determines social security, a condition that lent authority to the man in a middle-class couple. In *Die Illusion vom Grossen Paar*, Ulrike Prokop summed up:

In the analyses of Horkheimer, Marcuse, and Habermas, the modern middle-class lifestyle was judged to be a requirement for individualization and thus ultimately as progress for both genders. This ignores the fact that the condition of women in work and culture began to deteriorate in the eighteenth and nineteenth centuries. The shift from a home-based economy to a market economy, the exclusion of women from new economic centres of power (management and market-oriented production), their isolation from professional life as the result of the social differentiation of careers, and the containment of women's work and social relations to the family (processes affecting the upper middle-class), have all been retrograde for women. In terms set forth by the new middle classes, women lacked the socio-economic basis for becoming a self-fulfilled individual.[22]

Did this change in gender relations in the eighteenth and nineteenth centuries influence aesthetic sensibility to the build and physique of men and women? Does heterosexual attraction depend on who is short or who is entitled to be tall, to overwhelm, to dominate physically? The ideal spouse for the middle-class man is the woman who is shorter and thinner than he (anyone else would be threatening). The opposite—i.e. a short delicate man and a tall corpulent woman—is unbearable both for the couple and for others, because it is banned by the internal images and needs of middle-class society.

But this height ratio was not immediately internalized. For a long time social suitability remained a more important criterion for relationships than physical suitability, and the old estates system proved stronger than the new gender system. The *ancien régime* (even after it ended) determined who suited whom and who belonged with whom.

Figure 20 indicates that the pattern of a tall man and a short woman was insignificant in the last third of the eighteenth century. In Daniel Chodowiecki's 1774 drawing of a wedding procession, the couples portrayed are not generally acceptable in today's terms. Two of the four women clearly tower over their husbands, not only with their own physical bulk, but also with their high hairdos.[23] In the 'Göttinger Taschenkalender' of 1779 and 1780, Chodowiecki, a popular German illustrator of his day and a typical representative of middle-class self-consciousness,[24] also published two series of copper engravings titled 'Natural and Artificial Ways of Life' (Fig. 21). The first couple (right) is old-fashioned, i.e. artificial, powdered, and all dressed up. They are contrasted with a biblical couple representing humankind in nature.[25] 'Nature'—meaning 'naturalness'—is the motto of middle-class emancipation in the eighteenth century: It does not necessarily mean primeval or original but is juxtaposed to the 'unnatural' world of the aristocracy.[26]

Chodowiecki's didactic portrayal of social relations through stark contrasts are somewhat wooden and overdrawn. What interests us is that the height ratios of the two couples shift. The massive aristocratic woman, marked by a wig, a hat, high-heeled shoes, and gargantuan gown, and who seems to burst out of the picture, turns into a thin woman, who looks up to the taller man, a woman who is 'natural' and emotionally pure. The figure of the man also changes. The narrow-shouldered, affected aristocrat, who stands still and looks frail, is

FIG. 20. 'The Wedding Procession' by Daniel Chodowiecki, 1774. From Sabine Gieske (ed.), *Jenseits vom Durchshnitt. Vom Kleinsein & Großsein* (Marburg, 1998), 75. Courtesy of the Jonas Press, Marburg.

contrasted with an idealized man: broad-shouldered, taller than the woman, a man in motion, striding ahead vigorous and healthy.

Chodowiecki's graphic portrayals of 'artificial and natural ways of life' is one of many pictures with the same instructive and moral intent that were produced and distributed in the last third of the eighteenth century. Their (presumed) effect was reinforced by a flood of magazine articles, novels, poems, and treatises, in which the new ideal middle-class lifestyle and concomitant notions of female and male 'destiny' were phrased in conceptual or poetic terms. These publications also popularized the views of middle-class philosophers (for example, Rousseau, Kant, Fichte) on marriage, the duties of husband and wife, and the natural features of 'femininity' and 'masculinity' as the building blocks of their theoretical project for a new bourgeois age.[27]

Caricatures of the use of wigs were also popular in the late eighteenth century. In a moral tone, the middle-class illustrator made fun of old-fashioned women who wore sweeping baroque garb, high heels, and immensely high hairdos (Fig. 22). In the illustrations, these women appear massive, which, according to the spiteful critic,

Natur   *Afectation*

FIG. 21. 'Natural and Artificial Ways of Life' by Daniel Chodo-wiecki, 1779/1780. From Sabine Gieske (ed.), *Jenseits vom Durchschnitt. Vom Kleinsein & Großsein* (Marburg, 1998), 76. Courtesy of the Jonas Press, Marburg.

obstructs and endangers social life. Even more to the point, the big woman in Chodowiecki's illustration looks superior. She even throws the shorter man to the floor, nearly devouring him with her gigantic wig. She generates chaos, she looks threatening, she must no longer be allowed to indulge in foul play—at least from the point of view of the middle-class, progressive illustrator, commenting on the social questions of his time.

The transition to what is now considered an aesthetically pleasing form of a couple continued into the nineteenth century. For some time, the requirement that the man be taller was not absolute. In his 1842 *Die kleinen Leiden des Menschlichen Lebens*, Oskar L. B. Wolff, under the pseudonym of Pliny, the Younger, wrote about 'tall women': 'They also have a curious penchant for short men and like to marry them.'[28] And as late as 1889, in *Natural Inheritance*, Francis Galton stated that the choice of a spouse seems to be 'wholly independent of

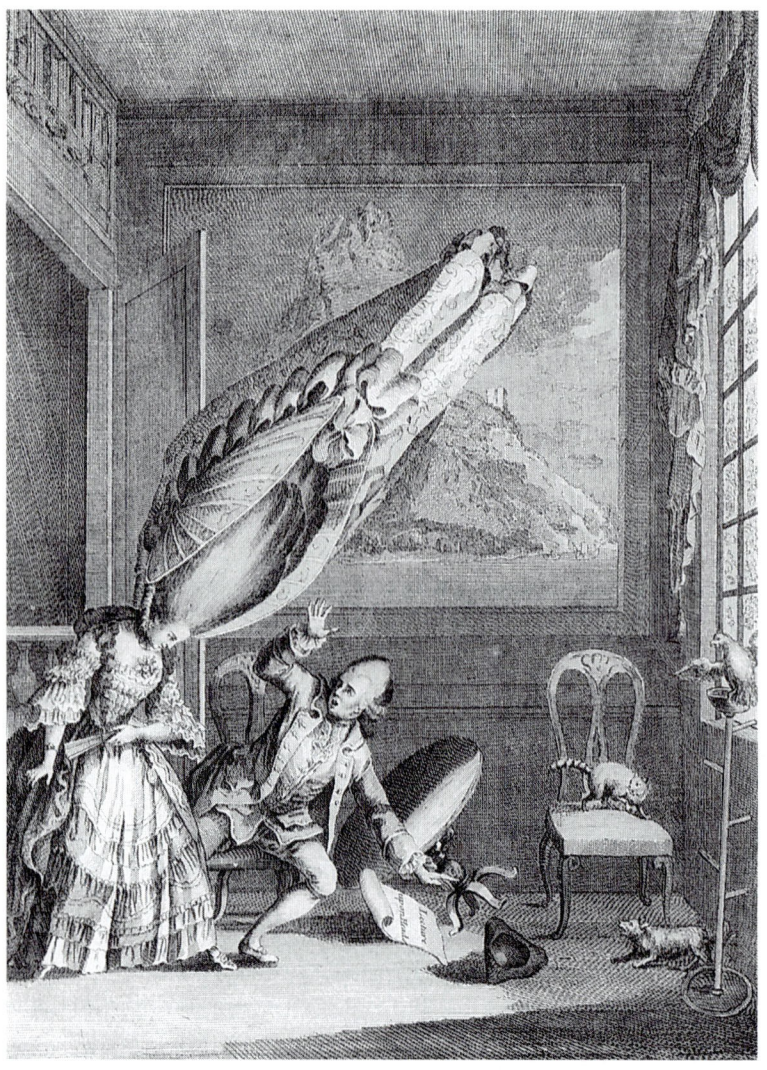

FIG. 22.   The French Lady, 1770. From Sabine Gieske (ed.), *Jenseits vom Durchschnitt. Vom Kleinsein & Großsein* (Marburg, 1998), 78. Courtesy of the Jonas Press, Marburg.

stature'.[29] The top hat, the mark of a financially secure middle-class man, should be studied more closely, for no other item of clothing was better suited to make the man look taller and the woman shorter, and thus to establish the modern gender ratio (Fig. 23). Moreover, the discussion of women's shoe fashion in the late eighteenth century is also instructive. While aristocratic women's high-heeled shoes were criticized harshly, flat shoes were seen as especially suitable for a stroll in the fresh air by the new middle classes.[30]

But the new model of the ideal couple was constantly being redesigned—especially visually. Guidebooks, entertainment magazines, and calendars—the expressions of cultural self-confirmation and reflections of contemporary ideals—proclaimed what was morally astute and what was reprehensible. 'Calendars, almanacs, magazines for the entertainment and edification of both sexes, especially the female, became very fashionable in the last decades of the eighteenth century. They were popular (like today's tabloids) and widely circulated. Like the photos in our journals, drawings and colored plates in almanacs, calendars, etc., served to impart its norms and rules graphically.'[31] These media presented not only descriptions, but also pictorial models of how the modern woman was to behave towards her intended, and how the couple was to appear. If she was not obviously shorter, obedient, and oriented towards the man, she was not chosen. An 1840 cartoon, entitled 'The Short Man and the Tall Woman',[32] showed polemically what happens if the reverse ratio were established: the tyrannical, tall female orders her degraded husband to do housework; she's the boss and gives herself airs in a ridiculous, upside-down, unacceptable world. The shorter woman, by contrast, was culturally acceptable and often shown leaning on a bigger, physically superior man. The body itself was presented as proof of this difference, not only visually in graphic media, but also in theoretical discussions of the bodies of women and men.[33]

Body Discourse

Throughout the eighteenth and nineteenth centuries, physicians and educators emphasized the physical differences of the sexes, defining normal and deviant, healthy and sick.[34] Thus, in 1787, Ernst Brandes wrote in *Über die Weiber*: 'Nature created man as lord of creation. Is the woman not made smaller, more delicate, weaker?'[35] The new

FIG. 23. An elegant couple in Hyde Park from the *Berliner Illustrirte Zeitung*, 20 (1906), 344.

middle-class epistemology concentrated on a connection between physical aptitudes and mental ability, making alleged distinctions between male and female anatomy. Since man has stronger muscles, tighter nerves, and more rigid sinews, he has a heavier skeleton, more courage, a bolder sense of enterprise, and a greater intellect.[36] The royal Privy Councillor of Fulda, doctor of medicine Konrad Anton Zwierlein, also supported this notion: 'A person of large build, especially when the height and breadth of the body are truly in proportion, has something superior and dignified in his nature. This refers mainly to men; the fair sex seems to be an exception to this point.'[37] Johannes Georg Krünitz advanced a remarkably similar argument in his Encyclopedia of 1797, under the entry, 'Body—size, height, breadth, scale, proportion, stature'.

A man is always displeasing, no matter how splendid the form of his body, if he is not suitably tall. A man of tall build, especially when the height and breadth of the body are in a true ratio, has something superior, dignified, and majestic in his nature, that evokes esteem for him in everyone else. This applies mainly to men. . . . Females in general are not only shorter than males, but they also have a smaller head, a longer neck, more compact shoulders, and chest. . . . Their muscles are less visible, their contours are more even and indistinct, and their movements are more delicate.[38]

In the words of sociologist Claudia Honegger: 'Strong organs and hard reason for men; weakness and sensitivity for women.'[39] She is referring here to the physiologist Ph. Fr. Walther, who wrote in the early nineteenth century: 'In general, the human female is shorter, less developed, softer in all parts and has a softer, smaller skeleton. Her breathing is more limited, the thorax less spacious, the vascular system is less developed, the muscles less taut, rough, sensitive. The whole irritable system is more predominant in men than in women.'[40] 'Female nature' must be adapted 'to its proper role.'[41] The male body also had to conform to the demands of the modern age. The ideal man of the nineteenth century had to be strong-willed and exercise strength and vigour in pursuit of his professional goals. In 1864, Carl Reclam wrote: 'Daily struggle and battle inhibits man's sensibility to feeling. . . . The daily practice of his trade hardens man and must harden him because otherwise he could not maintain himself in a flourishing state of health and vigor.'[42] Reclam's readers are told very clearly: 'Man is armed by nature for his harder life work. From the moment of birth, the majority of newborn boys are longer and heavier than the majority of newborn girls. This difference continues until the ages from 2 to 12 when boys and girls have the least difference in body size and height, as well as intellect. After age 12 their growth patterns as well as their intellectual development is again different.'[43] These conclusions were prevalent in medical circles and also appeared in popular magazines, many of which were aimed at a female audience. From the late eighteenth century on, male and female bodies were portrayed asymmetrically, and were soon perceived that way in fact: they now carried a new 'burden' as social historian Ulrike Döcker put it.[44]

The Middle-Class Man Grown Up

To sum up the physical difference of the two sexes at a glance, we can say that, in the man, irritability predominates, in the female body elasticity and sensibility predominate. In the former, everything is calculated for a greater external effect, in the latter for internal formation and acceptance of external influence. . . . The man's aspiration is aimed outward at a broad field, the woman cares for the narrow circle of the family. Man's mind is creative, woman's mind is receptive and preserving. . . . Man's ideal is greatness, his sense of beauty is directed almost exclusively at woman. She admires the greatness in man, her only sphere is found in the world of beauty.[45]

Thus, a picture of a man and a woman emerges: he stands erect and straight, his form is strong (as, for example, the portraits of soldiers and athletes in the nineteenth and early twentieth century).[46] The middle-class woman expects greatness from her husband, who has grown up, has become tall and will soon be upwardly mobile. But to fulfil this requirement, he needs a shorter woman. She mirrors him, reflects that he is tall and that there is someone who is even weaker than he is. And she, the delicate one, acting within the family, needs him, the successful man, who has chosen her alone, the man the world looks up to; and thus she is also exalted.

This image was developed at the same time as other notable advances in the eighteenth and nineteenth centuries. It was an age of inventions, achievements, and discoveries that produced the middle-class period and the modern age. Amid these vast cultural upheavals, the middle-class man—who was actively involved in them as engineer, scientist, or merchant—felt small, threatened, and powerless; yet at the same time, he appeared tall. But such feelings of inadequacy were

FIG. 24. 'How to Grow Taller'. An advertisement from the Cartilage Company of Paris from the *Berliner Illustrirte Zeitung*, 47 (1906), 757.

channelled, concealed, and averted, displaced to relations of couples, where they could still be controlled and determined. Thus, the popular size distinction in the relation of couples can be seen as the result of a collective fantasy of modernity, as an expression of pride in progress, and as a sign of the anxiety and threat it entails. The spirit of the modern age settled indiscriminately in the body.

Summary

'In view of the esteem for the "appropriate" ratio of size and the care that is taken to produce male physical superiority in society, it is no wonder that "reverse" ratios are extremely rare. In these cases, elaborate tricks (such as elevator shoes for men) were manufactured to simulate a standard height relation.'[47] In the late nineteenth century, advertisements for many dubious products responded to the desire among men to seem taller. Stretching cartilage, for example, was supposed to make one taller, at least that was the promise of 'The Cartilage Company' of Paris in 1906 (Fig. 24). According to the ad, through this procedure, the body gained two to ten centimetres height. 'If you want to increase your height, to be able to look over a crowd of people or not feel anxious next to tall people, and to enjoy the advantages that tall persons always have, write for our brochure, How to Grow.'[48] Today the discomfort of being too short or too tall can be avoided by taking hormones that claim to achieve the desired change in the body invisibly, privately, and painlessly, thus not only satisfying social pressures, but also stabilizing the regime of perception.

The middle-class gender design of the 'ideal couple' is reproduced even now. 'It is social exclusion that makes physical difference a problem and conspicuous. And conversely, it is physical failure vis-à-vis the power of the average that entails social exclusion. . . . In this respect, the desire to be like everyone else is a desire for social integration in the positive sense.'[49] This is why men and women (especially those who consider themselves liberated) agree to a design that limits and determines their choice of partner. And this is where the circle of the unconsciously embedded presentation of the couple is closed. Only when the internalized structure is recognized as such can it be renegotiated and remade. As long as it remains unconscious, the 'abnormal couple' reaps only smiles of aversion. But times are obviously changing and so are standards.

Notes

This is a revision of my essay 'Großer Mann und Kleine Frau. Das ideale Paar im bürgerlichen Entwurf', in Sabine Gieske (ed.), *Jenseits vom Durchschnitt. Vom Kleinsein & Großsein* (Marburg, 1998), pp. 61–94.

1. See Esther Fischer-Homberger, *Hunger – Herz – Schmerz – Geschlecht. Brüche und Fugen im Bild von Leib und Seele* (Bern, 1997), pp. 79 and 83.
2. See Günther G. Bauer, 'Hofriesen und Schauriesen des 17. und 18. Jahrhunderts', in *Riesen*, ed. by Roland Floimair and Lucia Luidold (Salzburg, 1996), pp. 111–14.
3. See Erving Goffman, *Gender Advertisements* (New York, 1979), p. 28.
4. Ibid.
5. See Gail Tom and Jeanne Shevell, 'The Height of Success', *Sociology and Social Research*, 71 (1986), p. 15; see also Tuvia Melamed and Nicholas Bozionelos, 'Managerial Promotion and Height', *Psychological Reports*, 71 (1992), pp. 587–93.
6. Rudolf Stichweh, 'Der Körper des Fremden', in Michael Hagner (ed.), *Der falsche Körper. Beiträge zu einer Geschichte der Monstrositäten* (Göttingen, 1995), p. 183. Among other things, he also refers to the fact that one of the classical topoi in the description of the Jews, at the turn of the twentieth century, was physical height. Jews, it was stated, were shorter than Germans, Russians, Anglo-Saxons, and Scandinavians, but just as tall as Southern Europeans. See p. 182.
7. Ellen Berscheid and Elain H. Walster, 'Physical Attractiveness', in Leonard Berkowitz (ed.), *Advances in Experimental Social Psychology*, vol. 7 (New York, 1974). Quoted in John S. Gillis and Walter E. Avis, 'The Male-Taller Norm in Mate Selection', *Personality and Social Psychology Bulletin*, 6 (1980), p. 396.
8. See Michel Foucault, *Der Wille zum Wissen. Sexualität und Wahrheit*, 1 (Frankfurt/ a.M., 1977).
9. Goffman, *Gender Advertisements*, p. 28. In this connection, see the studies on 'male-taller norm' by Hugo G. Beigel, 'Body Height in Mate Selection', *Journal of Social Psychology*, 39 (1954), pp. 257–68.
10. Irene Dölling, *Der Mensch und sein Weib. Frauen- und Männerbilder. Geschichtliche Ursprünge und Perspektiven* (Berlin, 1991), p. 47.
11. Ibid., p. 55.
12. Gitta Mühlen Achs, *Wie Katz und Hund. Die Körpersprache der Geschlechter* (Munich, 1993), p. 47. Studies should be made to determine whether size patterns also play a role in homosexual relations; and if so, what role they play in the relational structure.
13. Bruni Kobbe, 'Das schräge Bild der alten Eva. Stupide, geistlos, unterwürfig. So zeichneten Forscher die urgeschichtliche Frau. Sie irrten', *Die Zeit*, Nr. 1: 26 (December 1997), p. 34.
14. This indicates that people who historically did not suffer from malnutrition were bigger on average than their contemporaries, and refers particularly to socially higher classes who were physically larger than their subordinates or servants. See John Komlos, *Ernährung und wirtschaftliche Entwicklung unter Maria Theresia und Joseph II. Eine anthropometrische Geschichte der Industriellen Revolution in der Habsburgmonarchie* (Vienna, 1991); Astrid

Schumacher, 'Zur Bedeutung der Körperhöhe in der menschlichen Gesellschaft', Diss. (Hamburg, 1980). This social difference in size can be traced in the 1960s. Studies made in 1961 showed that workers' children were on average almost five centimetres shorter than children from an academic milieu. See Georg Kenntner, 'Die Veränderungen der Körpergröße des Menschen. Eine biogeographische Untersuchung', Diss. (Karlsruhe, 1963), p. 161.

15. We shall discuss below the understanding of the process of change concerning the ratio of size in relations of couples, at least roughly, and of 'nobility' and 'middle class'. Obviously it is problematic to think here at any given moment of a homogeneous culture, closed in itself and impervious to external culture; the same also applies to peasant culture. Moreover, different criteria had to be considered according to regional, religious, gender, and mentality levels. Emphasizing the French Revolution as the historical turning point between the *ancien régime* and middle-class society, should also be avoided. See Lynn Hunt, *Symbole der Macht. Die Französische Revolution und der Entwurf einer politischen Kultur* (Frankfurt/ a.M., 1989), pp. 11–30.

16. See Martin Scharfe, 'Die Volkskunst und ihre Metamorphose', *Zeitschrift für Volkskunde*, 70 (1974), p. 233.

17. See Ulrike Prokop, *Die Illusion vom Großen Paar*, Vol. 2: *Das Tagebuch der Cornelia Goethe. Psychoanalytische Studien zur Kultur*, ed. by Alfred Lorenzer (Frankfurt/ a.M., 1991), pp. 104 ff.

18. Ibid., pp. 137–9.

19. Ibid., p. 7.

20. Ibid.

21. Family does not mean the middle-class nuclear family, but the household with the patriarch and matriarch, servants, children, and relatives. See Heidi Rosenbaum, *Formen der Familie. Untersuchungen zum Zusammenhang von Familienverhältnissen, Sozialstruktur und sozialem Wandel in der deutschen Gesellschaft des 19. Jahrhunderts* (Frankfurt/ a.M., 1982), p. 116.

22. Prokop, *Die Illusion*, vol. 2, p. 123. And see Joan Landes (ed.), *Feminism, the Public and the Private* (Oxford, 1998).

23. There is a question of which rules (of seeing and aesthetic reception) relations of size in rural societies or worker classes were subject to and when changes were indicated here. Different criteria may be assumed to be relevant here than in middle-class relations, since the gender relations here offered women different areas of power, or that other economic necessities prevailed. The aspect of physical capacity for work, physical strength, and physical presence is certainly quite important as well.

24. See *Bürgerliches Leben im 18. Jahrhundert. Daniel Chodowiecki 1726–1801. Zeichnungen und Druckgraphik*, rev. by Peter Märker, ed. by Klaus Gallwitz and Margret Stuffmann, Städelsches Kunstinstitut und Städtische Galerie, Frankfurt/ a.M., 8 June 1978 to 20 August 1978 (Frankfurt/ a.M., 1978), pp. 15, 17.

25. See Dölling, *Der Mensch*, pp. 107 f.

26. See *Bürgerliches Leben*, p. 121.

27. Dölling, *Der Mensch*, p. 110.

28. Pliny the Younger [Oskar L. B. Wolff], *Die kleinen Leiden des Menschlichen Lebens* (Leipzig, 1842), p. 142.

29. Quoted in Gillis and Avis, 'The Male-Taller Norm', p. 396. These authors,

however, assumed that Galton 1889 reached this result erroneously from a sample that was too small, because by 1903, Pearson and Lee came to opposite conclusions. See Karl Person and A. Lee, 'On the Laws of Inheritance in Man: I. Inheritance of Physical Characters', *Biometrika*, 2 (1903), pp. 357–97; Francis Galton, *Natural Inheritance* (London, 1889). In their study of 1,000 couples, they postulate a connection between selection of partner and physical size. Galton's study may not have been based on an incorrect observation, but on an observation at a historical turning-point. Between the end of the nineteenth century and the beginning of the twentieth was precisely when that conspicuous shift of partner choice concerning gender-specific height patterns seems to have taken place. But Gillis and Avis did not take account of that in their 1980 publication.

30. See Joanna Schulz, '"Bei dem Gange gehörig gerade". Zur Ambivalenz des Haltungsideals für Bürgerinnen', in *Der aufrechte Gang. Zur Symbolik einer Körperhaltung*, Projekt des Ludwig-Uhland-Instituts für empirische Kulturwissenschaft der Universität Tübingen (Tübingen, 1990), pp. 26 ff.

31. Dölling, *Der Mensch*, p. 104.

32. See '"Der kleine Mann und die große Frau." Neuruppiner Druckgraphik um 1840', in Werner Hirte (ed.), *Die Schwiegermutter und das Krokodil. 111 bunte Bilderbogen für alle Land- und Stadtbewohner soweit der Himmel blau ist* (Munich, 1969), p. 13.

33. See Barbara Duden, *Geschichte unter der Haut. Ein Eisenacher Arzt und seine Patientinnen um 1730* (Stuttgart, 1987).

34. See the studies of the cultural history of the body by Duden, *Geschichte*; Thomas Laqueur, *Auf den Leib geschrieben. Die Inszenierung der Geschlechter von der Antike bis Freud* (Frankfurt/ a.M., 1992); Londa Schiebinger, *Schöne Geister. Frauen in den Anfängen der modernen Wissenschaft* (Stuttgart, 1993), ch. 8.

35. Ernst Brandes, *Über die Weiber* (Leipzig, 1787). Quoted in Claudia Honegger, *Die Ordnung der Geschlechter. Die Wissenschaften vom Menschen und das Weib 1750–1850* (Frankfurt/ a.M., 1991), p. 54.

36. See Honegger, *Die Ordnung*, pp. 65 ff.

37. Konrad A. Zwierlein, *Der Arzt für Liebhaberinnen der Schönheit* (Heidelberg, 1789), p. 16.

38. Johann G. Krünitz, *Oekonomisch-technologische Encyklopädie oder allgemeines System der Stats-Stadt-Haus-und Land-Wirthschaft, und der Kunst-Geschichte, in alphabetischer Ordnung* (Berlin, 1797), pp. 85 ff.

39. Honegger, *Die Ordnung*, p. 159.

40. Ph. Fr. Walther, *Physiologie des Menschen mit durchgängiger Rücksicht auf die comparative Physiologie der Thiere*, 2 vols. (Landshut, 1807–8), vol. 2, pp. 375 ff. Quoted in Honegger, *Die Ordnung*, p. 189. The term 'irritable' means stimulation, excitement. Brockhaus's *Conversations-Lexikon* of 1884, said: 'The term of I. was introduced into medicine by the English physician Glisson (1597–1677) and by Gorter in Harderwijk (1688–1762), but was developed particularly by A. von Haller, whose special contribution consisted of inserting an autonomous I. into the muscles, independent of the influence of the nerves, a view that has often been challenged by the latest research, but has been guaranteed.' Brockhaus' *Conversations-Lexikon. Allgemeine deutsche Real-Encyklopädie. Dreizehnte vollständig umgearbeitete Auflage*, 9 (Leipzig, 1884), p. 674.

41. Prokop, *Die Illusion*, vol. 2, p. 126.
42. Carl Reclam, *Des Weibes Gesundheit und Schönheit. Ärztliche Rathschläge für Frauen und Mädchen* (Leipzig and Heidelberg, 1864), p. 4 ff.
43. Ibid., p. 5 ff.
44. See Ulrike Döcker, *Die Ordnung der bürgerlichen Welt. Verhaltensideale und soziale Praktiken im 19. Jahrhundert* (Frankfurt/ a.M., 1994), p. 85.
45. Karl Ernst von Baer, *Vorlesungen über Anthropologie für den Selbstunterricht bearbeitet*, vol. 1 (Königsberg, 1824), p. 513. Quoted in Honegger, *Die Ordnung*, p. 210.
46. See Bernd J. Warneken, 'Biegsame Hofkunst und aufrechter Gang. Körpersprache und bürgerliche Emanzipation um 1800', in *Der Aufrechte Gang: Zur Symbolik einer Körperhaltung*, Projekt des Ludwig-Uhland-Instituts für empirische Kulturwissenschaft der Universität Tübingen (Tübingen, 1990), pp. 11–23.
47. Achs, *Wie Katz*, p. 47.
48. *Berliner Illustrirte Zeitung*, 47 (1906), p. 832.
49. Frigga Haug (ed.), *Sexualisierung der Körper*, Argument-Sonderband 90 (Göttingen, 1991), pp. 73 ff.

Part V. **Restrained Bodies**

14 The Anthropometry of Barbie

Unsettling Ideals of the Feminine Body in Popular Culture

Jacqueline Urla and Alan C. Swedlund

It is no secret that thousands of healthy women in the United States perceive their bodies as defective. The signs are everywhere: from potentially lethal cosmetic surgery and drugs to the more familiar routines of dieting, curling, crimping, and aerobicizing, women seek to take control over their unruly physical selves. Every year at least 150,000 women undergo breast implant surgery (Williams 1992), while Asian women have their noses rebuilt and their eyes widened to make themselves look 'less dull' (Kaw 1993). Studies show that the obsession with body size and the sense of inadequacy start frighteningly early; as many as 80 percent of 9-year-old suburban girls are concerned about dieting and their weight (Bordo 1991: 125). Reports like these, together with the dramatic rise in eating disorders among young women, are just some of the more noticeable fallout from what Naomi Wolf calls 'the beauty myth'. Fueled by the hugely profitable cosmetic, weight-loss, and fashion industries, the beauty myth's glamorized notions of the ideal body reverberate back upon women as 'a dark vein of self hatred, physical obsessions, terror of aging, and dread of lost control' (Wolf 1991: 10).

It is this conundrum of somatic femininity that female bodies are never feminine enough, that they must be deliberately and oftentimes painfully remade to be what 'nature' intended—a condition dramatically accentuated under consumer capitalism—that motivates us to focus our inquiry into deviant bodies on images of the feminine ideal. Neither universal nor changeless, idealized notions of both masculine and feminine bodies have a long history that shifts considerably across

From Jacqueline Urla and Alan C. Swedlund, 'The Anthropometrey of Barbie: Unsettling Ideals of the Feminine Body in Popular Culture', in *Deviant Bodies: Critical Perspectives on Difference in Science and Popular Culture*, eds. Jennifer Terry and Jacqueline Urla (Indiana University Press, 1998), pp. 240–87.

time, racial or ethnic group, class, and culture. Body ideals in twentieth-century North America are influenced and shaped by images from classical or 'high' art, the discourses of science and medicine, and increasingly via a multitude of commercial interests, ranging from mundane life insurance standards to the more high-profile fashion, fitness, and entertainment industries. Each have played contributing, and sometimes conflicting, roles in determining what will count as a desirable body in the late-twentieth century United States. In this essay, we focus our attention on the domain of popular culture and the ideal feminine body as it is conveyed by one of pop culture's longest lasting and most illustrious icons: the Barbie doll.

Making her debut in 1959 as Mattel's new teenage fashion doll, Barbie rose quickly to become the top-selling toy in the United States. Thirty-four years and a woman's movement later, Barbie dolls remain Mattel's best-selling item, netting over one billion dollars in revenues worldwide (Adelson 1992), or roughly one Barbie sold every two seconds (Stevenson 1991). Mattel estimates that in the United States over 95 percent of girls between the ages of 3 and 11 own at least one Barbie, and that the average number of dolls per owner is seven (E. Shapiro 1992). Barbie is clearly a force to contend with, eliciting over the years a combination of critique, parody and adoration. A legacy of the postwar era, she remains an incredibly resilient visual and tactile model of femininity for prepubescent girls headed straight for the twenty-first century.

It is not our intention to settle the debate over whether Barbie is a good or bad role model for little girls or whether her unrealistic body wrecks havoc on girls' self-esteem. Though that issue surrounds Barbie like a dark cloud, such debates have too often been based on literal-minded, decontextualized readings of popular culture. We want to suggest that Barbie dolls, in fact, offer a much more complex and contradictory set of possible meanings that take shape and mutate in a period marked by the growth of consumer society, intense debate over gender and racial relations, and changing notions of the body. Building on Marilyn Motz's (1983) study of the cultural significance of Barbie, and fashion designer extraordinaire BillyBoy's adoring biography, *Barbie, Her Life and Times*, we want to explore not only how it is that this popular doll has been able to survive such dramatic social changes, but also how she takes on new significance in relation to these changing contexts.

We begin by tracing Barbie's origins and some of the image makeovers she has undergone since her creation. From there we turn to an experiment in the anthropometry of Barbie to understand how she compares to standards for the 'average American woman' that were emerging in the postwar period. Not surprisingly our measurements show Barbie's body to be thin—very thin—far from anything approaching the norm. Inundated as our society is with conflicting and exaggerated images of the feminine body, statistical measures can help us to see that exaggeration more clearly. But we cannot stop there. First, as our brief foray into the history of anthropometry shows, the measurement and creation of body averages have their own politically inflected and culturally biased histories. Standards for the 'average' American body, male or female, have always been imbricated in histories of nationalism and race purity. Secondly, to say that Barbie is unrealistic seems to beg the issue. Barbie *is* fantasy: a fantasy whose relationship to the hyperspace of consumerist society is multiplex. What of the pleasures of Barbie bodies? What alternative meanings of power and self-fashioning might her thin body hold for women/girls? Our aim is not, then, to offer another rant against Barbie, but to clear a space where the range of her contradictory meanings and ironic uses can be contemplated: in short, to approach her body as a meaning system in itself, which, in tandem with her mutable fashion image, serves to crystallize some of the predicaments of femininity and feminine bodies in late twentieth-century North America.

A DOLL IS BORN

> Parents thank us for the educational values in the world of Barbie. . . . They say that they could never get their daughters well groomed before—get them out of slacks or blue jeans and into a dress . . . get them to scrub their necks and wash their hair. Well, that's where Barbie comes in. The doll has clean hair and a clean face, and she dresses fashionably, and she wears gloves and shoes that match.
>
> (Ruth Handler, 1964, quoted in Motz)

Legend has it that Barbie was the brainchild of Mattel owner Ruth Handler, who first thought of creating a three-dimensional fashion doll after seeing her daughter play with paper dolls. As an origin story this one is touching and no doubt true. But Barbie was not the first doll of her kind, nor was she just a mother's invention. Making sense of Barbie requires that we look to the larger sociopolitical and cultural milieu that made her genesis both possible and meaningful. Based on a German prototype, the 'Lili' doll, Barbie was from 'birth' implicated in the ideologies of the Cold War and the research and technology exchanges of the military-industrial complex. Her finely crafted durable plastic mold was, in fact, designed by Jack Ryan, well known for his work in designing the Hawk and Sparrow missiles for the Raytheon Company. Conceived at the hands of a military-weapons-designer-turned-toy-inventor, Barbie dolls came onto the market the same year that the infamous Nixon–Krushchev 'kitchen debate' took place at the American National Exhibition in Moscow. Here, in front of the cameras of the world, the leaders of the capitalist and socialist worlds faced off, not over missile counts, but over 'the relative merits of American and Soviet washing machines, televisions, and electric ranges' (May 1988: 16). As Elaine Tyler May has noted in her study of the Cold War, this much-celebrated media event signaled the transformation of American-made commodities and the model suburban home into key symbols and safeguards of democracy and freedom. It was thus with fears of nuclear annihilation and sexually charged fantasies of the perfect bomb shelter running rampant in the American imaginary, that Barbie and her torpedo-like breasts emerged into popular culture as an emblem of the aspirations of prosperity, domestic containment, and rigid gender roles that were to characterize the burgeoning post-war consumer economy and its image of the American Dream.

Marketed as the first 'teenage' fashion doll, Barbie's rise in popularity also coincided with, and no doubt contributed to, the postwar creation of a distinctive teenage lifestyle. Teens, their tastes, and their behaviors were becoming the object of both sociologists and criminologists as well as market survey researchers intent on capturing their discretionary dollars. While J. Edgar Hoover was pronouncing 'the juvenile jungle' a menace to American society, retailers, the music industry, and moviemakers declared the 13 to 19-year-old age bracket 'the seven golden years' (Doherty 1988: 51–2).

Barbie dolls seemed to cleverly reconcile both of these concerns by personifying the good girl who was sexy, but didn't have sex, and was willing to spend, spend, spend. Amidst the palpable moral panic over juvenile delinquency and teenagers' new-found sexual freedom, Barbie was a reassuring symbol of solidly middle-class values. Popular teen magazines, advertising, television, and movies of the period painted a highly dichotomized world divided into good (i.e., middle-class) and bad (i.e., working-class) kids: the clean-cut, college-bound junior achiever versus the street-corner boy; the wholesome American Bandstander versus the uncontrollable bad seed (cf. Doherty 1988; and Frith 1981, for England). It was no mystery where Barbie stood in this thinly disguised class discourse. As Motz notes, Barbie's world bore no trace of the 'greasers' and 'hoods' that inhabited the many B movies about teenage vice and ruin. In the life Mattel laid out for her in storybooks and comics, Barbie, who started out looking like a somewhat vampy, slightly Bardot-esque doll, was gradually transformed into a ' "soc" ' or a "frat"—affluent, well-groomed, socially conservative' (Motz 1983: 30). In lieu of backseat sex and teenage angst, Barbie had pajama parties, barbecues, and her favorite pastime, shopping.

Every former Barbie owner knows that to buy a Barbie is to lust after Barbie accessories—that pair of sandals and matching handbag, canopy bedroom set, or country camper. Both conspicuous consumer and a consumable item herself, Barbie surely was as much the fantasy of U.S. retailers as she was the panacea of middle-class parents. For every 'need' Barbie had, there was a deliciously miniature product to fulfill it. As Paula Rabinowitz has noted, Barbie dolls, with their focus on frills and fashion, epitomize the way that teenage girls and girl culture in general have figured as accessories in the historiography of postwar culture; that is as both essential to the burgeoning commodity culture as consumers, but seemingly irrelevant to the central narrative defining Cold War existence (Rabinowitz 1993). Over the years, Mattel has kept Barbie's love of shopping alive, creating a Suburban Shopper Outfit and her own personal Mall to shop in (Motz 1983: 131). More recently, in an attempt to edge into the computer game market, we now have an electronic 'Game Girl Barbie' in which (what else?) the object of the game is to take Barbie on a shopping spree. In 'Game Girl Barbie', shopping takes skill, and Barbie plays to win.

Perhaps what makes Barbie such a perfect icon of late capitalist constructions of femininity is the way in which her persona pairs endless consumption with the achievement of femininity and the appearance of an appropriately gendered body. By buying for Barbie, girls practise how to be discriminating consumers knowledgeable about the cultural capital of different name brands, how to read packaging, and the overall importance of fashion and taste for social status (Motz 1983: 131–2). Being a teenage girl in the world of Barbie dolls becomes quite literally a performance of commodity display, requiring numerous and complex rehearsals. In making this argument, we want to stress that we are drawing on more than just the doll. 'Barbie' is also the packaging, spin-off products, cartoons, commercials, magazines, and fan club paraphernalia, all of which contribute to creating her persona. Clearly, as we will discuss below, children may engage more or less with those products, subverting or ignoring various aspects of Barbie's 'official' presentation. However, to the extent that little girls *do* participate in the prepackaged world of Barbie, they come into contact with a number of beliefs central to femininity under consumer capitalism. Little girls learn, among other things, about the crucial importance of their appearance to their personal happiness and to their ability to gain favor with their friends. Barbie's social calendar is constantly full, and the stories in her fan magazines show her frequently engaged in preparation for the rituals of heterosexual teenage life: dates, proms, and weddings. A perusal of Barbie magazines, and the product advertisements and pictorials within them, shows an overwhelming preoccupation with grooming for those events. Magazines abound with tips on the proper ways of washing hair, putting on makeup, and assembling stunning wardrobes. Through these play scenarios, little girls learn Ruth Handler's lesson about the importance of hygiene, occasion-specific clothing, knowledgeable buying, and artful display as key elements to popularity and a successful career in femininity.

Barbie exemplifies the way in which gender in the late twentieth century has become a commodity itself, 'something we can buy into . . . the same way we buy into a style' (Willis 1991: 23). In her insightful analysis of the logics of consumer capitalism, cultural critic Susan Willis pays particular attention to the way in which children's toys like Barbie and the popular muscle-bound 'He-Man' for boys link highly conservative and narrowed images of masculinity and femininity with commodity consumption (1991: 27). In the imaginary world of Barbie and teen advertising, observes Willis, being or becoming a teenager, having a 'grown-up' body, is inextricably bound up with the acquisition of certain commodities, signaled by styles of clothing, cars, music, etc. In play groups and fan clubs (collectors are a whole world unto themselves), children exchange knowledge about the latest accessories and outfits, their relative merit, and how to find them. They become members of a community of Barbie owners whose shared identity is defined by the commodities they have or desire to have. The articulation of social ties through commodities, is, as Willis argues, at the heart of how sociality is experienced in consumer capitalism. In this way we might say that playing with Barbie serves not only as a training ground for the production of the appropriately gendered woman, but also as an introduction to the kinds of knowledge and social relations one can expect to encounter as a citizen of a post-Fordist economy.

BARBIE IS A SURVIVOR

A field trip in 1991 to Evelyn Burkhalter's Barbie Hall of Fame, located just above her husband's eye, ear, and throat clinic in downtown Palo Alto, California, revealed a remarkable array of Barbie dolls from across the globe. With over 1,500 dolls on display, several thousand more in storage, and an encyclopedic knowledge about Barbie's history, Mrs. Burkhalter proudly concluded her tour of the dolls with an emphatic, 'Barbie is a survivor!' Indeed! In the past three decades, this popular children's doll has undergone numerous changes in her fashion image and 'occupations' and has acquired a panoply of ethnic 'friends' and analogues that have allowed her to weather the dramatic social changes in gender and race relations that arose in the course of the 1960s and 1970s.

As the women's movement gained strength in the 1970s, the media and popular culture felt the impact of a growing self-consciousness about sexist imagery of women. The toy industry was no exception. Barbie, the ever-beautiful bride-to-be, became a target of some criticism and concern for parents who worried about the effects such a toy would have on their daughters. Barbie buffs like BillyBoy describe the 1970s as the doll's dark decade, a time when sales dipped, quality worsened as production was transferred from Japan to Taiwan, and Barbie was lampooned in the press (BillyBoy 1987). Mattel responded by trying to give Barbie a more diversified wardrobe and a more 'now' image. A glance at Barbie's resumé, published in *Harper's* magazine in August 1990, while incomplete, shows Mattel's attempt to expand Barbie's career options beyond the original fashion model:

Positions Held

1959–present	Fashion model
1961–present	Ballerina
1961–4	Stewardess (American Airlines)
1964	Candy striper
1965	Teacher
1965	Fashion editor
1966	Stewardess (Pan Am)
1973–5	Flight attendant (American Airlines)
1973–present	Medical doctor
1976	Olympic athlete
1984	Aerobics instructor

1985	TV news reporter
1985	Fashion designer
1985	Corporate executive
1988	Perfume designer
1989–present	Animal rights volunteer

It is only fitting, given her origin, to note that Barbie has also had a career in the military and aeronautics space industry: she has been an astronaut, a marine, and, during the Gulf War, a Desert Storm trooper. Going from pink to green, Barbie has also acquired a social conscience, taking up the causes of UNICEF, animal rights, and environmental protection. According to Mattel, the doll's careers are chosen to 'reflect the activities and professions that modern women are involved in' (quoted in *Harper's*, 2 August 1990, p. 20). Ironically, former Mattel manager of marketing Beverly Cannady noted that the doctor and astronaut uniforms never sold well. As Cannady candidly admitted to *Ms.* magazine in a 1979 interview, 'Frankly, we only kept the doctor's uniform in line as long as we did because public relations begged us to give them something they could point to as progressive' (Leavy 1979: 102). Despite their efforts to dodge criticism and present Barbie as a liberated woman, it is clear that glitz and glamour are at the heart of the Barbie doll fantasy. Motz reports, for example, that in 1963 only one out of sixty-four outfits on the market was job-related. There is no doubt that Barbie has had her day as astronaut, doctor, rock star, and even presidential candidate. She can be anything she wishes to be, although it is interesting that the difference between occupation and outfit has never been entirely clear. As her publicists emphasize, Barbie's purpose is to let little girls dream. And that dream continues to be fundamentally about leisure and consumption, not production.

For anyone tracking Barbiana, it is abundantly clear that Mattel's marketing strategies are sensitive to a changing social climate. Just as Mattel has sought to present Barbie as a career woman with more than air in her vinyl head, they have also tried to diversify her otherwise lily-white suburban world. About the same time that Martin Luther King was assassinated and Detroit and Watts were burning in some of the worst race riots of the century, Barbie acquired her first black friend. 'Colored Francie' appeared in 1967, failed, and was replaced the following year with Christie, who also did not do terribly well on the market. In 1980, Mattel went on to introduce Black Barbie, the first doll with Afro-style hair. She, too, appears to have suffered from a low advertising profile and low sales (Jones 1991). Nevertheless, the 1980s saw a concerted effort on Mattel's part to 'go multicultural', coinciding with a parallel preference in the pages of high-fashion magazines, such as *Elle* and *Vogue*, for racially diverse models. With the expansion of sales worldwide, Barbie has acquired multiple national guises (Spanish Barbie, Jamaican Barbie, Malaysian Barbie, etc.). In addition, her cohort of 'friends' has become increasingly ethnically diversified, as has Barbie advertising, which now regularly features Asian, Hispanic, and African American little girls playing with Barbie. Today, Barbie pals include a smattering of brown and yellow plastic friends, like Teresa, Kira, and Miko, who appear in her adventures and, very importantly, can share her clothes. This diversification has not spelled an end to reigning Anglo beauty norms and body image. Quite the reverse. When we line the dolls up together, they look virtually identical. Cultural difference is reduced to surface variations in skin tone and costumes that can be exchanged at will. Like the concomitant move toward racially diverse fashion models, 'difference' is remarkably made over into sameness, as ethnicity is tamed to conform to a restricted range of feminine beauty.

Perhaps Mattel's most glamorous concession to multiculturalism is their latest creation, Shani. Billed as tomorrow's African American woman, Shani, whose name, according to Mattel, means 'marvelous' in Swahili, premiered at the 1991 Toy Fair with great fanfare and media attention. Unlike her predecessors, who were essentially 'brown plastic poured into blond Barbie's mold', Shani, together with her two friends, Asha and Nichelle (each a slightly different shade of brown), and boyfriend, Jamal, created in 1992, were decidely Afro-centric, with outfits in 'ethnic' fabrics rather than the traditional Barbie pink (Jones 1991). The packaging also announced that these dolls' bodies and facial features were meant to be more like those of real African American women, although they too can interchange clothes with Barbie.

A realization of the growing market share of African American and Hispanic consumers has no doubt played a role in the changing face of Barbie. However, as *Village Voice* writer Lisa Jones has pointed out, there is a story other than simple economic calculus here. On the one hand, Mattel's social consciousness reflects the small but significant inroads black women have made into the company's top-level employee structure. It also underscores the growing authority of and recourse to expert knowledge, particularly psychological experts, as a way of understanding the social consequences of popular culture. As it turns out, Mattel product manager Deborah Mitchell and the principal fashion designer, Kitty Black-Perkins, are both African

American (Barbie's hair designer is also non-Anglo). Both women had read clinical psychologist Dr. Darlene Powell-Hopson's *Different and Wonderful: Raising Black Children in a Race-Conscious Society* (Powell-Hopson and Hopson 1990), and, according to Jones's account, became interested in creating a doll that could help give African American girls a positive self-image. Mattel eventually hired Powell-Hopson as a consultant, signed on a public relations firm with experience in targeting black consumers, and got to work creating Shani. Now, Mattel announced, 'ethnic Barbie lovers will be able to dream in their own image' (*Newsweek* 1990: 48). Multiculturalism cracked open the door to Barbie-dom, and diversity could walk in, so long as she was big-busted and slim-hipped, had long flowing hair and tiny feet, and was very very thin.

'The icons of twentieth-century mass culture,' writes Susan Willis, 'are all deeply infused with the desire for change,' and Barbie is no exception (1991: 37) In looking over the course of Barbie's career, it is clear that part of her resilience, appeal, and profitability stems from the fact that her identity is constructed primarily through fantasy and is consequently open to change and reinterpretation. As a fashion model, Barbie continually creates her identity anew with every costume change. In that sense, we might want to call Barbie the prototype of the 'transformer dolls' that cultural critics have come to see as emblematic of the restless desire for change that permeates postmodern capitalist society (Wilson 1985: 63). Not only can she renew her image with a change of clothes, Barbie also is seemingly able to clone herself effortlessly into new identities—Malibu Barbie; Totally Hair Barbie; Teen Talk Barbie; even Afrocentric Barbie/Shani—without somehow suggesting a serious personality disorder. Furthermore, Barbie's owners are at liberty to fantasize any number of life choices for the perpetual teenager; she might be a high-powered fashion executive, or she just might marry Ken and 'settle down' in her luxury condo. Her history is a barometer of changing fashions and changing gender and race relations, as well as a keen index of corporate America's anxious attempts to find new and more palatable ways of selling the beauty myth and commodity fetishism to new generations of parents and their daughters. The multiplication of Barbie and her friends translates the challenge of gender inequality and racial diversity into an ever-expanding array of costumes, a new 'look' that can be easily accommodated into a harmonious and illusory pluralism that never ends up rocking the boat of WASP beauty.

What is striking, then, is that, while Barbie's identity may be mutable—one day she might be an astronaut, another a cheerleader—her *hyper-slender, big-chested body has remained fundamentally unchanged over the years*—a remarkable fact in a society that fetishizes the new and improved. Barbie did acquire flexible arms and legs, as we know, and her hair, in particular, has grown by leaps and bounds, making Superstar Barbie the epitome of a 'big-hair girl'. Collectors also identify three distinctive changes in Barbie's face: the original cool, pale look with arched brows, red, pursed lips, and coy sideways glance gave way in the late 1960s to a more youthful, straight-haired, teenage look. This look lasted about a decade, and in 1977 Barbie acquired the exaggerated, wide-eyed, smiling look associated with Superstar Barbie that she still has today (Melosh and Simmons 1986). But her measurements, pointed toes, and proportions have not altered significantly in her thirty-five years of existence. We turn now from Barbie's 'persona' to the conundrum of her body and to our class experiment in the anthropometry of feminine ideals. In so doing, our aim is deliberately subversive. We wish to use the tools of calibration and measurement—tools of normalization that have an unsavory history for women and racial or ethnic minorities—to destabilize the ideal. In this way, our project represents a strategic use of scientific measurement and the authority it commands against the powerfully normative image of the feminine body in commodity culture. We begin with a very brief historical overview of the anthropometry of women and the emergence of an 'average' American female body in the postwar United States, before using our calipers on Barbie and her friends.

THE MEASURED BODY: NORMS AND IDEALS

> The paramount objective of physical anthropology is the gradual completion, in collaboration with the anatomists, the physiologists, and the chemists, of the study of the normal white man under ordinary circumstances.
>
> (Ales Hrdlicka, 1918)

As the science of measuring human bodies, anthropometry belongs to a long line of techniques of the eighteenth and nineteenth centuries concerned with measuring, comparing, and interpreting variability in different zones of the human body: craniometry, phrenology, physiognomy, and comparative anatomy. Early anthropometry shared with these an understanding and expectation that the body was a window into a host of moral, temperamental, racial, or gender characteristics. It sought to distinguish itself from its predecessors, however, by adhering to rigorously standardized methods and quantifiable results that would, it was hoped, lead to the 'complete elimination of personal bias' that anthropometrists believed had tainted earlier measurement techniques (Hrdlicka 1939: 12). Although the head (especially cranial capacity) continued to be a source of special fascination, by the early part of this century physical anthropologists, together with anatomists and medical doctors, were developing a precise and routine set of measurements for the entire body that would permit systematic comparison of the human body across race, nationality, and gender.

Under the aegis of Earnest Hooton, Ales Hrdlicka, and Franz Boas, located respectively at Harvard University, the Smithsonian, and Columbia University, anthropometric studies within U.S. physical anthropology were utilized mainly in the pursuit of three general areas of interest: identifying racial and or national types; the measurement of adaptation and 'degeneracy'; and a comparison of the sexes. Anthropometry was, in other words, believed to be a useful technique in resolving three critical border disputes: the boundaries between races or ethnic groups; the normal and the degenerate; and the border between the sexes.

As is well documented by now, women and non-Europeans did not fare well in these emerging sciences of the body (see the work of Blakey 1987; Gould 1981; Schiebinger 1989, 1993; Fee 1979; Russett 1989); measurements of women's bodies, their skulls in particular, tended to place them as inferior to or less intelligent than males. In the great chain of being, women as a class were believed to share certain atavistic characteristics with both children and so-called savages. Not everything about women was regarded negatively. In some cases it was argued that women possessed physical and moral qualities that were superior to those of males. Above all, woman's body was understood through the lens of her reproductive function; her physical characteristics, whether inferior or superior to those of males, were inexorably dictated by her capacity to bear children. These ideas, none of which were new, were part of the widespread scientific wisdom that reverberated throughout the development of physical anthropology and informed a great deal of what many of its leading figures had to say about the shape and size of women's bodies. Hooton, in his classic *Up From the Ape* (1931) was to regularly compare women, and especially non-Europeans of both sexes, to primates. Similarly, Hrdlicka's 1925 comparative study of male and female skulls went to rather extra-ordinary lengths to explain how it could be that women's brains (and hence, intelligence) were actually smaller than men's, even though his measurements showed females to have a cranial capacity relatively larger than that of males. Boas stood alone as an exception to this trend toward evolutionary ranking and typing. Although he did not address sex differences per se, his work on European migrants early in the century served to refute existing hypotheses on the hereditary nature of perceived racial or ethnic physical differences. As such, his work pushed physical anthropology and anthropometry toward the study of human adaptability and variation rather than the construc-tion of fixed racial, ethnic, or gender physical types.

It is striking that, aside from those studies specifically focused on the comparison of the sexes, women did not figure prominently in physical anthropology's attempt to quantify and typologize human bodies. In the studies of race and nationality, anthropometric studies of males far outnumbered those of females in the pages of the *American Journal of Physical Anthropology*. And where female bodies were measured, non-European women far outnumbered white women as the subjects of the calipers. Although Hrdlicka and others considered it necessary to measure both males and females, textbooks reveal that more often than not it was the biologically male body that stood in as the generic and ideal representative of the race or of humankind. It is, in fact, somewhat unusual that, in the quote from Hrdlicka given above, he should have called attention to the fact that physical anthro-pology's main object of study was indeed the 'white male', rather than the 'human body'. With males as the unspoken prototype, women's bodies were frequently described (subtly or not) as deviations from the norm: as subjects, the measurement of their bodies was occasion-ally risky to the male scientists, and as bodies they were variations from the generic or ideal type (their body fat 'excessive', their pelvises maladaptive to a bipedal [i.e., more evolved] posture, their muscu-lature weak). Understood primarily in terms of their reproductive capacity, women's bodies, particularly their reproductive organs, geni-talia, and secondary sex characteristics, were instead more carefully scrutinized and measured within 'marital adjustment' studies and in the emerging science of gynecology, whose practitioners borrowed liberally from the techniques used by physical anthropologists.

In the United States, an attempt to elaborate a scientifically sanc-tioned notion of a normative 'American' female body, however, was taking place in the college studies of the late nineteenth and early twentieth centuries. By the 1860s, Harvard and other universities had begun to regularly collect anthropometric data on their male student populations, and in the 1890s comparable data began to be collected from the East Coast women's colleges as well. Conducted by departments of hygiene, physical education, and home economics, as well as physical anthropology, these large-scale studies gathered data on the elite, primarily WASP youth, in order to determine the dimensions of the 'normal' American male and female. The data from one of the earliest cosexual studies, carried out by Dr. Dudley Sargent, a professor of physical education at Harvard, were then used to create two life-sized statues that were exhibited at the Chicago World's Fair in 1893 and put on display at the Peabody Museum. Effectively excluded from these attempts to define the 'normal' or average body, of course, were those 'other' Americans—descendants of African slaves, North American Indians, and the many recent European immigrants from Ireland, southern Europe, and eastern Europe—whose bodies were the subject of racist, evolution-oriented studies concerned with 'race crossing', degeneracy, and the effects of the 'civilizing' process (see Blakey 1987).

Standards for the average American male and female were also being elaborated in a variety of domains outside of academia. By the early part of the twentieth century, industry began to make widespread commercial use of practical anthropometry: the demand for standardized measures of the 'average' body manifested in everything from Taylorist designs for labor-efficient workstations and kitchens to standardized sizes in the ready-to-wear clothing industry (cf. Schwartz 1986). Certainly, one of the most common ways in which individuals encountered body norms was in the medical examination required for life insurance. It was not long before such companies as Metropolitan Life would rival the army, colleges, and prisons as the most reliable source of anthropometric statistics. Between 1900 and 1920, the first medicoactuarial standards of weight and height began to appear in conjunction with new theories linking weight and health. The most significant of these, the Dublin Standard Table of Heights and Weights, developed in 1908 by Louis Dublin, a student of Franz Boas and statistician for Metropolitan Life, became the authoritative reference in every doctor's office (cf. Bennett and Gurin 1982: 130–8). However, what began as a table of statistical averages soon became a means of setting ideal norms. Within a few years of its creation, the Dublin table shifted from providing a record of statistically 'average' weights to becoming a guide to 'desirable' weights that, interestingly enough, were notably below the average weight for most adult women.

In her history of anorexia in the United States, Joan Brumberg points to the Dublin table, widely disseminated to doctors and published in popular magazines, and the invention of the personal, or bathroom, scale as the two devices most responsible for popularizing the notion that the human figure could be standardized and that abstract and often unrealistic norms could be uniformly applied (1988: 232–5).

By the 1940s the search to describe the normal American male and female bodies in anthropometric terms was being conducted on many fronts. Data on the average measurements of men and women were now available from a number of different sources, including surveys of army recruits from World War I, the longitudinal college studies, sample measurements from the Chicago World's Fair, actuarial data, and extensive data from the Bureau of Home Economics, which had amassed measurements to assist in developing standardized sizing for the garment industry. Between the two wars, nationalist interests had fueled eugenic interests and provoked a deepening concern about the physical fitness of the American people. Did Americans constitute a distinctive physical 'type'; were they puny and weak as some Europeans had alleged, or were they physically bigger and stronger than their European ancestors? Could they defend themselves in time of war? And who did this category of 'Americans' include? Questions such as these fed into an already long-standing preoccupation with defining a specifically American national character and, in 1945, led to the creation of one of the most celebrated and widely publicized anthropometric models of the century: Norm and Norma, the average American male and female. Based on the composite measurements of thousands of young people, described only as 'native white Americans', across the United States, the statues of Norm and Norma were the product of a collaboration between obstetrician-gynecologist Robert Latou Dickinson, well known for his studies of human reproductive anatomy, and Abram Belskie, the prize student of Malvina Hoffman, who had sculpted the Races of Mankind series. Of the two, Norma received the greatest media attention when the Cleveland Health Museum, which had purchased the pair, decided to sponsor, with the help of a local newspaper, the YWCA, and several other health and educational organizations, a contest to find the woman in Ohio whose body most closely matched the dimensions of Norma. Under the catchy headline, 'Are You Norma, Typical Woman?' the publicity surrounding this contest instructed women in how to measure themselves at the same time that it extolled the virtues of Norma's body compared to those of her 'grandmother', Dudley

Sargent's composite of the 1890s woman. Within ten days, 3,863 women had sent in their measurements to compete for the $100 prize in U.S. War Bonds that would go to the woman who most resembled the average American girl.

Although anthropometric studies such as these were ostensibly descriptive rather than prescriptive, the normal or average and the ideal were routinely conflated. Nowhere is this more evident than in the discussions surrounding the Norma contest. Described in the press as the 'ideal' young woman, Norma was said to be everything an American woman should be in a time of war: she was fit, strong-bodied, and at the peak of her reproductive potential. Commentators waxed eloquent about the model character traits—maturity, modesty, and virtuousity—that this perfectly average body suggested. Curiously, although Norma was based on the measurements of living women, only about one percent of the contestants came close to her proportions. Harry Shapiro, curator of physical anthropology at the American Museum of Natural History, explained in the pages of *Natural History* why it was so rare to find a living, breathing Norma. Both Norma and Norman, he pointed out:

> . . . exhibit a harmony of proportion that seems far indeed from the usual or the average. One might well look at a multitude of young men and women before finding an approximation to these normal standards. We have to do here then with apparent paradoxes. Let us state it this way: the average American figure approaches a kind of perfection of bodily form and proportion; the average is excessively rare. (Shapiro 1945: 51)

Besides bolstering the circulation of the *Cleveland Plain Dealer*, the idea behind the contest was to promote interest in physical fitness. Newspaper articles emphasized that women had a national responsibility to be fit, if America was to continue to excel after the war. Commenting on the search for Norma, Dr. Bruno Gebhard, director of the Cleveland Health Museum, was quoted in one of the many newspaper articles as saying that 'if a national inventory of the female population of this country were taken there would be as many "4Fs" among the women as were revealed among the men in the draft' (Robertson 1945: 4). The contest provided the occasion for many health reformers to voice their concern about the need for eugenic marital selection and breeding. Beside weakening the 'American stock', Gebhard claimed, 'the unfit are both bad producers and bad consumers. One of the outstanding needs in this country is more emphasis everywhere on physical fitness' (ibid). Norma was presented to the public as a reminder to women of their duty to the nation, and, not incidentally, Norma could also serve as a hypothetical standard in women's colleges for the detection of faulty posture and 'so that students who need to lose or gain in spots or generally may have a mark to shoot at' (ibid).

Norma and Norman were thus more than statistical composites, they were ideals. It is striking how thoroughly racial and ethnic differences were erased from these scientific representations of the American male and female. Based on the measurements of white Americans, eighteen to twenty-five years old, Norm and Norma emerged carved out of white alabaster, with the facial features and appearance of Anglo-Saxon gods. Here, as in the college studies that preceded them, the 'average American' of the postwar period was to be visualized only as a youthful white body.

However, they were not the only ideal. The health reformers, educators, and doctors who approved and promoted Norma as an ideal for American women were well aware that her sensible, strong, thick-waisted body differed significantly from the tall, slim-hipped bodies of fashion models in vogue at the time. Gebhard and others tried through a variety of means to encourage women to ignore the temptations of 'vanity' and fashion, but they were ill equipped to compete with the persuasive powers of a rapidly expanding mass media that marketed a very different kind of female body. As the postwar period advanced, Norma would continue to be trotted out in home economics and health education classes. But in the iconography of desirable female bodies, she would be overshadowed by the array of images of fashion models and pinup girls put out by advertisers, the entertainment industry, and a burgeoning consumer culture. These idealized images were becoming, as we will see below, increasingly thin in the 1960s and 1970s while the 'average' woman's body was in fact getting heavier. With the thinning of the American feminine ideal, Norma and subsequent representations of the statistically average woman would become increasingly aberrant, as slenderness and sex appeal—not physical fitness—became the premier concern of postwar femininity.

THE ANTHROPOMETRY OF BARBIE: TURNING THE TABLES

As the preceding discussion makes abundantly clear, the anthropometrically measured 'normal' body has been anything but value-free.

Formulated in the context of a race-, class-, and gender-stratified society, there is no doubt that quantitatively defined ideal types or standards have been both biased and oppressive. Incorporated into weight tables, put on display in museums and world fairs, and reprinted in popular magazines, these scientifically endorsed standards produce what Foucault calls 'normalizing effects', shaping, in not altogether healthy ways, how individuals understand themselves and their bodies. Nevertheless, in the contemporary cultural context, where an impossibly thin image of women's bodies has become the most popular children's toy ever sold, it strikes us that recourse to the 'normal' body might just be the power tool we need for destabilizing a fashion fantasy spun out of control. It was with this in mind that we asked students in one of our social biology classes to measure Barbie to see how her body compared to the average measurements of young American women of the same period. Besides estimating Barbie's dimensions as if she were life-sized, we see the experiment as an occasion to turn the anthropometric tables from disciplining the bodies of living women to measuring the ideals by which we have come to judge ourselves and others. We also see it as an opportunity for students who have grown up under the regimes of normalizing science—students who no doubt have been measured, weighed, and compared to standards since birth—to use those very tools to unsettle a highly popular cultural ideal.

Initially, this foray into the anthropometry of Barbie was motivated by an exercise in a course entitled Issues in Social Biology. Since one objective of the course was to learn about human variation, our first task in understanding more about Barbie was to consider the fact that Barbie's friends and family do represent some variation, limited though it may be. Through colleagues and donations from students or (in one case) their children we assembled seventeen dolls for analysis. The sample included:

1 early 1960s Barbie
4 mid 1970s-to-contemporary Barbies, including a Canadian Barbie
3 Kens
2 Skippers
1 Scooter
Assorted Barbie's friends, including Christie, Barbie's 'black' friend
Assorted Ken's friends

To this sample we subsequently added the most current versions of

Barbie and Ken (from the 'Glitter Beach' collection) and also Jamal, Nichelle, and Shani, Barbie's more recent African American friends. As already noted, Mattel introduced these dolls (Shani, Asha, and Nichelle) as having a more authentic African American appearance, including a 'rounder and more athletic' body. Noteworthy also are the skin color variations between the African American dolls, ranging from dark to light, whereas Barbie and her white friends tend to be uniformly pink or uniformly suntanned.

Normally, of course, before undertaking the somewhat invasive techniques of measuring people's bodies, we would have written a proposal to the Human Subjects Committees of the department and the university; submitted a written informed consent form detailing the measurements to be taken and seeking permission to use the data, by name, in subsequent reports; and, finally, discussed the procedures and the importance of the research with each of the subjects. However, since our subjects were unresponsive, these protocols had to be waived.

Before beginning the actual measurements, we discussed the kinds of data we thought would be most appropriate. Student interest centered on height and chest, waist, and hip circumference. Members of the class also pointed out the apparently small size of the feet and the general leanness of Barbie. As a result, we added a series of additional standardized measurements, including upper arm and thigh circumference, in order to obtain an estimate of body fat and general size.

After practising with the calipers, discussing potential observational errors, and performing repeated trial runs, we began to record. All the measurements were taken in the Physical Anthropology Laboratory at the University of Massachusetts under clean, well-lit conditions. We felt our almost entirely female group of investigators would no doubt

TABLE 1. Measurements of Glitter Beach Barbie, African American Shani, and the average measurements of the 1988 U.S. Army women recruits[a]

Measurements	Barbie	Shani	U.S. Army 'Norma'
Height	5'10"	5'10"	5'4"
Chest circum.	35"	35"	35.7"
Waist circum.	20"	20"	31"
Hip circum.[b]	32.50"	31.25"	38.10"
Hip breadth	11.6"	11.0"	13.49"
Thigh circum.	19.25"	20.00"	22.85"

[a] 'Norma' is based on 2,208 army recruits, 1,140 of whom were white, 922 of whom were black.
[b] Hip circumference is referred to as 'buttock circumference' in anthropometric parlance.

have pleased Hrdlicka, since he believed women anthropometrists to be more skilled at the precise, small-scale measurements our experiment required. In scaling Barbie to be life-sized, the students decided to translate her measurements using two standards: (a) if Barbie were a fashion model (5'10") and (b) if she were of average height for women in the United States (5'4"). We also decided to measure Ken, using both an average male stature, which we designated as 5'8", and the more 'idealized' stature for men, 6'.

For the purposes of this chapter, we took measurements of dolls in the current Glitter Beach and Shani collection that were not available for our original classroom experiment, and all measurements were retaken to confirm estimates. We report here only the highlights of the measurements taken on the newer Barbie and newer Ken, Jamal, and Shani, scaled at their ideal fashion-model height. For purposes of comparison, we include data on average body measurements from the standardized published tables of the 1988 Anthropometric Survey of Army Personnel. We have dubbed these composites for the female and male recruits Army 'Norma' and Army 'Norm', respectively.

Barbie and Shani's measurements reveal interesting similarities and subtle differences. First, considering that they are six inches taller than 'Army Norma', note that their measurements tend to be considerably less *at all points*. 'Army Norma' is a composite of the fit woman soldier; Barbie and Shani, as high-fashion ideals, reflect the extreme thinness expected of the runway model. To dramatize this, had we scaled Barbie to 5'4", her chest, waist, and hip measurements would have been 32"-17"-28", clinically anorectic to say the least. There are only subtle differences in size, which we presume intend to facilitate the exchange of costumes among the different dolls. We were curious to

TABLE 2. Measurements of Glitter Beach Ken, African American Ken, and the average measurements of the 1988 U.S. Army male recruits[a]

Measurements	Ken	Jamal	U.S. Army 'Norm'
Height	6'0"	6.0"	5'9"
Chest circum.	38.4"	38.4"	39.0"
Waist circum.	28.8"	28.8"	33.1"
Hip circum.[b]	36.0"	36.0"	38.7"
Hip breadth	12.2"	12.2"	13.46"
Thigh circum.	20.4"	20.04"	23.48"

[a] 'Norm' is based on 1,774 males, 1,172 of whom were white and 458 of whom were black.
[b] Hip circumference is referred to as 'buttock circumference' in anthropometric parlance.

see the degree to which Mattel had physically changed the Barbie mold in making Shani. Most of the differences we could find appeared to be in the face. The nose of Shani is broader and her lips are ever so slightly larger. However, our measurements also showed that Barbie's hip circumference is actually larger than Shani's, and so is her hip breadth. If anything, Shani might have thinner legs than Barbie, but her back is arched in such a way that it tilts her buttocks up. This makes them appear to protrude more posteriorly, even though the hip depth measurements of both dolls are virtually the same (7.1″). Hence, the tilting of the lumbar dorsal region and the extension of the sacral pelvic area produce the visual illusion of a higher, rounder butt. This is, we presume, what Mattel was referring to in claiming that Shani has a realistic, or ethnically correct, body (Jones 1991).

One of our interests in the male dolls was to ascertain whether they represent a form closer to average male values than Barbie does to average female values. Ken and Jamal provide interesting contrasts to 'Army Norm', but certainly not to each other. Their postcranial bodies are identical in all respects. They, in turn, represent a somewhat slimmer, trimmer male than the so-called fit soldier of today. Visually, the newer Ken and Jamal appear very tight and muscular and 'bulked out' in impressive ways. The U.S. Army males tend to carry slightly more fat, judging from the photographs and data presented in the 1988 study.

Indeed, it would appear that Barbie and virtually all her friends characterize a somewhat extreme ideal of the human figure, but in Barbie and Shani, the female cases, the degree to which they vary from 'normal' is much greater than in the male cases, bordering on the impossible. Barbie truly is the unobtainable representation of an imaginary femaleness. But she is certainly not unique in the realm of female ideals. Studies tracking the body measurements of *Playboy* magazine centerfolds and Miss America contestants show that between 1959 and 1978 the average weight and hip size for women in both of these groups have decreased steadily (Wiseman et al. 1992). Comparing their data to actuarial data for the same time period, researchers found that the thinning of feminine body ideals was occurring at the same time that the average weight of American women was actually increasing. A follow-up study for the years 1979–88 found this trend continuing into the 1980s: approximately 69 percent of *Playboy* centerfolds and 60 percent of Miss America contestants were weighing in at 15 percent or more below their expected age and height category In short, the majority of women presented to us in the media as having desirable feminine bodies were, like Barbie, well on their way to qualifying for anorexia nervosa.

OUR BARBIES, OUR SELVES

> I feel like Barbie; everyone calls me Barbie; I love Barbie. The main difference is she's plastic and I'm real. There isn't really any other difference.
>
> (Hayley Spicer, winner of Great Britain's Barbie-Look-Alike competition)

On the surface, at least, Barbie's strikingly thin body and the repression and self-discipline that it signifies would appear to contrast with her seemingly endless desire for consumption and self-transformation. And yet, as Susan Bordo has argued in regard to anorexia, these two phenomena—hyper-thin bodies and hyper-consumption—are very much linked in advanced capitalist economies that depend upon commodity excess. Regulating desire under such circumstances is a constant, ongoing problem that plays itself out on the body. As Bordo argues:

[In a society where we are] conditioned to lose control at the very sight of desirable products, we can only master our desires through a rigid defense against them. The slender body codes the tantalizing ideal of a well-managed self in which all is 'in order' despite the contradictions of consumer culture. (1990: 97)

The imperative to manage the body and 'be all that you can be'—in fact, the idea that you can *choose* the body that you want to have—is a pervasive feature of consumer culture. Keeping control of one's body, not getting too fat or flabby—in other words, conforming to gendered norms of fitness and weight—are signs of an individual's social and moral worth. But, as feminists Bordo, Sandra Bartky, and others have been quick to point out, not all bodies are subject to the same degree of scrutiny or the same repercussions if they fail. It is women's bodies and desires in particular where the structural contradictions—the simultaneous incitement to consume and social condemnation for overindulgence—appear to be most acutely manifested in bodily regimes of intense self-monitoring and discipline. 'The woman who checks her make-up half a dozen times a day to see if her foundation has caked or her mascara run, who worries that the wind or rain may spoil her hairdo has become just as surely as the inmate of the Panopticon, a self-policing subject, a self committed to a relentless self surveillance'(Bartky 1990: 80). Just as it is women's appearance that is subject to greater social scrutiny so it is that women's desires, hungers, and appetites are seen as most threatening and in need of control in a patriarchal society.

This cultural context is relevant to making sense of Barbie and the meaning her body holds in late consumer capitalism. In dressing and undressing Barbie, combing her hair, bathing her, turning and twisting her limbs in imaginary scenarios, children acquire a very tactile and intimate sense of Barbie's body. Barbie is presented in packaging and advertising as a role model, a best friend or older sister to little girls. Television jingles use the refrain, 'I want to be just like you', while look-alike clothes and look-alike contests make it possible for girls to live out the fantasy of being Barbie. And, finally, in the pages of the *National Enquirer*, where cultural fantasies have a way of becoming nightmare reality, we find the literalization of becoming Barbie, thanks to the wonders of modern medical technology. In short, there is no reason to believe that girls (or adult women) separate Barbie's body shape from her popularity and glamour.

This is exactly what worries many feminists. As our measurements show, Barbie's body differs wildly from anything approximating 'average' female body weight and proportions. Over the years her wasp-waisted body has evoked a steady stream of critique for having a negative impact on little girls' sense of self esteem: While her large breasts have always been a focus of commentary it is interesting to note that, as eating disorders are on the rise, her weight has increasingly become the target of criticism. For example, the 1992 release of a Barbie aerobics workout video for girls was met with the following angry letter from an expert in the field of eating disorders:

I had hoped these plastic dolls with impossible proportions would have faded away in this current health-conscious period; not at all . . . Move over Jane Fonda. Welcome again, ever smiling, breast-thrusting Barbie with your stick legs and sweat-free aerobic routines. I'm concerned about the role model message she is giving our young. Surely it's hard to accept a little cellulite when the culture tells you unrelentingly how to strive for thinness and the perfect body. (Warner 1992)

There is no doubt that Barbie's body contributes to what Kim Chernin (1981) has called 'the tyranny of slenderness'. But is repression all her hyper-thin body conveys? Looking once again to Susan

Bordo's work on anorexia, we find an alternative reading of the slender body—one that emerges from taking seriously the way anorectic women see themselves and make sense of their experience:

> For them, anorectics, [the slender ideal] may have a very different meaning; it may symbolize not so much the containment of female desire, as its liberation from a domestic, reproductive destiny. The fact that the slender female body can carry both these (seemingly contradictory) meanings is one reason, I would suggest, for its compelling attraction in periods of gender change. (Bordo 1990: 103)

Similar observations have been made about cosmetic surgery: women often explain their experience as one of empowerment, taking charge of their bodies and lives (Balsamo 1992; Davis 1991). What does this mean for making sense of Barbie? We would suggest that a subtext of agency and independence, even transgression, accompanies this pencil-thin icon of femininity. One could argue that, like the anorectic body she resembles, Barbie's body displays conformity to dominant cultural imperatives for a disciplined body and contained feminine desires. As a woman, however, her excessive slenderness also signifies a rebellious manifestation of willpower, a visual denial of the maternal ideal symbolized by pendulous breasts; rounded stomach and hips. Hers is a body of hard edges, distinct borders, self-control. It is literally impenetrable. Unlike the anorectic, whose self-denial renders her gradually more androgynous in appearance, in the realm of plastic fantasy Barbie is able to remain powerfully sexualized, with her large, gravity-defying breasts, even while she is distinctly nonreproductive. Like the 'hard bodies' in fitness advertising, Barbie's body may signify for women the pleasure of control and mastery, both of which are highly valued traits in American society and predominantly associated with masculinity (Bordo 1990: 105). Putting these elements together with her apparent independent wealth can make for a very different reading of Barbie than the one we often find in the popular press. To paraphrase one Barbie-doll owner: she owns a Ferrari and doesn't have a husband— she must be doing something right!

Invoking the testimonies and experiences of women caught up in the beauty myth is not meant to suggest that playing with Barbie, or becoming like Barbie, is a means of empowerment for little girls. But it is meant to signal the complex and contradictory meanings that her body has in contemporary American society. Barbie functions as an ideological sign for commodity fetishism and a rather rigid gender ideology. But neither children nor adult consumers of popular culture are simply passive victims of dominant ideology. It is sensible to assume that the children who play with Barbie are themselves creative users, who respond variously to the messages about femininity encoded in her fashions and appearance. Not only do many children make their own clothes for their Barbie dolls, but anecdotes abound of the imaginative uses of Barbie. In the hands of their owners, Barbies have been known to occupy roles and engage in activities anathema to the good-girl image Mattel has carefully constructed. In the course of our research in the past year, we have heard or read about Barbies that have been tattooed, decapitated, and had their flowing locks shorn into mohawks. Knowing only the limits of a child's imagination, Barbies have become amazon cave warriors, dutiful mommies, evil axe-murderers, and *Playboy* models. As *Mondo Barbie*, a recent collection of Barbie-inspired fiction and poetry, makes clear, the possibilities are endless, and sexual transgression is always just around the corner (Ebersole and Peabody 1993; see also Rand, 1994).

It is clear that a next step we would want to take in the cultural interpretation of Barbie is an ethnographic study of Barbie-doll owners. In the meanwhile, we can know something about these alternative appropriations by looking to various forms of popular culture and the art world. Barbie has become a somewhat celebrated figure among avant-garde and pop artists, giving rise to a whole genre of Barbie satire, known as 'Barbie Noire' (Kahn 1991). According to Peter Galassi, curator of *Pleasures and Terrors of Domestic Comfort*, an exhibit at the Museum of Modern Art in New York, 'Barbie isn't just a doll. She suggests a type of behaviour—something a lot of artists, especially women, have wanted to question' (quoted in Kahn 1991: 25). Perhaps the most notable sardonic use of Barbie dolls to date is the 1987 film *Superstar: The Karen Carpenter Story*, by Todd Haynes and Cynthia Schneider. In this deeply ironic exploration into the 1970s, suburbia, and middle-class hypocrisy, Barbie and Ken dolls are used to tell the tragic story of Karen Carpenter's battle with anorexia and expose the perverse underbelly of the popular singing duo's candy-coated image of happy, apolitical teens. It is hard to imagine a better casting choice to tell this tale of femininity gone astray than the ever-thin, ever-plastic, ever-wholesome Barbie.

For Barbiana collectors it should come as no surprise that Barbie's excessive femininity also makes her a favorite persona of female impersonators, alongside Judy, Marilyn, Marlene, and Zsa Zsa. Appropriations of Barbie in gay camp culture have tended to favor the early, vampier Barbie look: with the arched eyebrows, heavy black eye-liner, and coy sideways look—the later superstar version of Barbie, according to BillyBoy, is just *too* pink. But new queer spins are con-stantly popping up. For example, multiple layers of meaning abound in this image of Barbie that appeared in an advertisement for a hair salon in the predominantly gay South End neighborhood of Boston. Here, superstar Barbie, good-girl teenage fashion model, is presented to us tattooed, dressed in leather, and painted over with heavy makeup. She is accompanied by lyrics from drag-queen performer Ru Paul's hit song, 'Supermodel'. A towering seven feet of gender-bending beauty in heels, Ru Paul opens and closes this wonderfully campy song about runway modeling, with the commanding phrase: 'I have one thing to say: you better work!' In the song, 'work' or 'work your body', refers simultaneously to the work of moving down the runway with 'savoir-faire' and to the work of illusion, the work of producing a perfect feminine appearance, a 'million-dollar derrière'. In the advertise-ment, of course, it is Barbie, with her molded-by-Mattel body, who stands in for the drag queen, commanding the spectator, whip in hand, to *work her body*. Barbie, in this fantasy-scape, becomes the mistress of body discipline, exposing simultaneously the artifice of gender and the feminine body.

In the world of Barbie Noire, the hyper-rigid gender roles of the toy industry are targeted for inversion and subversion. While Barbie is transformed into a dominatrix drag queen, Ken, too, has had his share of spoofs and gender bending. Barbie's somewhat dull steady boy-friend has never been developed into much more than a reliable escort and proof of Barbie's appropriate sexual orientation and popularity. In contrast to that of Barbie, Ken's image has remained boringly con-stant over the years. He has had his 'mod', 'hippie' and Malibu-suntan days, and he has gotten significantly more muscular. But for the most part, his clothing line is less diversified, and he lacks an independent fan club or advertising campaign. In a world where boys' toys are G.I. Joe-style action figures, bent on alternately saving or destroying the world, Ken is an anomaly. Few would doubt that his identity was primarily another one of Barbie's accessories. His secondary status vis-à-vis Barbie is translated into emasculation and/or a secret gay iden-tity: cartoons and spoofs of Ken have him dressed in Barbie clothes, and rumors abound that Ken's seeming lack of sexual desire for Barbie is only a cover for his real love for his boyfriends, Alan, Steve, and Dave.

Inscrutable with her blank stare and unchanging smile, Barbie is thus available for any number of readings and appropriations. What we have done here is examine some of the ways she resonates with the complex and contradictory cultural meanings of femininity in post-war consumer society and a changing politics of the body. Barbie, as we, and many other critics, have observed, is an impossible ideal, but she is an ideal that has become curiously normalized. In a youth-obsessed society like our own, she is an ideal not just for young women, but for all women who feel that being beautiful means look-ing like a skinny buxom, white 20-year-old. It is this cultural impera-tive to remain ageless and lean that leads women to have skewed perceptions of their bodies, undergo painful surgeries, and punish themselves with outrageous diets. Barbie, in short, is an ideal that constructs women's bodies as hopelessly imperfect. It has been our intention to unsettle this ideal and, at the same time, to be sensitive to other possible readings, other ways in which this ideal body figures and reconfigures the female body. For example, implicit in the various strategies of technologically mediated body-sculpting and surveillance that women engage in to meet these ideals is not only a conception of the female body that is inherently pathological, but an increasingly imaginary body of malleable, replaceable parts. Fueled by an ideology demanding limitless improvement and an increasingly popular cybor-gian science fiction, the modern paradigm of 'body as machine', says Susan Bordo, is giving way to an understanding of the body as 'cul-tural plastic'. The explosion in technologically assisted modifications through cosmetic surgery, piercing, aerobics, and nautilus all point to a conception of the body as raw material to be fragmented into parts, molded, and reshaped into a more perfect form. Lacking any essential truth, the body has become, like Barbie, all surface, a ground for sta-ging cultural identities.

What to make of this apparent denaturalizing of the feminine body is not clear. Feminists call attention to the way women use these tech-niques to take control over their bodies, while others are hesitant to join with current trends in cultural studies that would celebrate these as empowered acts of resistance. Our concern is not to decry the corruption of a fictitious 'natural' body, but to underscore how these acts of self-re-creation are inflected by power and desire. 'Fashion surgery', as Balsamo calls it, liberates one from the body one is born with, and as Nan Goldin's 1993 photo essays of transvestites and trans-sexuals make apparent, advances in this medical technology have made possible new permutations of the gendered body. What they do *not* do is erase the larger cultural matrix and power relations that propel women to undertake certain kinds of body transformations instead of others. The different matrices of power in which individuals are located make it such that, while all body transformations in some way treat the gender and the body as cultural plastic, they do not have similar meanings. Further, the potential to surgically alter bodies may challenge the naturalness of gender and the determining power of biological sex itself, but it has not unsettled the notion that gender is fundamentally located in the physical body, rather than in language, gesture, or other performative displays. Indeed, a variety of social and cultural forces conspire to make body modification so normal, so necessary, that 'electing *not* to have cosmetic surgery is sometimes interpreted as a failure to deploy all available resources to maintain a youthful, and therefore socially acceptable and attractive, body appearance' (Balsamo 1992: 216). It is the complexity of this terrain that leads artists such as Barbara Kruger to describe the body, particularly the body of the socially disempowered, as a *battleground*, a terrain of multiple sites of conflict and resistances, where histories of racial, national, and gender inequalities come to bear upon the 'choices' individuals make with regard to their bodies.

We have explored some of the battleground upon which the serious play of Barbie unfolds. If Barbie has taught us anything about gender, it is that femininity in consumer culture is a question of carefully performed display, of paradoxical fixity and malleability. One outfit, one occupation, one identity can be substituted for another, while Barbie's body has remained ageless, changeless, untouched by the ravages of age or cellulite. She is always a perfect fit, always able to consume and be consumed. Mattel has skillfully managed to turn the challenges of feminist protest, ethnic diversity, and a troubled multiculturalism to a new array of outfits and skin tones, annexing these to a singular anorectic body ideal. Cultural icon that she is, Barbie nevertheless cannot be permanently located in any singular cultural space. Her meaning is mobile as she is appropriated and relocated into different cultural contexts, some of which, as we have seen, make fun of many of the very notions of femininity and consumerism she personifies. As we consider Barbie's many meanings, we should remember that Barbie is not only a denizen of subcultures in the United States, she is also world traveler. A product of the global assembly line, Barbie dolls owe their existence to the internationalization of the labor market and global flows of capital and commodities that today characterize the toy industry, as well as other industries in the postwar era. Designed in Los Angeles, manufactured in Taiwan or Malaysia, distributed worldwide, Barbie™ is American-made in name only. Speeding her way into an expanding global market, Barbie brings with her some of the North American cultural subtext we have outlined in this analysis. How this teenage survivor then gets interpolated into the cultural landscapes of Mayan villages, Bombay high-rises, and Malagasy towns is a rich topic that begs to be explored.

References

Adelson, Andrea (1992). 'And Now, Barbie Looks Like a Billion'. *New York Times*, 26 November: sec. D, p. 3.

Balsamo, Anne (1992). 'On the Cutting Edge: Cosmetic Surgery and the Technological Production of the Gendered Body'. *Camera Obscura*, 28: 207–38.

Banner, Lois W. (1983). *American Beauty*. New York: Knopf.

Bartky, Sandra Lee (1990). 'Foucault, Femininity, and the Modernization of Patriarchal Power', in *Femininity and Domination: Studies in the Phenomenology of Oppression*, pp. 63–82. New York: Routledge.

Bennett, William, and Gurin, Joel (1982). *The Dieter's Dilemma: Eating Less and Weighing More*. New York: Basic Books.

BillyBoy (1987). *Barbie, Her Life and Times, and the New Theater of Fashion*. New York. Crown.

Blakey, Michael L. (1987). 'Skull Doctors: Intrinsic Social and Political Bias in the History of American Physical Anthropology'. *Critique of Anthropology*, 7 (2): 7–35.

Bordo, Susan R. (1990). 'Reading the Slender Body', in *Body/Politics: Women and the Discourses of Science*, ed. Mary Jacobus, Evelyn Fox Keller, and Sally Shuttleworth, pp. 83–112. New York: Routledge.

—— (1991). 'Material Girl: The Effacements of Postmodern Culture', in *The Female Body: Figures, Styles, Speculations*, ed. Laurence Goldstien, pp. 106–30. Ann Arbor, University of Michigan Press.

Breckenridge, Carol A. (1990). 'Editor's Comment: On Toying with Terror'. *Public Culture*, 2 (2): i–iii.

Brumberg, Joan Jacobs (1988). *Fasting Girls: The History of Anorexia Nervosa*. Cambridge, Mass.: Harvard University Press. Reprint, New York: New American Library.

Chernin, Kim (1981). *The Obsession: Reflections on the Tyranny of Slenderness*. New York: Harper and Row.

Cogan, Frances B. (1989). *All-American Girl: The Ideal of Real Woman-hood in Mid-Nineteenth-Century America*. Athens and London: University of Georgia Press.

Davis, Kathy (1991). 'Remaking the She-Devil: A Critical Look at Feminist Approaches to Beauty'. *Hypatia*, 6 (2): 21–43.

Doherty, Thomas (1988). *Teenagers and Teenpics: The Juvenilization of American Movies in the 1950s.* Boston: Unwin Hyman.

duCille, Ann (1995). 'Toy Theory: Blackface Barbie and the Deep Play of Difference', in *The Skin Trade: Essays on Race, Gender, and the Merchandising of Difference.* Cambridge, Mass.: Harvard University Press.

Ebersole, Lucinda, and Peabody, Richard (eds.)(1993). *Mondo Barbie.* New York: St Martin's Press.

Fee, Elizabeth (1979). 'Nineteenth-Century Craniology: The Study of the Female Skull'. *Bulletin of the History of Medicine,* 53: 415–33.

France, Kim (1992). 'Tits 'R' Us'. *Village Voice,* 17 March, 22.

Frith, Simon (1981). *Sound Effects: Youth, Leisure, and the Politics of Rock'n' Roll.* New York: Pantheon.

Goldin, Nan (1993). *The Other Side.* New York: Scalo.

Gould, Stephen Jay (1981). *The Mismeasure of Man.* New York: Norton.

Halberstam, Judith (1994). 'F2M: The Making of Female Masculinity', in *The Lesbian Postmodern,* ed. Laura Doan, pp. 210–28. New York: Columbia University Press.

Hohmann, Delf Maria (1985). 'Jennifer and Her Barbies: A Contextual Analysis of a Child Playing Barbie Dolls'. *Canadian Folklore Canadien,* 7 (1–2): 111–20.

Hopson, Darlene Powell, and Hopson, Derek (1990). *Different and Wonderful: Raising Black Children in a Race-Concious Society.* New York: Prentice Hall.

Hrdlicka, Ales (1925). 'Relation of the Size of the Head and Skull to Capacity in the Two Sexes', *American Journal of Physical Anthropology,* 8: 249–50.

—— (1939). *Practical Anthropometry.* Philadelphia: Wistar Institute of Anatomy and Biology.

Jones, Lisa (1991). 'Skin Trade: A Doll is Born'. *Village Voice,* 26 March: 36.

Kahn, Alice (1991). 'A Onetime Bimbo Becomes a Muse'. *New York Times,* 29 September.

Kaw, Eugenia (1993). 'Medicalization of Racial Features: Asian American Women and Cosmetic Surgery'. *Medical Anthropology Quarterly, 7 (1):* 74–89.

Lawson, Carol (1993).'Toys Will Be Toys: The Stereotypes Unravel'. *New York Times,* 11 February: sec. C, pp. 1, 8.

Leavy, Jane (1979). 'Is There a Barbie Doll in Your Past?' *Ms,* September: 102.

Lord, M.G. (1994). *Forever Barbie: The Unauthorized Biography of a Real Doll.* New York: William Morrow.

May, Elaine Tyler (1988). *Homeward Bound: American Families in the Cold War Era.* New York: Basic Books.

Melosh, Barbara, and Simmons, Christina (1986). 'Exhibiting Women's History', in *Presenting the Past: Essays on History and the Public,* ed. Susan Porter Benson, Stephen Brier, and Roy Rosenzweig, pp. 203–21. Philadelphia: Temple University Press.

Morse, Margaret (1987). 'Artemis Aging: Exercise and the Female Body on Video'. *Discourse*, 10 (1987/8) 20–53.

Motz, Marilyn Ferris (1983). 'I Want to Be a Barbie Doll When I Grow Up: The Cultural Significance of the Barbie Doll', in *The Popular Culture Reader*, 3rd edn., ed. Christopher D. Geist and Jack Nachbar, pp. 122–36. Bowling Green: Bowling Green University Popular Press.

Newsweek (1990). 'Finally. Barbie Dolls Go Ethnic'. *Newsweek*, 13 August: 48.

Rabinowitz, Paula (1993). 'Accessorizing History: Girls and Popular Culture'. Discussant Comments, Panel #150: Engendering Post-war Popular Culture in Britain and America. Ninth Berkshire Conference on the History of Women. Vassar College, 11–13 June 1993.

Rand, Erica (1994). 'We Girls Can Do Anything, Right Barbie? Lesbian Consumption in Postmodern Circulation', in *Lesbian Postmodern*, ed. Laura Doan, pp. 189–209. New York: Columbia University Press.

Robertson, Josephine (1945). 'Theatre Cashier, 23, Wins Title of "Norma"'. *Cleveland Plain Dealer*, 21 September: 1, 4.

Russett, Cynthia Eagle (1989). *Sexual Science: The Victorian Construction of Womanhood*. Cambridge, Mass.: Harvard University Press.

Schiebinger, Londa (1989). *The Mind Has No Sex?: Women in the Origins of Modern Science*. Cambridge, Mass.: Harvard University Press.

—— (1993). *Nature's Body: Gender in the Making of Modern Science*. Boston: Beacon.

Schwartz, Hillel (1986). *Never Satisfied: A Cultural History of Diets, Fantasies and Fat*. New York: Free Press.

Shapiro, Eben (1992). ' "Totally Hot, Totally Cool." Long-Haired Barbie Is a Hit'. *New York Times*, 22 June: sec. D, p. 9.

Shapiro, Harry L. (1945). *Americans: Yesterday, Today, Tomorrow*. Man and Nature Publications. (Science Guide No. 126). New York: American Museum of Natural History.

Spencer, Frank (1992). 'Some Notes on the Attempt to Apply Photography to Anthropometry during the Second Half of the Nineteenth Century', in *Anthropology and Photography, 1860–1920*, ed. Elizabeth Edwards, pp. 99–107. New Haven: Yale University Press.

Sprague Zones, Jane (1989). 'The Dangers of Breast Augmentation'. *The Network News* (July/August): 1, 4, 6, 8. Washington, DC: National Women's Health Network.

Stevenson, Richard (1991). 'Mattel Thrives as Barbie Grows'. *New York Times*, 2 December.

Terry, Jennifer C. (1992). 'Siting Homosexuality: A History of Surveillance and the Production of Deviant Subjects (1935–1950)'. Ph.D. diss. (University of California at Santa Cruz).

Warner, Patricia Rosalind (1992). Letter to the editor. *Boston Globe*, 28 June.

Williams, Lena (1992). 'Woman's Image in a Mirror: Who Defines What She Sees?' *New York Times*, 6 February: sec. A, p. 1, sec. B, p. 7.

Willis, Susan (1991). *A Primer for Daily Life*. London and New York: Routledge.

Wilson, Elizabeth (1985). *Adorned in Dreams: Fashion and Modernity*. London: Virago.

Wiseman, C., Gray, J., Mosimann, J., and Ahrens, A. (1992). 'Cultural Expectations of Thinness in Women: An Update'. *International Journal of Eating Disorders*, 11 (1): 85–9.

Wolf, Naomi (1991). *The Beauty Myth: How Images of Beauty Are Used against Women*. New York: William Morrow.

15 Foot-Binding in Neo-Confucian China and the Appropriation of Female Labor

C. Fred Blake

This essay interprets foot-binding in Neo-Confucian China as a voluntary ordeal undertaken by mothers to inform their daughters of how to succeed in a world authored by men. The ordeal entailed an exchange between mothers and daughters on many levels. Outwardly, it informed a daughter of the necessity of sacrificing the products of her body in service to the Neo-Confucian family system. Inwardly, the ordeal embodied for a woman—at the deepest level of her being—the lived memory of her mother. In foot-binding women both supported and subverted the Neo-Confucian way of being civilized.

Most of the writing on foot-binding in China highlights its erotic aspect. The only comprehensive work in English, *Chinese Footbinding: The History of a Curious Erotic Custom* by Howard Levy (1966), follows and reaffirms the traditional notion that foot-binding was a feminine mystique designed to please men. This notion provided grist for theories of human behavior, notably those of Thorstein Veblen and Sigmund Freud. Veblen ([1899] 1934) cited foot-binding as an example of conspicuous waste in which women surrendered their usefulness as a gesture to signify status in a male world. Freud ([1927] 1961) held that by mutilating their feet, women allayed the castration anxieties of men. Among more recent interpreters, Mary Daly (1978) cites foot-binding as an example of female victimhood in the grip of patriarchy, while Patricia Ebrey (1990) seeks an explanation in the Song dynasty reconstruction of gender, which aimed to accentuate the contrast between the genteel Chinese and their uncouth northern rivals. In Ebrey's view, foot-binding made women even more delicate, reticent, and stationary than they already were to allow men to refine themselves without seeming too 'effeminate'.

From C. Fred Blake, 'Foot-binding in neo-Confucian China and the Appropriation of Female Labor', *Signs*, 19 (1994), 1–25, 93–5. Reprinted with permission.

Each of these interpretations offers insight into the nature of foot-binding, but they also beg some crucial questions. For instance, how was this custom enforced in the everyday lives of women? Foot-binding was, after all, a discipline that required a constant vigilance not only during the initial stages of a little girl's coming of age but also through the better part of her adulthood. Also, what was the intended object of this discipline? The traditional interpretations beg the question of how these economies of male desire (for sexual diversion, domination, and social status) related to the economies of female work. Foot-binding was not simply a fashion marking the leisured life of 'upper-class' women. Foot-bound women labored in biological reproduction; depending on historical, regional, and family circumstance, many labored also in economic production. The physical labor women provided requires us to rethink the relationship between the more or less systematic mutilation of women's feet during the past millennium and the recruitment of their bodies into a particular system of economic production and reproductive activity.

In seeking answers to these questions I adopt the three-dimensional model of the 'mindful body' suggested by medical anthropologists Nancy Scheper-Hughes and Margaret Lock (1987). This model urges us to begin with the 'mindful' nature of our lived bodies—in the case at hand, to comprehend foot-binding as the advent of a 'feminine' self-consciousness. From this initial angle, we see foot-binding as a protracted discipline that mothers brought to bear upon their daughters in the name of a mother's love and a daughter's virtue. Both mother and daughter turned the brute resistance of a little girl's body into a body mindful of its fate in a world of sacrifice and duty and mindful as well of its ability to exercise control over itself and others. The embodiment of a self-consciousness also presupposes a discourse community in which the body is objectified, in this case as a thing of feminine mystery. As a rhetoric of gender, foot-binding became the critical moment in the historical domination of women's bodies by 'man made language' (to borrow from Spender [1980]). Foot-binding entailed intense, protracted physical pain, and, according to Elaine Scarry (1985: 4), 'whatever pain achieves, it achieves in part through its unsharability, and it ensures this unsharability through its resistance to language.' If this painful discipline between mothers and daughters became a 'tradition', it was represented and recorded almost entirely in the male voice. I will argue that foot-binding was the muted voice of women in contention with the dominant discourse of Neo-Confucian values and definitions of reality. Or, to paraphrase

Karl Marx ([1885] 1968: 97), foot-binding was a world that women made, but they did not make it under circumstances they themselves chose.

This suggests in the final analysis that the mindful body, an object of discourse, is also an object of incorporation or, in the words of Scheper-Hughes and Lock (1987), a 'body politic'. The body politic of foot-binding was its relationship to the historical system of production and reproduction during the late imperial period from the tenth century to the beginning of the twentieth century. Foot-binding was a mechanism for recruiting women's bodies into a labor-intensive economy by capturing their 'uterine power' to reproduce biological units of labor but also by commandeering their 'labor power' to produce economic goods. It facilitated what Hill Gates (1989) calls the 'commoditization of Chinese women' in which the products of women's labor were appropriated by male kinsmen in the patriarchal family system. Foot-binding was an integral part of that family system, and although it was a symbol of the leisured classes to which people in all walks of life generally aspired, it did not necessarily exempt women from participation in the system of economic production. Foot-binding simply changed the terms of that participation such that the labor women contributed was veiled by a mystique of women as sexual and maternal but otherwise worthless.

THE EMBODIED SELF AND THE MANAGEMENT OF SPACE

Gender differences are not only biologically determined, culturally constructed, or politically imposed, but also ways of living in a body and thus of being in the world. The existential condition of living in a body mediates our perceptual experience of the world (see Merleau-Ponty 1986). Perceptual experience is fundamentally spatial in nature if for no other reason than that our bodies are situated in space. This sense of space in which we feel and perceive our bodies to be located, extending into, and moving through is fundamental to our ability to perceive a world and embody a self. The spatial underpinning of our existence is simply taken for granted until our bodies are impaired and we are forced to reorganize our senses and the way we perceive the world. Foot-binding radically modified the bodily means by which a person perceived, experienced, and extended herself into the world. Radical and painful reduction in the size of the foot just as the foot was growing and extending a little girl's body into space brought the girl face-to-face with the spatial underpinning of her world.

For little girls whose feet were bound, the ground of perceptual experience was radically restructured beginning between the ages of 5 and 7. The period of trauma and enforced discipline generally lasted until she was between 13 and 15. In Western terms this is the period of latency when the child, having experienced a sense of finality in its libidinal attachment to mother, begins the transition into the discipline of the social world (Erikson 1963: 258). Chinese do not, however, conceptualize discreet stages of development linked to any one particular organic or cognitive agency. Also, the domestic discipline of little girls could begin earlier than 5 or 6 depending on the material disposition of the family or the whim of particular parents. Chinese, nevertheless, recognize that around the age of 5 or 6, the child's ability 'to understand things' (*dongshi*) is sufficient for the child to begin in earnest the discipline of the mindful body. The Confucian tradition refers to this discipline as *xiushen*, literally 'to cultivate the body', or, in the words of Tu Wei-Ming (1990), 'to become one's body'.

The Confucian body is not a discrete somatic object. It does not pose a world in terms of a radical or epistemological otherness. The Confucian body is rather a focal point of relationships centered around the primary relationship between parents and child. The child's ability 'to understand things' presupposes a 'body' that is inherently 'mindful' of others beginning with its parents. In recognizing the moral significance of its most intimate others, however, the child begins to cultivate its 'mindful body' as the basis of its relationship to less intimate others and the world at large. (In Chinese this innate capacity to be mindful of others is referred to as *ren*.) Both little girls and little boys experienced the awakening consciousness to the moral significance of others as a definite transition into the discipline of becoming their mindful bodies. The discipline of the mindful body facilitated a serious program of instruction in the Neo-Confucian fundamentals of industry and virtue.

For a little boy between the ages of 5 and 7, the discipline required that he relinquish the indulgence of his mother's world in order to embody the world of his father. Where the material means of the family allowed separate living quarters, this often entailed the physical removal of a boy from the interior rooms of his mother to the outer rooms of his father (Levy 1968: 75; Ko 1992). His father's world was anchored in the power of Neo-Confucian texts and genealogy that, with state sanction, formed the foundation of the family, the ancestral hall, and the school. This outer world invested power in mastery of the word, first the spoken word, later the written word. Resistance to the power of the word could be answered with a sound beating, usually the first and most memorable beating in a little boy's life (Yang 1945: 203–4; Lin 1947: 52). 'Fathers who used to be affectionate become distant, with a tendency to lecture' (Wolf 1972: 67). 'The father assumes a more dignified attitude toward [his son] and is frequently severe' (Yang 1945: 58). Mothers were torn between their need to support the discipline of their sons and their desire to provide a refuge from the harshness of that discipline. The boy's experience at this age is described with such words as 'abrupt', 'bewildering', 'drastic' 'radical and painful' (Wolf 1970; Solomon 1972: 47). In these terms the boy's experience with his father was not unlike the little girl's experience with her mother.

Yet the experiences of little boys and little girls were fundamentally different. While little boys, having begun to discover their bodies in their mother's world, were suddenly constrained to rediscover their bodies from inside their father's world, little girls, sharing their mother's sense of embodied space, learned to embody a culturally defined orientation toward the relative otherness of the male world. Neo-Confucian thought was long on its demand that females become virtuous and industrious while bending to the will of male authority, but it was short on how this process of 'becoming her body' should be accomplished. The process was left to the people, and by default to women, to figure out and act upon. Given the culturally exaggerated sense of a woman's body as mediating space and given the cultural necessity that a woman properly orient her body—that is, bend it to the will of male authority—it is reasonable that a girl's way of signifying her womanhood should be conceptualized in bending the organs that control space, spatial extension, and motion. Through the bending, twisting, and compressing of the feet, a girl's sense of managing space was radically modified and a mother delivered her daughter into a world where 'becoming one's body' led to moral and spiritual self-improvement.

Entry into the social world required little girls and boys to develop a sense of industry aimed at some level of morally informed self-realization. But this is where the similarity between their experiences ended. Boys' and girls' modes of self-realization, of becoming their respective bodies in relationship to others, differed completely. The boy's self-realization focused on the locutionary and literary power of the word. The girl's self-realization required her not merely to become, but to 'overcome her body' by restricting the space it filled. The difference was spelled out in proverbs like 'Teng er bu teng xue, teng nu bu teng jiao' (If you care for your son, care not if he suffers in his studies; if you care for your daughter, care not if she suffers in her feet). The difference was dramatized in innumerable little ways. In the western part of Henan, for instance, a daughter having her feet bound might receive a writing brush from her mother. The writing brush was a powerful symbol of masculinity and the world of civil affairs. But unlike her brother, the little girl did not receive the brush with the hope that she might learn how to shape literary discourse. Instead, she grasped the 'point' of the brush in the hope that her feet might acquire its 'pointed' shape (Yao 1936: 2:61). Once her feet acquired the shape of the brush, she might give the pattern of her small shoe (now invested with the mystical power of her self-sacrifice) to the man in her life to use as a bookmark and an antidote against bookworms (Yao 1936: 1:65–6).

For purposes of binding, 5 to 7 were the optimum years in a girl's physical, mental, and social maturation. Her prepubescent bones were still flexible. She was at the age of understanding (*dongshi*) and could be made to appreciate the necessity of a bodily discipline that entailed severe and protracted pain. Although largely filtered through male voices during the period when foot-binding was under attack, testimonies of foot-bound women attempted to find words for the kind of pain experienced in binding—burning, throbbing feet swallowing the body in fire—from severe traumas that created months, even years, of oozing sores, bandages stiff with dried pus and blood, and sloughed-off gobs of flesh. These accounts tell of girls losing appetites and sleep, running away, hiding, surreptitiously attempting to loosen their bandages, and enduring beatings while trying to comply with their mothers' demands. But the girls were still at a tender enough age to retain the basic trust in the implicit goodness of their mothers' intentions. The 'tradition' could not have passed from mothers to daughters if not for mothers' credibility as 'caring'. The conundrum of a mother's care consciously causing her daughter excruciating pain is contained in a single word, *teng*, which in the proverb cited above refers to 'hurting', 'caring', or a conflation of both in the same breath. The Chinese sometimes say of a parent beating a child, 'Da shi teng, ma shi ai' (Beating is caring, scolding is loving). In the ordeal of foot-binding, the exchange of *teng* between a mother and daughter was highly reflexive. In his novel about a late nineteenth century family of antique dealers, Feng Jicai (1986: 9) uses the word *teng* to describe the physical 'pain' little Xiang Lian felt when her grandmother began binding her feet. Later in the same paragraph, he uses *teng* to refer to the conflation of 'care' with 'pain' that Xiang Lian began to perceive behind her grandmother's grim countenance.

Howard Levy (1966: 249–51) asked each of his elderly informants how she thought her mother felt when she first began binding her daughter's feet. Most recalled that their mothers showed no pity, but some recited the old refrain that a mother could not love both her daughter and her daughter's feet at the same time. By separating the mother's care for her daughter from the vessel of her daughter's worldly existence, the mother and daughter shifted their cares to a different order of consciousness. In the struggle with her mother over the painful, bloody, and terrifying labor of making the brute nature of her feet materialize into an object of beauty, mystery, and discipline, the daughter formed a new self-consciousness based outwardly on a sense of dependency and attachment to a male-dominated world and inwardly on an ability to exercise some control over her own destiny and that of the persons to whom she was attached.

Mothers constantly informed their daughters that binding was necessary in order to find a good family into which to marry. Mothers impressed upon their daughters that the mark of a woman's attraction resided more in her character as revealed in the bind of her feet than in the face or physique with which nature had endowed her. Her selection in marriage was the task of her prospective mother-in-law, whose criterion for a good daughter-in-law was the discipline that the bound foot represented (Pruitt 1967: 22). Concomitantly, binding was a matter of both families' reputations—or 'face' in the Neo-Confucian contest for social status, which I will discuss. A daughter learned that she carried her family's reputation or face in the bind of her feet, and that her family's face, whether that of her own family or the one into which she married, belonged to the male heads.

A woman's dependency on her family was made manifest in her disabled feet. A popular saying was that in her lifetime, a woman leaned on three men, her father, her husband, and her son. But if a woman's bound feet made her appear weak and vulnerable and thus dependent on men, it also veiled her inner sense of managing those appearances and thus being able to exercise considerable control over herself and those to whom she was attached. The microcosm of her mindful body thus learned to mirror the organization of the family in which women (as *neiren* or 'inside persons') managed the 'internal' (*nei*) and 'secret' (*si*) affairs of the domicile while men managed the 'external' (*wai*) and 'public' (*gong*) affairs of society. This separation of spheres, originally formulated in the ancient *Book of Rites*, was embedded in Neo-Confucian doctrine and the organization of daily life (Ko 1992).

Before she reached puberty, at around 14 the girl's severe bouts with pain subsided and she began to bind her feet without help from her mother. Her feet having achieved the desired proportion and shape, the young woman concentrated her attention on maintaining or even refining the bind of her feet in preparation for the ultimate payoff, marriage. The sexual design of the compressed foot, to which most of the literature on the subject is devoted, however, veils other aspects of its meaning. The years of pain and suffering—or, in the words of an old refrain, 'the two cisterns of tears' a young woman shed around attempts to win control of her bodily space—ended as she entered the world of menstruation and birthing. Foot-binding was a prelude to, even a preparation for, the sexual maturation of her body. In this sense, foot-binding helped her to 'overcome her body' in two important ways.

First, in mastering her embodiment of space the young woman with bound feet became acutely attuned to the significance of exercising control over reproductive functions, which were both a threat to and highly valued by the male-dominated family system. Emily Ahern (1975) points out that because of their reproductive activities, Chinese women are constituted as a threat to social and cosmic order and therefore 'powerful' and 'dangerous'. In realizing her primary function of reproduction, the young woman transgressed critical social boundaries by the taint of uterine discharge and the introduction of her and her babies' unfamiliar, unsocialized bodies into the family to which she was married. The products of her body—especially male infants—were highly valued. Even her menstrual blood was positively valued as a sign of regularity and fertility (Furth 1986). But because these bodily products tend to 'disrupt' and thus to 'defile' the established sense of familial order, they were also 'dangerous' and subject to rigorous self-control.

Second, foot-binding prepared the young woman for the aggravation, pain, and dread associated with menstruation, sexual consummation, pregnancy, and birthing. The act of sexual consummation, which by custom took place on the marriage night, could be attended with considerable anxiety, especially by the bride. Although we know very little about sexuality in Neo-Confucian China, I think it is fair to assume that the anxiety surrounding the initial experience, given the necessity of proving female virginity, was not altogether different from what we can glean from modern studies of 'traditional' village women and marriage rites. Some of these reports are especially graphic. In Margery Wolf's report from a Taiwan village of the late 1950s, 'every woman our assistants could pry information out of insisted, and often with bitterness, that their mothers had told them nothing to prepare them for their first sexual experience. [Mothers] wish to spare [their daughters] any knowledge of what is at least by custom an unpleasant act' (Wolf 1972: 139).

But if mothers failed to verbally instruct their daughters in how to deal with the 'unpleasant customs' of married life, which included the necessity of submitting her body to a man's will, foot-binding obviated the need to verbalize these matters with any specificity. Foot-binding provided mothers with an effective means for instructing their daughters in how to handle all kinds of bodily insult. Foot-binding at least informed the mindful body of its fate in patriarchal society and armed it with an effective discipline, not only—as the conventional wisdom taught—to maintain the strict rule of chastity, but also to deal with the travails of becoming a wife, daughter-in-law, and mother.

We will probably never know what kinds of verbal instructions mothers gave daughters in the actual ordeal of binding. My point is that the *teng* exchanged between mothers and daughters in the ordeal of binding was itself a lesson about language. I think foot-binding pointed to the notion of women as 'muted' (Ardener 1975). That women were muted does not refer to women's incapacity to make utterances, to be clever or entertaining with words, or even to be aggressive with words, but rather to their incapacity to give their words the weight of authority, to have their words taken seriously, or to make decisive statements worthy of public credibility. References to 'women's talk' assumed that women were naturally inclined to 'talk too much' (*duoyin*). Every little girl stood warned against developing a 'fast mouth and sharp tongue' (*(kuai-zui lishe)*, as this was grounds for a husband to divorce a wife ('The Shrew' 1973). While village women typically engaged in 'loud and gossipy talk' (Wolf 1972: 40), those who aspired to be women 'with class' were taught to refine their natural instincts for 'chatter' in ways that compared to the 'pleasing sounds' of 'swallows twittering' (*yanyu*). Conversely, a man's speech act, which presumed the weight of authority and substance, could be ridiculed by comparing it to the 'the endless chatter' of old women or worse, by comparing it to an 'old lady's footbinding cloth, smelly as it is long'. This common idiom, which ridiculed a man's speech act by linking it to women's foot-bindings, can be found embedded in political discourse as well as everyday talk: 'Some of our comrades love to write long articles with no substance very much like the "footbindings of a slattern, long as well as smelly"' (Mao [1942] 1965: 56, 57).

Becoming a mother was of course the critical event of a woman's life as it entailed the act of birthing. Although no conclusive evidence exists about the relative risks to a woman's health in birthing and foot-binding, I think it is fair to suggest that birthing must have been by far the graver risk. It is unlikely in my opinion that Neo-Confucian society would have committed so much of its resources to a rite of passage that would entail a significant risk of mortality to its adolescent population. If female infants were expendable, those who were raised to adolescence became important to the reproductive and productive capacity of a labor-intensive economy, a point I will discuss. Still, the pain of reproductive labor, often agonizing, was not as protracted as the agony of twisting and compressing the lower extremities. Of course, the pain associated with birthing was considered entirely natural, while the protraction of pain and suffering associated with foot-binding was altogether an intentional act of the human will.

The suggestion that foot-binding was a prelude to menstruation and birthing posits a phenomenological connection between the foot and the womb, or 'seed chamber' (*zigong*), that anchored femininity in motherhood. The ostensible theory of binding focused attention on the physiological benefits of constricting the lower limbs to concentrate the blood in the upper parts of the legs and groin where sexual sensitivity was increased and the fertility of the 'seed chamber' was enhanced (Yao 1936: 2: 343). Binding was compared to pruning trees to quicken and concentrate the sap in bearing fruit. I will interpret this theory as a mystification that hid the labor power of women. But a deeper connection existed between the foot and the womb. Essentially, by controlling that which the culture invested in her feet, a woman controlled that which the culture invested in her womb and defined as the key to her femininity—her ability to bear children.

We can take this one step further: by controlling that which the culture invested in her feet, a woman might mediate at some level of intentionality the reproduction of male sexual characteristics. In other words, while the womb is an exclusively female organ, it has the power to obtrude and extrude male bodies. Feet, on the other hand, are possessed equally by females and males, but the obtrusive and intrusive nature of feet allows them to represent the male organ. (In China this representation was realized when a bride intruded her foot into the groom's shoe as a ritual prelude to the act of sexual consummation or when she left her small shoes in the Guan Yin temple to betoken her most earnest desire to become pregnant [Fortune 1857: 39–40].) Thus the obtrusive and extrusive nature of the womb and the obtrusive and intrusive nature of the feet each possess androgenic properties. Female gender therefore represents an essential androgynous ambiguity, one internal, the other external. In bending and compressing her external obtrusion, the female manipulated and rearranged the external features of her sex and won control over the reproduction of sexual characteristics. In the process, she made a world in which males intrude as husbands and obtrude and extrude as sons. If the rewards for bearing sons were basic to a woman's position in the family and her own sense of fulfillment, then foot-binding, insofar as it increased a woman's sense of control over her body, also contributed to her sense of reproductive competence.

Past the age of puberty and for the better part of her adult life, a woman remained vigilantly focused on the state of her feet to maintain the desired effect. The reduction of the foot was not accomplished in a onetime operation, which a radical excision of the toes would have made possible. Some Chinese writers recognized, at least for rhetorical purposes, that if 'chopping off the foot' to make it 'smaller' and 'weaker' was feasible, it was not what binding actually intended (Yao 1936: 1: 63). Although these writers were never so specific, simple amputation would have defeated the whole point of binding as a 'voluntary', protracted, and disciplined self-sacrifice.

The question of volition brings us finally to the historical tensions between feminine and masculine consciousness. Foot-binding provided the means by which men appropriated custody over uterine property by distancing themselves from the toil of production. 'Production' is used here in its widest possible sense, from production of the small feet to other forms of labor, including the labor of biological reproduction. The key to this process of appropriation was in the mystification of the sexual object. While they fawned over the appearance and disappearance of the tiny feet in an erotic context, men contorted themselves in combinations and permutations of disgust, contempt, anger, rage, and pity for anything that associated them with the actual production of tiny feet. As we shall see, the taboo against male contact with the binding process also helped to preserve the integrity of Neo-Confucian teaching that opposed self-mutilation.

For their part, women attended to their feet in the strictest privacy. They often washed their feet separately from other parts of their bodies. Separate washing also shielded other parts of their bodies and other persons' bodies from possible contamination. In recalling the bathing customs in his village in southern Jiangsu, a modern writer tells how his happiest memories were of bathing with his grand-mother, who 'had a pair of small pointed feet like rice dumplings. Without removing her binders or her shoes, she propped her feet on the edge of the tub while immersing the rest of her body in the water. In this position she splashed water on her grandson and rubbed his back' (Ding 1989: 6). Even in the artistic depiction of an erotic bathing scene, a woman is depicted stepping from her bath with the help of her paramour clad only in her little red shoes (*Erotic Art of China* 1977: pl. 21). While 'erotic representations depict(ed) women stark naked,' according to one authority, 'I have never seen or read about a picture that showed a woman's uncovered bound foot' (Van Gulik 1961: 218). In reciting the story of her life, Lao Ning tells of a man who cursed his daughters for disturbing his peace of mind with their crying while having their feet bound. His wife carried the basin of water with its bandages into the presence of her husband and threw the basin at his feet. 'The water splashed and strips of foot-binding cloth flew in all directions. Because he was an official and because there were so many of us around he dared not to beat her nor to revile her. But his anger was too great for him to bear. He jumped up, seized his whip, and began to thrash and beat the dogs' (Pruitt 1967: 79). Other women were not so lucky: when a woman in San Francisco accidentally splashed water on a man while washing her feet, the man flew into a rage. She tried to assuage his anger with offers of sweets as a prophy-lactic against the contamination, but his rage was too great and he shot her to death (*St. Louis Globe-Democrat 1881: 11*) A common curse was to accuse a man of 'carrying his old lady's foot water', thus suggesting his abject devotion to his wife. (The same phrase could be framed as a compliment).

There were of course exceptions to the prohibition of male contact with the sacraments of the deformed foot. These exceptions were in situations that were outside the routines of everyday life, such as ill-ness and sexual arousal. These occurred in highly restricted settings in which social conventions are generally suspended and taboos may be overturned (see Turner 1969; Douglas 1970). In the process, that which is normally considered dangerous and polluting may become a powerful healing agent. A variety of male impairments were treatable by various female body fluids including menstrual blood (Furth 1986: 47; see also Seaman 1981), while illnesses such as cholera, malaria, strep throat, and faint were treated with various paraphernalia of the bound foot. Similarly, rigorous restrictions on male contact with the uncovered foot facilitated a reversal of the gender hierarchy when men abandoned all their normal pretensions, debased themselves, handled the wash, drank from the tiny shoe, fondled, sniffed, and mouthed the deformity itself.

The dread of contamination and the lengths to which women went to control the threat of defilement from their hidden deformity was, as I have suggested, linked to their fecundity and childbearing activities. A woman presented the products of her labor (her tiny feet and male infants) to be re-presented as precious objects in the world of male discourse. As we shall see in the next section, men made up all kinds of names for the tiny feet. And it was the male heads of households who by ancient custom named and entitled the newborn males (Chai and Chai 1967: 476). Whether female infants received a name was less important, and in some village traditions, at least, they remained unnamed (R. Watson 1986). The act of naming and entitling is a distancing device as it empowers the namer to endow the named with an objective social existence. By naming the products of women's bodies, men distanced themselves from the production process and appropriated, among other assets, the male members of their descent lines, which in turn provided them their sense of immortality.

Men lived in a world mediated not by the weapons of oppression, which remained ominously in the background, but by the civil functions and power of language. The distance a man bore from the practical act of oppression and degradation of the other was his measure of civilization. In Neo-Confucian China the ideal of masculinity did not countenance, indeed found disreputable, attributes of direct confrontation, bravado, machismo, or aggressiveness. These traits were ascribed to 'the uneducated' and 'barbarians' such as the Mongols and Manchus, who, it so happened, rejected foot-binding. The ideal masculine image among Neo-Confucian Chinese was modeled on the dignified bearing of the scholar-gentleman. It was the disposing of the word—rational and careful management of the communicative act, made manifest in the concern for 'face' and in verbal and literary competence—that was the measure of manliness. Foot-binding allowed men to appear genteel, but not simply to cover the threat of appearing too effeminate as Ebrey (1990: 221) argues. Foot-binding, in my opinion, exempted men from direct responsibility for dominating and degrading women. It allowed women to deflect the brutality and threat of direct male domination by taking that domination into their own bodies and making it their own.

THE SOCIAL BODY AND THE RHETORIC OF FOOT-BINDING

The embodied self is also a 'social body' in that it is an organic object of discourse. It mirrors, reflects, projects, and represents the language-based categories that society takes to be the 'natural' order of things (Douglas 1973; Needham 1973). In China the (male-dominated) discourse invested the sexual and reproductive functions of the female body with special cosmic significance and in the process defined its place in the gender hierarchy. But foot-binding dramatically altered and in some ways subverted this 'natural' order of things in a way that revealed the willful and artificial aspects of the feminized body. The symbolic contradictions, the tensions between representations of 'nature' and 'culture', and the contest of words around the mysteries of gender and the feminized body constituted what I, paraphrasing Kenneth Burke ([1950] 1969), would call a 'rhetoric of foot-binding'.

There are many terms, trite terms and spontaneous expressions, descriptive and euphemistic, metaphoric and metonymic, vulgar and literary, for the small bound foot and its sartorial aspects. These myriad terms were used to depict the bound foot under different conditions from subtle changes in lighting and mood as well as with historical and regional differences in style. The term that packed in the most symbolic meaning across time and space was *jinlian*, usually translated as 'golden lotus'. The *lian* is the central symbol of Buddhism (Ch'en 1972: 445). The life structure of the *lian* evoked the life world of women, even more so the life world of foot-bound women. The flower floating on the surface of the water, a symbol of purity, was sustained by an umbilical cord rooted in the mud and slime of pond and lake bottoms. Just as the lily appeared to float, the woman was supposed to evoke an illusion of floating by ambulating in short, mincing steps, a style referred to as 'Bu bu sheng lian hua' (Each step bears a lotus blossom). The bound feet supposedly forced a woman to walk in this manner. This style of ambulating, according to Chinese writers, also made the sexual anatomy and physiology more voluptuous and sensitive. The *lian* suggested both spiritual detachment and carnal desire. In other contexts the *lian* was directly associated with its hidden connection to reproductive functions. Not only was its seed-filled pod emblematic of the womb (*zigong* or 'seed chamber') and thus of childbearing (Williams 1974: 258), but other parts of the plant, including the flower and the root, provided a rich pharmacopoeia for disorders associated with childbearing. In one particular remedy, the petal of the lotus flower eased a difficult birth when inscribed with a man's 'literary style' and swallowed by his wife during labor (Smith and Stuart [1911] 1969: 280).

Although *lian* was modified with any number of prefixes, the most popular prefix was *jin*. Translated as 'golden', 'precious', or 'gilded', *jin* reinforced the worldly aspect of bound feet as objects of commodity value: 'Such small and delicate feet as these are worth a thousand ounces of gold' was a common adage. The symbol of the *jinlian* thus combined the transcendental motif of Buddhist metaphysics with the worldly symbol of material value.

The rhetoric of the bound (and unbound) foot played on and mystified the tensions and contradictions between the mundane things of this world and the spiritual essences of cosmic order. Much of the rhetoric found expression in folk songs collected by Chinese folklorists during the early part of the twentieth century (see Liu [1925] 1971). Songs about bound-footed women could point to the irony and hypocrisy of living in a world of appearances, deception, and manipulation that finally showed itself as a pitiful state of uselessness. In a song from Henan (Liu [1925] 1971: 204–5), a 'madam big feet' jealously scorns another woman with bound feet: 'You've never eaten rice or noodles, how can you put on such airs? You are like two trowels that can't mix the mud; you are like two threshing boards that can't thresh the grain; you are like two pieces of dried meat that can't be sacrificed at the temple; you are like two softshell turtle lamps that set upon the stage cannot shine brightly!'

The theme of deception, mystification, and realization (as 'useless') is expressed in a song from Suzhou described in the novel *Sancun Jinlian* (Feng 1986: 21). The song describes a woman who is seen binding her feet by her doting paramour. As he strolls about the room, the man describes her tiny delicacies with poetic hyperbole as 'bamboo shoots in winter', 'like three-cornered rice dumplings in May, so fragrant so sweet', 'the fragrant chestnuts of June', 'so delicate, so pointed'. The woman blushes and says, 'Your cheap talk only betrays how horny you are. Tonight we will sleep end-to-end so my golden lotus is next to your mouth. Then I'll ask you how sweet it is and how you like the taste of the winter bamboo shoots!' In this case, the self-deception and mystification is revealed in the man's talk and it is the woman binding her feet who suddenly throws a cold light on his effort to mystify her feet. Although this song is recited in a modern novel in which the author places its performance in a small-foot contest around the turn of the twentieth century, it reflects two interesting aspects of foot-binding. The lyrics give some idea of the kinds of allusions used and the way they were used to mystify the bound foot, but they also suggest how women exercised control over the whole process.

The cosmic consequences of not binding the feet were told in another folk song from Henan province (Liu [1925] 1971: 205–7). The song recounts how a 'madam big feet' orders a new pair of shoes to go to the temple, tries to stuff her oversized feet in the shoes, knocks down walls, and desecrates temples and engulfs six towns when she stops to urinate before returning home 'with big strides'. The 'big-footed' woman was not simply clumsy; everything about her body was gross and it threatened the cosmic order. Her body represented the unstrung forces of nature that in earthquake and flood bring ruin, engulfment, contamination, and death. Social life on the scale envisioned by the Neo-Confucian imperial order was possible only if these elemental forces were restrained and mediated by human effort and individual initiative (see Weber 1968: 28).

The relationship between the size of women's feet and the cosmic order as mediated in the rhetoric of foot-binding can be enlarged upon by observing that natural-sized feet were attributed to two distinct classes of women. One was the ordinary woman (as depicted in the previous song) who lived in the carnal world and whose 'big feet' were compared to 'lotus ships'. As we have seen, the 'big feet' of ordinary women were demeaned as clumsy and crude and as a disaster to the natural foundation—the productivity—of the civilized world. The other kind of woman with natural-sized unbound feet was extraordinary in her power to confer benefit on the world. These women took a variety of extraordinary forms. They included legendary paragons of filial devotion (e.g., Meng Jiangnu), women warriors (e.g., Hua Mulan), goddesses of mercy (e.g., Tian Hou), and the Buddhist redeemer of humankind (Guan Yin). These women were extraordinary because each in her own way freely chose, or successfully resisted, or, in the last instance, transcended the ordinary world of marriage and reproduction. Tian Hou and Guan Yin were models of maternity and sacrifice. They were powerful deities to whom women especially appealed for help to become pregnant. The Buddhist Guan Yin provided her worldly supplicants with the karma of reproductivity and bore constant witness to their agony. Guan Yin was some times depicted bending forward as if to offer her supplicant the infant in her arms (Fortune 1857: 39–40), but when she and the other female members of her extraordinary class were depicted with exposed feet, they were always 'natural feet' (*tianjiao*). It was inconceivable that these legendary and powerful women would be seen in bound feet. Bound feet were artificial things invented by ordinary people. Indeed, during the era of foot-binding, women with 'big feet' could be damned with faint praise by comparing their feet with the natural feet of Guan Yin (see Levy 1966: 118).

Foot-binding thus mediated these images of femininity. The refined woman with bound feet was obviously like the common woman with 'big feet'. Both were ordinary and worldly and caught in the karma of reproductive activity. But the woman with bound feet also possessed something of the power and the virtue of the extraordinary woman. The two images of femininity were mediated in the act of self-mutilation. The woman with bound feet was crude, carnal, and potentially defiling, and yet she was true, pure, loyal, and compassionate. In the first instance, foot-binding was a vulgar act of self-exaltation. It intended self-beautification, aimed at mystification and manipulation. In the second instance, foot-binding was an act of self-mortification. It aimed to transcend carnal desire through self-mutilation. In her state of exaltation, the woman with small feet was moved by cunning, avarice, envy, and jealousy, the excesses of which were especially lamented in women (Furth 1986: 58). These motives were glossed as 'narrow heartedness' (*shao-xinyen*), a word widely used to describe the common vice of women. But in mortifying her feet, the ordinary woman of the world revealed a capacity for self-sacrifice that surpassed the maternal instinct observable in nature. The self-conscious sacrifice of her body to redeem her world in its otherness (i.e., to bear the progeny and thus sustain the male ancestral line of the family into which she married) raised her feminine instincts to the level of compassionate being that is intended in the figure of Guan Yin or ideal 'motherhood'. Thus binding the feet created a feminine mystique that enabled an ordinary woman to exalt her carnal body in the quest for spiritual perfection!

Although Buddhist symbols were culturally and historically associated with women and foot-binding, I am not aware of any references to foot-binding in formal Buddhist doctrines. The same holds true in reference to other formal doctrines of imperial law and Neo-Confucian philosophy. Neo-Confucianism was institutionalized as the dominant discourse from around the tenth century to the end of the nineteenth century, the period within which foot-binding was more or less popular. That we cannot cite Neo-Confucian doctrine on the virtue of foot-binding has enabled such modern-day Confucian scholars as Tu Wei-Ming (1985: 143) to claim that 'no convincing evidence has been found to establish a causal relationship between the rise of Neo-Confucian culture in the tenth century and the prevalence of such appalling social customs as footbinding' (1985: 143). But this way of dismissing a relationship because it fails the test of simple causality not only obfuscates the genius of Chinese civilization, it begs the obvious question: How do we describe what was indeed a complex relationship? Although foot-binding was not 'caused' by Neo-Confucian doctrine, foot-binding was women's way of participating in the community of Neo-Confucian discourse. Foot-binding affirmed the Neo-Confucian claim on moral superiority, but not without also contradicting it and in some ways subverting that claim. This 'affirmation' can be seen in the way foot-binding was embedded in the most cherished institution of Neo-Confucian culture, the family.

Neo-Confucianism was 'pro-family', to put it mildly. The doctrine of Neo-Confucianism anchored the interests of the state in the affairs of the family. The distinctions of gender and generation that shaped the family were given legal sanction and backed by the threat of draconian punishments. Parricide (the killing of a parent) was a crime of treason punishable by the 'slow and painful death' of a thousand cuts; striking a parent was also a capital crime, punishable by beheading (Staunton [1810] 1966). The severe and arbitrary application of imperial law, however, was mediated to the extent that the power to enforce its moral code was transferred to local kinship organizations (Hsiao 1972: 342; Potter and Potter 1990: 10). But local sanctions could also be harsh and were especially brutal in punishing the 'sexual misconduct' blamed on women. With foot-binding, women extended the logic of self-regulation and self-control to themselves and thus obviated the dreadful consequences of regulation through external sanctions.

But in obviating the threat of severe external sanctions, foot-binding violated an ironclad dogma that pervaded the whole of the Confucian (and Neo-Confucian) doctrine—the rule that prohibits physical abuse of the body. This prohibition was conveyed in dicta of Confucian-inspired texts that every child memorized in the course of learning the language. At my request, one young Chinese woman (reared for most of her life in the United States) wrote the apothegm with only a moment's hesitation:

'Dare not injure even the hair and skin of the body received from mother and father.' Its reasoning was a fundamental precept of Confucian ethical thought. 'Our bodies . . . are not our own possessions pure and simple; they are sacred gifts from our parents and thus laden with deep ethicoreligious significance' (Tu 1985: 118). Confucians recognized that this was the genesis of a system of ethics and that its logic required little more than an awareness of and appreciation for the facts of life that are realized by the simplest reflection: this body in which I locate my being is not mine entirely. It also belongs to those who brought it, and thereby me, into existence; and as those others (my mother and father) gave me life by giving me part of themselves, I will in turn give them life (as ancestral beings) by giving up part of myself to yet another generation (my children). This ethic is taught to youngsters on a daily basis and in no uncertain terms. 'A link in a chain' has been used to describe this sense of being in the world (Wolf 1979: viii). As a 'link' in a chain of being, the individual's purpose in life was to reproduce himself or herself on behalf of his or her forebears. Not to bear progeny was a sacrilege that violated the first principle of filial piety. One's body belonged to the chain of being, to one's parents and to their parents. Any kind of physical abuse that posed a threat or that was even extraneous to the well-being of the temporal linkage (e.g., excising pieces of flesh or marking the flesh with tattoos) was forbidden as a desecration. There were occasions, however, when people violated the rule against self-mutilation in order to dramatize the rule of filial piety. Gestures as grand as boiling a piece of flesh excised from one's own body to cure a morbidly ill parent received popular acclaim and dramatized the meaning of self-sacrifice in service to one's parents until it was discountenanced by imperial edict in 1729 (deGroot [1892] 1972: 458–9, 793). Of course the most widely acclaimed exception to the rule against self-mutilation was foot-binding.

But the rule against self-mutilation applied less to women because they did not constitute, in the dominant discourse, a formal link in the agnatic chain of descent through fathers and sons. They constituted instead a formal 'break' in that linkage. Women as a custom did not remain in the domicile where they were born. They did not, as a rule, reproduce their parents progeny. If a woman's purpose in life was to reproduce herself, it was on behalf of her husband's forebears, not her own. Women were, strictly speaking, not allowed to fulfill the highest ideal of filial piety, an ideal that was ironically grounded in the virtue attributed to ideal womanhood—the maternal instinct to care and be cared for. Foot-binding ostensibly separated a daughter from her parents to serve and to immortalize a stranger's parents. The stigma of separation was the wound to her feet, which Richard Solomon (1972:

36) compares to the tragic Greek figure Oedipus ('wounded foot'), whose separation from parents was similarly marked. But in my opinion the likeness extends no further than the symbol of separation implied in the 'wounded foot'. The foot-bound daughter was more like a Chinese Persephone in that the bind of her feet bonded her to her mother in the process of wedding her reproductive powers to a strange man. Although the daughter's 'wounded foot' marked her separation from her mother in marriage it did not mark her as having forgotten who her mother was. Indeed, it was the embodiment of her mother's memory and all the 'knowledge that is subliminal, subversive, and preverbal . . . between two alike bodies' in the long ordeal of binding (to borrow from Adrienne Rich [1976, 220]).

With marriage the bride carried the mark of her mother with the painfully acquired knowledge embodied therein into the custody of a mother-in-law and her son. As preparation for, and coupled with, the rites of marriage, foot-binding mystified the bride's movement through space from one domicile to another to become the link between families in order to bear the (male) link between generations. If she herself was not a 'link' in the formal sense, it was she who 'made' the formal links (between fathers and sons) that constituted Chinese society. Her role was like Guan Yin's. She sacrificed her body to help others (in this case, men) procreate themselves. As she embodied the cosmic order of things, she stretched, bent, rent, and mended and thus rendered the sinews of the social order much as she did the sinews of her body in binding and birthing. The marriage ritual was enveloped from start to finish in symbols of fecundity and childbearing. Through her reproductive labor, she would make what was otherwise a mere abstraction, a list of men's names and entitlements—a genealogy of fathers and sons—into immediate, tangible, temporal reality. As in the Greek myth of Persephone and Demeter, the fertility of the earth and the continuity of human life are tied to the endurance of the bond between mother and daughter.

Family reputation and imperial law required that a daughter keep her virtue intact, even to the point of death. This was the extent of her formal commitment to her natal family (Ng 1987). From birth she was already thought of as 'belonging to another family' and generally treated accordingly. The ever-growing realization that her presence in the family was merely temporary intensified the anxieties between parents and daughters. Popular references to daughters as *peiqianhuo* (goods on which money is lost) made daughters feel they used household resources that they could never repay. Daughters could compare themselves to the family-owned draft animal and note that fathers were less willing to part with their animals than they were with their own daughters (e.g., see Blake 1979a). In poor families daughters might be sold outright or sent into marriage sooner rather than later, while in elite families fathers (Levy 1968: 180) and, even more so, mothers (Ko 1992) with certain dispositions might just as readily indulge their daughters and encourage them to develop their literary talents in line with Neo-Confucian ideals of self-cultivation.

By whatever means parents chose to fill the void between themselves and their daughters, ordinary mothers were saddled with the task of teaching their daughters the realities of a woman's domestic roles. In the ordeal of foot-binding and in its complex symbolism of separation and identification, mothers surpassed—and in some ways subverted— the expectations of their institutional role by establishing a powerful and enduring relationship with their daughters. The Neo-Confucian family system and the civilization that rested upon it veiled its own ontology in the 'uterine family' (Wolf 1972). Margery Wolf suggested the concept of the uterine family to show how Chinese women exercise influence over their sons in order to contest loyalties in the family system. The other source of 'uterine' strength was hidden in the ostensible separation of mothers from daughters that foot-binding signified.

THE BODY POLITIC AND THE INCORPORATION OF FEMALE LABOR

While the mindful body of women was mystified in the rhetoric of foot-binding, its biological power to produce and reproduce things of material value was incorporated in a 'body politic'. Foot-binding cannot be fully explained without reference to the historical system of material production in which the sexual, reproductive, and economic products of women's labored bodies were systematically appropriated to make possible a Neo-Confucian way of being civilized. The remarkable fact about foot-binding is that while the modern world has relegated it to a historical curiosity, it exchanged untold amounts of human energy on a daily basis without direct force of law—even in violation of imperial edicts—and it lasted for a thousand years across generations, centuries, and dynasties. The origins of its popularity are located in the Song dynasty (Van Gulik 1961: 253; Levy 1966: 188).

C. FRED BLAKE

The Song dynasty was a period of social ferment in which Neo-Confucian doctrine was institutionalized and a commercial economy developed in an increasingly labor-intensive system of agricultural production. This late imperial social configuration lasted into the late nineteenth and early twentieth centuries, when it disintegrated under the challenge of European-based capitalism. That the popularity of foot-binding flowed and ebbed with the rise and fall of China's late imperial order suggests that the custom of disabling women's feet and the system of production bear more than a fortuitous relationship one with the other.

My historical analysis does not attempt to correlate fine-grained historical changes in the structure of the political economy with the popularization of foot-binding. My aim here is to correlate the essential features of the late imperial political economy with the sociocultural mystification of foot-binding that I have described. The critical features of the late imperial political economy included a labor-intensive grain-agriculture-based adaptation, a (Neo-Confucian) ideologically driven political system, and a growing urban-centered commercial economy based on what Gates (1989) calls 'petty capitalism'. These features were integrated in the family system in which the contributions of women's labor (biological and economic) was essential and in which women took part as both pawns and players in the increasingly fluid system of social climbing. Also, my data on the economic contributions of foot-bound women are necessarily limited and come mainly from the period when the mystification of foot-binding was penetrated by the reified world of European capitalism in which foot-binding became scorned as irrational and economically wasteful. This led, however, to the view that handicapping women removed a large segment of the population from the process of economic production and made women practically worthless. Few observers were able to see through either the Chinese mystification or the European reification to appreciate the economic contributions that foot-bound women made. I believe that the impairment of women's feet served more to mask than to completely restrict their participation in economic production.

During the late imperial society, the family-based system of production was increasingly integrated by an expanding 'petty capitalist' economy. The 'self-exploitation' (Chayanov 1986) of family labor increasingly transferred the value of women's labor to male kinsmen. Referring to this process as the 'commoditization of women', Gates (1989) points to the many ways in which Neo-Confucian discourse called upon women to sacrifice themselves for the material welfare of the family coupled with 'petty capitalist' references to the commodity value of their bodies. The material contributions that women made to the family were indeed substantial. They included women's traditional handiwork—making items like clothes and shoes—as well as their biological contributions in making sons for the labor-intensive economy. As I have discussed here, foot-binding helped to mystify the process by which these products of women's bodies were appropriated by the male-dominated family system. Foot-binding facilitated the process of appropriation by a mystification that entailed masking a woman's labor power. The ritual-like idiom of 'masking' women's labor power—of forcefully making it 'disappear'—reflected the intentional, artificial, and arbitrary nature of foot-binding. The bound foot was seen and appreciated for the mystification that was intended. This way of mystifying social relationships, I believe, was unique to the 'petty capitalist' system of family-based self-exploitation of late imperial China. It was altogether different from the reification of consciousness by which the European-based capitalist mode of commoditization totally rationalizes a life world and ridicules devices such as foot-binding as wasteful.

As noted, foot-binding mystified more than the handiwork of domestic labor; it also helped to mediate the reproductive strategy of the family-based system of labor-intensive agriculture. The labor-intensive economy generated high levels of output that required and supported a huge and growing population (Banister 1987: 3). A number of economic historians have noted the long-term critical relationship between agricultural production and labor intensification in China's late imperial society. The system achieved high levels of productivity but was 'trapped' (Elvin 1973), 'stagnated' (Chao 1986), or 'involuted' (Huang 1990) in its own demographic success. But the agricultural countryside with its pronatalist values was also (and increasingly) integrated into the urban commercial ('petty capitalist') classes of artisans, traders, merchants, and financiers. Networks of rural markets operating outside the imperial system of tributary extraction facilitated trade in raw materials, handicrafts, and small-scale industrial goods on at least a regional basis (Skinner 1977). Small-scale industry and commerce were established as a way of life as early as the thirteenth century (Gernet 1970) and became increasingly characteristic of late imperial Chinese society. But the petty capitalist classes with all their commercial power and clamor for social recognition did not seek cultural or political autonomy. Instead they tied their aspirations for recognition and status to respect and support for the traditional bureaucratic-based elites (Balazs 1972). In turn, the ideologically steered bureaucratic system that was based on surplus extraction from family-based farming extended its traditional consumption values to the nontraditional gentrified elites (Lippit 1987). The system required a vast army of agricultural producers, at the same time allowing petty capitalism to flourish everywhere and people from all walks of life to climb the many and varied rungs of the social ladder. Women embodied this manifold synthesis, which required their labor as incubators, producers, and game players.

As I have shown, the structure of opportunities in late imperial society was predicated on self-improvement in the management of human relationships. The family system was not only the training ground for this social orientation toward life but also the principal organization within which this life was produced, biologically and economically. Family relationships were systematized, rationalized, codified, and ritualized sufficiently to be employed in a variety of productive enterprises of small and medium scale. The family system provided the social training and the bodies for the pool of labor necessary to sustain the increased outputs that the system required. Women involved themselves at every level of this recruitment process. They harnessed themselves to the pronatalist systemwide drive for biological reproduction while they managed individual strategies for limiting the strain on family resources by the widespread although proscribed practice of female infanticide (Ho 1959: 58–62; Doolittle [1865] 1966: 203–9). It was in this manifold of labor-intensive production and social climbing that foot-binding's popularity acted as a mechanism for hiding the labor power of women. Foot-binding impaired women's feet; it gave a woman and those who possessed her a claim on social status by mystifying her sexual and reproductive powers, but it did not immobilize her to the point that she could not walk.

Feet that were bound so small that the woman had to be carried disqualified her as a legitimate player in the great contest to appear sexually interesting. Many men, indeed, even connoisseurs of the small foot, warned against the mania for 'smallness'. Some argued the matter in terms of aesthetics while others argued the matter in personal terms. One man became alarmed after his sister's daughter exhausted herself in a contest with her cousins. Every night she added a length of cloth to her binding in order to make it tighter. He warned her to stop lest she impair her ability to walk. He cynically offered to save her further anguish by simply slicing off her feet. (It is relevant to my thesis that this same uncle had instigated his niece's nightmarish quest in the first place by another thoughtless remark in which he compared her 'big feet' with those of her cousins. 'Uncle's remark hit me like an ice cold downpour and a clap of thunder and I could not find a space small enough to hide in' [Yao 1936: 1: 258–60]). A woman might even use her opponent's mania for smallness as a strategy to overcome the latter's challenge in a small foot contest. In the novel *Sancun Jinlian* (Feng 1986: 46), Xiang Lian, the protagonist, faces an opponent whose reputation for tiny feet exceeds only her own. Xiang Lian's winning strategy is to invite her opponent to view the peach blossoms in the courtyard. As Xiang Lian leads the way her opponent falls and is unable to pick herself up. Sensing victory, Xiang Lian scolds rhetorically: 'Are not feet that are unable to stand a person on her own truly wasted?' It was in the walking that the whole project of compressing the twisted foot was realized. We have already noted that the gait of foot-bound women was construed by men not only as sexually provocative, but also as transferring a woman's strength from her feet to her thighs and groin. This added to the perception of women as weak on their small feet but strong in their reproductive and sexual functions. Yet, because foot-bound women retained physical mobility, their labor power was left intact to a considerable degree.

In most areas of central and north China, women of all social classes bound their feet. The motive, even among working-class women, was the dream, the hope, or the real expectation of upward mobility through marriage. But even without the promise or hope for upward mobility, women bound their feet to signify their claim on the dignity accorded those who embodied refinement and a 'sense of class'. For many working-class women the hope of upward social mobility through marriage never materialized, but they did not stop binding their own feet or those of their daughters. There was always the possibility of a change in personal or family fortune. Many continued to bind despite their poverty because it was more painful, they said, to let out the feet or, worse, to let them out only to begin binding again when the circumstances improved (McCunn 1981: 30–1). Even when the possibility of upward mobility through self-improvement and marriage proved personally elusive, there was still the dignity of one's own family and its male heads to uphold and there was the next generation of daughters who if properly informed with a sense of class might entertain the hope and promise of mobility through self-improvement and marriage.

The life story of Lao Ning is especially poignant in this regard, in part because in the end she did not even have the dignity of her husband and his family to uphold. Still, she bound her feet only to spend a lifetime laboring as a domestic servant and begging in the city streets in turn-of-the-century northern Shandong (Pruitt 1967). Sometimes her attempts to bind were frustrated by her absolute lack of means to purchase the materials for bandages and shoes. While bound feet hindered Lao Ning they did not prevent her from walking about the city all day and doing physical labor. Nor did she hesitate to bind her daughter's feet when the time came.

As mentioned at the outset, the evidence for the actual work performed by foot-bound women comes from observations made toward the very end of the period in which the custom had been popular. The end of the nineteenth and beginning of the twentieth centuries was a period of industrialization and penetration by European-based capitalism. This certainly facilitated the struggle to end foot-binding, but it is not clear how far we can extrapolate these observations to comprehend the labor of women in previous centuries and dynasties. Still, the observations I offer are mostly of foot-bound women working in traditional commercial and agricultural occupations.

The traditional work of women based on ancient patterns that preceded the advent of foot-binding was domestic-based sewing, spinning, embroidering, and weaving. The domestic task of women was to provide shoes and clothes for their families. While suffering through the early phases of binding, all little girls were impressed with the traditional need to develop a sense of industry at the very least by learning how to sew and spin. This domestic industry along with other small- and medium-scale industries employed the labor of women in various places and times throughout the late imperial period. Data from the waning years of the dynastic period show that women's labor in the traditional handicraft industry, especially in spinning and weaving, was the cheapest form of labor (McCunn 1981: 30; Huang 1990: 85). But these data also show that the household incomes produced by women in home and community-based rural handicraft industries, although defined as 'supplementary', were sometimes comparable to or even greater than the income from male-managed agricultural or industrial production (see, e.g., Fei 1939: 170; Gamble 1968: 15).

From the early part of the twentieth century, women were increasingly recruited into the modernizing sector of the textile industry, and some reports indicate that the women workers in these urban industries were still binding their feet. A sociological survey of female labor in the city of Chefoo, where Lao Ning lived and worked for much of her life, reported in 1926 that 'practically all women and girls who examine nets have bound feet. I have found some who walk to and from work covering a distance of three miles every day. Forewomen in most shops and factories have unbound feet. The custom of binding feet is still fixed among these working people' (Rietveld 1926: 560). In this case women from the lower class of workers were still binding their feet while the 'forewomen' had begun to quit the practice. Beyond this, women with bound feet performed all kinds of menial labor; they worked as serving women in large houses and in city shops. In the countryside of north and central China, women with bound feet often contributed their labor to agricuitural production, especially during critical labor shortages. Here again it was defined mostly as 'supplementary' work or relegated to 'cheaper' divisions of the agricultural labor force.

The 1926–33 social survey of Ting Hsien, a network of rural villages in the Central Plain north of the Yellow River, provides data that allow us to account for some of the productive labor done by foot-bound women (Gamble 1968: 46–9, 52, 60). Based on a sample of 1,736 women in 515 families, Gamble calculated that 99.2 percent of the village women born before 1890 had bound feet. Although Gamble made no direct reference to the relationship between foot-binding and the work women did, the data make plain the fact that agricultural and cottage production were dependent in significant ways on the contributions of women with bound feet. Between 1926 and 1933, 71 percent of women over the age of 10 had bound feet, and 80 percent of women over the age of 12 reported farm work as their primary occupation. This means that at least half of the agriculture done by women was done by women with bound feet. Only fifteen women (1.3 percent), a smaller percentage than that of men, reported that they were idle or had no occupation. Of working women, 95 percent reported spinning and weaving as their primary or secondary occupation. Fifty-seven percent of Ting county's industrial income was from cottage industry, which employed mostly women. Forty-three percent of the industrial income was from 'wage labor', which employed mostly men (Gamble 1968: 15). At least during the period of Gamble's observations, the foot-bound women of Ting Hsien made substantial contributions to the local economy.

South of the Central Plain in the Yangtze River region, women were also recruited for agricultural labor in varying degrees depending on the structure of local and regional economies. Here the custom of foot-binding was widely practiced. Observers from the mid-nineteenth century such as Robert Fortune pointed out that 'in the central and eastern provinces . . . [foot-binding] is almost universal—the fine ladies who ride in sedan-chairs, and the poorer classes who toil from morning till evening in the fields, are all deformed in the same manner' (1857: 248–9). In the Yangtze delta village of Huaiyang-qiao, studied by Philip Huang, the task of planting rice in wet paddy was assigned to men while women were assigned tasks that had much lower prestige and only one-third the material reward. Huang believes that 'this division of labor rested on an irrational assumption' that women were not as capable as men in the physical labor of rice agriculture: 'That a fiction could be so completely and irrationally maintained attests to the power of custom and ideology. In this case, I think that the association of planting with men might have been linked to the custom of footbinding, which made sloshing around in the flooded fields a rather messy business. Wu Xiaomei . . . , for example, said she never did any farm work for which she had to take off her shoes, which rules out not only planting but also weeding whenever the fields were wet' (Huang 1990: 56). While working in wet fields was obviously inconvenient and sometimes risky (see below), Wu Xiaomei's refusal to do work that required her to remove her shoes in public must have been dictated by the same sense of modesty that caused Ding Wenyuan's grandmother (mentioned earlier) to keep her shoes on while bathing in the tub with her grandson. Huang (1990: 56) goes on to say that economic imperatives could overcome the fiction of women's unfitness for farm work: in other areas of the Yangtze delta, growing cotton was strictly women's work, although Huang does not indicate whether the women responsible for cotton production bound their feet. Some women with bound feet did take charge of certain outdoor tasks and oversee the labor of others. William Lockhart, a missionary physician in mid-nineteenth-century Shanghai, treated an elderly woman who suffered a compound fracture when she slipped, 'owing to her crippled feet', while supervising the spring cutting of bamboo shoots (Fortune 1857: 251).

In regions where women were systematically recruited into the production of wet rice, as in the south coastal provinces of Fujian and Guangdong, foot-binding was not generally accepted (and in a few places actively rejected as utterly impractical and pretentious). In these areas, foot-binding marked differences between women who spent their working lives in rice paddy agriculture and women whose lifestyles did not require them to remove their shoes in public and labor in their bare feet. Although foot-binding was not prevalent in the agricultural villages of the far south, it was practised in the towns and cities where distinct local and regional styles were noted by traveling votaries as late as the end of the nineteenth century (see Yao 1936: 1: 133). For now, anecdotes such as these provide the most direct evidence of the relationship between foot-binding and agricultural labor.

Our small stock of anecdotes on the agricultural work done by foot-bound women suggests that these women were capable of providing significant physical labor to the system of economic production. They could exempt themselves from labor in wet fields that required workers to remove their shoes, but they could be called upon to labor in the 'lighter' and 'cheaper' divisions of farm work. We also know that there is no inherent contradiction between agricultural work on the part of women and high rates of female fertility (Banister 1987: 133–4). The physical labor that women could and I believe did provide coupled with years devoted to bearing and rearing children describes the world of Chinese women. This world was enmeshed in the historical synthesis of a system of political economy that coupled a systemwide strategy of labor-intensive production with family-based strategies for social climbing. Foot-binding played a role in every aspect of this system.

During the past century the regional system of imperial hegemony that gave foot-binding its raison d'être began to give way to the challenge of the global capitalist system. Although the Manchu dynasty (1644–1911) had forbidden foot-binding by imperial fiat from the beginning of its reign, it was the modern revolution with its patriotic appeal to 'unleash the labor power of women' that finally spelled its end. But the end of foot-binding was no more abrupt than the attempt to reconstruct a modern society was decisive. The practice lingered in the cities for twenty years into the Republican period and held on in remote parts of the countryside until the Great Leap Forward of 1958–60. The Great Leap mobilization and collectivization of the Chinese countryside aimed to transform radically the traditional forces and relations of production. The total mobilization of women into the labor force was a key component in this effort (see Blake 1979*b*). In some areas foot-bound women who were mobilized under a piece-work system surprised observers by outproducing their male co-workers. A young man from Ningxia province told me how his grandmother, who still has three-inch bound feet, gained a local reputation for harvesting three times as much wheat per day as her male co-workers. If the end of foot-binding helped to 'unleash the labor power of women', which it undoubtedly helped to do, it certainly made the question of what women's labor is worth a point of lively debate. My argument that foot-binding did more to mask than to completely cripple the labor power of women is of course contestable. I offer it in the spirit of a hypothesis that a renewed interest in research on foot-binding might find challenging.

CONCLUSION

An adequate theory of foot-binding must move beyond the conventional typifications of Chinese culture in order to see how they are constituted in, and help to inform, individual experience, social relationships, and the historical process. The three-dimensional model of the 'mindful body' that was suggested by Scheper-Hughes and Lock (1987) for medical anthropology seems particularly appropriate for comprehending the complex questions surrounding foot-binding.

The first dimension of the mindful body is its intentional embodiment of a world. Mindful of its experience in the 'real world', the body becomes a 'self'. Here is where an intentional mutilation of a woman's body induced the (re)embodiment of her self by the application of a rigorous discipline designed to inflict and to overcome protracted physical pain. This embroiled a mother and daughter in a project that embodied the world in its spatial modality. It dramatized the ontological tension between femininity and masculinity such that 'femininity' came from the muted management of body and space while 'masculinity' came from the articulation of language and time.

The second dimension of the mindful body is its organically based system of symbols. The body provides a means by which a socially constituted reality represents itself as 'natural'. Foot-binding was an intentional manipulation of a 'natural (female) body' that was predefined in the Neo-Confucian-dominated domain of male-managed discourse. The rhetoric of foot-binding thus connected human effort and individual initiative on the part of women to the ontology of femininity, to the idea of self-control over individual fates, to the social roles that allowed women to fulfill their purpose as bearers of sons, to the fertility of the earth, and to the cosmic order of things. It was a powerful mystification of gender because it entailed the mindful bodies of women overcoming intense, protracted physical pain in a drama that coupled 'self-sacrifice' with 'self-exaltation'.

The third dimension of the mindful body is the socialized object that presents itself to the incorporation process of the body politic. This is where women's bodies came under the full weight of male-managed discourse in a historical system of production and reproduction. The relative objectification of erotic desire—the crippling of the foot, the natural symbol of labor power—made the labor of foot-bound women less visible and easier to appropriate than it otherwise might have been. The everyday reality of foot-binding masked the work of women by defining incubation more as 'nature' than as 'labor' and by defining foot-bound women's labor as worthless in view of the obvious, if indeed artificial, disability of their bodies.

The three bodies discussed here are, according to Scheper-Hughes and Lock (1987), a 'matrix' of interpersonal emotional experiences. In foot-binding the nexus of interpersonal emotional experience began in the 'interuterine' intersubjectivity of mothers and daughters. Over a period of months and years mothers and daughters shared an indescribable, hurtful, caring, mystifying experience. Foot-binding did not 'cause' this exchange. Foot-binding was merely the mute idiom of expression by which mothers chose to inform daughters about 'the world out there'. Verbal instructions and didactic manuals on how to behave, how to succeed in the real world, simply paled by the side of two mindful bodies locked in a manifold of contesting wills and fused intentions rigorously and relentlessly unmaking and remaking the world. Foot-binding embodied the memory of mothers in their daughters. Men were excluded from this province of memory just as men excluded women from their written genealogies that recorded for posterity the names of their fathers. The embodiment of mothers in daughters was cultural ontology: it intended the world in its ostensibility, that is, in making the world appear other than what it is. The writing of genealogy was cultural authority. It intended the world in its objectivity, that is, in describing the world just as it appears to be. The world of cultural authority upheld the gender hierarchy as natural. The world of cultural ontology as revealed in the ordeal of foot-binding showed the gender hierarchy to be unnatural and altogether artificial. Foot-binding was the way women in China supported, participated in, and reflected on the Neo-Confucian way of being civilized.

References

Ahern, Emily (1975). 'The Power and Pollution of Chinese Women', in *Women in Chinese Society*, ed. Margery Wolf and Roxane Witke, 193–214. Stanford, Calif.: Stanford University Press.

Ardener, Edwin (1975). 'Belief and the Problem of Women', in *Perceiving Women*, ed. Shirley Ardener, 1–28. London: Malaby.

Ayscough, Florence (1937). *Chinese Women Yesterday and Today*. Cambridge, Mass.: Riverside.

Balazs, Etienne (1972). *Chinese Civilization and Bureaucracy*. New Haven, Conn.: Yale University Press.

Banister, Judith (1987). *China's Changing Population*. Stanford, Calif.: Stanford University Press.

Blake, C. Fred (1978). 'Death and Abuse in Marriage Laments: The Curse of Chinese Brides'. *Asian Folklore Studies*, 27 (1): 13–33.

—— (1979*a*). 'The Feelings of Chinese Daughters towards Their Mothers as Revealed in Marriage Laments'. *Folklore*, 90 (1): 91–7.

—— (1979*b*) 'Love Songs and the Great Leap: The Role of a Youth Culture in the Revolutionary Phase of China's Economic Development'. *American Ethnologist*, 6 (1): 41–54.

Burke, Kenneth [(1950) 1969]. *A Rhetoric of Motives*. Berkeley: University of California Press.

Chai, Ch'u, and Chai, Winberg (eds.)(1967). *Li Chi, Book of Rites*, trans. James Legge. New Hyde Park, NY: University Books.

Chao, Kang (1986). *Man and Land in Chinese History: An Economic Analysis*. Stanford, Calif.: Stanford University Press.

Chayanov, A. V. (1986). *The Theory of Peasant Economy*. Madison: University of Wisconsin Press.

Ch'en, Kenneth K. S. (1972). *Buddhism in China: A Historical Survey*. Princeton, NJ: Princeton University Press.

Daly, Mary (1978). *Gyn/Ecology: The Metaethics of Radical Feminism*. Boston: Beacon.

deGroot, J. J. M. [(1892) 1972]. *The Religious System of China*, vol. 2. Taipei: Cheng Wen.

Ding, Wenyuan (1989) 'Yuguoli di Xiangqing' (Nostalgia for the hometown bathtub). *People's Daily*, 21 January: 6, overseas edition.

Doolittle, Justus [(1865) 1966] *Social Life of the Chinese*, vol.2. Taipei: Cheng Wen.

Douglas, Mary (1970). *Purity and Danger: An Analysis of Concepts of Pollution and Taboo*. Baltimore: Penguin.

—— (1973). *Natural Symbols: Explorations in Cosmology*. New York: Vintage.

Ebrey, Patricia (1990). 'Women, Marriage, and the Family in Chinese History', in *Heritage of China: Contemporary Perspectives on Chinese Civilization*, ed. Paul S. Ropp, 197–223. Berkeley and Los Angeles: University of California Press.

Elvin, Mark (1973). *The Pattern of the Chinese Past.* Stanford, Calif.: Stanford University Press.

—— (1989). 'Tales of *shen* and *xin*: Body-Person and Heart-Mind in China during the Last 150 Years', in *Fragments for a History of the Human Body*, pt. 2, ed. M. Feher, 267–88. New York: Urzone.

Erikson, Erik (1963). *Childhood and Society.* New York: Norton.

Erotic Art of China (1977). New York: Crown.

Farquhar, Judith (1991) 'Objects, Processes, and Female Infertility in Chinese Medicine', *Medical Anthropology Quarterly*, n.s., 5 (4): 370–99.

Fei Hsiao-Tung (1939). *Peasant Life in China: A Field Study of Country Life in the Yangtze Valley.* London: Kegan Paul, Trench, Trubner.

Feng Jicai (1986). 'Sancun Jinlian' (The three-inch golden lotus). *Shouhuo*, 59 (May): 4–73.

Fortune, Robert (1857). *A Residence among the Chinese: Inland, on the Coast, and at Sea.* London: Murray.

Freud, Sigmund [(1927) 1961]. 'Fetishism', in *The Standard Edition of the Complete Psychological Works of S. Freud*, vol. 21, ed. and trans. J. Strachey 149–57. London: Hogarth.

Furth, Charlotte (1986). 'Blood, Body and Gender: Medical Images of the Female Condition in China, 1600–1850'. *Chinese Science*, 7 :43–66.

Gamble, Sidney D. (1968). *Ting Hsien: A North China Rural Community.* Stanford, Calif.: Stanford University Press.

Gates, Hill (1989). 'The Commoditization of Chinese Women', *Signs: Journal of Women in Culture and Society*, 14 (4): 799–832.

Gernet, Jacques (1970). *Daily Life in China on the Eve of the Mongol Invasion, 1250–1276.* Stanford, Calif.: Stanford University Press.

Gilligan, Carol (1982). *In a Different Voice: Psychological Theory and Women's Development.* Cambridge, Mass.: Harvard University Press.

Heider, Karl G. (1970). *The Dugum Dani: A Papuan Culture in the Highlands of West New Guinea.* New York: Viking Fund Publications in Anthropology.

Ho P'ing-Ti (1959). *Studies on the Population of China: 1368–1953.* Cambridge, Mass.: Harvard University Press.

Hsiao Kung-chuan (1972). *Rural China: Imperial Control in the Nineteenth Century.* Seattle: University of Washington Press.

Hsu, Francis L. K.(1949). 'Suppression versus Repression: A Limited Psychological Interpretation of Four Cultures'. *Psychiatry*, 12 (3): 223–42.

Huang, Philip C. C. (1990). *The Peasant Family and Rural Development in the Yangzi Delta, 1350–1988.* Stanford, Calif.: Stanford University Press.

Huang Shu-min (1989). *The Spiral Road.* Boulder, Colo.: Westview.

Ko, Dorothy (1992). 'Pursuing Talent and Virtue: Education and Women's Culture in Seventeenth- and Eighteenth-Century China'. *Late Imperial China*, 13 (1): 9–39.

Levy, Howard S. (1966). *Chinese Footbinding: The History of a Curious Erotic Custom.* New York: Walton Rawls.

Levy, Marion J., Jr. (1968). *The Family Revolution in Modern China*. New York: Atheneum.

Lin Yueh-hwa (1947). *The Golden Wing: A Sociological Study of Chinese Familism*. London: Kegan Paul, Trench, Trubner.

Lippit, Victor D. (1987). *The Economic Development of China*. Armonk, NY: Sharpe.

Liu Ching-an [(1925) 1971]. *Geyao yu Funu* (Folk songs and women). Taipei: Orient Cultural Service.

Liu, Daling, et al. (eds.) (1992). *Zhongguo Dangdai Xingwenhua: Zhongguo Liangwanli 'Xingwenming' Diaocha Baogao* (Sexual behavior in modern China: A report of the nationwide 'Sex Civilization' survey on 20,000 subjects in China). Shanghai: Sanlian Shudian.

Lukacs, Georg (1971). *History and Class Consciousness*, trans. Rodney Livingstone. Cambridge, Mass.: MIT Press.

McCunn, Ruthanne Lum (1981). *Thousand Pieces of Gold*. San Francisco: Design Enterprises of San Francisco.

Mao Zedong [(1942) 1965]. 'Oppose Stereotyped Party Writing', in *Selected Works of Mao Tse-Tung*, vol. 2. Beijing: Foreign Languages.

Marx, Karl [(1885) 1968]. 'The Eighteenth Brumaire of Louis Bonaparte', in *Selected Works*. New York International Publishers.

Merleau-Ponty, Maurice (1986). *The Phenomenology of Perception*, trans. Colin Smith. London: Routledge & Kegan Paul.

Needham, Rodney (ed.) (1973). *Right and Left: Essays on Dual Symbolic Classification*. Chicago: University of Chicago Press.

Ng, Vivien W. (1987). 'Ideology and Sexuality: Rape Laws in Qing China'. *Journal of Asian Studies*, 46 (1): 57–70.

Ots, Thomas (1990). 'The Angry Liver; the Anxious Heart and the Melancholy Spleen: The Phenomenology of Perceptions in Chinese Culture'. *Culture, Medicine and Psychiatry*, 14: 21–58.

Potter, Sulamith Heins, and Potter, Jack M. (1990). *China's Peasants: The Anthropology of a Revolution*. Cambridge: Cambridge University Press.

Pruitt, Ida (1967). *A Daughter of Han: The Autobiography of a Chinese Working Woman*. Stanford, Calif.: Stanford University Press.

Rawski, Evelyn S. (1985). 'Economic and Social Foundations of Late Imperial Culture', in *Popular Culture in Late Imperial China*, ed. David G. Johnson, Andrew J. Nathan, and Evelyn S. Rawski, 3–33. Berkeley and Los Angeles: University of California Press.

Rich, Adrienne (1976). *Of Woman Born: Motherhood as Experience and Institution*. New York: Norton.

Rietveld, Harriet (1926) 'Women and Children in Industry in Chefoo'. *Chinese Economic Monthly*, 3 (12): 559–62.

Ropp, Paul S. (1976). 'The Seeds of Change: Reflections on the Conditions of Women in the Early and Mid-Ching'. *Signs*, 2 (1): 5–23.

St. Louis Globe-Democrat (1879).'They Would Go into a Salt Cellar'. *St Louis Globe-Democrat*, 14 September: 11.

—— (1881) 'Fatal Results of Chinese Superstition'. *St Louis Globe-Democrat*, 16 November: 11.

Sangren, P. Steven (1983). 'Female Gender in Chinese Religious Symbols: Kuan Yin, Ma Tsu, and the "Eternal Mother"'. *Signs*, 9 (1): 4–25.

Scarry, Elaine (1985). *The Body in Pain: The Making and Unmaking of the World.* New York and Oxford: Oxford University Press.

Scheper-Hughes, Nancy, and Lock, Margaret (1987). 'The Mindful Body: A Prolegomenon to Future Work in Medical Anthropology'. *Medical Anthropology Quarterly*, 1 (1): 6–41.

Seaman, Gary (1981). 'The Sexual Politics of Karmic Retribution', in *The Anthropology of Taiwanese Society*, ed. Emily M. Ahern and Hill Gates, 381–96. Stanford, Calif.: Stanford University Press.

'The Shrew' (1973), in *Chinese Literature: Popular Fiction and Drama*, ed. H. C. Chang, 23–55. Edinburgh: Edinburgh University Press.

Skinner, G. William (1977). 'Introduction: Urban Development in Imperial China', in *The City in Late Imperial China*, ed. G. William Skinner, 3–32. Stanford, Calif.: Stanford University Press.

Smith, F. Porter, and Stuart, G. A. [(1911) 1969]. *Chinese Materia Medica: Vegetable Kingdom*, revised by Ph. Daven Wei, 2nd edn. Taipei: Ku T'ing Book House.

Solomon, Richard (1972). *Mao's Revolution and the Chinese Political Culture.* Berkeley: University of California Press.

Spender, Dale (1980). *Man Made Language.* London: Routledge & Kegan Paul.

Staunton, George Thomas (trans.) [(1810) 1966]. *Ta Tsing Leu Lee: Being the Fundamental Laws and a Selection from the Supplementary Statutes of the Penal Code of China.* Taipei: Cheng Wen.

Turner, Victor W. (1969). *The Ritual Process: Structure and Anti-Structure.* Chicago: Aldine.

Tu Wei-Ming (1985). *Confucian Thought: Selfhood as Creative Transformation.* Albany, NY: SUNY Press.

—— (1990). 'Perceptions of the Body'. Paper presented at the East–West Center Workshop, Honolulu, 5 November.

Van Gulik, Robert (1961). *Sexual Life in Ancient China.* Leiden: Brill.

Veblen, Thorstein [(1899) 1934]. *The Theory of the Leisure Class.* New York: Modern Library.

Watson, James L. (1985). 'Standardizing the Gods: The Promotion of T'ien Hou along the South China Coast, 960–1960', in *Popular Culture in Late Imperial China*, ed. David G. Johnson, Andrew J. Nathan, and Evelyn S. Rawski, 292–324. Berkeley and Los Angeles: University of California Press.

—— (1986). 'Anthropological Overview: The Development of Chinese Descent Groups', in *Kinship Organization in Late Imperial China, 1000–1940*, ed. Patricia Buckley Ebrey and James L. Watson, 274–92. Berkeley and Los Angeles: University of California Press.

Watson, Ruby (1986). 'The Named and the Nameless: Gender and Person in Chinese Society'. *American Ethnologist*, 13 (4): 619–31.

Weber, Max (1968). *The Religion of China*, trans. Hans H. Gerth. New York: Free Press.

Williams, C. A. S. (1974). *Outlines of Chinese Symbolism and Art Motives*. Rutland, Vt.: Tuttle.

Wolf, Arthur P., and Chien-Shan, Huang (1980). *Marriage and Adoption in China, 1845–1945*. Stanford, Calif.: Stanford University Press.

Wolf, Margery (1970). 'Child Training and Chinese Families', in *Family and Kinship in Chinese Society*, ed. Maurice Freedman, 37–62. Stanford, Calif.: Stanford University Press.

—— (1972). *Women and the Family in Rural Taiwan*. Stanford, Calif.: Stanford University Press.

—— (1979). 'Introduction', in *Old Madam Yin: A Memoir of Peking Life*, by Ida Pruitt, v–ix. Stanford, Calif.: Stanford University Press.

Yang, Martin, (1945). *A Chinese Village: Taitou, Shantung Province*. New York: Columbia University Press.

Yao Ling-hsi (1936). *Cai Fei Lu* (A record of gathering radishes). 2 vols. Tianjin: Shidai Gongsi Yinxing.

Zhao Liming (1990). '"Nushu" di Wenzi Xue Jiazhi' (The literary value of studying 'Nushu'). *Xinhua Wenzhai*, 3: 168–71.

Zhou Shuyi (1990). 'Jiangyong County's Unique "Women's Language"'. *China Today*, North American edn., 39 (9): 53–5.

and illuminates the patterns of Western modernity through the prism of Islamism. Therefore, this is a study of the 'embeddedness of gender'[4] in the elaboration of Islamism on the one hand and modernism on the other. I argue that contemporary Islamism cannot be adequately understood in isolation from the local constructs of Western modernity in which women have an edificatory role.

This research project originated from an observable event—namely, the Islamist veiling movement of university students in Turkey during the post-1983 period. The issue of veiling became a decisive force in the radicalization of the Islamist movement once female students wearing headscarves were banned from universities, causing veiled students to mobilize, organizing sit-ins and demonstrations. Veiling can thus be considered as a social movement in that Muslim female students articulate their claims collectively and publicly and define the objectives of their action autonomously. The veiling movement has become a source of political conflict and polarization between secularists and Islamists, one that has engaged intellectuals, university faculty, mass media, and political parties in a fierce debate. But, foremost, this debate has revealed the deep social and cultural cleavages between secularists and Islamists in general and among women in particular. In a way it has revealed the extent to which 'Westernist' versus 'Islamist' confrontations have been reproduced among the citizens of the same nation, religion, and gender. This study also stems from the 'personal uneasiness' of the author, both as a Turkish intellectual and woman. First, the most cherished master-narrative in Turkey relating education and modernization to woman's emancipation is contested by educated Islamist women. The premise that traditions and religion disappear with the advent of modernity, an evolutionary progression that is often taken as a natural consequence of secular scientific education, no longer holds. Second, as a product of Turkish secular modernism, I am puzzled by the advent of this new Islamic figure of women sharing the same classes, educations, and professions but asserting at the same time their aspirations for an Islamic ideal and identity; the commonality or communicability between these two figures of women, of the same religion, nationality and gender, thus becomes one of the questions underlying this study. Hence, I will attempt to establish a link between 'private uneasiness' and 'public issue', between 'biography' and 'history' thereby trying to translate my own experience into Mill's term, through a 'sociological imagination'.[5]

The first section of this chapter describes the sociological significance of the political debate about Islamist veiling. The second section discusses the methodological difficulties stemming from the shift in master-narratives to a focus on agency and meaning within a highly polarized and politicized context. These difficulties are not limited in time and space to the conditions of the research; instead, they lead to the more general question of the location of the researcher, to the connection between intellectuals and knowledge, culminating in the age-old social science problem of the connection between analysis and engagement, between structure and agency. The third section reviews the Turkish mode of modernization (initiated by the Ottoman-Turkish elites during the nineteenth century and resulting in the foundation of a secular nation-state by Mustafa Kemal Ataturk in 1923) in order to grasp from a historical perspective the contemporary dispute between Islamists and modernists. It reexamines the transformative impact of modernization on the private and public spheres, on the relations between men and women, and on the self-definition of Turks. It problematizes the cherished concept of 'universal civilization' held by Turkish reformist elites, which came to be synonymous with Western European culture. The fourth section argues for the need to locate and study women as the symbols and central agents of this 'civilizing project'. The fifth section discusses the significance of Islamist movements, and more particularly that of Islamist veiling, in terms of the intrigue and resistance it offers to the civilizing project. It depicts the ways in which the Islamist identity, constructed in gendered terms, cannot be separated from the perceptions and constructions of Western modernity. The sixth section studies Islamism's implicit critique of secular ways of life, one that results in women's corporal and public invisibility. Finally, the last section examines the conflicting agendas Islamist women acquire through participating in secular education and religious politics, on the one hand, and adhering to the principles of Islamist communalism, on the other. While the former necessitates public participation, the latter sanctions the individuation of women, whose morality specifically involves community and Islamist authenticity.

SOCIOLOGICAL SIGNIFICANCE OF VEILING

Islamic veiling is a political issue in both Muslim and Western European countries: it highlights the tension within the core values of society, ranging from secularism of the public space, the place of religion in education, and individual rights to multiculturalism and multiconfessionalism.[6] The veiling of Islamist students appears as a controversial issue because it is the most visible reminder of religiosity and traditional roles of women in modern social contexts, such as university campuses, urban centers, political organizations, and industrial workplaces. This veiling is commonly perceived as a force of 'obscurantism' and is often identified with women's subservience; as such, it is interpreted as blurring the clear-cut oppositions between religion and modernity and as an affront to contemporary notions of 'gender emancipation' and 'universal progress.' Hence, the revival of Islamist movements throughout the Muslim world is often interpreted as a challenge to Western modernity, which is built upon a uni-directional notion of evolutionary progress conceived in terms of binary oppositions between religion and secularism, the private and public spheres, and particularism and universalism. Also Islamist veiling embodies the battleground for the two competing conceptions of self and society, Western and Islamist. Metaphorically, women's covered bodies revitalize contemporary Islamist movements and differentiate them from the secularist project.

I therefore argue here not only that the question of veiling is not an auxiliary issue for Islamist movements but, on the contrary, highlights the centrality of the gender issue to Islamist self-definition and implied Western criticism. Hence, veiling is a discursive symbol that is instrumental in conveying political meanings. Accordingly, the significance of contemporary Islamist movements can only be understood, I believe, in terms of their problematic relation to Western modernity, a relation that takes shape and acquires sense through women's bodies and women's voices. Islamism, therefore, is shaped by a selective reconstruction of identity rather than by an unchanging—that is, a historical and context-free, or fixed—identity. Consequently, it can be said that the questions of women, modesty, and sexuality are discerned and problematized by the contemporary Islamist movements more as a result of critical dependence on modernity rather than of loyalty to Islamic religion.

In its contemporary form veiling conveys a political statement of Islamism in general and an affirmation of Muslim women's identity in particular. In this respect it is distinct from the traditional Muslim woman's use of the headscarf.[7] While the latter is confined within the boundaries of traditions, handed down from generation to generation and passively adopted by women, the former is an active reappropriation by women that shifts from traditional to modern realms of life and conveys a political statement. Veiling is a political statement of women, an active reappropriation on their behalf of Islamic religiosity and way of life rather than its reproduction by established traditions. In this respect veiling does not express passive submission to prevalent community norms but, instead, affirms an active interest in Islamic scripture. Educated lower- and middle- class women claim to know the 'true' Islam and hence differentiate themselves from traditional uneducated women; these young Islamist women reject foremost the model provided by their mothers, who are perpetuating traditions and traditional religion within their domestic lives without any claim to knowledge and praxis. Paradoxically, the veiled students, who owe their newly acquired class status and social recognition to their access to secular education, also empower themselves through their claim on Islamic knowledge and politics. Veiled students, as new female actors of Islamism, acquire and aspire for 'symbolic capital'[8] of two different sources: religious and secular. Their recently acquired visibility, both on university campuses and within Islamist movements, indicates women's appropriation of this new symbolic capital and the emergence of a new figure, the female Islamist intellectual.

Furthermore, women's participation in Islamist movements symbolizes a new sense of belonging and ushers in a new community of believers. If the traditional way of covering oneself changes from one Muslim country to another in terms of the form of 'folk' dresses, the contemporary Islamist outfit is similar in all Muslim countries: it is through the symbolism of women's veiling that a commonality of identity and the Muslim community (*umma*) is reconstructed and reinvented at the transnational level.

In sum, the veiling of women is not a smooth, gradual, continuous process growing out of tradition. On the contrary, it is the outcome of a new interpretation of Islamic religion by the recently urbanized and educated social groups who have broken away from traditional popular interpretations and practices and politicized religion as an assertion of their collective identity against modernity. The veiling of women is the most salient trait of the contemporary Islamist movement, which is grounded on this focal tension among Islamism, traditionalism, and modernism.

The difference between the two profiles of Muslim women and the new significance of veiling were sharply underlined by the 'headscarf dispute', which erupted in 1984 in Turkey.[9] The use of the traditional headscarf by lower-middle-class women living on the fringes of modern city life has gone almost unnoticed, considered as a residual practice of traditionalism, while the adoption of the Islamist outfit by a group of university students, a phenomenon of the post-1983 period, is considered as a manipulative tool of the rising Islamist fundamentalist movement and, consequently, has provoked a very polarized political dispute between secularists and Islamists.

To distinguish the Islamist from the traditional way of covering, the Turkish mass media labeled the new veiling of women as the 'turban movement'.[10] The word *turban*, coined for the new veiling movement, is polysemic and hints at the transgressive nature of women's participation in Islamist movements. *Turban*, originally from Turkish *tü(i)bend* and Persian *düband*, refers to a head-dress of Muslim origin, consisting of a long linen, cotton, or silk scarf. In modern Turkish, however, it is employed as a French word, one denoting a fashion of head-dress, itself adopted from Ottoman Turkish. As such, the word *turban* itself has witnessed the transmission, interplay, and interpenetration of words and practices between the Ottoman Turkish heritage and the West. Labeling the female Muslim students' movement of veiling as the 'turban movement' differentiates it from the use of the traditional headscarf and suggests fashion and a change—that is, a modern way of appropriation as opposed to simply perpetuating tradition. On the other hand, *turban* also means a religious garment used by men—that is, *sarik* in Turkish—and, though it recalls the power of Islamic *ulema* (religious classes), it is used to distinguish women's religious politicization and empowerment. Hence, the label 'turban' itself reveals the hybrid and transgressive character of the Islamist veiling movement grounded on the power dynamics between East and West, traditionalism and modernity, men and women. Contrary to the traditional practice of Islamic veiling, or the Islamic headscarf, which conveys a specific meaning—that is—'return to traditions', 'return to fundamentalism', 'subservience of women'—and suggests binary oppositions such as 'Islam is essentially different from the West', and 'Westernization is a condition of women's emancipation', the label of 'turban' represents the hybrid and transgressive nature of Islamism in general and women's participation in the Islamist movement in particular.

AGENCY, CONFLICTUALITY AND SELF-REFLEXIVITY

Theories of modernization have forced us to seek, and find, symmetrical and linear lines of development that occur almost independently of historical and geographical context. Today the epistemological pendulum is swinging from evolutionary reasoning and methodological positivism to the question of agency and the subsequent analysis of particularistic, context-bound interpretations of modernity and self. Such a shift has an undeniably liberating potential on the study of 'non-Western' countries. The distancing from the universalistic master-narratives of modernization and emancipation opens up the space for the examination of subjective constructions of meanings, cultural identities, and social conflicts; in short, it enables the examination of the specific articulations between modernity and the local fabric.

The move toward new forms of interpenetration and hybridization between the particularistic and universalistic, the local and global, situates us on a terrain that is not designated, determined, and mediated by social scientific language. This new positioning leads to the common tendency of naming all sorts of puzzling hybridizations and paradoxes as either parochial signs of a 'pathology of backwardness' or, on the other extreme, as culturalist essentialism[11] or simply as postmodern relativism. Hence, such a move brings with it a problem of conceptualization and necessitates new linkages between empirical and theoretical languages. It also necessitates a repositioning of the researcher with respect to 'local' context and native language. The weakening of the master-narratives of modernization and emancipation changes the role of intellectuals from transmitters and defenders of universal values to that of interpreters of hybrid, paradoxical, multidirectional social realities.[12] This new attention can be translated into a sociological awareness only through the mediation of a context-bound language.

The original Turkish title of this chapter and the book from which it was drawn, *Modern Mahrem*, for instance, articulates the difficulties in crossing cultural boundaries and national languages: *Mahrem* literally refers to intimacy, domesticity, secrecy, women's space, what is forbidden to a foreigner's gaze; it also means a man's family. In the text the concept of 'mahrem' illuminates the difficulty of rendering this 'private' space, private in the modern Western sense. The concept becomes more than a simple question of translation, an analytic category, a key for understanding the issues of intimacy, sexual segregation, and communal morality in a Muslim society. Using the Western concept of 'private sphere' instead of *mahrem* would have led to the suppression of the distinctiveness of the domestic sphere in a Muslim context. Understanding the particularities from within therefore requires a broad sociological consciousness and conceptualization.

Similarly, the word *modern* literally indicates the Western set of values of individualism, secularism, and equality produced by the Enlightenment, the industrial revolution, and pluralistic democracy; it does not have a synonym in Turkish.[13] But it does have its own life and meaning in Turkish history. Rather than *modernity*, which refers to a given consciousness of the present time and state of development, the concept of 'modernization'—expressing a political will to 'become modern'— is used in the Turkish vocabulary of history and politics. Modernization can be interpreted as a continuous endeavor to over-come the lag in scientific, economic, and political development. As a consequence, Western modernity as a construct is an intrinsic part of Turkish intellectual and political life that changes over time and according to the ideological climate. No accurate reading of history, social agency, and social conflict is possible without taking into con-sideration this dependence on the concept of modernity. Public narra-tives (e.g., about Westernization, progress, the emancipation of women) as well as counter-narratives of oppositional movements (e.g., leftism, Islamism) carry the imprint of a problematic relation to modernity. If we want to understand and account for practices of social actors, ontological narratives, collective identities, and political conflicts, we must first recognize that at the very level of individual and collective identities it is necessary to decode local constructs of modernity. In other words, modernity cannot be conceived as a geo-graphical entity, external to the Turkish experience but, on the con-trary, needs to be elaborated as part of the historical experience, social setting, and the identity of social actors. It is in non-Western settings that modernity needs to be examined for its part in shaping public narratives, collective identities, and social practices.

Thus, the title *Modern Mahrem* combines two distinct yet inter-penetrating civilizational categories: one intimate, gendered, and secret and the other public, universalist, and manifest. It alludes to the unsolicited encounter and interaction between Islam and modernity throughout Turkish history and hints more particularly at the Islamist woman's critical assessment of modernity and her forbidden participation in the public sphere. This chapter highlights and interprets local constructions of self and modernity, hybrid conjunctions, and asymmetrical social realities as forms of social practice and not as deviations from the evolutionary trajectory predicted by modernization theories. On the contrary, discrepancies between social theories and social practices are taken as a starting point of inquiry instead of simply juxtaposing ready-made social scientific language with a given social phenomenon and thereby suppressing the particulars.

Typical social theories explain Islamist movements in general and veiling in particular by assigning priority either to sociopolitical factors or to the essence of the religion itself, presumed to be alien to a series of transformations, such as reformism and secularism, that took place in the West. The first stand leads to causal explanations in which economic stagnation, political authoritarianism, rural exodus, and urban anomie are enumerated as the causes of radical Islamism. Even though such approaches describe the social environment within which oppositional movements are rooted, they fail to explain how and why Islam acquires such an appeal for cultural and political empowerment. The approaches that give priority to the Islamic essence, on the other hand, suppose an immutable nature of Islam and therefore take the religion out of its historical and political context. In this perspective Islamism and Islamic veiling appear as a 'deviation' or a 'pathology'; the underlying assumption of the argument is that, if modernization and secularization were successful, such 'anamolous reactions' would not occur. And for both the political as well as the cultural approaches, the veiled women are an extension of a wider phenomenon: either as subsidiary militants of the fundamentalist political movement or as passive transmitters of traditional values. Such analyses, giving priority to the determinism of the system and structure, ignore the questions of agency and the formation of an Islamic actor and their contribution to the (re)production of the Islamic social order.

Fortunately, recent social theories have focused increased attention on questions of agency and cultural reproduction. Within the perspective of the sociology of action, social movements are not explained as anomalies but, instead, as contributing to the 'self-production of society'.[14] Social movements express the struggle for the control of historicity, that is, for the control of a cultural model that is not distinct from social conflict, through which society produces and reproduces itself.[15] Therefore, focusing on the questions of agency and identity becomes integral to an understanding of how the society reproduces and also transforms itself.

473

Emphasizing the questions of agency and identity and locating women at the center of our analysis will lead to new perspectives in the understanding of the discord between Islamism and modernism as it opens up a new territory of conflict. As such, focusing on Islamist women will not be limited to an examination of women's narratives and practices but will be enlarged to gain an understanding of how women's agency (or lack of agency) acts upon Islamist social movements. The focus of this analysis will thus be situated at the interplay between gender identities and political ideologies, at the intersection of ideological configurations and everyday power relations between the sexes. As Michel Foucault puts it, the body is the locus of all struggles of power, which works by the organization and division of space. The analysis of power, not as established domination (as something people hold) but, rather, as a 'set of open strategies', will enable us to investigate the significance of the veiling movement at the level of the subjectivity of social actors and their social practice.[16] Consequently, identities (Islamist and women's) are not taken as fixed, essentialist categories but as reconstructions over time as well as accommodations to power relations. Thus, the formation of a collective identity for Islamist women can only be elaborated upon in their relationships with secularist women, with Islamist men, and also with other Muslim women. Once the narratives of women are historicized and contextualized, the fixity of identities can be transcended.[17]

A rethinking of Islamist veiling in terms of agency requires a sociological approach and method that render narratives of social actors in terms of their relationships and conflicts. Therefore, the choice of the sociological method is not independent from the general sociological approach. Although both deep interviews and group discussions are used in this study, the latter are privileged to the extent that they provide a relational setting and reinforce the self-reflexivity of actors. The group discussions were conducted according to the principles of 'sociological intervention', elaborated by Alain Touraine as a method of the sociology of action.[18] Discussions in groups render narratives of Islamist women within a relational setting—that is, both in relation to one another and to the sociologist but also in relation to adversaries or spokepersons of the movement invited to the group. For instance, the encounter of the group first with one of the powerful male Islamist intellectuals defending women's rights from within Islam, and a later one with a secularist, Westernized feminist differed radically in both the terms of the debate as well the self-definition of the group. Sociological intervention gives a chance to analyze the formation of a collective identity of women in its shifting boundaries, in its conflictuality, and in its accommodation to power relations. The group members meeting with one another, the sociologist, and the invited outsiders discover their differences and acquire a sense of 'memory' and of 'history'. As the research situation creates a distancing from the imperatives of collective action and identity, it increases in turn the self-reflexivity of the actors, urging them to go beyond given opinions. Sociological intervention gives priority to the production of meaning and knowledge as an interactive process between the actors and the analyst. The label 'intervention' recalls this need to interact with actors of the movement in order to produce meanings of narratives, agency, and conflictuality. In a sense there is a place for 'surprise' built into a research process, making space for the agency and self-analysis of actors, which would engender unanticipated meanings and conflicts, requiring new conceptualizations in turn. In other words, in a research situation the researcher faces aspects of the studied subject matter that do not fit, which are dissonant or simply meaningless in the context of the theoretical framework developed beforehand. Disregarding these aspects can be disabling for the understanding of the social dynamics. Thus, at a given stage of the research process the 'disempowerment' of the researcher—that is, recognition of the rigidity of the theoretical framework in respect to the experiential, thus malleable, empirical world—is necessary if new territory of the meaning of social action and history is to be explored.

To sum up, the trajectory of the sociological analysis presented here can be described as a move from an observable micro-level event (the headscarf dispute) to focused research (of veiled Islamist students), which in turn highlights the general problem of Islamism and modernism from the point of view of women. The book begins by giving the historical setting, although the historical analysis is not intended as an explanation of contemporary events. On the contrary, the contemporary discord between Islamists and modernists ushers in a new reading of history and a new reading of the Turkish mode of modernization, which in turn leads us to explore territories of exclusion and domination and to examine silenced representations of those stigmatized as 'backward', 'uncivilized', and 'irrational'— that is, the 'dark side' of modernization.

....

WEAK HISTORICITY AND WESTERN MODERNITY

....

Unlike Western modernity, which was forged by the forces of indus-trialization, production, and class conflict, Turkish modernization was an outcome of the Westernism and secularism of reformist elites for whom women's emancipation from the traditional Islamic way of life would pave the way to Westernization and secularization for the larger society. In almost all Muslim countries, due to the historical specificity of the mode of modernization, women's issues have been pivotal in constructing both Islamism and modernism.

The modernization project takes a very different turn in a non-Western context in that it imposes a political will to 'Westernize' cul-tural codes, lifestyles, and gender identities. The terms *Westernization* and *Europeanization,* widely used by nineteenth- and twentieth-century reformers prior to the more recent emphasis on the structural and universal traits of modernization, overtly express the willingness to borrow institutions, ideas, and manners from the West. The history of Turkish modernization as almost a civilizational conversion can be considered a most radical example of such a cultural shift. In their attempts to penetrate the lifestyle, public manners, gender behavior, body care, and the daily customs of the people, as well as their willing-ness to change their self-conception as Turks, Kemalist reforms[19] extended far beyond the modernization of the state apparatus and the transition from a multi-ethnic Ottoman Empire to a secular republican nation-state. With the contemporary renewal of Islamist movements both in Turkey and elsewhere, it is crucial to return to history, to a reconsideration of this civilizational shift in order to understand the emotional, personal charge, '*habitus*' in Pierre Bourdieu's terms, which mark the realm of conflict between today's Islamists and modernists.

As Western experience and culture basically define the terms of civilization, attempts at modernization in a Muslim country center around Westernization. Once Western history, from the Renaissance through the Enlightenment to industrialization and currently to the Information Age, became the terrain of innovation and the reference point of modernity, non-Western experiences faded and lost their power as world historymakers. Instead, these experiences have been defined as residuals, without specific names, viewed only in relation to the 'West' as 'non-Western', as 'the rest', and, as today, as 'Islamic', as 'essentially' distinct from Western civilization. These non-Western societies which can no longer participate in the 'carnival of change', are excluded from being bearers of history and knowledge; they are, instead, left on the periphery of Western civilization with a 'mutilated outlook'.[20] This exclusion in turn results in the formation of societies with 'weak historicity', that is with a weak capacity to generate modernity as a societal 'self-production' as an indigenous development grounded in the interaction of cultural fabric and social praxis. Weak historicity is the sociological counterpart of the economic definition of the developmental lag between the 'core countries' and the 'periphery' in terms of the level of economic production, cultural creativity, and scientific knowledge. Weak historicity represents the repercussions of this lag on the discursive constructs and praxis of social and political actors. In societies in which modernity does not emerge from the indigenous culture and its development, history-making becomes a continuous effort toward modernization and Westernization (or later the refusal of it) shaped by the will of the society's political and intellectual elites.

The encounter of the East with the West has not produced a reciprocal exchange between the two cultures but, instead, has resulted in the decline of Islamic identity. The term *civilization* does not refer to the historical relativism of each culture, as in 'Islamic', 'Arabic' or 'French' civilization, but, rather, designates the historical superiority of the West as the holder of modernity. The concept of 'civilization' blankets many themes, ranging from technology and worldview to rules of decent manners to the provision of a conglomeration of the phenomena that make the West distinct from other contemporary yet 'primitive' societies.[21] In other words, the concept of civilization is not a neutral, value-free concept; to the contrary, it specifies the superiority of the West and attributes universality to a specifically Western cultural model. In opposition to the German notion of 'kultur', which stresses national differences, the concept civilization has a universal claim: it plays down the national differences, emphasizing instead what is imagined to be common among peoples.[22] More precisely, it expresses the self-image of the European upper class, who see themselves as the 'standard-bearers of expanding civilization', thus serving as a counterpart to the other tendency in society, that of barbarism.[23] The concept of civilization encompasses the idea of progress, referring to something that is constantly in motion, moving forward. Not only does it solely designate a given state of development (of Western countries), but it also encapsulates an ideal to be reached (by the non-Western world). The ideology of positivism underlying the concept of civilization attributes universality to Western civilization and suggests its applicability at anywhere any time. Thus, the main objective of modernization, as Turkish modernists have stipulated perfectly well, is to 'reach the level of contemporary civilization' (*muasir medeniyet seviyesine erişmek*), as defined by the West.

CIVILIZING 'KEMALIST' WOMEN

The debate among Ottoman elites on the ways and limits of Westernization at the end of the nineteenth century reveals the significance of the unsettled link of modernization with cultural identity and, in turn, with gender. While the Ottoman Westernists argued that only the emancipation of woman from the religious constraints and traditional ties would bring on the civilizing process, the Ottoman conservatives conjectured that according such freedoms to women would inevitably lead to the breakdown of the moral fabric of society. Consequently, veiling, a sign of 'backwardness' for the proponents of Westernization, because it implied the separation of women from 'civilized human beings', became, on the contrary, a sign of 'virtue' for conservatives protecting both women's sexuality and their social morality.[24] In this debate the Westernists eventually triumphed over the conservatives with the success of the Turkish Kemalist Revolution in 1923.

An exploration of the Islamic religion as a constituent element of a social organization based on the segregation of sexes is vital to understanding how and why the civilizing process of the Kemalist revolution was constituted and culturally coded in gendered terms. The most resistant antagonism between the Islamic and the modern Western civilizations can therefore be grounded in the normative definitions of gender relations, in the religious-cultural code of honor and modesty, and in the organization of interior and exterior spheres, as well as on the separation of male and female.[25] Unlike most national revolutions, which redefine the attributes of an 'ideal man', the Kemalist revolution celebrated an 'ideal woman'. Within the emerging Kemalist paradigm,[26] women became bearers of Westernization and carriers of secularism, and actresses gave testimony to the dramatic shift of civilization.[27]

The grammar of Turkish modernization can best be grasped by the implied equation established between national progress and women's emancipation. More than the construction of citizenship and human rights, it is the construction of women as public citizens and women's rights that are the backbone of Kemalist reforms.[28] Kemalist reforms aiming at the public visibility of women and the social mixing of the sexes thus implied a radical change in the definitions of the private and public spheres and of Islamic morality based on female 'modesty' and 'invisibility'. According to the Kemalist project, women's visibility and the social mixing of men and women thus endorse women's existence in the public sphere. The taking off of the veil by women (1924 onwards), the establishment of compensatory coeducation for girls and boys (1924), the granting of political rights such as eligibility for political offices, women's suffrage (1934), and, finally, the abolition of the 'Sharia', the Islamic law, and the subsequent adoption of the Swiss Civil Code (1926) were all measures undertaken to guarantee the public visibility and citizenship of women in the new Turkish nation-state.

The celebration and acquisition of women's visibility both in their corporality and in their public roles as models for emulation thus made the secularization of social life in Turkey possible. Photographs of women unveiled, of women in athletic competitions, of female pilots and professionals, and photographs of men and women 'miming' European lifestyles depicted the new modernist interpretations of a 'prestigious' life in the Turkish nation-state.[29] Novels of the Turkish republic, focused on this new 'civilized' way of life—on its decor, goods, and clothing—celebrated the ideal attributes and rituals of a 'progressivist and civilized' republican individual: tea saloons, dinners, balls, and streets were defined as the public spaces for the socializing sexes; husbands and wives walking hand in hand, men and women shaking hands, dancing at balls, or dining together, reproduced the European mode of encounter between male and female. Among the cast of characters of the new republic, the serious, hard-working, professional women devoted to national progress, appeared as a touchstone set apart from a 'superficial' and mannered claim for Europeanness.[30] Against Ottoman cosmopolitanism, Kemalist female characters endorsed seriousness, modesty, and devotion and accommodated the presumed (preIslamic) Anatolian traditional traits—thus, they represented the nationalist project.

In summary, the civilizing process intruded upon Turkey, and other non-Western contexts, bringing with it new ethical and aesthetic values for self-definition, representations of the body, gender relations, lifestyles and spatial divisions; it labored in the minds and bodies through the establishment of education for women and their new corporal and public visibility; and it introduced the principles of equality and the social interaction of the sexes, two very problematic notions for Islamic social imagery and world. 'Women's visibility, women's mobility, and women's voices'[31] were and continue literally and symbolically to form the stakes of the battle between the modernists and the Islamists in Turkey and elsewhere in the non-Western world.

'ISLAM IS BEAUTIFUL': THE QUEST FOR A NEW DISTINCTION

The concept of civilization, once problematized, loses its value-free status and refers, instead, to the power relations between those who appropriate civilized manners and the 'barbarian', 'primitive' others. In the context of Turkish modernization the distinction between 'civilized' and 'uncivilized' manners has been highly scrutinized: the *alla franca* (European) mode and behavior are praised, whereas everything associated with the *alla turca* (Turkish) mode acquires a negative meaning. It is curious that Turks themselves currently label their own habits as if European eyes are watching over their daily lives and employ the foreign word *alaturka* to represent a sort of ideal self.[32]

The reappropriation of the term *alaturka* symbolizes that changes in lifestyles and aesthetic values are not innocuous, in that they reiterate the civilizational shift from an Islamic to a Western one, a shift in which taste, as a social marker, establishes cultural distinctions and social stratification among classes.[33] Hence, cultural codes and lifestyles define the stakes in the implicit power struggle between the Westernist and the parochial elites. Rather than employing the concept of social classes, which emphasizes economic exploitation for some, it is more useful to refer to this stratification through the concept of *habitus* that encompasses lifestyles.[34] One can argue that upper-middle-class Kemalist women, who acquire education and a professional career and who cultivate their bodies and ways of life in a 'secular' manner, form a distinct social status group. Thus, the acquisition of social distinction and social status, rooted in the exclusion of the Islamic world, forms the main focus of social and political discord between the secularists and the Islamists. The Turkish experience can be considered unique among its Muslim counterparts in this 'epistemological break', its radical discontinuity between traditional self-definitions and Western constructs. Although the majority of the Turkish population easily create hybrid forms to integrate their daily practice of religion, their traditional conservatism, and their aspirations to modernity, modernist elites, with their implicit value reference to binary oppositions, situate themselves squarely on the side of the 'civilized', 'emancipated' and 'modern'. Consequently, Kemalist women, liberated from the religious or cultural constraints of the intimate sphere and fully participating in public life, are faced with a radical choice: they can be either culturally Western or Muslim. It is this choice, and the polarity embedded within it, that generates cognitive dissonance around the separate value systems of the elites and the rest of the populace and thus raises the issue of competing legitimacies.[35] The subsequent discourse of the elites is formulated from the Kemalist center of power, fortified by state support against all 'other' forms of opposition.

Contemporary Islamist ideology reveals this power struggle in an aggravated form by challenging the equation established by the modernist elites between 'civilized' and 'Westernized'. It promotes the return of the Muslim actors to the historical scene in terms of their own religious morality. The Islamist body politic conveys a distinct sense of self (*nefs*) and society (*umma*), one in which the central issue becomes the control of sexuality through women's veiling and the segregation of the sexes in public life. Hence, the veiling of women emerges as the most visible symbol of this Islamization of the self and society. But, in addition, other social forms, such as beards worn by men and taboos regarding chastity, promiscuity, homosexuality, and alcohol consumption, define a new consciousness of the Islamic self and the Islamic way of life.

Thus, through the political radicalization of Islam, Muslim identity makes itself apparent and seeks to acquire legitimacy in a modern political idiom. Islamism emphasizes Muslim identity and reconstructs it in the modern world. Islamic faith and the Islamic way of life become a reference point for the re-ideologization of seemingly simple social issues such as the veiling of Muslim students at the university, allocation of prayer spaces in public buildings, construction of a mosque at the center of Istanbul, segregation of the sexes in the public transportation system, censorship of erotic art, and discouragement of alcohol consumption in restaurants. Activism regarding these issues demonstrates the Islamist problematization of a universalistic construction of Western civilization and criticism of the 'secular way of life'.

The problematization of social issues demonstrates that the Islamic movements share the same critical sensitivity with contemporary

Western social movements vis-á-vis Enlightenment modernity. The Islamist movements are thus similar to civil rights, feminist, environmental, and ethnic movements[36] in that all display the force of the repressed identities and issues (race, gender, nature, ethnicity, and religion, respectively) when challenged by industrial modernity, and all reawaken latent memories and identity politics. Identity politics, particularism, and localism against the uniformity of abstract universalism are common features of the postmodern condition. In this regard, Islamism is similar to feminism. While feminism questions the universalistic and emancipatory claims of the category of 'human being' and asserts, instead, women's difference, Islamism problematizes the universalism of the notion of civilization, and asserts Islamic difference. While women reinforce their identity by labeling themselves feminists, Muslims emphasize theirs by naming themselves Islamists. Civil rights activists also asserted the primacy of difference through the use of the motto 'Black Is Beautiful', thus rejecting the equation of emancipation with *white* and *Western*. Difference, therefore, becomes the source of empowerment for contemporary social movements and the content of identity politics. It is in this context of the rejection of the universalism of Enlightenment modernity and the assertion of difference that the motto 'Islam Is Beautiful' has gained credence among Muslims.

Islamist movements therefore render visible a realm of conflict, one whose terms are defined by different discursive constructs and normative values of self, gender relations, and lifestyles. It is situated at the crossroads of religion and politics. The self-definition of the Muslim subject acquires meaning only through the religious idiom, and radical politics empowers and conveys this meaning to society at large. This realm can be approached through Alain Touraine's analytical framework of the struggle for the control of historicity—that is, for a cultural model tied directly to class conflict.[37] In this formulation, Islamic movements do not solely express a reaction to a given situation of domination but also present both a countercultural model of modernization and a new paradigmatic self-definition.

AUTHENTICITY AND OCCIDENTALISM

The quest for differentiation from the Westernized self, for an 'authentic' Islamic way of life, has engendered a critical alertness for both traditional Islam and the contemporary forms of Western modernity imposed by the globalization of culture and lifestyles. Islamism conveys a tacit resistance to Western modernity and its four basic principles and forces of change: secularism, equality, individualism, and confessionalism. The most significant experiences of Western modernity (and the most alien to a Muslim context) are the expansion of secularization into new realms of life and the extension of the principle of equality into new social relations.[38] The principle of equality confers legitimacy on a continuous societal endeavor that aims to overcome all social differences that have been traditionally accepted as insurmountable and natural. As such, equality among citizens, races, nations, workers, and, last, the sexes defines the historical and progressive itinerary of Western societies. Similarly, as more realms of life are socially and historically reconstructed and delivered from transcendental natural definitions, the process of secularization deepens. This synergy between secularism and equality is the ultimate product of the modern Western individual, epitomizing his or her self-reflexivity as a mental and corporal being.

Another touchstone of differences between Westernized and Islamist individuals is in approaches to the human body. The semiotics of body care by the secularized individual and the religious one indicates asymmetric conceptions of the self and body. For the modern individual the body liberates itself progressively from the hold of natural and transcendental definitions and enters into the spiral of secularization; it penetrates that realm in which human rationality exercises its will to tame and master the human body through science and knowledge. Genetic engineering and biological reproduction are two examples of highly scientific interventions. Also, the phobia about cholesterol, taboos on smoking, and obsessions with fitness demonstrate how the word *healthy* has become magical in defining new lifestyles inspired by the scientific (rather than the erotic) desire of body cultivation. 'Bodysculpting', fitness, sveltness, thinness, and cultivating energy emerge as the ideals of a modern individual witnessing the leveling of differences between the sexes (the interchangeability of roles and outlook) and ages (that all can be young forever), culminating in the replacement of lifecycles with lifestyles.[39] The Islamist discourse and practice, conversely, scrupulously reinforces and creates hierarchies of the differences among sexes (e.g., veiling for women, beards for men) and among generations (for women, e.g., the stages of virgin, married woman, then mother). Any blurring between the feminine and masculine roles, especially the physical masculinization of women, is considered a transgression. In this respect veiling is considered a trait of feminity, of feminine modesty and virtue. The conception of the body conveys a different set of meanings in which the body is transformed as it mediates devotion to religious sacraments and purified through religious practices such as ablutions, fasting, and praying. In addition to the physical body, the Islamic spiritual being (*nefs*) is mastered and purified through submission to a divine will. But, again, this comparison is not simply a matter of difference between the secular and religious conceptions of self that could be valid for almost all religions; instead, in this context it is a civilizational matter. The genealogy of self and body conceptions in the West and within Islam also helps explain the almost compulsory resistance of contemporary Islamists to secularism and equality. The Islamic self and community are reconstructed in opposition to the premises of the modern individual; the Islamic identity searches for its 'authenticity', and for its distinction from the Western identity in defining its essential foundational roots.[40]

Another distinctive drive of modern societies is the drive to 'confess', to 'tell the truth about sexuality', as Michel Foucault put it. According to Foucault, the emergence of modernity can only be understood within the context of this urge, which stems from earlier religious practices; it entails a desire to confess the most intimate experiences, desires, illnesses, uneasiness, and guilt in public in the presence of an authority who may judge, punish, forgive, or console the confessor.[41] This explains why everything rooted in the private sphere, long considered the most difficult to reveal, becomes public, political, and a source of knowledge once it is confessed. In a sense the motto of the feminist movement, 'The Personal is Political', contributes to this movement of exit from and transparency in the public sphere in that it focuses on the process through which intimate relations of domination are transformed into relations of power. Abortion rights, sexual harassment, and date rape, witness, by the novelty of their labeling, this movement from the private to the public. Talk shows and popular trials have, for example, become a center of debate in U.S. society, mediating this urge to confess, to intrude on personal relations to detect the truth about the relations of power inherent in the private sphere.

While modern Western society reveals the truth about revelations in the private sphere, it also transforms truth into a matter of individual conscience.[42] In contrast to the interiorizing of truth in the modern West, Islam privileges the community (*cemaat*) in guiding the individual through life and giving oneself up to God. Veiling symbolizes the primacy of the community and conveys the forbidden, intimate sphere and the confinement of the self and sexuality within the limits of the private, thus shielding the private sphere from disclosure to the public gaze.[43]

The Islamic subject thus reveals him- or herself and quests for authenticity by refusing to assimilate into Western modernity and by rediscovering religion and memory, often repressed by Western secular rationalism and universalism. The reappropriation of Islam as a way of life provides a new anchor for the self and thereby recreates an 'imagined political community'; it reinforces social ties among those individuals who do not know one another but who yearn for profound horizontal attachment.[44] Islamism works as an imagined community, one that is forged and reinforced by and within the realm of the sacred.[45] The Islamic way of life, as a selective construct of religion facing modernity, traces the borders of this imagined community and resists the trivialization of chosen lifestyles by the impact of the commodity logic and hedonistic individualism characteristic of modern times.

WOMAN'S PERSONALITY OR ISLAMIC COMMUNALISM

The quest for difference and authenticity, though necessary for an initial elaboration of identity, nevertheless has its own limits. For instance, who decides such questions as What is really Islamic? and Who is a real Muslim? can lead to essentialist definitions and exclusionary standards, thereby turning the imagined community into an 'oppressive communitarianism'. The 'return of the repressed' can in turn be repressive. This totalitarian tendency can be easily triggered by a search for the 'total' Islamic identity, one freed from the 'corrupt and dominating' effects of Western modernity. The more one reinforces the relationship between the 'pure' self and 'total' community, the more Islamic politics becomes an imposed lifestyle, veiling a compulsory emblem, and women the moral guardians of the Islamic identity and community. This would entail the control of the public sphere by means of canceling out individual choices in determining lifestyles, monopolizing the cultural code, and instituting an Islamist form of the colonization of the self.

In other words, the totalitarian dimension emerges from the utopian ideal of a single identity for the collectivity. The pairing of a fixed identity and a utopian community presupposes a harmonious, a historical nondifferentiated social order that avoids all subversive conflict. Women are the touchstones of this Islamic order in that they become, in their bodies and sexuality, a *trait d'union* between identity and community. This implies that the integrity of the Islamic community will be measured and reassured by women's politically regulated and confined modesty and identity (such as compulsory veiling, restricted public visibility, and the restrained encounter between the sexes). Traditional gender identities and roles thus underlie Islamic authoritarianism.

Overpoliticization of the Islamic utopia—a yearning for the holistic implementation of the three pillars of Islam, the three Ds of '*din, dünya ve devlet*', (faith-religion, life-world, and state)[46]—leads to totalizing conceptions of identity, religion, and state. This utopian dimension is one of the distinctive traits of Islamist movements and distinguishes them from other 'new social movements'. Only self-limitation of Islamism as a political project and the autonomization of Islam within civil society will open up Islam as a space for diversity and pluralism. However, gender questions will remain the yardstick by which Islamic pluralism is measured.

Thus, the direction in which Islamist movements evolve will hinge significantly on the elaboration and recognition of women's identity and agency. Paradoxically, Islamic politics can delimit women's individuality and visibility at the same time that the politicization of women within Islamic politics empowers Muslim women. Muslim women are empowered by an Islamism that assigns them a 'militant', 'missionary', political identity, and by secular education, which provides them a 'professional', 'intellectual' legitimacy. The new public visibility of Muslim women, who are outspoken, militant, and educated, brings about a shift in the semiotics of veiling, which has long evoked the traditional, subservient domestic roles of Muslim women. The new veiling represents the public and collective affirmation of women who are searching for recognition of their Muslim identity through its expression—that is, through Islamism. Islamism, as the 'politics of recognition',[47] empowers Muslim women, providing intellectual and political tools for asserting their long-silenced difference, by questioning the equation between *civilized* and *Westernized*. It also empowers them by supplying a collective identity. Furthermore, Islamic politics provides a realm of opportunities for their 'self-realization' and becomes a vehicle for their public and professional visibility. Women now work as columnists and journalists; they attend political meetings, militate for the Islamist party, write best-selling novels, and make films—all inspired by the new language of Islamism. An elite cadre of Islamic women is thus emerging from within Islamism.

Woman's participation in Islamism has had unintended consequences; a latent individuation of women is at work. Women, once empowered by their public and professional visibility, continue to follow and develop personal life strategies. At the same time, while never forgetting the primacy of their identities as mothers and wives, women confront and criticize the Islamist ideology. The exit from the *mahrem* (domestic intimate) sphere forces women to question traditional gender identities and male definitions of 'licit' and 'illicit' behavior, thereby unveiling relations of power between Islamist men and women. Criticizing the 'pseudoprotectionism' of Muslim men, veiled women claim their right to 'acquire personality'—that is, 'a life of their own'—and, consequently, provoke disorder in Islamic gender definitions and identities.

Hence, the title *Forbidden Modern* conveys a double meaning. First, *forbidden* refers to the gendered construct of the private sphere (*mahrem*). The moral psychology of the domestic sphere depends on woman's controlled sexuality, which is guaranteed by woman's corporal and behavioral modesty (exemplified by veiling) as well as by limiting encounters between the sexes (social segregation of the sexes). Veiling suggests modesty, the forbidden woman. Yet, through Islamism and modernism, women have acquired public forms of visibility, sharing with men the same urban, political, and educational territories. Again under the veil, a new profile of Muslim woman is emerging, which in turn constitutes a threat to the moral psychology of gender identifications. Second, *Forbidden Modern* points to this encounter of women with modernity which is taking place in practice but is 'forbidden' in principle by Islamism. It is these paradoxes, ambiguities, and unfolding tensions—between the politicization of Islam and the individuation of woman, between the Islamist utopia and personal strategies for women, between the quest for authenticity and that for native modernity—that lie at the core of the veiling movement in Turkey.

Notes

1. The terms *Islamic movements* and *radical Islamism* are used interchangeably in this text, designating the contemporary Islamic movements as a collective action whose ideology was shaped during the late 1970s by Islamist thinkers all over the Muslim world (such as Abu-l Maudoodi in India, Sayyid Qutb in Egypt, Ali Shariati in Iran, and Ali Bulaç in Turkey) and by the Iranian Revolution. The term *radicalism* is used in the sense that there is a return to the origins, to the fundamentals of Islam, to address a critique to Western modernity on the one hand and a desire to realize a systemic change, to create an Islamic society, on the other. The terms *Islamic* and *Islamism* are not differentiated, although the latter term refers to the project of transforming society through political and social empowerment and the former to Muslim culture and religion in general.

2. The contemporary form of Islamist veiling is a head scarf that completely covers the hair, throat, and upper part of the chest and a long, loose-fitting gown in a discreet color that reaches to the heels.

3. The categories of East and West, rather than pointing to fixed geographical entities, refer to cultural representations that are not independent from the relations of domination between developed Western countries and the non-Western ones (labeled differently according to ideological epochs, such as Eastern, Third World, and recently Islamic). As the East was an integral part in the development of Western identity, similarly the West has been woven into the history making of the East. For the problematization of these categories, see Edward Said, *Orientalism* (New York: Pantheon, 1978).

4. Micaela di Leonardo,(ed.), *Gender at the Crossroads of Knowledge: Feminist Anthropology in the Postmodern Era* (Berkeley: University of California Press, 1991), 30–1.

5. C. Wright Mills, *The Sociological Imagination* (London: Penguin Books, 1971).

6. For instance, as *Le Monde* indicate, France witnessed a very passionate 'juridical, political and quasi-philosophical dispute' around the rights of female Muslim students covering their heads in French high schools, often referred to in the media as '*l'affaire du foulard*'. Recently, legislation ('la circulaire Bayrou', 20 September 1994) prohibited the utilization of all 'ostentatious signs' in high schools. For the debate on 'La France et l'Islam', see *Le Monde*, 13 October 1994.

7. For an analysis of the differentiation between traditional, pious and Islamist styles and the symbolic fanctions of covering in Egypt, see Andrea B. Rugh, *Reveal and Conceal: Dress in Contemporary Egypt* (Syracuse: Syracuse University Press, 1986); and Sherifa Zuhur, *Revealing Reveiling: Islamist Gender Ideology in Contemporary Egypt* (New York: State University of New York Press, 1992).

8. Pierre Bourdieu, *La Distinction: Critique sociale du jugement* (Paris: Editions du Minuit, 1979) (pub. in English as *Distinction* [Cambridge, Mass.: Harvard University Press, 1984]).

9. For the analysis of the events and the terms of this debate, see Emelie A. Olson, 'Muslim Identity and Secularism in Contemporary Turkey: The Headscarf Dispute', Anthropological Quarterly, 58: 6 (Oct.1985).

10. Islamists themselves do not use the word *turban* and are critical of this label,

preferring to use *tessettür*, that is, covering of women, or in modern Turkish *başörtüsü* (headscarf).

11. For a criticism of culturalist essentialism and its relation to postmodernism, see Aziz Al-Azmeh, *Islams and Modernities* (London: Verso, 1993).

12. Zygmunt Bauman, *Legislators and Interpreters: On Modernity, Post-Modernity and Intellectuals* (Ithaca, NY: Cornell University Press, 1987).

13. The Turkish word *çağdaş* (contemporary) used synonymously with *modern*, is conceptualized more in terms of the 'coming future', progress; it does not invoke the 'present' time. It is interesting to note that the concepts 'modernity' and 'modernism' have been widely used in Turkish for a decade because of Turkish intellectuals' interest in the postmodern debate. But its use as such differentiates more and more the 'modern' from the 'West'.

14. Alain Touraine, *Production de la société* (Paris: Editions du Seuil, 1973) (pub. in English as *The Self-Production of Society*, trans. Derek Coltman [Chicago: University of Chicago Press, 1977]).

15. Alain Touraine, *La Voix et le regard* (Paris: Editions du Seuil, Paris, 1978) (pub. as *The Voice and the Eye: An Analysis of Social Movement* [Cambridge: Cambridge University Press, 1981]). Touraine speaks of historicity as the creation of a historical experience and not of a position in historical evolution. He defines historicity as 'the set of cultural models a society uses to produce its norms in the domains of knowledge, production, and ethics', Alain Touraine, *Critique of Modernity* (Cambridge: Blackwell, 1995), 368–9.

16. Michel Foucault, *L'Usage des plaisirs*, vol.2: *Histoire de la sexualité* (Paris: Editions Gallimard, 1984) (pub. as *The History of Sexuality*, vol.1, trans. Robert Hurley [New York: Vintage Books, 1990]).

17. Somers and Gibson argue forcefully for 'joining narrative to identity', a condition, according to the authors, of introducing 'time, space, and analytic relationality' and thereby circumventing 'essentialist' approaches to identity. For the methodological implications of the concept of narrative for the social theory of action, see Margaret G. Somers and Gloria D. Gibson, 'Reclaiming the Epistemological "Other": Narrative and the Social Constitution of Identity', in *Social Theory and the Politics of Identity*, ed. Craig Calhoun (Cambridge, Mass.: Blackwell, 1994), 65 and 37–90.

18. For a detailed discussion of these principles, see Touraine, *La Voix*. The author has participated with Alain Touraine in studying the new social movements, particularly the feminist movement, in France. Apart from studying Islamist veiled women, she has also conducted research according to the principles of sociological intervention on leftist engineers in Turkey. The discussion of sociological intervention accounts for her own interpretation of the method in relation to the research conducted on Islamist women.

19. Mustafa Kemal Pasha (1881–1938), the Ottoman general who led the War of Independence (1919–22), was given the name Atatürk (father of Turks) when he established the Republic of Turkey in 1923. Kemalism refers to his modernist and secularist ideology that has continued to influence Turkish society to this day. Kemalist reforms refer to the transition from the multiethnic Ottoman Empire to the foundation of a secular republican nation-state (1923). The abolition of the Caliphate (1924), the abolition of Sharia, the

adoption of the Swiss civil code (1926), and the abandonment of the Arabic script, replacing it with the Latin alphabet (1928), were among the Kemalist reforms. See Bernard Lewis, *The Emergence of Modern Turkey* (London: Oxford University Press, 1968).

20. Daryush Shayegan, *Le Regard mutilé* (Paris: Albin Michel, 1989).

21. Norbert Elias, *The History of Manners: The Civilizing Process*, vol.1 (New York: Pantheon, 1978).

22. Ibid. 5.

23. Ibid. 50.

24. There is abundant recent research in different contexts on the centrality of the 'woman question', ranging from investigations of modernization, colonialism, and Islamism to the gendered nature of power relations. See Leila Ahmed, *Women and Gender in Islam* (New Haven: Yale University Press, 1992). In the Iranian context the 'women's issue' has held a privileged position in the writings of the ideologues of the constitutional revolutionary period (1905–11); see, for instance, Farzaneh Milani, *Veils and Words: The Emerging Voices of Iranian Women Writers* (Syracuse: Syracuse University Press, 1992); Deniz Kandiyoti (ed), *Women, Islam and the State* (Philadelphia: Temple University Press, 1991), 48–76.

25. For a discussion on the Islamic worldview, encompassing the Koran, its interpretations and the images of the Prophet's wives as models for emulation (sources of *sunna*) defining female righteousness based on segregation and modesty, see Barbara F. Stowasser, *Women in the Qur'an, Traditions, and Interpretation* (Oxford: Oxford University Press, 1994). For a discussion of the cultural code of honor and modesty from an anthropological point of view, see Lila Abu-Lughod, *Veiled Sentiments: Honor and Poetry in a Bedouin Society* (Berkeley: University of California Press, 1986).

26. The Kemalist paradigm involves the ideological and intellectual premises of Turkish modernity, which can be summarized as the master-narrative of secularism, republicanism, and gender equality.

27. I am therefore not in agreement with the prevalent Turkish 'feminist' reading of the relation between Kemalism and women's rights, highlighting the 'given' and not 'taken' nature of women's rights, the absent or limited agency on the part of women, and the predominance of patriarchal nationalism. These ready-made formulations of the Western feminist discourse fall short, in my view, as explanations of the particular relationships between Kemalism and women in Turkey.

28. This explains why, in contemporary Turkey, the violation of women's rights and secularism hurts the feelings of the elite more than does the violation of human rights and democracy.

29. Sarah Graham-Brown, *Images of Women: The Portrayal of Women in Photography of the Middle East, 1860–1950* (New York: Columbia University Press, 1988).

30. For a critique of the superficial Westernization of Turkish novels and their male characters, see Şerif Mardin, 'Super Westernization in the Ottoman Empire in the Last Quarter of the Nineteenth Century', in *Turkey: Geographic and Social Perspectives*, ed. P. Benedict et al. (Leiden: E. J. Brill, 1974).

31. Milani, *Veils and Words*, 238.

32. It is to be observed that *alaturka* is still used currently in the everyday

language of Turkish elites, with a negative connotation—although in recent years there is an appreciation, a sort of nostalgic romantic feeling, for Ottoman and early Turkish taste, cooking, furniture, etc.

33. Milani, *Veils and Words*.

34. For the elaboration and use of the concept of *habitus* in relation to lifestyles, see Bourdieu, *La Distinction*, 189–248.

35. In terms of historical classification and political experience, the Democrat Party legacy, which characterized the transition to political pluralism in Turkey in 1946, is of crucial importance. The Democrat Party, considered by state elites to be too liberal on religious and economic issues, gave voice to those segments of society that were not part of the bureaucratic Kemalist state; it therefore created political mediation, a buffer between the state and society. We can even claim that, instead of Turkey's secularism, which has been imitated to a certain extent in a majority of Muslim countries, it is the Democrat Party legacy that defines Turkish 'specificity'. I omit the Democrat Party legacy argument here because I think that rather than creating a new intellectual legacy, it provided a political representation of Muslim identity.

36. Calhoun, *Social Theory*.

37. Touraine, *The Voice and the Eye* (trans. from French, *La Voix et le regard* [Paris: Editions Seuil, 1978]).

38. Alexis de Tocqueville, *De la démocratie en Amérique* (Paris: Garnier-Flammarion, 1981), with preface by François Furet, 1: 32–5.

39. Anthony Giddens, *Modernity and Self-Identity: Self and Society in the Late Modern Age* (Palo Alto, Calif.: Stanford University Press, 1991), 80–1.

40. The notion of the 'Islamic self' is not appropriate, in the sense in which *self* implies secularism and separation from the community. But this ambiguity is helpful precisely in understanding how community is reconstructed and reinterpreted through Islamism and the Muslim self. The politicization of religion thus paradoxically engenders self-reflexivity on the Islamic self.

41. Foucault, *History of Sexuality*, 1: 61.

42. C. A. O Van Nieuwenhuijze, *The Lifestyles of Islam* (Leiden: E. J Brill, 1955), 144.

43. Milani argues that the absence in Persian literature of autobiography as a literary genre demonstrates the 'reluctance to talk publicly and freely about the self', a condition confined not only to women, who are 'privatized', but also to men, who are expected to be 'self-contained'. See Milani, *Veils and Words*, 201–2.

44. Here I employ Anderson's analysis of nationalism in the context of Islamism. See Benedict Anderson, *Imagined Communities: Reflections on the Origin and Spread of Nationalism* (Thetford: Thetford Press, 1983).

45. Durkheim long ago pointed out that the two distinct realms—the sacred and profane—are both indispensable for the establishment and reproduction of social ties.

46. Nazih Ayubi, *Political Islam: Religion and Politics in the Arab World* (London and New York: Routledge, 1991), 63.

47. Charles Taylor, *Multiculturalism: Examining the Politics of Recognition*, ed. and intro. Amy Gutman (Princeton: Princeton University Press, 1994).

Further Reading

ANGIERS, NATALIE, *Woman: An Intimate Geography* (Boston: Houghton Mifflin, 1999).

BAECQUE, ANTOINE DE, *Body Politic: Corporeal Metaphor in Revolutionary France, 1770–1800*, trans. Charlotte Mandell (Stanford, Calif.: Stanford University Press, 1997).

BORDO, SUSAN, *Unbearable Weight: Feminism, Western Culture, and the Body* (Berkeley: University of California Press, 1993).

BRUMBERG, JOAN JACOBS, *Fasting Girls: The Emergence of Anorexia Nervosa as a Modern Disease* (Cambridge, Mass.: Harvard University Press, 1988).

—— *The Body Project: An Intimate History of American Girls* (New York: Random House, 1997).

BULLOUGH, VERN, and BULLOUGH, BONNIE, *Cross Dressing, Sex, and Gender* (Philadelphia: University of Pennsylvania Press, 1993).

BUSH, BARBARA, *Slave Women in Caribbean Society, 1650–1838* (Bloomington: Indiana University Press, 1990).

BUTLER, JUDITH, *Bodies that Matter* (New York: Routledge, 1993).

BYNUM, CAROLINE, *Jesus as Mother: Studies in the Spirituality of the High Middle Ages* (Berkeley and Los Angeles: University of California Press, 1982).

—— *Holy Feast and Holy Fast: The Religious Significance of Food to Medieval Women* (Berkeley: University of California Press, 1987).

—— *Fragmentation and Redemption: Essays on Gender and the Human Body in Medieval Religion* (New York: Zone Books, 1991).

—— 'Why All the Fuss about the Body? A Medievalist's Perspective', *Critical Inquiry*, 22 (1995), 1–33.

CADDEN, JOAN, *Meanings of Sexual Differences in the Middle Ages: Medicine, Natural Philosophy, and Culture* (Cambridge: Cambridge University Press, 1993).

CAVALLARO, DANI, *The Body for Beginners* (New York: Writers and Readers, 1998).

COAKLEY, SARAH, *Religion and the Body* (Cambridge: Cambridge University Press, 1997).

COOPER, FREDERICK and STOLER, ANN (eds.), *Tensions of Empire: Colonial Cultures in a Bourgeois World* (Berkeley: University of California Press, 1997).

DOUTHWAITE, JULIA, *Exotic Women: Literary Heroines and Cultural Strategies in Ancien Régime France* (Philadelphia: University of Pennsylvania Press, 1992).

DUDEN, BARBARA, 'A Repertory of Body History', in Michel Feher,

Romana Naddaff, and Nadia Tazi (eds.), *Fragments of a History of the Human Body*, iii (New York: Zone, 1989).

—— *The Woman Beneath the Skin: A Doctor's Patients in Eighteenth-Century Germany*, trans. Thomas Dunlop (Cambridge, Mass.: Harvard University Press, 1991).

—— *Disembodying Women*, trans. Lee Hoinacki (Cambridge, Mass.: Harvard University Press, 1993).

FAUSTO-STERLING, ANNE, *Myths of Gender: Biological Theories about Women and Men* (New York: Basic Books, 1985).

—— 'The Five Sexes' *The Sciences* (March/April 1993), 20–5.

FEATHERSTONE, MIKE HEPWORTH, and TURNER, BRYAN, (eds.), *The Body: Social Process and Cultural Theory* (London: Sage, 1991).

FEE, ELIZABETH, and KRIEGER, NANCY, (eds.), *Women's Health, Politics, and Power: Essays on Sex/Gender, Medicine, and Public Health* (Amityville, NY: Baywood Publishing, 1994).

FEHER, MICHEL, NADDAFF, ROMANA and TAZI, NADIA, (eds.), *Fragments of a History of the Human Body* (New York: Zone, 1989).

FILDES, VALERIE, *Breasts, Bottles and Babies: A History of Infant Feeding* (Edinburgh: Edinburgh University Press, 1986).

—— *Wet Nursing: A History from Antiquity to the Present* (Oxford: Basil Blackwell, 1988).

—— (ed.), *Women as Mothers in Pre-Industrial England* (London: Routledge, 1990).

FISCHER-HOMBERGER, ESTHER, *Krankheit Frau und andere Arbeiten zur Medizingeschichte der Frau* (Bern: Hans Huber Verlag, 1979).

FOUCAULT, MICHEL, *The History of Sexuality*, trans. Robert Hurley (New York: Pantheon Books, 1986).

—— *Discipline and Punish: The Birth of the Prison*, trans. Alan Sheridan (New York: Pantheon Books, 1977).

GELPI, BARBARA, *Shelley's Goddess: Maternity, Language, Subjectivity* (New York: Oxford University Press, 1992).

GERO, JOAN and CONKEY, MARGARET (eds.), *Engendering Archaeology: Women and Prehistory* (Oxford: Blackwell, 1991).

GILBERT, SCOTT, 'Resurrecting the Body: Has Postmodernism Had Any Effect on Biology.', *Science in Context*, 8 (1995), 563–77.

GILMAN, SANDER, *Difference and Pathology: Stereotypes of Sexuality, Race and Madness* (Ithaca, NY: Cornell University Press, 1985).

—— *Sexuality: An Illustrated History* (New York: John Wiley & Sons, 1989).

GUTWIRTH, MADELYN, *The Twilight of the Goddesses: Women and Representation in the French Revolutionary Era* (New Brunswick: Rutgers University Press, 1992).

HAGER, LORI, (ed.), *Women in Human Evolution*, (New York: Routledge, 1997).

HILLMAN, DAVID and MAZZIO, CARLA (eds.), *The Body in Parts: Fantasies of Corporeality in Early Modern Europe* (New York: Routledge, 1997).

HOLLANDER, ANNE, *Seeing Through Clothes* (New York: Penguin Books, 1975).
—— *Sex and Suits* (New York: Knopf, 1994).

HONEGGER, CLAUDIA, *Die Ordnung der Geschlechter: Die Wissenschaften vom Menschen und das Weib* (Frankfurt: Campus, 1991).

HUBBARD, RUTH, *The Politics of Women's Biology* (New Brunswick: Rutgers University Press, 1990).

HUNT, LYNN, *Politics, Culture, and Class in the French Revolution* (Berkeley and Los Angeles: University of California Press, 1984).

—— *The Family Romance of the French Revolution* (Berkeley: University of California Press, 1992).

—— (ed.), *Eroticism and the Body Politic* (Baltimore: Johns Hopkins University Press, 1991).

—— (ed.), *The Invention of Pornography: Obscenity and the Origins of Modernity, 1500–1800.* (New York: Zone, 1993).

JACOBUS, MARY, KELLER, EVELYN FOX, and SHUTTLEWORTH, SALLY, (eds.), *Body/Politics: Women and the Discourses of Science* (New York: Routledge, 1990).

JORDANOVA, LUDMILLA, *Sexual Visions: Images of Gender in Science and Medicine between the Eighteenth and Twentieth Centuries* (Madison: University of Wisconsin Press, 1989).

KELLER, EVELYN FOX and LONGINO, HELEN, (eds.), *Feminism and Science* (Oxford: Oxford University Press, 1996).

KESSLER, SUZANNE, 'The Medical Construction of Gender: Case Management of Intersexed Infants', *Signs: Journal of Women in Culture and Society,* 16 (1990), 3–26.

KEVLES, BETTYANN KEVLES, *Females of the Species: Sex and Survival in the Animal Kingdom* (Cambridge, Mass.: Harvard University Press, 1986).

LANDES, JOAN, *Women and the Public Sphere in the Age of the French Revolution* (Ithaca, NY: Cornell University Press, 1988).

—— (ed.), *Feminism, the Public and the Private* (Oxford: Oxford University Press, 1998).

LAQUEUR, THOMAS, *Making Sex: Body and Gender from the Greeks to Freud,* (Cambridge, Mass.: Harvard University Press, 1990).

LAWRENCE, CHRISTOPHER and SHAPIN, STEVEN (eds.), *Science Incarnate: Historical Embodiments of Natural Knowledge.* (Chicago: University of Chicago Press, 1998).

LOCK, MARGARET, *Encounters with Aging: Mythologies of Menopause in Japan and North America* (Berkeley: University of California Press, 1993).

MACLEAN, IAN, *The Renaissance Notion of Woman: A Study in the Fortunes of Scholasticism and Medical Science in European Intellectual Life* (Cambridge, England, 1980).

MAINES, RACHEL, *The Technology of Orgasm: 'Hysteria', the Vibrator, and Women's Sexual Satisfaction* (Baltimore: Johns Hopkins University Press, 1999).

MARTIN, EMILY, 'The Egg and the Sperm: How Science has Constructed a Romance Based on Stereotypical Male–Female Roles', *Signs: Journal of Women in Culture and Society*, 16 (1991), 485–501.

—— *The Woman in the Body: A Cultural Analysis of Reproduction*, 2nd edn. (Boston: Beacon, 1992).

MASTROIANNI, ANNA, FADEN, RUTH, and FEDERMAN, DANIEL, (eds.), *Women and Health Research*, ii (Washington, DC: National Academy Press, 1994).

MELZER, SARA, and NORBERG, KATHRYN (eds.), *From the Royal to the Republican Body: Incorporating the Political in Seventeenth- and Eighteenth-Century France* (Berkeley: University of California Press, 1998).

MENDELSOHN, KATHLEEN, NIEMAN, LINDA, ISAACS, KRISTA, LEE, SOPHIA, and LEVISON, SANDRA, 'Sex and Gender Bias in Anatomy and Physical Diagnosis Text Illustrations', *Journal of the American Medical Association*, 262 (26 October 1994), 1267–70.

MERCHANT, CAROLYN, *The Death of Nature: Women, Ecology, and the Scientific Revolution* (San Francisco: Harper and Row, 1980).

MOSCUCCI, ORNELLA, *The Science of Woman: Gynaecology and Gender in England, 1800–1929* (Cambridge: Cambridge University Press, 1990).

MOSSE, GEORGE, *The Image of Man: The Creation of Modern Masculinity* (New York: Oxford University Press, 1996).

NUSSBAUM, FELICITY, *Torrid Zones: Maternity, Sexuality, and Empire in Eighteenth-Century English Narratives* (Baltimore: Johns Hopkins University Press, 1995).

NYE, ROBERT A., *Masculinity and Male Codes of Honor in Modern France* (New York: Oxford University Press, 1993).

—— (ed.), *Sexuality* (Oxford: Oxford University Press, 1999).

OUDSHOORN, NELLY, *Beyond the Natural Body: An Archeology of Sex Hormones* (London: Routledge, 1994).

OUTRAM, DORINDA, *The Body and the French Revolution: Sex, Class, and Political Culture* (New Haven: Yale University Press, 1989).

PETHERBRIDGE, DEANNA, and JORDANOVA, LUDMILLA, *The Quick and the Dead: Artists and Anatomy* (Berkeley: University of California Press, 1997).

POOVEY, MARY, *Making of a Social Body: British Cultural Formation, 1830–1864* (Chicago: University of Chicago Press, 1995).

PRATT, MARY LOUISE, *Imperial Eyes: Travel Writing and Transculturation* (London: Routledge, 1992).

RAMASWAMY, SUMATHI, 'Virgin Mother, Beloved Other: The Erotics of Tamil Nationalism in Colonial and Post-Colonial India', *Thamyris*, 4 (1997), 9–39.

ROPER, M., and TOSH, J. (eds.), *Manful Assertion: Masculinities in Britain since 1800* (London: Routledge, 1991).

Russett, Cynthia, *Sexual Science: The Victorian Construction of Woman-hood* (Cambridge, Mass.: Harvard University Press, 1989).

Ruzek, Sheryl, Clarke, Adele, and Olesen, Virginia (eds.), *Women's Health: Complexities and Differences* (Columbus: Ohio State University Press, 1997).

Schiebinger, Londa, *The Mind Has No Sex? Women in the Origins of Modern Science* (Cambridge, Mass.: Harvard University Press, 1989).

—— *Nature's Body: Gender in the Making of Modern Science* (Boston: Beacon Press, 1993).

—— 'Lost Knowledge, Bodies of Ignorance, and the Poverty of Taxonomy as Illustrated by the Curious Fate of *Flos Pavonis*, an Abortifacient' in Caroline Jones and Peter Galison (eds.), *Picturing Science, Producing Art* (New York: Routledge, 1998).

Sinha, Mrinalini, *Colonial Masculinity: The 'Manly Englishman' and the 'Effeminate Bengali' in the Late Nineteenth Century* (Manchester: Manchester University Press, 1995).

Spanier, Bonnie, *Im/partial Science: Gender Ideology in Molecular Biology* (Bloomington: Indiana University Press, 1995).

Squier, Susan, *Babies in Bottles: Twentieth-Century Visions of Reproductive Technology* (New Brunswick: Rutgers University Press, 1994).

Stepan, Nancy Leys, *The Idea of Race in Science: Great Britain 1800–1960* (Hamden, Conn.: Archon Books, 1982).

—— 'Race and Gender: The Role of Analogy in Science', *Isis*, 77 (1986), 261–77.

Terry, Jennifer, and Urla, Jacqueline (eds.), *Deviant Bodies: Critical Perspectives on Difference in Science and Popular Culture* (Bloomington: Indiana University Press, 1995).

Tuana, Nancy (ed.), *The Less Noble Sex: Scientific, Religious, and Philo-sophical Conceptions of Woman's Nature* (Bloomington: Indiana University Press, 1993).

Theweleit, Klaus, *Male Fantasies* (Minneapolis: University of Minnesota Press, 1989).

Yalom, Marilyn, *A History of the Breast* (New York: Knopf, 1997).

Index

abortion 247–50, 251
Ackermann, Jakob 31, 36, 38
Aegineta, Paulus 75–6
Airy, George Biddell 317–18, 327
Albinus, Bernhard 32–4, 36, 37, 38
Amazons 184, 272, 273–6, 284, 286
animals 13
anorexia 418–20, 421
anti-Semitism 355, 359, 360
apes 12, 212, 219–20, 222
Appel, Toby 209
Apollo 30, 34
Ardener, Edwin 275
Artemis 273
Aristotle, 27, 32, 44, 74, 89, 133, 135, 169, 170
Athena 16
Auerbach, Elias 360–1, 364
Austen, Jane 324
Aveling, James 123–4

Bantock, G. G. 123–4
Barbin, Abel (née Alexina) 118–22, 124, 125, 127, 129, 130, 131, 136, 144
Barclay, John 36, 38–40, 41
Barnes, Fancourt 122–3, 127, 131, 136
Bartholdi, Auguste 265–6, 268
Bartmann, Sarah (Saartje Bartman) 203–4, 207, 211–14
Bauhin, Gaspar 34, 69
Beller, Steven 357
Bence Jones, Henry 322, 323
Berliner, Bernhard 355
Bettes, John 164
Bidloo, Godfried 29, 30, 34
Billy Boy 398, 403
Bischoff, T. L. W. von 45, 50
bisexual 61, 64, 82
Boas, Franz 408, 409
body
  Christian 355
  Confucian 432
  Jewish 355
  mindful 16, 430, 458–9
  politic 3, 7–11, 182, 187–8, 276, 355–6, 359, 431, 449
  social 442

Bonaparte, Marie 67, 82
Bordo, Susan 2, 418, 420
Bourdieu, Pierre 476
Brandes, Ernst 385
breasts 30, 205, 212, 219, 223, 269, 272, 273, 276, 278–81, 283, 284, 286, 287, 400, 406, 419, 420
Brown-Séquard, Charles Edouard 90–1, 92, 96
Brumberg, Joan 411
Burke, Kenneth 442
Butler, Judith 129

Cadden, Joan 133
Callow, Nancy 100
Callow, Robert 100
Cameron, Julia Margaret 335–6, 337
Camper, Pieter 220
cap, red liberty 183–4, 267, 268, 270, 271–2, 273, 275, 287
Case, John 166–9, 170, 177
Chapman, John 321–2
Charcot, Jean Martin 365–6, 367
Charity 272, 276, 278, 279, 280, 284
chastity 158, 164, 274
Cherin, Kim 419
Cheselden, William 29, 34
child, women compared to 42–4
Chodowiecki, Daniel 381–3
Clark, Andrew 323
clitorectomy, 16, 75–7, 82
clitoris 5, 16, 58–82, 137, 139 , 140, 223
clothing 8–9, 17, 36, 77, 119, 144, 155, 159, 162, 163–4, 165, 166, 171, 182–91, 269, 270–4, 283, 328, 334, 375, 382, 401, 410, 465–87
  uniforms 192–3, 194
  top hat 383, 385, 386
cockade 182–4, 186, 192
codpiece 157
colonial bodies 10, 204, 226, 234–55
Colp, Ralph 318, 324
Columbus, Renaldus 5, 58, 62, 65–74
Comte, Auguste 48
Conkey, Margaret 378
Copernican universe 169, 170–1
corset 36

DATE DUE

NOV 1 3 2003			
DEC. 0 3 2005			
GAYLORD			PRINTED IN U.S.A.